D0533366

Advanced Statistics

Michael Hugill

Bell & Hyman

First published in 1985 by
Bell & Hyman Limited
Denmark House
37–39 Queen Elizabeth Street
London SE1 2QB

British Library Cataloguing in Publication Data

Hugill, Michael J.
Advanced statistics.
1. Mathematical statistics
I. Title
519.5   QA276

ISBN 0 7135 2491 X

Phototypeset by Advanced Filmsetters (Glasgow) Ltd
Printed and bound in Great Britain by
M. & A. Thomson Litho Ltd, East Kilbride, Scotland

Contents

Preface

The scope of A-level Statistics syllabuses has greatly increased in recent years and it has not been easy for students to find suitable guidance in current elementary textbooks. This book is intended to fill the gap, by covering in a systematic and comprehensive manner the statistical mathematics needed by students taking the subject at any of the various levels.

No previous knowledge of the subject is assumed and the approach, especially in the early chapters, is informal with the aim of providing a text which can be easily read, enabling the student to move fairly quickly through the basic concepts. Mathematical ideas that are intuitively reasonable but whose formal proofs are best avoided on a first reading are given extended treatment in Appendices to the chapters. In this way it is hoped that the book will be found to be more genuinely comprehensive than most elementary books while at the same time not allowing the reader to be held up in the early stages by too much theory. The reader will nevertheless have ready access to the theory when he needs it.

Each new idea is followed immediately by straightforward exercises in the techniques involved, with further, more varied, examples following a little later. Since the usefulness of a book of this kind depends on there being sufficient suitable examples for the student to practise on, an unusually large number are provided, graded in approximate order of difficulty, with the more demanding ones marked with an asterisk (★). There are therefore more than enough for the purpose of revision.

The bulk of the book deals with the basic work common to most syllabuses. The remainder is intended for the student taking the more advanced papers—variously called 'Further Mathematics', 'Mathematics (Statistics)', 'Pure Mathematics with Statistics'—and provides a generous coverage of topics required for these papers. Much of this work is handled rather cursorily in current A-level texts, while the treatment in more advanced books is not suitable at this level. The chapters (or sections of chapters) dealing with these topics are also marked with an asterisk.

Students may find it helpful to be given some guidance on tackling the work on a 'first reading', and the following outline offers a suggested plan for an initial course.

Suggestions for a first reading

Chapter 1	Introduction.
Chapters 2, 3	Frequency distributions. Averages and measures of spread.
Chapters 4, 5	Probability and probability laws, up to page 77. Subsequent work in statistics depends mainly on an elementary understanding of the basic laws. It is not advisable to spend time on complicated problems at this stage.
Chapter 6	Probability distributions. The idea of a 'model' is central. The theoretical results in this chapter are important but the detailed proofs may be postponed.
Chapter 7	Probability generating functions. This chapter is not essential, but there are considerable advantages in mastering pages 112–15 at least.
Chapter 8	The binomial and geometric distributions, omitting the negative binomial.
Chapter 9	The Poisson distribution. This may be taken at any time after the student has met the exponential function in pure mathematics. It should not be tackled too early in the course.
Chapter 10	Continuous probability distributions, up to page 179, and the theoretical results on page 187.
Chapter 11	The normal distribution to page 206, and the statement of the central limit theorem (pages 214–15).
Chapter 12	Sampling distributions and the idea of estimation, without dwelling on consistency and efficiency at this stage (pages 227–36). The nature of the sampling distribution of the mean (pages 241–5).
Chapter 13	Hypothesis tests and confidence interval for means (large samples).
Chapter 15	Tests for proportions, to page 300.
Chapter 18	Correlation.
Chapter 19	Linear regression.
Chapter 23	Chi-squared tests for goodness of fit and contingency, up to page 461.

Further reading

The more advanced topics are placed, for convenience, in what seems to be the logical position in this book but students may take these at any appropriate time. They include:

Chapter 5	Bayes' theorem on probability.
Chapter 7	Further applications of the probability generating function.
Chapter 8	The negative binomial distribution.
Chapter 10	The exponential distribution.
Chapter 11	The normal approximation to the binomial and Poisson distributions. Fitting a normal distribution to given data.
Chapter 12	Consistency and efficiency of estimators.
Chapter 14	Small sample hypothesis tests and confidence intervals.

I am grateful to the following G.C.E. Examination Boards for permission to use questions from past papers:

The Associated Examining Board (AEB)
Joint Matriculation Board (JMB)
Oxford and Cambridge Schools Examination Board (O & C), who also publish papers set by Mathematics in Education and Industry (MEI) and the School Mathematics Project (SMP)
Southern Universities' Joint Board (SUJB)
University of Cambridge Local Examinations Syndicate (C)
University of London School Examinations Board (L)
University of Oxford Delegacy of Local Examinations (O)

I am also indebted to the following for permission to reprint parts of tables:

Cambridge University Press for four tables from *New Cambridge Elementary Statistical Tables* by D. V. Lindley and W. F. Scott, and for one table from *SMP Advanced Tables*.

Finally my thanks are due to the staff of Bell & Hyman, particularly to Antonia Murphy for much encouragement and expert advice on the shaping of this book, to Sandra Wegerif for the enthusiasm she brought to the task of preparing it for publication, and to Anne Hollifield for her editorial help and advice.

M.J.H.

The author and publishers would like to thank the following for their permission to reproduce copyright material: Victor Gollancz Limited for an extract from *The White Hotel* by D. M. Thomas; William Heinemann Limited for an extract from *The Military Philosophers* by Anthony Powell; an extract from *Schindler's Ark* by Thomas Keneally, reprinted by permission of Hodder & Stoughton Limited.

List of symbols

$$\sum_{i=1}^{N} x_i$$ the sum of N observations $x_1 + x_2 + \ldots + x_N$

X	a random variable (i.e. capital letters for random variables)
x	a particular value of a random variable
$\bar{x}$	the mean of a set of observations, sample mean
$\mathrm{var}(x)$	the variance of a set of observations; $\mathrm{var}(x) = \Sigma(x-\bar{x})^2/N$
$\bar{X}$	the random variable representing sample means
$E(X), \mu$	the expectation (or mean) of the random variable X, the population mean
$E[f(X)]$	the expectation of the function $f(X)$
$\mathrm{Var}(X), \sigma^2$	the variance of the random variable X, the population variance
$\hat{\theta}$	unbiased estimator of the population parameter θ
s^2	unbiased estimate of the population variance from a sample; $s^2 = \Sigma(x-\bar{x})^2/(N-1)$
π	the proportion in a population
$P(A)$	the probability of event A
$n(A)$	the number of occurrences of event A
$P(A \cup B)$	the probability of event A or event B (or both)
$P(A \cap B)$	the probability of event A and event B
$P(A \mid B)$	the conditional probability of A given B
${}^nP_r$	the number of permutations of n objects taken r at a time
$n!$	factorial n
${}^nC_r, \binom{n}{r}$	the number of selections of r objects that may be made from n objects
p_i	the probability of the value x_i of the discrete random variable X
p_{ij}	the probability of the values (x_i, y_j) in the joint distribution of X and Y
$f(x)$	the probability density function of the continuous random variable X
$F(x)$	the cumulative distribution function of the continuous random variable X

$G(t)$	the probability generating function of the discrete random variable X; $G(t) = E(t^X)$
$M(t)$	the moment generating function of the random variable X (discrete or continuous); $M(t) = E(e^{tX})$
μ'_r	the rth moment about zero; $\mu'_r = E(X^r)$
μ_r	the rth moment about the mean; $\mu_r = E[(X - \mu)^r]$
$B(n, p)$	binomial distribution with parameters n and p
$N(\mu, \sigma^2)$	normal distribution with mean μ and variance σ^2
$N(0, 1)$	standard normal distribution
Z	the standardised normal variable
$\phi(z)$	the probability density function of Z
$\Phi(z)$	the cumulative distribution function of Z
H_0	the null hypothesis
H_1	an alternative hypothesis
Type I error	rejecting H_0 when it is true
Type II error	accepting H_0 when it is false
t	Student's t statistic
v	the number of degrees of freedom
t_v	the critical value of t for v degrees of freedom
T^-, T^+	the sum of the ranks in the Wilcoxon signed-rank test
W	the sum of the ranks for the smaller sample in the Wilcoxon rank-sum test
$L(\theta)$	the likelihood of a parameter θ for a distribution with a single parameter
$L(\theta_1, \theta_2)$	the likelihood of the parameters θ_1, θ_2 for a distribution with two parameters
S_{xx}, S_{yy}	$\Sigma(x - \bar{x})^2, \Sigma(y - \bar{y})^2$ calculated from a set of sample observations (x, y)
S_{xy}	$\Sigma(x - \bar{x})(y - \bar{y})$
$\mathrm{cov}(x, y)$	the covariance of sample observations (x, y); $\mathrm{cov}(x, y) = S_{xy}/n$
$\mathrm{Cov}(X, Y)$	the covariance of the random variable (X, Y)
r	the product-moment correlation coefficient of a sample; $r = S_{xy}/\sqrt{S_{xx}S_{yy}}$
r_s	Spearman's rank correlation coefficient
ρ	the population product-moment correlation coefficient; $\rho = \mathrm{Cov}(X, Y)/\sqrt{\mathrm{Var}(X).\mathrm{Var}(Y)}$
α, β	population parameters of a regression line
a, b	least-squares estimates of α and β; $a = \bar{y} - b\bar{x}, b = S_{xy}/S_{xx}$
χ^2	a chi-squared distribution
χ^2_v	the chi-squared distribution with v degrees of freedom

$F(v_1, v_2)$ the F-distribution with v_1 and v_2 degrees of freedom

SS_b, SS_w, SS_t the sums of squares for between-samples, within-samples and total
                 variations (single-factor Anova)

s_b^2, s_w^2 between-samples and within-samples estimates of σ^2

SS_c, SS_r, SS_e the sums of squares for between-columns, between-rows and
                 residual variations (two-factor Anova)

s_c^2, s_r^2, s_e^2 between-columns, between-rows and residual estimates of σ^2

1 Introduction

1.1 Statistical mathematics

The kind of problem that statistical mathematics is designed to deal with, and the kind of answer that it produces, are best illustrated by two examples. These will also be used to introduce some important terms that will be used throughout this book.

EXAMPLE 1 A sample of 200 people were interviewed by an opinion-polling organisation and 80 people in the sample declared their intention of voting for the Radical Party in the next local council elections. The rest said they would vote for the Ratepayers' Party. There were no other parties in the field. What conclusion could be drawn about the voting intentions of the population of the locality as a whole?

Answer: If the pollsters conclude that the proportion of the voting population intending to vote 'Radical' is between 33% and 48%, the chance that they are correct is 95%.

EXAMPLE 2 In a large American university it has always been the custom to find the weights of each of a sample of 50 students from the new year's intake. The average weight in 1942 was 70.4 kg. Now, the average over many previous years of the whole of each year's intake had been estimated to be 67.2 kg. The authorities suspected that the increase in average weight of the 1942 sample might imply that there had been a general increase in weight, possibly reflecting improved diet since the Depression of the early 1930s. Is the increase in average weight of the sample from the 1942 intake really significant, or might it, on the contrary, have arisen by chance?

Answer: Using the figures mentioned (and some others), the authorities concluded that it was highly probable that the average weight of the whole 1942 intake *was* significantly greater than in previous years.

1.2 Samples

The first thing to notice is that in each of these examples the data available were from a sample only.

This is the case in most statistical investigations. It would, for instance, be impracticable and expensive to obtain the voting intentions of the whole voting population of a town. In example 2, the size of the annual intake was such that it was, presumably, out of the question to weigh more than a sample of 50. In some scientific investigations, samples are often all that can be available. A researcher, for example, who is investigating the average size of water-rats in various parts of the country, would clearly not be able to measure all the water-rats in the various areas, only those that he or she managed to trap.

It seems likely that the larger the sample, the more reliable the conclusion, and indeed we shall prove later that this is generally the case. Nevertheless, the nature of the investigation may impose the use of quite a small sample. Testing industrial products 'to destruction' is a case in point. If the manufacturer of TV tubes wanted to check that their average life was up to some required standard, the test would involve running tubes until they 'die'. Naturally, he would not want to sacrifice too many.

1.3   Sample statistics

The next stage in the process is for the investigator to calculate, from the observations, some numerical quantity. In example 1 it was the proportion of the sample who intended to vote Radical; in example 2 it was the average weight of the student sample. Each of these numerical measures is called a *statistic* (in the singular).

Any quantity calculated from a sample is called a *sample statistic*. It describes, or summarises, in numerical form, some property of the sample. To say that the average weight of 50 people is 70.4 kg is one way of summarising simply the results of weighing each of them.

1.4   Populations

If we now consider the conclusion of example 1, it will be seen that it referred to the proportion of the *whole* voting population intending to vote Radical. Similarly, in example 2 the conclusion referred to the average weight of the *whole* 1942 intake. The use of the word 'whole' leads us now to define the word population as it is used in statistical theory.

Population is the name given to the complete set of items being investigated. In example 1 the population is the whole set of voting intentions of each member of the voting public. In example 2 it is the complete set of weights of all students admitted in 1942. These examples differ, of course, in that while 'weight' is a numerical measurement, 'voting intention' is an attribute. One is quantitative, the other is qualitative. We will now give further examples of both kinds.

Quantitative examples

1.   The complete set of the numbers of rooms in each household in Wigan.

2.   The complete set of heights of all 21-year-old males in Scotland.

3. The complete set of weights of all babies born in the United Kingdom from midnight 31 December 1980 to midnight 31 December 1981.

4. The complete set of lengths (body plus head) of all water-rats in Skye at a given time.

Qualitative examples

1. The complete set of data giving the sex of each baby born in the United Kingdom from midnight 31 December 1980 to midnight 31 December 1981.

2. The complete set of data stating whether or not each general practitioner in Surrey is a smoker.

Infinite populations

In the above examples, the populations are all *actual* and *finite*, although in the case of the water-rats in Skye the population can only be imagined. Such a population is '*hypothetical*' and is not accessible to the investigator, who can only be aware of a sample. In the other examples, each member of the population could, in theory, be actually observed.

Later in this book we shall come across many populations which are *infinite*; here we shall just mention one example. Suppose an investigation is being carried out to determine whether or not a die is biased. The experimenter records the number of 'sixes' after 1000 throws. This could be regarded as a sample from the population of the complete set of numbers of 'sixes' in all possible repetitions of the experiment. This is obviously a hypothetical concept as 'all possible repetitions of the experiment', being infinite in number, could not be carried out.

1.5 Probability

The last point to be noted about examples 1 and 2 is that the conclusions reached are given in terms of *probability*.

In example 1: '... the chance that they are correct is 95 %'; in example 2: '... the authorities concluded that it was highly probable that ...'.

A moment's thought will show that this is inevitable. In no case like these could we be certain of a conclusion about a whole population, since the observations are obtained from samples only. In example 1 it might well be the case that the sample, by chance, contained an unusually high proportion of people of one political inclination, and in example 2 the sample of students might also, by chance, have been uncharacteristic (as far as weight is concerned) of the year's intake as a whole. In other words, chance is built into the whole process. In spite of this fact, careful sampling, together with various refinements of the mathematical analysis, can reduce these uncertainties to a minimum.

1.6 The statistical process

The process outlined in these two examples is typical of statistical analysis. It is designed *to discover what we can about the nature of a population by examining a sample drawn from it.*

In example 1, the aim is to investigate the population of the voting intentions of all the town's voters. For a sample, the statistic (the proportion who intend to vote Radical) is calculated. The conclusion of the analysis shows that if the pollsters say that the proportion in the population with this intention is between 37.8% and 42%, the probability that they are right is 95%.

In example 2, the aim is to investigate the population of the weights of a whole year's university intake. For a sample, the statistic (average weight) is calculated. The conclusion of the analysis is that it is highly probable that the average population weight is significantly higher than expected. (In practice this probability is given in numerical form.)

We have not said anything at this stage about the method by which the conclusions were reached. This is called *inference* and is the subject of later chapters. Meanwhile, as a necessary preliminary, we must examine methods of calculating certain important statistics (these are the subject of chapters 2 and 3) and outline the theory of probability on which the whole of statistical mathematics depends (this is dealt with in chapters 4 and 5).

1.7   Random sampling

It was mentioned above that a sample may, by chance, be unrepresentative; it is difficult to guard against this. In addition, there is the possibility that the method of sampling is defective and will inevitably result in erroneous conclusions. If, for example, the views on some issue are obtained by using a sample of people picked from a telephone directory, part of the population at large has no chance at all of being selected – those who have no telephone, perhaps because they cannot afford one. The classic instance of this kind occurred during the 1932 American Presidential election. As a result of a very large sample of two million voters selected by mail (the fact that mailing lists represent a fairly narrow section of the population being overlooked), Governor Landon was tipped to win comfortably. In fact Roosevelt received about 60% of the vote.

For correct sampling, it is essential that every member of the population should have an *equal chance of being selected* and that a member's chance of being chosen is independent of previous selections. This is called *random sampling*, or strictly *simple random sampling*.

The method may be pictured (or actually carried out) by representing each member of the population by a numbered ticket, mixing the tickets in a large bowl and drawing out a sample of the required size, the tickets having been thoroughly mixed after each draw. For true random sampling, a ticket must be replaced after it has been drawn, so that it may well happen that a member appears more than once in a sample. Should this happen, it is a matter of choice, depending on the nature of the investigation, whether a duplicated member is used; if the investigation concerns heights of people it is sensible to use samples of different people.

Random numbers

A better method, and one which is mathematically equivalent, is to use a table of *random numbers.* This consists of the digits 0 to 9 in random order. As an example, here is part of a table, in which the numbers are printed in pairs·

58	91	05	97	96	86	90	64	77	84
25	02	71	93	90	31	12	14	73	36

Suppose we wished to pick a sample of four days from the days of the year. These are first numbered from 1 to 365, in the natural order. (If we were picking a sample of four from 365 people, the way in which they were numbered would not matter.) Starting anywhere in the table, and proceeding in any systematic manner, the first four numbers that occur are chosen, a number greater than 365 being discarded. In this case we might start at 25 and read off the digits in threes, working along the rows. This would give

$$250 \quad 271 \quad 031 \quad 121$$

or 7 August, 28 August, 31 January and 1 May.

However the sampling is done, it is sometimes not practicable to replace a member after it has been chosen. This is called sampling without replacement. If the population is fairly small, it is clear that once a member has been chosen and not replaced the chance of the rest is slightly altered. For a large population the difference between sampling with and without replacement is not significant. In this book it will be assumed that the sampling is either genuinely random or that the population is large enough for the sampling to be effectively random.

2 Frequency distributions: averages

In this chapter and the next one we shall be concerned with observations obtained from samples, and with ways of:

1. displaying the data *graphically*, since sets of figures on a page convey no image and therefore tell us little;

2. *describing* the data in some concise mathematical form, in other words calculating '*statistics*' (which, as we stated in chapter 1, are quantities which represent or summarise some aspect of the data).

To take a trivial example, if out of 900 housewives chosen at random 387 preferred cooking by gas to cooking by electricity, the nature of the information suggests that the relevant statistic is just the *proportion* (0.43) who prefer gas cooking. If, on the other hand, the housewives were asked to give the amount of their last quarterly electricity bill, this information could be summarised by calculating the *average* bill for those who cooked by gas, and the average for those who cooked by electricity. In this chapter we shall be concerned with averages, and although it will be assumed that the observations concerned come from samples, the methods described will apply to any sets of data (including those from whole populations, provided these are not too large).

2.1 Discrete and continuous variables

Data from samples can be of two kinds, discrete and continuous, and the difference between them is illustrated in the following examples.

In a traffic enquiry designed to compare the contributions of private and public transport to road congestion, 100 successive cars travelling along a certain road were observed and the number x of passengers (excluding the driver) was noted. The results were:

x	Frequency (f)
0	40
1	29
2	18
3	11
4	2
	100

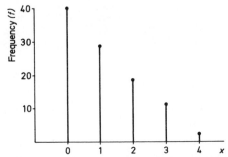

6

The information in this table is called a *frequency distribution* (in some books you may see the frequencies written $f(x)$ and the set of frequencies called the frequency function). The frequency distribution may be shown graphically by a frequency chart (or line graph) where the *height* is proportional to the frequency. The vertical axis is therefore marked 'frequency'.

An example like this, where the 'variable' x takes distinct values is called a *discrete* distribution, and x is a discrete variable. Discrete variables are often integers, although they need not be. For example, x could be $3\frac{1}{2}, 4, 4\frac{1}{2}, \ldots$, if the data concerned were shoe sizes.

Suppose, now, that the data were the results of measuring the lengths of a sample of 90 ears of barley, in millimetres to the *nearest millimetre*. In spite of being given to the nearest millimetre, the length could in theory be anywhere in a given range, so that 24.136 is a possible theoretical value. Such a variable is *continuous*.

To tabulate the data in a practicable way, the observations are divided into *classes* by grouping, so that, for example, recorded lengths of 25, 26, 27, 28, 29, which represent measured lengths anywhere between 24.5 and 29.5, might be put in a class labelled 24.5–29.5.

The result is that, if we have decided to group the lengths into classes 14.5–19.5, 19.5–24.5, ..., say, the numbers 14.5, 19.5, 24.5 make suitable *end-points* because no *recorded length* can be 14.5 or 19.5 or 24.5, etc. There is therefore no doubt as to which class any particular recorded length belongs, and there is *no break* between neighbouring classes. This last point is important when we come to graphical representation of the data. (If a measured length comes out as 19.5, the usual convention is to put the recorded length in the class above, as we shall see.)

If, then, the recorded lengths to the nearest millimetre are collected in the form shown in table 1, the appropriate classification for graphical representation is as shown in table 2.

Table 1			Table 2		
x	f		x	f	
15–19	1		14.5–19.5	1	
20–24	4		19.5–24.5	4	
25–29	22		24.5–29.5	22	
30–34	35		29.5–34.5	35	
35–39	20		34.5–39.5	20	
40–44	8		39.5–44.5	8	
	90			90	

Notice that:

1. This kind of distribution is sometimes shown with classes 15– , 20– ,..., where '15– ' means '15 up to but excluding 20'.

The end-points are still 14.5, 19.5, ..., and if a measured length comes out at 19.5 it goes naturally into the class above.

2. A class is sometimes given in an open-ended manner, such as '40 or more'. In this sort of case, the unspecified upper end-point is arbitrary. An intelligent decision must therefore be made, and there is usually no particular difficulty about this. For example, a distribution of the ages of women in the UK with the last class given as

'85 or over' suggests naturally an upper end-point of, say, 100.

3. The *class width* is the difference between successive endpoints, so that in our example it is 5.

In practice, the end-points and class width are chosen arbitrarily, subject to the following considerations:

1.   that there should not be too few nor too many classes to be useful,

2.   that no *recorded* observation should fall on an end-point. This is ensured by choosing the end-points to $\frac{1}{2}$ *a unit* beyond the accuracy of measurement.

In our example, since the measurements were taken to the nearest integer, the units for the end-points were taken to 0.5. If the measurements had been taken to the nearest 1 decimal place, the classes might be:

$$\text{For a class width of 5} \qquad 14.45\text{–}19.45,\ldots$$

$$\text{For a class width of 2.5} \qquad 14.45\text{–}17.95,\ldots.$$

2.2   Histograms

In the example we have just described, the data were given in what is called a *grouped* frequency distribution. Such a distribution is best displayed graphically by means of a *histogram*, like the one opposite, which displays the distribution of the lengths of ears of barley given on page 7.

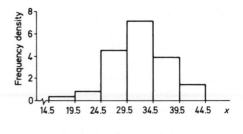

The important points to notice about a histogram are:

1.   the *area* of the rectangle drawn on top of each class interval is proportional to the frequency;

2.   the vertical axis cannot therefore be labelled 'frequency'. It is, in fact, marked 'frequency density', which is the frequency per unit of class width. (In this example the unit has been taken as 1.)

Frequency density is therefore *frequency ÷ class width*, so that, for example, the frequency density for the class interval 24.5–29.5 is $22/5 = 4.4$, and this is the height of the rectangle.

If you are wondering why it is necessary to use frequency density rather than frequency, the answer is that if, for any reason, the distribution does not give the frequencies for *equal* class intervals, it may still be represented by a histogram.

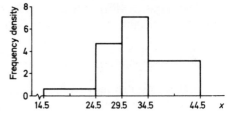

The histogram here shows the data of our example, in which the first two and the last two classes are combined.

The frequency density for the class interval 24.5–29.5 is $22/5 = 4.4$, as before, but for the interval 34.5–44.5 the frequency density is $28/10 = 2.8$.

Discrete data, too, may be given in grouped frequency form, particularly if the sample is a large one.

Here we have a histogram for the following distribution of marks gained by 150 candidates in an examination.

Marks	f
1–10	6
11–20	22
21–30	60
31–40	51
41–50	11
	150

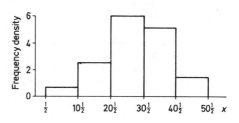

As for the continuous case, the classes have end points $\frac{1}{2}$, $10\frac{1}{2}$, etc.

2.3   Frequency polygons

Although the best way of representing a frequency distribution is the histogram, there is an alternative which is particularly useful when we want to compare the shapes of two distributions. It is called a *frequency polygon*. If a histogram has already been drawn, the mid-points of adjacent rectangles are joined by straight lines (the mid-points of the end rectangles being carried on down to zero.)

Here is a frequency polygon derived from the first histogram on page 8.

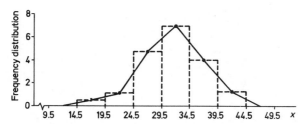

To draw a frequency polygon without first drawing a histogram, all that is needed is a simple frequency chart (see page 6) on which neighbouring points are joined. We shall adopt this procedure in the next example, which shows how to compare two frequency distributions effectively.

EXAMPLE 1   The novels of two contemporary writers, Anthony Powell and the late Evelyn Waugh, are broadly similar in being ironic comedies about twentieth-century society in peace and war. As part of an analysis of their styles of writing a descriptive passage of 100 words is taken from a novel by each, and the number of letters in the words of these samples is counted. The two extracts are:

> *Towards morning the teleprinter's bell sounded. A whole night could pass without a summons of that sort, for here, unlike the formations, was no responsibility to wake at four and take dictation – some brief unidentifiable passage of on the whole undistinguished prose – from the secret radio, calling*

and testing in the small hours. Sleep was perfectly attainable when no raid intervened, though recurrent vibration from one or both machines affirmed next door the same restlessness of spirit that agitated the Duty Officer's room, buzzing all the time with desultory currents of feeling bequeathed by an ever changing tenancy. Endemic as . . .

Anthony Powell, The Military Philosophers

They had lined the aisle; then while the register was being signed, had formed up along this path which led from the door to the motor car. Their finery had excited cries of admiration. As the organ sounded the first notes of the Wedding March they had drawn their swords and held them in a posture for which no drill book has a name, forming an arch over the wedded couple. The bride had smiled right and left looking up at each of them in the eyes, thanking them. The bridegroom held his top hat in his hand and greeted . . .

Evelyn Waugh, Unconditional Surrender

The letter count for these two authors is:

Letters per word	1	2	3	4	5	6	7	8	9	10	11	12	13	14	15
Powell	2	14	15	23	10	4	12	6	5	4	0	2	0	2	1
Waugh	2	13	28	20	18	8	7	2	0	2	0	0	0	0	0

If we tried to compare these two distributions graphically in one diagram by superimposing one histogram on the other the effect would be very confusing. Using a frequency polygon for each, on the other hand, brings out the differences in shape of the two very clearly.

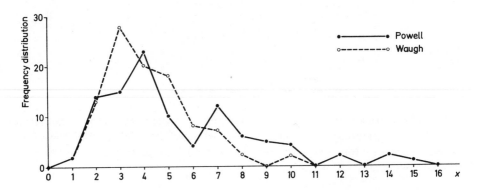

On the strength of these admittedly small samples it would seem that Powell may be inclined to use a greater range of word lengths than Waugh, whose distribution is more compact. To say anything more precise than this would involve comparing the average word length for the two authors, and possibly making some other calculations (see page 18, question 13). Nevertheless, this short discussion of histograms and frequency polygons will serve as an introduction to the possible shapes of some frequency distributions, and we shall end this section by giving the names that are used to describe some special features of distributions.

A *symmetrical* distribution is one whose histogram (or frequency polygon) is symmetrical about a vertical axis.

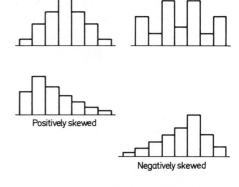

Positively skewed

Negatively skewed

A distribution which is not symmetrical is said to be *skewed*. A distribution which is positively skewed has a 'tail' to the right, a negatively skewed distribution has a 'tail' to the left.

In the Powell–Waugh diagram, it is clear that both distributions are positively skewed and it looks as though the Powell one is more skewed than the Waugh. However, we have not yet defined a measure for skewness.

There are other methods of representing statistical observations graphically such as pie charts and bar charts, but these play no part in advanced work.

EXERCISE 2a

1. One hundred oranges were tested for their vitamin C content, with the following results: (the amount of vitamin C is given to the nearest milligram):

Amount of vitamin C (mg)	25	26	27	28	29	30	31	32	33	34	35
Number of oranges	1	6	4	13	18	21	16	12	6	2	1

Draw a histogram and frequency polygon for this distribution.

2. The number of sparking plugs issued each day from stock in a garage, for 30 working days, were:

```
75    72    61    72    83    74    76    81   101    92
91   108    83    62   100    99    98    76    58    73
102    56    68    57    83    81    79   102    97    69
```

Using classes 55–59, 60–64, etc., draw up the frequency distribution table, and draw a histogram.

Hint

55–59	
60–64	|
65–69	
70–74	|||
75–79	||
80–84	||
⋮	

Reading along the figures in the order given, insert a tick (tally-mark) in the appropriate class. When a fifth mark is made, draw a stroke across the previous four ⃦𝙃𝙃 (rather than \| \| \| \| \|). Then count up the ticks in each class.

3. The figures below show individual marks (out of 50) in an examination paper in Economics taken by sixty candidates:

```
26   17   42   28   23   17   39    3   38   10   44   25    4   10   33
11   38   31   38   23   42   29   40   37   18    6   22   15   15   32
32   15   12   37   38   37   12   29   12   18   27   30   30   12   40
45    4   26   24   22   33   17    8    3    7   42   39    6   41   20
```

Group these marks into classes 3–7, 8–12, etc. and draw up a frequency distribution table. Draw a histogram and frequency polygon.

4.   The following table shows the number of football teams who scored stated numbers of goals on a certain Saturday.

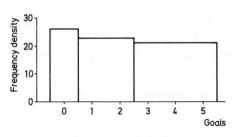

Number of goals	Number of teams
0	27
1 or 2	44
3, 4 or 5	21

The diagram is supposed to be a histogram of the data, but is, in fact, incorrect. Draw a correct diagram.
(MEI)

5.   When the first families had moved into New Town, a survey of the ages of children gave the following results:

Age (years)	Under 1	1 and under 2	2 and under 3	3 and under 5	5 and under 7	7 and under 10
Number	20	24	18	60	64	60

Draw a histogram for this distribution.

6.   The ages of cars belonging to a car-hire firm are classified in the following table:

Age (months)	0–5	6–11	12–17	18–23	24–29	30–35	36–47	48–59	60–83
Number of cars	12	16	28	9	26	26	71	31	26

Draw a histogram. What would the histogram look like if the figures were re-classified as below?

Age (months)	0–23	24–35	36–83

7.   One hundred acres of land growing potatoes in two different counties, A and B, give yields shown in the following distributions:

Yield (tons/acre)	9.4	9.5	9.6	9.7	9.8	9.9	10.0	10.1	10.2	10.3
Frequency A	4	6	14	16	17	13	11	9	7	3
B	2	7	10	11	15	16	16	12	8	3

Draw frequency polygons for each county on the same diagram and comment on the differences between the two distributions.

8.   The following passages are extracts from two books about World War II which are partly fact and partly fiction. Each passage contains 85 words. Do an analysis of the lengths of words used by the two authors (as on page 10), draw frequency polygons, and comment on the results.

He awoke, for about the tenth time that night, and groaned to himself when he realized it still was not dawn. He listened to the sounds of rustling in the wall. He would never hear those sounds again. His mouth was dry with excitement;

he wanted to command the sun to rise, so that they could start out on their journey. The first time he had 'moved house', it had only been a matter of crossing the city; and a miserable change it was too.

D. M. Thomas, *The White Hotel*

With a small ration of bread, they climbed up into cattle wagons. None of the guards who loaded them admitted to knowing where they were going. They squatted on the floorboards in the galley slave fashion. They kept fixed before their minds the map of Central Europe, and made continual judgments about the passage of the sun, gauging their direction by glimpses of light through the small ventilators near the roofs of the wagons. The navigation experts claimed the train was travelling generally south east.

Thomas Keneally, *Schindler's Ark*

Do you think this is a valid way of comparing styles?

2.4   Averages

In comparing two distributions like the word-length examples on page 10 (example 1), it is natural to want to include in the comparison an 'average' value for each of the distributions.

In this context the word 'average' is used for any single statistic which can be taken to represent a 'centre' of a distribution. The three most commonly used are the *mean*, the *median* and the *mode*, and since the mean is the statistic most commonly used we shall deal with it first.

The mean

The mean, or arithmetic mean, is the sum of all the numbers making up the original set of data (before any grouping has taken place) divided by the total number of observations in the set.

Hence the mean of the five numbers in the set

$$0.8, 1.2, 2.8, 3.1, 4.6$$

is $(0.8 + 1.2 + 2.8 + 3.1 + 4.6) \div 5 = 2.5,$

and in general, for a discrete distribution, if the data consist of the numbers

$$x_1, x_2, x_3, \ldots, x_N,$$

The mean $\bar{x} = \dfrac{x_1 + x_2 + x_3 + \ldots + x_N}{N}.$

This formula is expressed more shortly as

$$\bar{x} = \frac{\sum_{i=1}^{N} x_i}{N}.$$

The Σ (sigma) notation will be used constantly from now on, and a full account of it will appear shortly. Meanwhile, you need only know that, for example,

$$\sum_{i=1}^{5} x_i \quad \text{means} \quad x_1 + x_2 + x_3 + x_4 + x_5.$$

Suppose, now, that the discrete data are given in the form of a frequency distribution so that, for example, the marks (out of 20) obtained by 14 students in a test were as follows:

Marks (x)	11	12	14	16
Number of students (f)	2	4	5	3

Since the mark 11 occurs twice, the mark 12 occurs four times and so on, we have

$$\text{the mean} = \frac{(2 \times 11) + (4 \times 12) + (5 \times 14) + (3 \times 16)}{2 + 4 + 5 + 3} = 13.4 \quad \text{to 1 dec. place.}$$

The general result for a frequency distribution when x takes values $x_1, x_2, x_3, \ldots, x_n$ with frequencies $f_1, f_2, f_3, \ldots, f_n$ is

$$\bar{x} = \frac{f_1 x_1 + f_2 x_2 + f_3 x_3 + \ldots + f_n x_n}{f_1 + f_2 + f_3 + \ldots + f_n}.$$

The short form for this expression is

$$\bar{x} = \frac{\sum\limits_{i=1}^{n} f_i x_i}{\sum\limits_{i=1}^{n} f_i} \quad \text{or} \quad \frac{\sum\limits_{i=1}^{n} f_i x_i}{N}$$

$$\text{where} \quad N = \sum_{i=1}^{n} f_i.$$

Note: Whenever there is no ambiguity, the formula may be written without the suffixes.

$$\bar{x} = \frac{\Sigma fx}{\Sigma f} \quad \text{or} \quad \frac{\Sigma fx}{N}, \quad \text{where } N = \Sigma f.$$

EXAMPLE 2   Find the mean of the number of passengers in each of a sample of 100 cars for the distribution on page 6.

The working in the following table is self explanatory.

x	f	fx
0	40	0
1	29	29
2	18	36
3	11	33
4	2	8
	$N = 100$	$\Sigma fx = 106$

The mean number of passengers per car

$$\bar{x} = \frac{\Sigma fx}{N} = \frac{106}{100} = 1.06.$$

The mean is a useful statistic for one sort of comparison between two distributions. If, for example, a table similar to this were compiled during a strike of public transport, the fact that the new mean turns out to be, say, 2.21 is more illuminating than a comparison which just presents the two distributions side by side.

Calculating the mean of a grouped frequency distribution

When the data are given in grouped form (discrete or continuous) the method is the same as above, but as single observations are not available we first have to find some number to represent each class interval. This number is the 'class centre' (or mid-mark), which, as its name suggests is at the centre of the interval and is defined as follows. The *class centre* is the mean of the end points.

In dealing with continuous distributions, care must be taken to get the class centres right as an incorrect choice vitiates all the subsequent working. When the data are fully described, that is to say when we are told to what degree of accuracy the measurements have been made, there is no difficulty, as the following examples show.

Observations	Recorded data	Class interval	Class centre
Lengths (to the nearest cm)	50–59 60–69, etc.	49.5–59.5 59.5–69.5	54.5 64.5
Masses (to the nearest 0.1 kg)	40– 43– , etc.	39.95–42.95 42.95–45.95	41.45 44.45
Marks (integers)	10–14 15–19, etc.	9.5–14.5 14.5–19.5	12 17

When the class centres have been found, they take the place of x in the formula for $\bar{x}$. It is important, however, to realise that the value for $\bar{x}$ found in this way will be *approximate*, since we have to work on the assumption that all the data in a given class interval are uniformly spread over the interval, which is not necessarily the case.

EXAMPLE 3   For the distribution on page 7 giving the lengths (to the nearest millimetre) of a sample of 90 ears of barley, find the approximate mean. (In this chapter most of the examples will contain fairly small sets of observations, so as to avoid obscuring the methods by an excess of computation.)

Class	Class centre (x)	f	fx
14.5–19.5	17	1	17
19.5–24.5	22	4	88
24.5–29.5	27	22	594
29.5–34.5	32	35	1120
34.5–39.5	37	20	740
39.5–44.5	42	8	336
		$N = \Sigma f = 90$	$\Sigma\, fx = 2895$

From the table $\bar{x} \approx \dfrac{\Sigma\, fx}{N} = \dfrac{2895}{90} = 32.2$ to 1 dec. place.

Coding

Students who have calculators will not find the computation of a mean very troublesome. Nevertheless, for frequency distributions involving large numbers much trouble is saved if the data can be simplified before starting on the calculation. Indeed it is not sensible to work with unsimplified data when this can be avoided. This process is called 'coding', and will be described in detail in the next chapter.

For the moment we shall confine ourselves to one kind of coding only, which involves subtracting some convenient number from each value of x. As an illustration, consider the following frequency distribution, alongside a coded table. Since $x = 230$ is near the centre of the distribution and corresponds to the largest frequency, we shall work with a new variable u, where $u = x - 230$.

x	f	fx	u	f	fu
215	2	430	-15	2	-30
220	4	880	-10	4	-40
225	17	3825	-5	17	-85
230	33	7590	0	33	0
235	16	3760	5	16	80
240	5	1200	10	5	50
245	3	735	15	3	45
	$\Sigma f = 80$	$\Sigma fx = 18420$		$\Sigma f = 80$	$\Sigma fu = 20$

Since each value of u is 230 less than the corresponding value of x, $\bar{u}$ will also be 230 less than $\bar{x}$:

$$\bar{u} = \bar{x} - 230 \Rightarrow \bar{x} = \bar{u} + 230.$$

Now $$\bar{u} = \frac{\Sigma fu}{\Sigma f} = \frac{20}{80} = 0.25 \Rightarrow \bar{x} = 0.25 + 230 = 230.25,$$

which agrees, of course, with the value $\dfrac{\Sigma fx}{\Sigma f} = \dfrac{18\,420}{80} = 230.25.$

EXERCISE 2b

1. A man firing at a target obtained the following scores with forty shots:

Score	0	1	2	3	4	5
Number of shots with this score	2	8	11	12	6	1

Calculate the mean score per shot.

2. The numbers of telephone calls received by a Samaritan office between midnight and 6 a.m. on 15 successive nights were as follows:

$$3 \quad 2 \quad 1 \quad 5 \quad 1 \quad 2 \quad 4 \quad 3$$
$$1 \quad 2 \quad 3 \quad 3 \quad 5 \quad 2 \quad 3$$

Draw up a frequency distribution table and calculate the mean.

3. The frequency distribution for the number of matches in a match box, for a sample of 30

of a particular make, is

Number in box	46	47	48	49	50	51	52
Frequency	1	2	5	10	7	4	1

Draw the histogram of the distribution, determine the mean and mark it in the diagram.

4.   After an earthquake in Italy, a random sample of owners of houses in a village unaffected by the disaster said that they were prepared to give refuge to children from damaged homes. In detail the offers of accommodation were as follows:

Number of children	0	1	2	3	4	5	6 or more
Number of house owners	18	16	28	9	2	2	0

Find the mean of this distribution.

The total number of houses that might be available for taking in children was known to be 281. Estimate the number of children that the village as a whole could take in.

5.   Which of the two areas in exercise 2a, question 7 gives the greater mean yield per acre?

6.   A hospital librarian recorded the number of books borrowed weekly by patients in one particular ward during 30 successive weeks. The results were:

Number borrowed	3–5	6–8	9–11	12–14	15–17
f	4	6	8	7	5

Draw a histogram, making clear what the end points are, and calculate an approximate mean.

7.   In an experimental farm, 25 hectare plots of land growing potatoes gave yields as follows, in tonnes per hectare, to the nearest 100 kg:

Yield	3.4–3.6	3.7–4.0	4.1–4.4	4.5–4.9
f	2	6	9	8

Draw a histogram, marking the end points clearly and estimate the mean yield. In about how many plots was the mean exceeded?

8.   The percentage of vitamin C (to the nearest 0.1%) in 45 samples of a proprietary medicine was found to be distributed as follows:

%	3.45–3.85	3.85–4.25	4.25–4.65	4.65–5.05	5.05–5.45
f	11	2	8	19	5

Estimate the mean.

9.   The masses in kg of 60 boys of a certain age were found to give the following frequency distribution:

Mass	35–	40–	45–	50–	55–	60–	65–	70–
Number of boys	3	7	10	17	13	6	3	1

Estimate the mean mass.

10.   The marks (out of 30) obtained by 25 pupils in a test were:

19	22	15	15	8	26	16	14	9	13	9	16	20
17	16	15	18	11	14	21	12	20	21	16	17	

(a) Without grouping, find the mean mark.
(b) Arrange the data in classes, $7\frac{1}{2}$–$12\frac{1}{2}$, $12\frac{1}{2}$–$17\frac{1}{2}$ etc., and calculate the approximate mean.

11.   A sample of 100 sentences was taken from a book and the number of words per sentence was counted. The results were as follows:

Number of words per sentence	1–5	6–10	11–15	16–20	21–25	26–30	31–35	36–40	41–45
Frequency	16	22	18	11	12	9	8	2	2

   Code these figures by subtracting from each mid-mark the mid-mark of the class interval 11–15 and estimate the mean number of words per sentence.

12.   A student musician 'busking' in an Underground station found that, on 20 successive days he made the following amounts (in £s, to the nearest 10p).

3.2	4.1	4.7	3.6	5.4	4.4	5.7	5.0	3.5	5.1
3.6	5.3	4.9	5.6	5.9	5.6	4.2	4.2	4.0	3.3

   Group these figures into 6 classes, the first class being 3.15–3.65 and estimate the mean.
   What will your estimate of the mean be if the figures are grouped into 10 classes, the first being 3.15–3.45?

13.   Use the frequency distributions given on page 10 to find the mean word length for each of the two passages from books by Anthony Powell and Evelyn Waugh.

14.   Find the mean word length for the two passages from books by Thomas Keneally and D. M. Thomas given on pages 12–13. (*Note:* The answers to questions 13 and 14 only show that the mean word length of *samples* from the two books differ. Whether or not these differences are significant for the books as a whole is a matter that will be looked into later.)

The median

The median of a set of observations is the *middle* value when the observations are arranged in order of size.
   Suppose 7 pupils gain the following marks in a test:

$$31 \quad 38 \quad 43 \quad 47 \quad 50 \quad 52 \quad 55$$

then the median mark is 47 (the 4th in order from either end).
   If, subsequently, an additional pupil takes the same test so that there are now 8 marks:

$$31 \quad 38 \quad 38 \quad 43 \quad 47 \quad 50 \quad 52 \quad 55$$

there is no middle value and the median is taken as the mean of the middle two, in this case $(43+47)/2 = 45$ (which can be thought of as the $4\frac{1}{2}$th in order from either end).
   In general, then, if the number of observations is N,

$$\text{the median value is the } \frac{N+1}{2}\text{th observation,}$$

where, for example, if $N = 12$, the $6\frac{1}{2}$th observation is the mean of the 6th and 7th.

Calculating the median of a frequency distribution

Finding the median for a discrete, ungrouped, distribution presents no problems. For the distribution below,

x	f
11	1
12	3
15	5
16	1
17	1
	$N = 11$

the median is the 6th value of x, if these were laid out in order:

$$11 \quad 12 \quad 12 \quad 12\ldots$$

and is clearly 15.

For a continuous distribution (or a grouped discrete distribution) it is only possible to make an estimate of the median. There are two possible methods, both of which use the idea of *cumulative frequency*.

EXAMPLE 4   Estimate the median length of an ear of barley from the frequency distribution, given on page 7.

The first step is to establish the class boundaries (1st column).

The cumulative frequencies are given in the 3rd column which shows that there are: $1+4 = 5$ ears whose length is less than 24.5 mm, $1+4+22 = 27$ ears whose length is less than 29.5 mm, and so on.

Class	Frequency (f)	Cumulative frequency (cf)
14.5–19.5	1	1
19.5–24.5	4	5
24.5–29.5	22	27
29.5–34.5	35	62
34.5–39.5	20	82
39.5–44.5	8	90

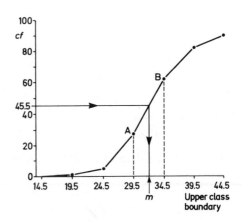

The procedure from now on is best understood if we draw a *cumulative frequency polygon*, plotting cumulative frequency against the *upper* class boundary:

1 against 19.5
5 against 24.5 etc.

These points are then joined by straight lines, since we have to make the assumption that the lengths are uniformly spread in the intervals.

Since $N = 90$, the median must be the $45\frac{1}{2}$th value. From the graph we see that it is the value m, since half the distribution will be less than m, and half greater. (With a large sample like this there is, in fact little point in taking the $45\frac{1}{2}$th rather than the 45th value as the method only gives an approximate answer based on the uncertain assumption of uniform spread in the class intervals.)

From the table and the graph, then, the median is somewhere in the class 29.5–34.5 and the diagram opposite shows the relevant portion of the cumulative frequency polygon, from A to B.

By similar triangles,

$$AQ = \frac{PQ}{BR} \times AR,$$

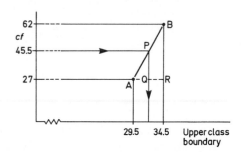

so that the median

$$\approx 29.5 + \frac{45.5 - 27}{62 - 27} \times 5$$

$$\approx 32 \, \text{mm}.$$

(*Note:* It is not strictly necessary to draw the whole cumulative frequency polygon to calculate the median. A sketch, like the one above, of the part that is needed will do.)

The estimated median in this example is close to the estimated mean (32.2). This will always be the case for a fairly symmetrical distribution.

The method just described is called *linear interpolation*, 'linear' since the points on the graph are joined by straight lines. An alternative method, which does not involve assuming an even spread of the observations in the class intervals, is to join the points on the cumulative frequency diagram by a smooth curve, and to estimate the median by drawing.

In this example, the part of the curve we are concerned with is so nearly straight that the approximate median comes out as the same value (32) as before.

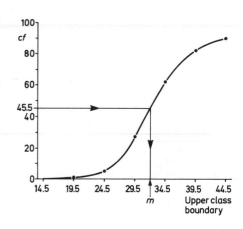

Cumulative frequency curves like this are not used solely for estimating medians. They have other uses, which will be described in the next chapter.

Mean and median

In this book we shall generally use the mean as centre since it is more reliable when statistical inferences are made (these being, as was pointed out in chapter 1, the principal objects of statistical mathematics). Nevertheless, the median is a useful measure for some purposes. It is, for example, preferable to the mean when a distribution contains values at the extremes which may be misleading in a particular context. Suppose a distribution of examination marks contained a few near 100, while the bulk of the distribution was of marks between 20 and 70. The mean would be calculated by using *all* the marks, and the marks at the top might have an effect out of proportion to their importance. The median, on the other hand, would not be much affected by these extreme values and would therefore be a better representative as a 'centre' or average.

The median also comes into its own in what is known as 'non-parametric' or 'distribution-free' estimation, which is the subject of chapter 15.

The mode

The least important of the three averages mentioned at the beginning of this section is the mode. As the adjective 'modish' (or fashionable) suggests, it is the value that occurs most often in a set of data.

For the discrete set 2.3, 3.6, 3.7, 4.8, 5.7 there is no mode.

For the discrete frequency distribution

x	2.3	3.6	3.7	4.8	5.7
f	3	8	4	2	1

the mode is clearly 3.6. If the frequency for 4.8 had been 8 instead of 2, there would have been two modes, 3.6 and 4.8.

A grouped frequency distribution has no identifiable mode and it is only possible to define a modal class (or modal classes). In example 4 above, the modal class is 29.5–34.5.

Since the modal class depends on the method of grouping, it is virtually useless as a central representative of the distribution. The word 'modal' does however serve a useful purpose in describing the general shape of distributions as being 'unimodal' (having one modal class) or bimodal (having two modal classes) and so on.

EXERCISE 2c

1.   Write down the median and mode of the following frequency distributions:

(a)

x	2	5	6	9	11
f	4	7	8	5	6

(b)

x	40	60	80	100	120
f	6	3	12	10	9

2.   Write down the median and mode of the frequency distribution given in exercise 2b, question 1.

3.   Write down the median and mode of the frequency distribution given in exercise 2b, question 3.

4.   For the following frequency distribution, draw a cumulative frequency polygon and use it to find the median.

x	2.5–3.5	3.5–4.5	4.5–5.5	5.5–6.5	6.5–7.5	7.5–8.5
f	7	31	33	17	11	1

5.   The lengths of 125 ears of barley (to the nearest millimetre) gave the following frequency distribution:

Length (mm)	15–19	20–24	25–29	30–34	35–39	40–44	45–49	50–54	55–59	60–64	65–69
Frequency	1	3	5	15	20	28	22	18	7	4	2

Taking as upper class limits 19.5, 24.5,..., draw a cumulative frequency polygon and estimate, by linear interpolation, the median.

6.   Use linear interpolation to find the median of the frequency distribution given in exercise 2b, question 6.

7.   Use linear interpolation to find the median of the frequency distribution given in exercise 2b, question 7.

8.   Using the data of question 5, draw a cumulative frequency *curve* and use it to find an estimate of the median, which does not assume a uniform spread of data in the class intervals.

2.5   Σ algebra

The Σ notation was briefly introduced on page 13 with the simple examples:

$$\sum_{i=1}^{N} x_i = x_1 + x_2 + \ldots + x_N,$$

$$\text{and}\ \sum_{i=1}^{n} f_i x_i = f_1 x_1 + f_2 x_2 + \ldots + f_n x_n.$$

Formula involving Σ can often be manipulated to give alternative forms, a process which is called Σ algebra. This process will often be used during the course of this book and you will need to be familiar with the few rules involved.
The basic rules are simple:

Rule 1

$$\sum_{i=1}^{n} (x_i + y_i) = \sum_{i=1}^{n} x_i + \sum_{i=1}^{n} y_i \quad \text{or} \quad \sum (x+y) = \sum x + \sum y.$$

This may be easily proved.

Rule 2

$$\sum_{i=1}^{n} ax_i = a \sum_{i=1}^{n} x_i \quad \text{or} \quad \sum ax = a \sum x$$

where *a* is a constant.

Proof: $$\sum_{i=1}^{n} ax_i = ax_1 + ax_2 + \ldots + ax_n$$

$$= a(x_1 + x_2 + \ldots + x_n) = a \sum_{i=1}^{n} x_i.$$

Rule 3

$$\sum_{i=1}^{n} c = nc$$

where c is a constant.

Proof: $\quad\quad \sum_{i=1}^{n} c = c+c+c+\ldots+c \quad$ (the sum of n c's)

$$= nc.$$

EXAMPLE 5 Use rules 1, 2 and 3 to write the following in an alternative form:

(a) $\Sigma\ 5x$,   (b) $\Sigma\ (ax+by)$,   (c) $\Sigma\ (2x+3)$,   (d) $\Sigma\ (ax+b)$,

where Σ means $\displaystyle\sum_{i=1}^{n}$.

(a) $\Sigma\ 5x = 5\ \Sigma\ x$.
(b) $\Sigma\ (ax+by) = \Sigma\ ax + \Sigma\ by = a\ \Sigma\ x + b\ \Sigma\ y$.
(c) $\Sigma\ (2x+3) = \Sigma\ 2x + \Sigma\ 3 = 2\ \Sigma\ x + 3n$.
(d) $\Sigma\ (ax+b) = \Sigma\ ax + \Sigma\ b = a\ \Sigma\ x + bn$.

In the next chapter we shall be using the following:

$$\sum_{i=1}^{n} x_i^2 = x_1^2 + x_2^2 + \ldots + x_n^2.$$

$$\sum_{i=1}^{n} f_i x_i^2 = f_1 x_1^2 + f_2 x_2^2 + \ldots + f_n x_n^2.$$

$$\sum_{i=1}^{n} (x_i-a)^2 = (x_1-a)^2 + (x_2-a)^2 + \ldots + (x_n-a)^2.$$

$$\sum_{i=1}^{n} f_i(x_i-\bar{x})^2 = f_1(x_1-\bar{x})^2 + f_2(x_2-\bar{x})^2 + \ldots + f_n(x_n-\bar{x})^2.$$

EXAMPLE 6 Find alternative forms for the following:

(a) $\Sigma\ (x^2-2x+4)$,   (b) $\Sigma\ (2x-3)^2$,   (c) $\Sigma\ (x-a)^2$,

where Σ means $\displaystyle\sum_{i=1}^{n}$.

(a) $\Sigma\ (x^2-2x+4) = \Sigma\ x^2 - \Sigma\ 2x + \Sigma\ 4 = \Sigma\ x^2 - 2\ \Sigma\ x + 4n$.
(b) $\Sigma\ (2x-3)^2 = \Sigma\ (4x^2-12x+9) = \Sigma\ 4x^2 - \Sigma\ 12x + \Sigma\ 9 = 4\ \Sigma\ x^2 - 12\ \Sigma\ x + 9n$.
(c) $\Sigma\ (x-a)^2 = \Sigma\ (x^2-2ax+a^2) = \Sigma\ x^2 - \Sigma\ 2ax + \Sigma\ a^2 = \Sigma\ x^2 - 2a\ \Sigma\ x + na^2$.

EXERCISE 2d

1.   Use rules 1, 2 and 3 to write the following in an alternative form:

(a) $\Sigma\ (3x+4y)$,   (b) $\Sigma\ (4x-5)$,   (c) $\Sigma\ (3x+1)$,

(d) $\Sigma\ (2x+c)$,   (e) $\Sigma\ (x+2y+3z+k)$.

2. Find alternative forms for

$$\text{(a) } \Sigma\ (x+5)^2, \quad \text{(b) } \Sigma\ (ax+b)^2.$$

3. Find alternative forms for

$$\text{(a) } \sum_{i=1}^{n} f_i(x_i-a), \quad \text{(b) } \sum_{i=1}^{n} f_i(x_i-a)^2.$$

4. Prove that $\Sigma\ (x-\bar{x})^2 = \Sigma\ x^2 - n\bar{x}^2$, where $\bar{x} = (\Sigma\ x)/n$.

EXERCISE 2e (miscellaneous)

1. The heights of 100 fourteen-year-old boys are given in the following table:

Height (cm)	145	146	147	148	149	150	151	152	153	154
Frequency	1	4	8	15	20	21	16	9	4	2

Calculate the mean height. (C)

2. The number of runs scored by 100 batsmen in cricket matches on a certain day was noted and the following frequency distribution obtained:

Number of runs	0–9	10–29	30–49	50–99
Frequency	48	34	8	10

Draw a histogram to illustrate this distribution.
Estimate the mean number of runs scored. (MEI)

3. The table below gives the number of shoots produced by 50 plants in a botanical research establishment.

Number of shoots	0–4	5–9	10–14	15–19	20–24	25–29	30–34	35–39	40–44	45–49
Frequency	1	1	1	6	17	16	4	2	1	1

Draw a histogram to illustrate these results, and estimate the mean number of shoots per plant. (C)

4. In a survey of the age distribution of teachers in schools the following data were obtained:

Age (in completed years)	20–29	30–39	40–49	50–59	60–69
Frequency (%)	24	28	26	19	3

State the modal class of this distribution and calculate the mean age, giving your answer to the nearest year. (MEI)

5. The average weekly earnings of women in 56 principal trades shortly after World War II were calculated, the distribution being shown in the frequency table below.

Earnings (shillings)	50–	55–	60–	65–	70–	75–80
Number of trades	1	4	19	17	13	2

(a) Calculate the mean of the average earnings in the 56 trades.
(b) Construct a cumulative frequency polygon from the data and estimate the median wage.
   In how many trades were the average earnings higher than the mean? (MEI)

6. The following table gives 72 values of a certain variable:

21	30	20	45	27	55	25	22	30	18	20	24
18	45	36	20	40	45	19	24	30	25	45	24
25	40	23	18	50	28	20	30	25	14	24	30
21	32	29	20	35	22	30	21	24	31	25	30
23	45	22	33	19	30	24	15	25	24	25	20
45	21	30	18	30	20	25	40	22	24	20	38

Using class intervals centred at 15, 20, ..., 55, draw up a frequency distribution for these values and draw a cumulative frequency curve.
Estimate the median of the distribution (a) from the curve, (b) by linear interpolation.
(MEI)

7. In order to estimate the mean length of leaves from a certain tree, a sample of 100 leaves was chosen and their lengths measured correct to the nearest millimetre. A grouped frequency table was set up and the results were as follows:

Mid-interval value (cm)	2.2	2.7	3.2	3.7	4.2	4.7	5.2	5.7	6.2
Frequency	3	5	8	12	18	24	20	8	2

(a) Display the table in the form of a frequency polygon and describe the distribution exhibited by this polygon.
(b) Calculate an estimate for the mean leaf length.
(c) What are the boundaries of the interval whose mid-point is 3.7 cm?
(d) Construct a cumulative frequency table and use it to estimate the sample median. (SUJB)

8. A random sample of 120 male computer operators had their diastolic blood pressure measured to the nearest millimetre. The table below summarises the results.

Blood pressure (mm)	55–59	60–64	65–69	70–74	75–79	80–84	85–89	90 and over
Frequency	1	4	10	21	35	29	13	7

Determine the median from a suitable graphical representation. (AEB 1982)

9. The following table gives an analysis by numbers of employees of the size of UK factories of less than 1000 employees manufacturing clothing and footwear.

Number of employees	11–19	20–24	25–99	100–199	200–499	500–999	Total
Number of factories	1500	800	2300	700	400	100	5800

Calculate as accurately as the data will allow the mean and median of this distribution. If 90 % of the factories have less than N employees, estimate N. (O & C)

3 Measures of spread

The mean is a statistic which, as we have seen, summarises a frequency distribution in a simple way. By itself, however, it is often not informative enough. If, for example, we wanted to compare the heights of 100 Italians and 100 Spaniards, it would certainly be important to know how the mean heights of the two samples compared, but we would get a fuller comparison if we also knew how the *spread* of the two distributions compared or, to put it another way, how each of the distributions as a whole varied.

Now the simplest measure of spread to calculate is the *range*. The range of a set of observations is merely the difference between the largest and smallest observations in the set. This is not a very useful statistic since it depends only on two observations and ignores the way the observations between these two are distributed. It also has the important disadvantage that the extreme values may be atypical. Suppose, for example, the distributions of the weekly wages of the non-medical staff in two hospital departments were identical apart from the highest and lowest wages – £100 and £40 in one case, £120 and £35 in the other. To use the ranges (£60 and £85 respectively) as a way of comparing these two distributions would be totally useless. Another more general disadvantage is that the range of a sample from a given population tends to increase as the sample size increases. There is a way of getting round the difficulty that the extreme values may be atypical, by calculating the range of the central portion of the distribution (see page 42). Nevertheless, it is of limited use and is not easy to develop mathematically.

3.1 Deviations from the mean

A more promising way forward is to use the *deviations* of each observation from the centre (usually the mean, but sometimes the median).

Mean deviation

If the distribution consists of the N observations $x_1, x_2, \ldots, x_N$ whose mean is $\bar{x}$, the deviations from the mean are $x_1 - \bar{x}, \ x_2 - \bar{x}, \ldots, x_N - \bar{x}$, and an *average* of these deviations would certainly represent the spread of the distribution. (The reason for saying 'an average' rather than 'the average' will become clear very shortly.)

Let us consider

$$\frac{\sum_{i=1}^{N} (x_i - \bar{x})}{N},$$

the mean of the deviations as a possible measure of the spread.

Suppose eight men have heights (in centimetres):

x: 168   168   170   172   172   172   173   173

with mean height 171 cm, so that the deviations from the mean are:

$x - \bar{x}$: -3 -3 -1 1 1 1 2 2

Clearly the mean of the deviations is 0, which is unhelpful. In fact the mean of the deviations will *always* be zero since

$$\Sigma (x - \bar{x}) = \Sigma x - N\bar{x} \Rightarrow \frac{\Sigma (x - \bar{x})}{N} = \frac{\Sigma x}{N} - \bar{x} = \bar{x} - \bar{x} = 0.$$

One way round this difficulty is to define the spread as the mean of the *absolute values* of the deviations, ignoring the signs, and use

$$\frac{\sum_{i=1}^{N} |x_i - x|}{N}$$

where $|-x| = |+x| = x$.

This statistic is called the *mean deviation* (or the *mean absolute deviation*).

For the heights of the eight men above, the mean deviation

$$= \frac{3 + 3 + 1 + 1 + 1 + 1 + 2 + 2}{8} = 1.75 \, \text{cm}.$$

Standard deviation

Although the mean deviation (or mean absolute deviation) is a useful statistic for comparing the spreads of two distributions, the modulus sign $|\,|$ makes it difficult to handle mathematically.

(Consider a simple piece of algebraic addition: $|x - 2| + |x - 3|$. If $x > 3$, its value is $x - 2 + x - 3 = 2x - 5$. If $2 < x < 3$, its value is $x - 2 + 3 - x = 1$. If $x < 2$, its value is $2 - x + 3 - x = 5 - 2x$.)

There is, however, another way of avoiding the difficulty over signs and that is to *square* each deviation before finding the average. This gives a statistic

$$\frac{\sum_{i=1}^{N} (x_i - \bar{x})^2}{N}$$

which is called the *mean squared deviation*, but as we need, as a measure of spread, a statistic which is in the *same units* as the original observations, we take the square root of the mean squared deviation and call it the *standard deviation*.

Hence:

$$\text{Standard deviation (s.d.)} = \sqrt{\text{mean squared deviation}}$$

$$= \sqrt{\left[\frac{\sum\limits_{i=1}^{N}(x_i - \bar{x})^2}{N}\right]}.$$

Thus, for the men's heights, the mean squared deviation

$$= \frac{9+9+1+1+1+1+4+4}{8} = 3.75\,\text{cm}^2$$

and

$$\text{s.d.} = \sqrt{3.75} = 1.94\,\text{cm}.$$

Calculating the mean and standard deviation for a frequency distribution

We shall again use the distribution of men's heights, but with the data given in a frequency distribution.

EXAMPLE 1   Calculate the mean and standard deviation for the distribution:

Height (cm)	168	170	172	173
Frequency	2	1	3	2

x_i	f_i	$f_i x_i$	$x_i - \bar{x}$	$(x_i - \bar{x})^2$	$f_i(x_i - \bar{x})^2$
168	2	336	-3	9	18
170	1	170	-1	1	1
172	3	516	1	1	3
173	2	346	2	4	8
	$N = 8$	1368			30

From the 3rd column, $\bar{x} = \Sigma\, fx/N = 1368/8 = 171$. This value is then used in the 4th and 5th columns. In the 6th column the squared deviations must be multiplied by the appropriate frequency.

$$\text{Thus}\quad \frac{\Sigma\, f(x - \bar{x})^2}{N} = \frac{30}{8}\quad \text{and}\quad \text{s.d.} = \sqrt{\frac{30}{8}} = 1.94\quad \text{(2 dec. places)}.$$

3.2   Summary of formulae for mean and standard deviation

We now have the following formulae for the mean and the standard deviation:
  When individual observations are listed:

x_1	x_2	$\cdots$	x_N

$$\bar{x} = \frac{\Sigma\, x}{N}\qquad \text{s.d.} = \text{standard deviation} = \sqrt{\left[\frac{\Sigma\,(x - \bar{x})^2}{N}\right]}.$$

For a discrete frequency distribution:

x_1	x_2	$\ldots$	x_n
f_1	f_2	$\ldots$	f_n

where $\Sigma f_i = N$,

$$\bar{x} = \frac{\Sigma fx}{N} \qquad \text{s.d.} = \text{standard deviation} = \sqrt{\left[\frac{f(x-\bar{x})^2}{N}\right]}.$$

For a continuous frequency distribution, the formulae are the same as those for a discrete distribution, where $x_1, x_2 \ldots$ are the class centres. The results will therefore be approximate.

Note: There is no generally accepted symbol for the standard deviation of a sample as defined above. The obvious candidate, '*s*', is not available as, by international agreement, it is used for another purpose. '*S*' is found in some text-books but as capital letters have a special significance in statistics work we will not use it.

3.3   Variance

Although we have been concerned with finding a measure of spread, the standard deviation, the mean squared deviation itself plays a very prominent part in much later work. It is given the name *variance*, so that

$$\text{Variance of } x = \text{var}(x) = \frac{\Sigma f(x-\bar{x})^2}{N} = (\text{s.d.})^2.$$

EXERCISE 3a

1.   A sample of 11 seed pods has the following number of seeds per pod:

$$4 \quad 5 \quad 5 \quad 6 \quad 8 \quad 9 \quad 12 \quad 14 \quad 15 \quad 19 \quad 24$$

Calculate the mean absolute deviation and the standard deviation.

2.   In eight international gymnastic competitions two regular competitors scored as follows in the parallel bars part of the competition:

Competitor A	25	25	26	20	27	30	24	23
Competitor B	23	26	25	25	$23\frac{1}{2}$	24	$26\frac{1}{2}$	27

Calculate the mean and standard deviation of the scores of each competitor. Which competitor was the more consistent performer on parallel bars?

3.   A man times two different routes from home to his place of work by car. On eight successive days the times for each, in minutes, are shown in the table

Route A	14	15	15	10	18	20	15	13
Route B	11	16	16	15	$15\frac{1}{2}$	13	$16\frac{1}{2}$	17

Calculate the mean and the standard deviation of these items for each of the two routes. On the basis of these results, which route seems the better?

4. Find the mean and the standard deviation for the following frequency distribution:

x	0	1	2	3	4	5	6
f	4	14	10	6	3	2	1

5. In a public examination, passes were awarded in six grades numbered from 1 to 6. The grades of 50 pupils were distributed as follows:

Grade	1	2	3	4	5	6
Number reaching grade	7	13	14	7	7	2

Calculate the mean and standard deviation of these pass grades.

6. Students in a hostel were asked how many baths they had taken during a given fortnight. The results were:

Number of baths	0	1	2	3	4	5	6	7	8	9	10
Frequency	0	2	3	8	4	4	3	6	3	2	1

Calculate the mean and standard deviation.

7. A record was kept of the number of goals scored in a season by the village football teams of Loose Chippings and Cumber Bottom. The results are shown in the table:

Loose Chippings	Number of goals	1	2	3	4	5	6	7
	Number of matches	1	4	6	8	6	4	1
Cumber Bottom	Number of goals	1	2	3	4	5	6	7
	Number of matches	2	3	4	4	4	3	2

Which of the two teams was the more consistent?

8. For the following grouped frequency distribution calculate approximate values of the mean and the standard deviation.

x	8–12	13–17	18–22	23–27	28–32
f	6	8	12	8	6

9. A research team from the Sociology Department at Neasden University observed 50 regular patrons at the *Goat and Compasses* with a view to establishing the structure of their drinking habits. To this end they carefully noted the time each man took to finish a pint of beer. The results were as follows:

Time (min)	0–	10–	20–	30–	40–50
Frequency	15	24	8	2	1

Estimate the mean and the standard deviation.

Draw a histogram and indicate, by drawing vertical lines, the proportion taking up to one, two, three standard deviations more, or less, than the mean time.

Hence complete a table like the one shown below, which appeared in the team's research report:

'Estimated number of patrons in a drinking situation taking less than the time shown to down a pint, measured in units of standard deviation from the mean.'

Time (min)	3	2	1	0	−1	−2	−3
Number	49	48	.	.	.	.	0

A simplified form for variance and standard deviation

Calculation of the standard deviation $= \sqrt{\text{var}(x)}$ using the formulae

$$\text{var}(x) = \frac{\Sigma \ (x - \bar{x})^2}{N} \quad \text{or} \quad \frac{\Sigma \ f(x - \bar{x})^2}{N}$$

presents no difficulty with a calculator, but there is one important disadvantage in using these expressions. If $\bar{x}$ is not an exact number it will have to be rounded off to a certain number of decimal places and the rounding error will be increased by the summation of the squared deviations. However these formulae may be manipulated into alternative forms which not only get round this snag but are also simpler to use.

The alternative simplified forms for variance are:

$$\text{var}(x) = \frac{\Sigma \ x^2}{N} - \bar{x}^2 \quad \text{or} \quad \frac{\Sigma \ fx^2}{N} - \bar{x}^2.$$

This expression may be proved using the basic rules of Σ algebra (page 22).

$$\Sigma \ (x - \bar{x})^2 = \Sigma(x^2 - 2\bar{x}x + \bar{x}^2)$$
$$= \Sigma \ x^2 - \Sigma 2\bar{x}x + \Sigma \ \bar{x}^2$$
$$= \Sigma \ x^2 - 2\bar{x} \ \Sigma \ x + N\bar{x}^2$$

since $2\bar{x}$ and $\bar{x}^2$ are constants for the summation.

Hence
$$\frac{\Sigma \ (x - \bar{x})^2}{N} = \frac{\Sigma \ x^2}{N} - 2\bar{x} \ \frac{\Sigma \ x}{N} + \bar{x}^2$$

$$= \frac{\Sigma \ x^2}{N} - 2\bar{x} \cdot \bar{x} + \bar{x}^2 \quad \left(\text{since} \ \frac{\Sigma \ x}{N} = \bar{x}\right)$$

$$= \frac{\Sigma \ x^2}{N} - 2\bar{x}^2 + \bar{x}^2$$

$$= \frac{\Sigma \ x^2}{N} - \bar{x}^2.$$

The proof that

$$\frac{\Sigma \ f(x - \bar{x})^2}{N} = \frac{\Sigma \ fx^2}{N} - \bar{x}^2$$

follows similar lines.

We therefore have the simplified formulae for standard deviation:

$$\text{s.d.} = \sqrt{\left(\frac{\Sigma x^2}{N} - \bar{x}^2\right)} \text{ for individual observations}$$

$$\text{or } \sqrt{\left(\frac{\Sigma fx^2}{N} - \bar{x}^2\right)} \text{ for a frequency distribution, where } N = \Sigma f.$$

As well as simplifying calculations, these results are of great importance in much theoretical work.

EXAMPLE 2 Calculate the mean, variance and standard deviation of the following numbers:

$$6 \quad 8 \quad 9 \quad 11 \quad 12 \quad 19$$

$$\bar{x} = \frac{\Sigma x}{N} = \frac{65}{6} = 10.833 \quad \text{to 3 dec. places.}$$

$$\frac{\Sigma x^2}{N} = \frac{36 + 64 + 81 + 121 + 144 + 361}{6} = 134.5.$$

Hence $\text{var}(x) = 134.5 - (10.833)^2 = 17.146$

and   s.d. $= \sqrt{17.146} = 4.14$ to 2 dec. places.

EXAMPLE 3   Find the mean and standard deviation of the following frequency distribution:

x	1	2	3	4	5
f	3	15	20	10	2

The calculation is set out in the following table:

x	f	fx	fx^2
1	3	3	3
2	15	30	60
3	20	60	180
4	10	40	160
5	2	10	50
	$N = 50$	143	453

← The final column is most easily calculated by multiplying the number in column 3 by the number in column 1.

$$\bar{x} = \frac{143}{50} = 2.86. \qquad \text{s.d.} = \sqrt{\left(\frac{453}{50} - 2.86^2\right)} = 0.94 \quad \text{to 2 dec. places.}$$

EXAMPLE 4   If a sample of 80 match-boxes containing an average of 48.2 matches per box and standard deviation 1.4 matches, is combined with a sample of 40 boxes containing an average of 49 matches per box and standard deviation 0.9 matches, what are the mean and standard deviation of the combined sample of 120?

This kind of problem can only be solved using the simplified form for standard deviation.

We shall call the numbers of matches in the individual boxes of the two samples respectively

$$x_1, x_2, \ldots, x_{80} \quad \text{and} \quad y_1, y_2, \ldots, y_{40}$$

with corresponding means and variances $\bar{x}$ and var(x), $\bar{y}$ and var(y).

Then $\bar{x} = \dfrac{\Sigma\ x}{80} \Rightarrow \Sigma\ x = 80 \times 48.2 = 3856$

$\bar{y} = \dfrac{\Sigma\ y}{40} \Rightarrow \Sigma\ y = 40 \times 49 = 1960$

so that $\Sigma\ x + \Sigma\ y = 5816$.

The mean of the combined sample is therefore

$$\frac{\Sigma\ x + \Sigma\ y}{120} = \frac{5816}{120} = 48.47 \quad \text{to 2 dec. places.}$$

Now

$$\text{var}(x) = \frac{\Sigma\ x^2}{80} - \bar{x}^2$$

$$\Rightarrow \Sigma\ x^2 = 80[\text{var}(x) + \bar{x}^2] = 80(1.4^2 + 48.2^2)$$

$$= 186\,016.$$

Similarly

$$\text{var}(y) = \frac{\Sigma\ y^2}{40} - \bar{y}^2$$

$$\Rightarrow \Sigma\ y^2 = 40[\text{var}(y) + \bar{y}^2] = 40(0.9^2 + 49^2)$$

$$= 96\,072.4.$$

Hence $\Sigma\ x^2 + \Sigma\ y^2 = 282\,088.4$ and the standard deviation of the combined sample is

$$\sqrt{\left(\frac{282088.4}{120} - 48.47^2\right)} = 1.18 \quad \text{to 2 dec. places.}$$

EXERCISE 3b

1. Find the mean and standard deviation of the numbers

$$1 \quad 3 \quad 4 \quad 6 \quad 10$$

(a) using the formula $\sqrt{\left[\dfrac{\Sigma\ (x - \bar{x})^2}{N}\right]}$ for standard deviation,

(b) using the formula $\sqrt{\left(\dfrac{\Sigma\ x^2}{N} - \bar{x}^2\right)}$.

2. Find the mean and standard deviation of the numbers:

$$2 \quad 3 \quad 5 \quad 6 \quad 8 \quad 9 \quad 10 \quad 12$$

3. The number of goals scored by an amateur football team during one season were as follows:

Goals per match	0	1	2	3	4 or more
Number of matches	8	7	4	6	0

Calculate the mean, standard deviation and variance of the number of goals per match.

4. Calculate the mean and standard deviation of the following frequency distribution:

x	1	2	3	4	5	6
f	12	15	11	30	22	10

5. The number of shoots produced by 50 plants in a botanical experiment were:

Number of shoots	0	1	2	3	4	5
Frequency	6	17	12	9	4	2

Fifty similar plants reared in specially treated soil gave the following distribution:

Number of shoots	0	1	2	3	4	5
Frequency	5	15	11	11	6	2

Calculate the mean and standard deviation for each of these distributions and comment on the results.

6. If k is any number, prove that

$$\text{var}(x) = \frac{\Sigma\ (x-\bar{x})^2}{N} = \frac{\Sigma\ (x-k)^2}{N} - (\bar{x}-k)^2.$$

(This shows that subtraction of an arbitrary number from the values of x makes no difference to the variance.)

7. Find the mean and standard deviation of the numbers

$$1 \quad 2 \quad 3 \quad 4.$$

Two tetrahedral dice have the figures 1,2,3,4 on their faces. When one of these dice is thrown, the score is the number on the face in contact with the table.

Write down all the possible *average* scores when the two dice are thrown together (there are sixteen of them) and calculate the mean and standard deviation of these averages.

8. The number of males in a sample of 810 litters of four cats are recorded in the table below:

Number of males	0	1	2	3	4
Number of litters	156	326	242	74	12

Calculate the mean and standard deviation of the number of males in this sample of litters.

(L)

9.   The degree of cloudiness of the sky is measured on an 11 point scale, with the value 0 corresponding to a clear sky and the value 10 corresponding to a completely overcast sky. Observations of the degree of cloudiness were recorded at a particular meteorological station at noon each day during the month of June over 10 consecutive years. The frequency distribution of the 300 observations was as follows:

Degree of cloudiness	0	1	2	3	4	5	6	7	8	9	10
Number of days	64	24	13	9	6	8	7	9	14	27	119

Calculate the sample mean and the sample variance.
Comment on the usefulness (or otherwise) of the sample mean as a measure of location of such a distribution.   (JMB)

10.   A teacher has calculated the mean test mark (out of 50) for 20 pupils to be 30.5 and the standard deviation of the marks to be 4.5. A pupil who was absent for the test did it on the following day and scored 32.
Calculate the mean and standard deviation for the marks of the 21 pupils.

11.   A sample of 10 observations with mean 15 and standard deviation 3 is combined with a sample of 40 observations with mean 18 and standard deviation 4.
Find the mean and standard deviation of the combined sample.

12.   In a survey of television viewing habits 350 householders were asked to keep a record of the number of hours their TV set was used during a given period. The mean turned out to be 12.4 hours with a standard deviation 2.1 hours. A second sample of 237 households produced a mean of 10.7 hours with standard deviation 1.8 hours.
Calculate the mean and standard deviation of both samples taken together.

13.★ If a sample of n_1 observations with mean $\bar{x}_1$ and standard deviation S_1 is combined with another sample of n_2 observations with mean $\bar{x}_2$ and standard deviation S_2, show that the mean of the combined sample is

$$\frac{n_1 \bar{x}_1 + n_2 \bar{x}_2}{n_1 + n_2}.$$

Show also that if S is the standard deviation of the combined sample

$$(n_1 + n_2)S^2 = n_1 S_1^2 + n_2 S_2^2 + \frac{n_1 n_2}{n_1 + n_2}(\bar{x}_1 - \bar{x}_2)^2.$$

3.4   Coding

Those of you who have calculators should not have found the computation encountered so far in this chapter very troublesome, especially as the data have been kept fairly simple.
It is, however, often advisable to simplify the data before embarking on the

calculations. This process is called *coding* and it is important to understand the process as it will be used in other contexts later in the book and has important theoretical consequences.

Coding involves one or other of the following: a change of origin, a change of scale or a change of origin and change of scale.

Change of origin

This involves subtracting some convenient number (usually one that is near the centre of the distribution) from each value of the variable x. We have already used this method in calculating the mean (see page 16).

To illustrate the idea we shall take these data:

x (class centre)	f
17	1
22	4
27	22
32	35
37	20
42	8

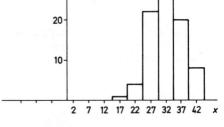

If we subtract 32 from each value of x and work with a new variable u where

$$u = x - 32,$$

we get a new distribution:

u	f
-15	1
-10	4
-5	22
0	35
5	20
10	8

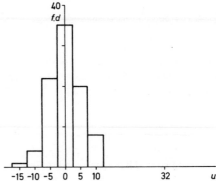

The effect is to change the 'origin' from 0 to 32, and, as the histograms show this results in the mean also being *reduced by 32*, so that

$$\bar{u} = \bar{x} - 32 \quad \text{or} \quad \bar{x} = \bar{u} + 32$$

whereas the *standard deviation (spread) and therefore the variance are unaltered*, so that

$$\text{var}(u) = \text{var}(x).$$

In general, then, if the values of x are $x_1, x_2, \ldots, x_n$ and the values of u are $u_1, u_2, \ldots, u_n$, where

$$u_i = x_i - k$$

then

$$\bar{u} = \bar{x} - k \Rightarrow \bar{x} = \bar{u} + k$$

and

$$\text{var}(x) = \text{var}(u).$$

EXAMPLE 5   Find the mean and standard deviation for the following distribution:

x (class centre)	61	64	67	70	73
f	5	18	42	27	8

We shall take 67 as the origin as it corresponds to the greatest value of f (the modal class, in fact). Setting the working out in tabular form:

x	f	$u = x - 67$	fu	fu^2
61	5	-6	-30	180
64	18	-3	-54	162
67	42	0	0	0
70	27	3	81	243
73	8	6	48	288
	$N = 100$		45	873

$$\bar{u} = \frac{45}{100} = 0.45 \Rightarrow \bar{x} = \bar{u} + 67 = 67.45.$$

$$\text{var}(x) = \text{var}(u) = \frac{873}{100} - 0.45^2 = 8.528 \Rightarrow \text{s.d.} = 2.92 \quad \text{(2 dec. places)}.$$

Note: Without changing the origin, the values of fx^2 would be large and unwieldy, so that the change of origin is worthwhile. As we shall see, further simplification can be made by changing the scale.

Change of scale

If the values of the variable x were, say, 1500, 2500, 3000, 3500 it would seem sensible to change the scale by using the new variable u, where

$$u = \frac{x}{500}$$

giving the values 3, 5, 6, 7.

For a general change of scale $u = x/c$, we would expect the mean and standard deviation to be reduced by the same factor c and the variance by a factor c^2. In other words:

$$\bar{u} = \frac{\bar{x}}{c} \Rightarrow \bar{x} = c\bar{u}$$

and

$$\text{var}(u) = \frac{\text{var}(x)}{c^2} \Rightarrow \text{var}(x) = c^2 \text{var}(u).$$

Similarly, if the values of x were 0.23, 0.27, 0.31, 0.36 putting $u = 100x$ would avoid the complications introduced by the decimal points.

Combining change of origin and change of scale

If, now, we changed the variable x by the relation

$$u = \frac{x-k}{c},$$

so that the origin is changed from 0 to k and the scale is reduced by the factor c, we would again expect the mean $\bar{u}$ to be related in the same way, and the variance to be reduced by the factor c^2, so that

$$\bar{u} = \frac{\bar{x}-k}{c} \Rightarrow \bar{x} = k+c\bar{u} \quad \text{and} \quad \text{var}(u) = \frac{\text{var}(x)}{c^2} \Rightarrow \text{var}(x) = c^2\text{var}(u).$$

A formal proof of these results can be found in the Appendix (page 48). For the present, we shall simply assume these are correct.

EXAMPLE 6   Using the data of example 5, find the mean and standard deviation by changing the origin to 67 and changing the scale by dividing by 3.

Note: The change of scale suggested is 3, since the class width is 3.

x (class centre)	f	$x-67$	$u = \dfrac{x-67}{3}$	fu	fu^2
61	5	-6	-2	-10	20
64	18	-3	-1	-18	18
67	42	0	0	0	0
70	27	3	1	27	27
73	8	6	2	16	32
	$N = 100$			15	97

$$\bar{u} = \frac{15}{100} = 0.15 \Rightarrow \bar{x} = 67+3\bar{u} = 67+3\times0.15 = 67.45.$$

$$\text{var}(u) = \frac{97}{100} - 0.15^2 = 0.9734$$

$$\Rightarrow \text{var}(x) = 9\times0.9734$$

$$\Rightarrow \text{s.d.} = 2.92 \quad \text{(2 dec. places)}.$$

In this example we used the coding

$$u = \frac{x-(\text{a class centre})}{\text{class width}}$$

a method which is strongly recommended when the values of x are in arithmetic progression, as often happens.

As a final illustration of coding we shall look at an example of a frequency distribution, starting from scratch with the observed data.

EXAMPLE 7   The weights, in pounds (to the nearest pound) of 95 rugby forwards were distributed as follows:

Weight (lb)	160–179	180–199	200–219	220–239	240–259	260–279
Frequency	6	30	35	16	7	1

Estimate the mean weight and the standard deviation.

The classes are 159.5–179.5 etc. so that the class width is 20. In the table that follows the origin is taken as 209.5 and the coding is

$$u = \frac{x - 209.5}{20}$$

so that $\bar{x} = 209.5 + 20\bar{u}$ and $\text{var}(x) = 400\,\text{var}(u)$.

Class	x (*class centre*)	f	$u = (x - 209.5)/20$	fu	fu^2
159.5–179.5	169.5	6	−2	−12	24
179.5–199.5	189.5	30	−1	−30	30
199.5–219.5	209.5	35	0	0	0
219.5–239.5	229.5	16	1	16	16
239.5–259.5	249.5	7	2	14	28
259.5–279.5	269.5	1	3	3	9
		$N = 95$		−9	107

$\bar{u} = -9/95 \Rightarrow \bar{x} = 209.5 + 20(-9/95) = 207.60$ (2 dec. places).

$$\text{var}(u) = 107/95 - (-9/95)^2 = 1.117$$

$$\Rightarrow \text{var}(x) = 400 \times 1.117 = 446.94$$

$$\Rightarrow \text{s.d.} = 21.14 \quad \text{to 2 dec. places.}$$

3.5   Summary of coding

	Change	Mean	Variance
Origin	$u = x - k$	$\bar{x} = k + \bar{u}$	$\text{var}(x) = \text{var}(u)$
Scale	$u = \dfrac{x}{c}$	$\bar{x} = c\bar{u}$	$\text{var}(x) = c^2\,\text{var}(u)$
Origin and scale	$u = \dfrac{x - k}{c}$	$\bar{x} = k + c\bar{u}$	$\text{var}(x) = c^2\,\text{var}(u)$

In later theoretical work it will be found useful to have expressions for $\bar{x}$ and var(x) when the origin is changed which are given explicitly in the following forms:

$$\bar{x} = \bar{u}+k = \frac{\Sigma\ fu}{N}+k = \frac{\Sigma\ f(x-k)}{N}+k.$$

$$\text{var}(x) = \text{var}(u) = \frac{\Sigma\ fu^2}{N}-\bar{u}^2 = \frac{\Sigma\ f(x-k)^2}{N}-(\bar{x}-k)^2.$$

EXERCISE 3c

1.   The heights of a 100 fourteen-year-old boys are given in the following table:

Height (cm to nearest cm)	145	146	147	148	149	150	151	152	153	154
Frequency	1	4	8	15	20	21	16	9	4	2

Using an origin of 150, calculate the mean and standard deviation.   (C)

2.   The following table gives data concerning the weights, in pounds, of 120 adult males.

Weight (mid-interval value)	100	120	140	160	180	200	220
Number of males	3	37	51	18	6	3	2

Estimate the mean and standard deviation of this distribution.   (MEI)

3.   The following table gives the distribution of systolic blood pressure, in millimetres of mercury, for 254 male workers aged 30 to 39 years.

Blood pressure (mid-interval value)	85	95	105	115	125	135	145	155	165	175
Number of men	1	2	12	47	77	65	32	13	4	1

Calculate the mean and standard deviation.   (MEI)

4.   The marks of a sample of 50 examination candidates were classified in intervals of 10 marks as follows:

Mid-interval value	4.5	14.5	24.5	34.5	44.5	54.5	64.5	74.5	84.5	94.5
Frequency	1	0	3	10	9	9	7	5	3	3

Estimate the mean and variance.

5.   Using an origin 159 and a scale factor of 3 estimate the standard deviation of the following distribution:

Class	152–154	155–157	158–160	161–163	164–166
Frequency	5	18	42	27	8

6.   A man who had to make weekly visits to a Social Security Office kept a record of the times he had to wait for service on 100 consecutive visits. The results were:

Time (mins)	5–9	10–14	15–19	20–24	25–29	30–34
Frequency	2	29	37	16	14	2

Calculate the mean and the standard deviation.

7.   The table below gives the masses in kilograms of 150 working coal miners of similar height.

Mass (kg)	50–54	55–59	60–64	65–69	70–74	75–79	80–84	85–89	90–94
Frequency	4	8	25	33	35	27	9	5	4

Estimate the mean and variance of this distribution.   (AEB 1980)

8.   The following data were collected in 100 ten-second counts of particles from a radio-active source.

Count	2600–2629	2630–2659	2660–2689	2690–2719	2720–2749	2750–2779	
Frequency	2	2	4	5	8	9	

Count	2780–2809	2810–2839	2840–2869	2870–2899	2900–2929	2930–2959	2960–2989
Frequency	14	12	12	8	12	8	4

Calculate the mean and variance for this sample.   (JMB)

9.   Find the variance of the following distribution of lengths, which were measured to the nearest centimetre.

Length (cm)	3.5–5.5	5.5–7.5	7.5–9.5	9.5–11.5	11.5–13.5
Frequency	2	3	17	43	62

Length (cm)	13.5–15.5	15.5–17.5	17.5–19.5	19.5–21.5	21.5–23.5
Frequency	81	70	61	18	3

10.   A roulette wheel designed for the domestic market contains pockets numbered $0, 1, 2, \ldots, 9$. The wheel is spun and the ball is introduced, coming to rest when the wheel stops. A wheel of this kind was spun 100 times, with the results shown below. Arrange the data in the form of a frequency distribution and calculate the mean and standard deviation.

```
5  6  2  6  4  8  5  2  5  8  3  3  5  9  0  2  4  7  4  6
4  6  5  9  0  3  8  7  5  5  2  4  9  4  2  1  5  6  2  9
5  4  5  1  0  6  9  8  7  3  9  7  6  8  2  4  4  3  4  6
8  5  5  5  9  1  6  7  7  2  0  3  7  3  2  5  4  5  6  6
3  9  9  4  5  4  7  2  5  4  5  5  8  2  4  3  3  6  9  6
```

3.6   The interquartile range

Suppose that the number of letters received by a particular department in a government Ministry on each of 15 successive working days is

165   138   253   133   151   170   98   180   174   172   141   303   131   134   162.

The mean of these numbers is 167.

However, if we arrange these numbers in order of size

98   131   133   134   138   141   151   162   165   170   172   174   180   253   303,

it is clear that the numbers 98, 253 and 303 at the ends are so different from the rest that they cannot be regarded as typical and must have occurred for some special reason. Possibly a one-day strike by civil servants accounts for the lower figure, and a build-up of mail after public holidays produced the higher figures. Whatever the reasons, the effect is that the mean, 167, cannot be taken as a satisfactory average and that it will be better to use the median (the middle number), which is 162.

How is the spread to be measured? Clearly we cannot use the standard deviation, which depends on deviations from the mean. What is needed is an equivalent measure which is associated with the median.

The usual method is to find the *range of the central half* of the distribution – not the whole range (303–98), since it includes the extreme values. Now just as the median divides a distribution into two equal parts, there are values which divided the distribution into four equal parts. These are called *quartiles*.

In our example there are seven numbers below the median and the central one of these is 134. This is the 1st quartile, Q_1. Similarly, the central number above the median is 174 and this is the 3rd quartile, Q_3. The 2nd quartile is, of course, the median. The 1st and 3rd quartiles are often called the *lower* and *upper* quartiles respectively.

The *interquartile range* is defined as $Q_3 - Q_1$, which in this case is 174–134 = 40.

As we said, it is the length of the interval which contains the central half of the distribution.

Finding the quartiles

We have already seen that if there are N observations, the median is the $\frac{1}{2}(N+1)$th observation, which, if N is large, may be replaced by the $\frac{1}{2}N$th (see page 19). Similarly

$$Q_1 \text{ is the } \tfrac{1}{4}(N+1)\text{th observation } (\tfrac{1}{4}N\text{th if } N \text{ is large}),$$

$$Q_3 \text{ is the } \tfrac{3}{4}(N+1)\text{th observation } (\tfrac{3}{4}N\text{th if } N \text{ is large}).$$

EXAMPLE 8   Estimate the interquartile range for the distribution given in example 3 on page 15.

We can follow the method used for finding a median.

First of all the cumulative frequencies are calculated.

Class	f	cf
14.5–19.5	1	1
19.5–24.5	4	5
24.5–29.5	22	27
29.5–34.5	35	62
34.5–39.5	20	82
39.5–44.5	8	90

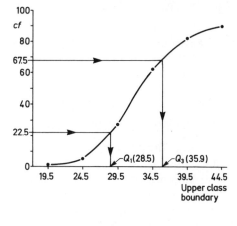

Since $N = 90$ and is large, Q_1 is the $\frac{1}{4} \times$ 90th or $22\frac{1}{2}$th observation and Q_3 is the $\frac{3}{4} \times$ 90th or $67\frac{1}{2}$th observation.

Q_1 and Q_3 may then be found by drawing, as accurately as possible, a cumulative frequency curve.

From the graph, $Q_1 = 28.5$ and $Q_3 = 35.9$ (the median $= 32$).

Hence the interquartile range is $35.9 - 28.5 = 7.4$.

Alternatively, the method of linear interpolation may be used, in which case only the relevant parts of the cumulative frequency polygon are needed.

By inspection of the cumulative frequencies, Q_1 must lie in the class 24.5–29.5.

Hence $Q_1 \approx 24.5 + \dfrac{22.5 - 5}{27 - 5} \times 5$

$= 28.5.$

Similarly, Q_3 must lie in the class 34.5–39.5, so that

$Q_3 \approx 34.5 + \dfrac{67.5 - 62}{82 - 62} \times 5$

$= 35.9.$

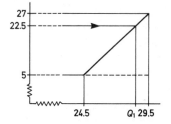

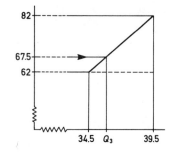

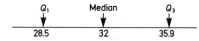

It is important to realise that the median is only half-way between the two quartiles if the distribution is symmetrical. In fact, the relative positions of Q_1, Q_3 and the median give a good indication of the skewness of the distribution. In our example, the interval between Q_1 and the median is slightly less than the interval between Q_3 and the median, which means that the distribution is slightly skewed to the right (positive skewness).

Deciles and percentiles

For very skew distributions, it is often useful to indicate where the extreme values lie by calculating *deciles* or *percentiles*, which as the names suggest, divide a distribution into 10 or 100 equal parts respectively.

The 3rd decile, for example, is the $\frac{3}{10}(N+1)$th observation and the 95th percentile is the $\frac{95}{100}(N+1)$th observation.

Deciles and percentiles may readily be found from a cumulative frequency curve.

The semi-interquartile range

Instead of the interquartile range, the semi-interquartile range $\frac{1}{2}(Q_3 - Q_1)$ is sometimes used as a measure of spread, since, for an approximately symmetrical distribution, it gives roughly the difference between a quartile and the median and is therefore a sort of average of the deviations from the median of all the observations.

EXERCISE 3d

1. In a certain experiment 100 seeds were planted in each of 80 pots. The number of seeds (x) which germinated in each pot was counted and recorded in the table below:

x	0–10	11–20	21–30	31–40	41–50	51–60	61 and over
f	8	16	20	18	12	6	0

Construct a cumulative frequency table for these data, and draw the corresponding cumulative frequency curve.

From your curve, estimate the interquartile range of this distribution.   (MEI)

2. The salaries of middle-grade administrative staff in a large industrial concern in 1984 are given in the following table, where the unit for x is £1000 to the nearest £100:

x	10.3–10.7	10.8–11.2	11.3–11.7	11.8–12.2	12.3–12.7	12.8–13.2	13.3–13.7	13.8–14.2
f	2	5	12	17	14	6	3	1

Estimate the interquartile range
(a) by drawing a cumulative frequency curve,
(b) by linear interpolation.

3. Find the interquartile range for the distribution of word lengths in the passage from Evelyn Waugh's *Unconditional Surrender*, whose distribution is given on page 10.

4. The table shows the marks, collected into groups, of 400 candidates in an examination. The maximum mark was 99.

Marks	0–9	10–19	20–29	30–39	40–49	50–59	60–69	70–79	80–89	90–99
f	10	26	42	66	83	71	52	30	14	6

Compile the cumulative frequency table and draw the cumulative frequency curve. Use your curve to estimate
(a) the median,
(b) the 10th and 90th percentiles,
(c) the 10 to 90 percentile range.   (C)

EXERCISE 3e (miscellaneous)

1. The number of candidates obtaining certain marks in an examination was as follows:

Marks	0–4	5–9	10–14	15–19	20–24	25–29	30–34	35–39	40–44
Number of candidates	2	8	7	11	10	8	6	5	2

(a) What was the modal group of marks?
(b) Calculate an estimate of the median mark.
(c) Draw a cumulative frequency graph of this distribution.
(d) From this, estimate the value of the semi-interquartile range.   (SUJB)

2. The following table shows the durations of 40 telephone calls from an office via the office switchboard.

Duration (min)	$\leqslant 1$	1–2	2–3	3–5	5–10	$\geqslant 10$
Number of calls	6	10	15	5	4	0

Obtain an estimate of the mean and standard deviation of the data.
Estimate the median, and the lower and upper quartiles.   (O & C)

3. The frequency distribution of a random sample of 250 rivets according to their head diameter, measured to the nearest 0.01 mm, is shown in the following table:

Diameter (mm)	*Number of rivets*
13.07–13.11	10
13.12–13.16	20
13.17–13.21	28
13.22–13.26	36
13.27–13.31	52
13.32–13.36	38
13.37–13.41	32
13.42–13.46	21
13.47–13.51	13

Construct a cumulative frequency diagram (in curve or polygon form) to illustrate these data.
Determine the value of
(a) the median,
(b) the mean,
(c) the standard deviation.   (MEI)

4. An inspection of 34 aircraft assemblies revealed a number of missing rivets as shown in the following table.

Number of rivets missing	0–2	3–5	6–8	9–11	12–14	15–17	18–20	21–23
Frequency	4	9	11	6	2	1	0	1

Draw a cumulative frequency curve. Use this curve to estimate the median and the quartiles of the distribution.   (O & C)

5. In September 1978, 100 children enrolled for 'beginners' swimming classes at the city pool. Their ages, to the nearest year, are given below

```
5  6  1  4  9  7  5  6  7  4  3  2  3  5  5  5  9  8  5  8
8  2  7  5  3  3  4  7  8  5  9  4  6  6  4  7  3  2  5  6
4  7  5  6  8  6  3  5  4  3  8  6  5  2  9  7  4  3  7  2
3  2  8  4  6  5  2  8  8  1  5  4  6  3  4  2  6  5  5  8
6  7  5  9  1  3  4  5  6  7  9  7  4  5  1  6  3  9  4  4
```

(a) Produce a frequency table for these data.
(b) Hence, or otherwise, calculate the mean and standard deviation of this sample.

(AEB 1980)

6. Show, from the basic definition, why the standard deviation of a set of observations $x_1, x_2, x_3, \ldots, x_i, \ldots, x_n$ with mean x may be found by evaluating

$$\sqrt{\left(\frac{\Sigma\, x_1^2}{n} - \bar{x}^2\right)}.$$

(a) Find, showing your work clearly and not using any preprogrammed function on your calculator, the standard deviation of the following frequency distribution:

x	25	26	27	28
f	2	0	15	11

(b) The average height of 20 boys is 160 cm, with a standard deviation of 4 cm. The average height of 30 girls is 155 cm, with a standard deviation of 3.5 cm. Find the standard deviation of the whole group of 50 children.

(SUJB)

7. In a fishing competition, the total catches of 50 anglers had masses (to the nearest 0.1 kg) as given in the following table.

Mass (kg)	0–0.2	0.3–0.7	0.8–1.2	1.3–1.7	1.8–2.2	2.3–3.7	3.8–5.2
Frequency	8	8	12	8	8	4	2

Draw a histogram to represent the frequency distribution.
    State formulae for the mean and standard deviation of a frequency distribution, explaining the symbols. Prepare a table for the given data, showing the mid-values of the class intervals and all the terms in the summations that have to be calculated to obtain the mean and variance. Calculate the mean, variance and standard deviation of the distribution.
    Given the additional information that all eight anglers placed in the first class interval caught nothing at all, obtain a revised value for the mean.

(JMB)

8. The pupils in each of three classes take the same examination. The number in each class and the mean and standard deviation of their marks are given in the table below.

Number in class	Mean	Standard deviation
25	60	8
35	40	9
40	55	10

Calculate the mean and standard deviation of the marks of the 100 pupils in the three classes taken together.

(C)

9. The total annual incomes of a random sample of 100 Camford families are summarised in the table below.

Annual income (£)	Number of families
$x < 10\,000$	0
$10\,000 \leqslant x < 14\,000$	42
$14\,000 \leqslant x < 18\,000$	37
$18\,000 \leqslant x < 24\,000$	11
$24\,000 \leqslant x < 30\,000$	8
$30\,000 \leqslant x < 40\,000$	2
$x \geqslant 40\,000$	0

(a) Estimate the mean annual income of this sample.
(b) Plot the data on a cumulative frequency graph.
(c) Estimate the 20th percentile and the 80th percentile of the family income distribution.

(C)

10. The table below shows the cumulative distribution of gross weekly earnings in £ among men aged 21 years and over in a survey period in 1977.

Earnings	30	35	40	45	50	55	60	65	70
% of men	0.5	1.1	2.4	5.5	10.8	18.2	27.0	36.6	45.7

Earnings	75	80	85	90	95	100	120	150	200
% of men	54.6	62.2	69.0	74.2	78.9	82.7	92.1	97.0	99.2

Find the proportion of men who earn more than £100 per week.
Estimate the median earnings and the semi-interquartile range.
Illustrate these data by drawing a histogram showing earnings in the ranges 0–40, 40–60, 60–80, 80–100, 100–120 and 120–200 £ per week. (O & C)

11. A man regularly runs round a fixed course. His times (to the nearest second) for 100 runs are distributed as shown in the following table.

Time	13 m 0 s –13 m 29 s	13 m 30 s –13 m 59 s	14 m 0 s –14 m 29 s	14 m 30 s –14 m 59 s	15 m 0 s –15 m 29 s
Frequency	2	3	16	23	25

Time	15 m 30 s –15 m 59 s	16 m 0 s –16 m 29 s	16 m 30 s –16 m 59 s	17 m 0 s –17 m 29 s
Frequency	18	7	4	2

Calculate the mean, variance and standard deviation, stating the units in each case.
Compile a cumulative frequency distribution and represent this graphically. Use your graph to estimate the median time. (JMB)

12. The following incomplete table shows the ages at which 987 men became widowers. Determine the numbers missing from the five places marked A, B, C, D, E. (There is no need to copy out the table.)

Age of man when wife died (years)	Mid-interval value: x_i	x_i'	Number of men: f_i	$f_i x_i'$	$f_i(x_i')^2$
18–22	20.5	−6	5	−30	180
23–27	25.5	−5	43	−215	1075
28–32	30.5	−4	108	−432	C
33–37	35.5	−3	107	−321	963
38–42	40.5	−2	130	−260	520
43–47	45.5	A	109	−109	109
48–52	50.5	0	115	0	0
53–57	55.5	1	88	88	88
58–62	60.5	2	76	152	304
63–67	65.5	3	82	246	738
68–72	70.5	4	59	B	944
73–77	75.5	5	41	205	1025
78–82	80.5	6	16	96	576
83–87	85.5	7	8	56	392
		Totals	987	D	E

Use your answers to estimate the mean and the standard deviation of this distribution. The mean deviation d, where d is defined by

$$d = \frac{f_i|x_i - \bar{x}|}{\Sigma f_i}$$

is sometimes used instead of the standard deviation as a measure of dispersion. Estimate d for the data in the table.

Suggest, for the general case, one possible advantage and one possible disadvantage in using d instead of the standard deviation as a measure of dispersion.   (MEI)

3.7   Appendix

Coding

If the variable x is replaced by u, where $u = (x - k)/c$, then
(a) $\bar{x} = k + c\bar{u}$,
(b) $\text{var}(x) = c^2 \text{var}(u)$.

Proof

(a) $x_i = k + cu_i$ $(i = 1, 2, \ldots, n)$

$$\Rightarrow \sum_{i=1}^{n} f_i x_i = \sum_{i=1}^{n} f_i(k + cu_i)$$

$$\Rightarrow \Sigma\, fx = k\,\Sigma\, f + c\,\Sigma\, fu \quad \text{(dropping the suffixes)}$$

$$= kN + c\,\Sigma\, fu$$

$$\Rightarrow \frac{\Sigma\, fx}{N} = k + c\frac{\Sigma\, fu}{N}$$

$$\Rightarrow \bar{x} = k + c\bar{u}.$$

(b)

$$\Sigma\, f_i(x_i - \bar{x})^2 = \Sigma\, f_i[(k + cu_i) - (k + c\bar{u})]^2$$
$$\Rightarrow \Sigma\, f(x - \bar{x})^2 = \Sigma\, f(cu - c\bar{u})^2$$
$$= \Sigma\, f[c(u - \bar{u})]^2$$
$$= c^2\, \Sigma\, f(u - \bar{u})^2$$
$$\Rightarrow \frac{\Sigma\, f(x - \bar{x})^2}{N} = c^2\, \frac{\Sigma\, f(u - \bar{u})^2}{N}$$
$$\Rightarrow \text{var}(x) = c^2\, \text{var}(u).$$

For change of origin only, $c = 1$ and

$$\bar{x} = k + \bar{u}, \qquad \text{var}(x) = \text{var}(u).$$

For change of scale only, $k = 0$ and

$$\bar{x} = c\bar{u}, \qquad \text{var}(x) = c^2\, \text{var}(u).$$

4 Probability

4.1 Defining probability

In chapter 1 we saw that the conclusions that are drawn from statistical analysis of data have to be given in terms of probability. How is probability to be defined? Of course some ideas about probability are intuitive, even before the word is defined, so that most people, for instance, would accept without argument that:

> The probability that, if a man jumps out of a window, he will land on the moon is zero;

> The probability that the next person to come through a door will be either male or female is 1;

provided, of course, that we attach the number 1 rather than, say, 100 to certainty.

Having established these two extremes, it is natural to say that the probability of any event lies somewhere between 0 and 1, so that, for example:

> The probability of finding at least one bacterium near the exit of a sewer pipe is high, and therefore somewhere at the top end of the scale;

> The probability that there will be at least three inches of rain in Malta on a given August Saturday is somewhere near the zero end of the scale.

The problem is to decide how to allocate definite numbers to probabilities. Now anyone who has given any thought to the matter might be inclined to say that there are some situations in which it is not difficult to allot probabilities; for example, the probability of getting a '5' when an unbiased die is thrown is 1/6. He might also say, though with slightly less confidence, that when a child is born the probability that it will turn out to be a girl is about 0.5. But even in the apparently simple example of the die there is a problem. What do we mean by an unbiased die? There are two possible answers:

1. that we can imagine a perfect (or *ideal*) die, so that, when it is thrown, there is no reason to expect one score rather than any other of the *possible* scores.

2. that we would describe a die as unbiased if, in a very *long run* each score turned up once in six throws. (This will be more precisely stated later.)

Now statistical experiments are concerned with the real world, so that a concept of probability that is to be of practical use cannot *start from* an assumption about some sort of ideal situation. Moreover, although the perfect die is not difficult to imagine, it is not easy to say what sort of corresponding ideal situation could be envisaged for dealing with an event like 'at least three inches of rain in Malta'.

In fact the situations for which a theory of probability is needed are so varied that the only way forward towards a *definition* of probability is to use the idea of long-run frequency mentioned in point 2 above.

Nevertheless, although we said that a definition cannot *start from* an assumption about an ideal situation, such situations have an important part to play in what follows and are particularly useful in demonstrating how to use probability.

Suppose we wanted to allocate some number between 0 and 1 to the probability that a child about to be born will be a girl. The obvious method would be to consult the records of the sex of new-born children over a long period of recent time. If in the last 40 000 births, the proportion of girls is 0.485 (which is in fact the current figure for England and Wales), we would use this with some confidence as the probability. Our confidence would be greater if, at a later time, the additional records confirmed that the proportion or relative frequency had settled down to 0.485.

It is this fraction, *long-term relative frequency*,

$$\text{or} \quad \frac{\text{number of girls so far in a long period of time}}{\text{number of births in the same period}},$$

that gives us the numerical value we allocate to the probability that a new-born child will be a girl. In other words, we have defined probability in this case as the number we judge to be appropriate in the light of a large number of observations. Although it is, in a sense, an estimate, it is one to which we are able to attach a strong degree of belief because it is based on a large number of *observations*.

The probability of getting a '5' by throwing a die is defined in a similar way, not by examining a large number of past records, but by carrying out the *experiment* a large number of times, and noting the *relative frequency* with which a '5' turns up. The results might be as shown in the figure. If we make the reasonable assumption that random fluctuations will decrease as the number of trials increases, the proportions of '5s' turning up can be expected to settle down to a 'limiting value'. This hypothetical limiting value is defined as the probability.

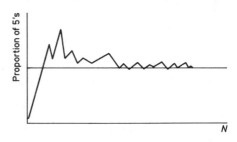

Writing the probability of an event A as $P(A)$, it is defined as follows:

$P(A)$ is the *long-run relative frequency* with which A occurs,

or, putting it in mathematical terms, if the number of trials is N and the event A occurs $n(A)$ times, then

$$P(A) = \lim \frac{n(A)}{N}$$

where 'lim' means the limiting value to which $n(A)/N$ settles down as N increases indefinitely.

What this means in practice is that we would take the relative frequency $n(A)/N$ after a large number of trials as an estimate of the probability, so that if, in the die-throwing experiment, 1672 '5s' turned up in 10 000 trials, the probability of a '5' would be taken as 0.1672.

Note: In the definition, the word 'trials' was used. In what follows it will be taken to mean either 'experiments' or 'observations'.

4.2   Outcome, possibility space, event

We shall now define some terms that will be used constantly in probability theory.

The result of any experiment (such as throwing a die) or trial (the name we might give to counting the number of girls in the records of 5000 new-born babies) is called an *outcome* and a set of *all distinct possible outcomes*, S, is called a *possibility space* (or sample space). As the following examples show, a given experiment may lead to more than one possibility space:

Experiment	*Possibility space*
Throwing a die and noting the score	$S = \{1, 2, 3, 4, 5, 6\}$ or $\{$score less than 3, score between 3 and 6 inclusive$\}$
Tossing a coin three times and noting the number of heads	$S = \{0, 1, 2, 3\}$
Tossing a coin twice and noting the occurrence of head or tail	$S = \{\text{HH, HT, TH, TT}\}$ or $\{$both the same, different$\}$
Tossing two dice and noting the individual scores, or the sum of the scores	$S = \{(1, 1), (1, 2), \ldots, (6, 6)\}$ or $\{2, 3, 4, 5, 6, 7, 8, 9, 10, 11, 12\}$
Counting the number of defective items in a sample of 7 taken from a production line	$S = \{0, 1, 2, 3, 4, 5, 6, 7\}$

We have already used the word 'event' without giving the formal definition:

Any subset of the possibility space is called an *event*.

In other words an event consists of one or more of the outcomes that make up the possibility space, so that in the first example above,

$\{1, 2\}$ is an event, and might be written as $\{$the score $< 3\}$;
$\{2, 4, 6\}$ is an event and is equivalent to $\{$the score is an even number$\}$.

The probability of *any* event, not just a simple event like getting a '5' with the throw of a die, is defined, as above, as the long-term relative frequency.

In each of the examples we have just given, the possibility space has been finite, consisting of a discrete number of outcomes, and we shall restrict ourselves to this sort of case here. Nevertheless a possibility space may be infinite, and it may be continuous, as the next two examples show:

Experiment	Possibility space
Throwing a die and noting the number of throws until a '4' turns up	$S = \{1, 2, 3, \ldots\}$
Testing cathode-ray tubes from a production line and noting their lengths of life (t)	$S = \{\text{all values of } t > 0\}$

Symmetry

On page 50 we stated that an unbiased die might be defined as one which was *assumed* to be so perfectly constructed and marked that there would be no reason to expect one score rather than any other to turn up on a single throw. In other words, *in advance of any experiment*, we would expect that the long-run relative frequency of the occurrence of any particular score would be 1/6.

In this case, the possible outcomes are taken to be *equally likely*, so that the number of '5s', say, in 2700 throws would be expected to be very near 450. This is an idealised experiment and we say that each outcome 1,2,3,4,5,6 is equally likely *on grounds of symmetry*.

In cases where symmetry may be reasonably assumed, the probability of an event A is defined simply as:

$$\frac{\text{The number of possible outcomes that count as } A}{\text{The total number of possible outcomes}} = \frac{n(A)}{n(S)}.$$

Most exercises in this chapter concern situations in which symmetry is assumed. The reason for this is that problems involving symmetry present the operation of the probability laws in the simplest light. If the wording of a problem includes the phrase '*at random*' it means that the problem *may* be treated as one in which the outcomes are *equally probable*. Care must be taken, however, to choose the correct possibility space (see example 2 below).

EXAMPLE 1   If a card is drawn at random from a normal pack, what is the probability that it is

(a) a court card,   (b) a black two?

(a) The total number of possible outcomes is 52 and they are equally probable, so $n(S) = 52$.

Twelve of these (jack, queen, king) are court cards, so that if A is the event {court card}, $n(A) = 12$.

$$\text{Hence}\quad P(A) = \frac{12}{52} = \frac{3}{13}.$$

(b) If B is the event {black two}, $n(B) = 2$, and

$$P(B) = \frac{2}{52} = \frac{1}{26}.$$

EXAMPLE 2   Two unbiased dice are thrown and the scores on each recorded. Find the probability that the sum of the scores is less than 5.

Since each individual score is equally likely, the 36 possible pairs of scores

$$(1,1), (1,2), \ldots, (6,5), (6,6)$$

are equally likely and they make up the sample space S.
  If $A = \{$the sum is less than 5$\}$, the following pairs count as A:

$$(1,1), (1,2), (1,3), (2,1), (2,2), (3,1)$$

and $n(A) = 6$.

$$\text{Hence}\quad P(A) = \frac{6}{36} = \frac{1}{6}.$$

Note: In this example, you might be tempted to take as possibility space the set of total scores 2, 3, 4, 5, 6, 7, 8, 9, 10, 11, 12 three of which, 2, 3, 4, are less than 5. It does not, however, follow that the required probability is 3/11, since the total scores listed are not equally probable.

EXERCISE 4a

1.   An unbiased die is thrown. What is the probability of the score being: (a) odd, (b) prime?

2.   A set of 50 cards have the numbers 1 to 50 marked on them. If one is picked at random, what is the probability that the number on it is a perfect square?

3.   Apart from the foreman, there are 22 men working on the night shift in a small factory. The 8 youngest are all apprentices and the 15 oldest are supporters of West Ham football team. If the foreman picks one at random and tells him to make the tea, what is the probability that the one chosen is an apprentice supporter of West Ham?

4.   A card is picked at random from a normal pack of 52. What is the probability that it is: (a) a king, (b) a heart, (c) a number less than 7, assuming that aces count 'high'?

5.   A computer generates random numbers ranging from 1 to 1000. What is the probability that the next one it produces will be a multiple of 23 or 29 or both?

6.   An unbiased coin is tossed three times. Write down a possibility space consisting of all the equally likely outcomes.
  If A is the event $\{$at least two tails$\}$, and B is the event $\{$head followed by a tail$\}$, find $P(A)$ and $P(B)$.

7.   Two people have packs of cards marked 1 to 10. If each selects a card at random, what is the probability that the sum of the numbers on the cards selected is prime?

8.   A symmetrical tetrahedral die has faces marked 1, 2, 3, 4, The score when it is thrown is the number on the face in contact with the table. If two such dice are thrown together, find the probability that: (a) the sum of the scores is less than 5, (b) the two scores are the same, (c) the sum of the scores is even.

9.   Two unbiased dice are thrown together. If C is the event $\{$the sum of the scores is 7 or 11$\}$, D is the event $\{$the sum of the scores is less than 8$\}$, E is the event $\{$the difference between the scores is 1$\}$, find $P(C)$, $P(D)$ and $P(E)$.

10.   If someone comes into the room where you are, what is the probability that he was born on the same day of the week as you?

Aids to counting

If a 'hand' of 13 cards is dealt from a pack of 52, what is the probability that it will contain, say, exactly 9 hearts? To answer this question we need to know how many possible hands there are and clearly any simple counting method is out of the question. The number of hands is in fact about 6.35×10^{11}. What is needed is some alternative method of 'counting' which will enable us to answer this kind of question without much difficulty. We start, however, with some much simpler questions.

EXAMPLE 3   There are three airfields in country A and five in country B. In how many ways can an airline fly one of its aircraft from A to B?

The answer is clearly 3×5, since for each of the 3 airfields in A there are 5 possible choices for B.

EXAMPLE 4   A coin is tossed 5 times. How many possible outcomes are there?

A typical outcome is HTHHT and counting would be laborious, so we argue like this. The first throw can give 2 possible results; for each of these there are 2 possible results for the second, giving a total of 2×2 for the first two throws. The answer for 5 throws is therefore $2 \times 2 \times 2 \times 2 \times 2$ or 2^5.

Examples 3 and 4 illustrate the general principle that if an event E_1 can occur in n_1 ways and an event E_2 in n_2 ways then the combined events E_1 and E_2 can occur in $n_1 \times n_2$ ways, a *multiplication rule* which can be extended to any number of events.

4.3   Permutations

An important consequence of this rule is exemplified in the next problem.

In how many ways can the letters H I P be arranged?

If we imagine three boxes into which cards with the letters printed on them are to be placed, the number of ways the first can be filled is 3. When this has been done there are only 2 letters available for the second box,

1	2	3
P	H	I

and when the first two boxes have been filled the letter left over must go into the third box. It follows that the total number of arrangements is $3 \times 2 \times 1$, and if the number of letters had been 5, the number of arrangements would have been $5 \times 4 \times 3 \times 2 \times 1$. In general, then:

The number of arrangements (or permutations) of n distinct objects is

$$n(n-1)(n-2) \times \ldots \times 3 \times 2 \times 1.$$

This is written $n!$ and read as 'factorial n'.

Suppose now that we had cards with the 7 letters BANQUET printed on them, 3 of which are to be placed in the three boxes. In how many ways could this be done?

Using the same argument as before, there are 7 possible choices for the first box, and for each of these there are 6 choices for the second box. When both of these have been filled there are 5 choices available for the third box. Hence the number of arrangements is $7 \times 6 \times 5$.

Such an arrangement (or permutation) of 7 objects, taken 3 at a time is written

$$^7P_3 = 7 \times 6 \times 5 \quad (\text{or } 7 \cdot 6 \cdot 5, \text{ with a } \cdot \text{ for } \times).$$

Other examples of this kind are:

1. The number of arrangements of 8 objects taken together is 8!
2. The number of arrangements of 8 objects taken 5 at a time is

$$^8P_5 = 8 \cdot 7 \cdot 6 \cdot 5 \cdot 4 \text{ which can be written as } \frac{8!}{3!}.$$

In general, then, $^nP_r = n(n-1)(n-2)\ldots(n-r+1) = \dfrac{n!}{(n-r)!}.$

Note: By the definition of ^nP_r, $^8P_8 = \dfrac{8!}{0!}$ and 0! cannot be defined in the usual way.

We get round this difficulty by observing that $^8P_8$ is the number of ways of arranging 8 objects, taking them all together, which is 8!. Hence

$$\frac{8!}{0!} = 8!$$

which only makes sense if we adopt the convention that $0! = 1$.

EXAMPLE 5   Write in factorial form: (a) $52 \cdot 51 \cdot 50 \cdot 49$, (b) $\dfrac{13 \cdot 12 \cdot 11}{1 \cdot 2 \cdot 3}$.

(a) $\dfrac{52!}{48!}$ (b) $\dfrac{13!}{10!3!}$.

EXAMPLE 6   How many changes can be rung with a peal of 6 bells, if the tenor bell is always rung last.

If the tenor bell comes last, there are only 5 bells to be arranged, giving $5! = 120$ changes.

EXAMPLE 7   How many arrangements can be made of the letters of
(a) the word DRAUGHT, if the vowels must not be separated,
(b) the word LIBELLOUS, where the letters L are indistinguishable?

(a) This amounts to arranging 6 objects D, R, G, H, T and AU where A and U may be switched round. Hence the number of arrangements is $6! \times 2 = 1440$.
(b) The number of ways of arranging the 9 *distinct* letters $L_1, L_2, L_3,$ I, B, E, O, U, S is 9!, but in fact the Ls are not distinct and could be re-arranged in 3! ways without making any difference. Hence the number of arrangements is

$$\frac{9!}{3!} = 60\,480.$$

EXAMPLE 8   An artist is offered 6 positions in a line to show 6 of his pictures at an exhibition. If he has 10 landscapes and 5 portraits available, in how many ways could he use these positions if 4 landscapes are to be shown together on the left with 2 portraits on the right?

The number of arrangements of the landscapes is $^{10}P_4$, and the number of arrangements of the portraits is $^5P_2$.

Using the multiplication rule, the overall number of arrangements is

$$^{10}P_4 \times {}^5P_2 = (10 \cdot 9 \cdot 8 \cdot 7) \times (5 \cdot 4) = 100\,800.$$

EXERCISE 4b

1.   Write in factorial form: (a) $^{11}P_4$, (b) $\dfrac{30 \cdot 29 \cdot 28 \cdot 27 \cdot 26}{1 \cdot 2 \cdot 3 \cdot 4 \cdot 5}$.

2.   In how many ways can the letters of the following words be arranged:
(a) FLYING,   (b) POSSUM (the Ss are indistinguishable),
(c) CONDITION,   (d) POSITIVISM.

3.   A woman has a collection of 10 china ornaments, all different, six in the shape of a shepherd and four in the shape of a dog. She arranges them on the mantlepiece, with all the shepherds together on the left and, in line with them, all the dogs together on the right. How many different arrangements are possible?   (C)

4.   How many integers are there between 12345 and 54321 which contain each of the integers 1,2,3,4,5 once and once only?

5.   In the Women's Institute competition for cake of the year, four entries are to be chosen in order of merit from sixteen finalists. How many ways are there of choosing them?

6.   In the Lord Mayor's procession, four costermonger's carts and four brewer's drays will move in line together, the costermonger's carts coming first. If there are eight carts and twelve drays available, in how many ways can the carts and drays be chosen?

7.   It was found that the first three (in order) to lead a rock climb could be chosen from a party in 210 ways. How many were there in the party?

8.   Simplify: (a) $\dfrac{n!}{(n-2)!}$,   (b) $\dfrac{(n+1)!}{n!}$,   (c) $\dfrac{(n+1)!}{(n-1)!}$.

4.4   Combinations (selections)

The calculation of the number of arrangements of n objects, taken r at a time, involves, as we have just seen, considering the *order* in which the objects are taken. In calculating probabilities, on the other hand, we are usually concerned with selections in which order is irrelevant. For instance, in the example mentioned earlier, if we want to know how many possible hands of 13 cards can be obtained from a pack of 52, the order in which they are dealt is of no consequence. The same is true in the simpler problem that follows.

EXAMPLE 9   Given the 7 letters BANQUET, how many selections of 3 letters can be made, if order does not matter?

To find out, we shall suppose that x is the number we are looking for.

Consider one of these selections, (B, A, T) say. This could be arranged *in order* in 3! ways and if we did this for each of the x selections we would end up with $x \times 3!$ arrangements.

However we know that the number of arrangements of 7 letters taken 3 at a time is $^7P_3$. Hence

$$x \times 3! = {}^7P_3 \Rightarrow x = \frac{^7P_3}{3!}.$$

In general, then, the number of selections or combinations of n distinct objects taken r at a time is

$$\frac{^nP_r}{r!} = \frac{n!}{(n-r)!r!}.$$

The symbol for this expression is $^nC_r \left[\text{or} \binom{n}{r} \right]$ so that

$$^nC_r = \frac{n!}{(n-r)!r!}.$$

Calculating ^nC_r

EXAMPLE 10   Find the number of ways that 4 people may be chosen from 13 to join an expedition.

The number of combinations $= {}^{13}C_4 = \dfrac{13!}{9!4!}$,

which is best calculated as

$$\frac{13 \cdot 12 \cdot 11 \cdot 10}{4!} = 715$$

with *four* integers moving downwards consecutively from 13 in the numerator, and 4! in the denominator.

It should be noted that $^{13}C_9 = \dfrac{13!}{9!4!} = {}^{13}C_4$, which is to be expected, since, when 4 people are selected for an expedition, the remaining 9 are automatically selected for being left behind.

The important general result is that

$$^nC_r = {}^nC_{n-r}.$$

This is a useful result when calculating $^{21}C_{19}$, for example, which is more easily evaluated as $^{21}C_2$.

EXAMPLE 11   A House of Commons select committee is to consist of 5 Conservative, 3 Labour and 2 SDP/Liberal MPs, chosen from the 11 Conservative, 7 Labour and 4 SDP/Liberal members who have offered themselves. In how many ways can this be done?

The number of possible combinations is

$$^{11}C_5 \times {^7C_3} \times {^4C_2} = \frac{11 \cdot 10 \cdot 9 \cdot 8 \cdot 7}{5!} \times \frac{7 \cdot 6 \cdot 5}{3!} \times \frac{4 \cdot 3}{2!}$$

$$= 97\,020.$$

EXAMPLE 12   How many possible hands of 13 cards can be selected from a pack of 52?

The number of hands $= {^{52}C_{13}} = \dfrac{52 \cdot 51 \cdot 50 \ldots 41 \cdot 40}{13!}$, which, after a few moments with a pocket calculator, gives the answer as approximately 6.35×10^{11}.

EXERCISE 4c

1.   Twenty mounted policemen are on call for duty to help with crowd control at a political rally. In the event the rally was poorly attended and only four were needed. In how many ways could they be chosen?

2.   In a People's Republic there are 14 ministerial posts to be filled from a praesidium consisting of 16 commissars. In how many ways can this be done?

3.   Six people assemble to await the arrival of a woman with a car who is going to give them a lift to a party. When she arrives it turns out that she only has room for three passengers at a time and will have to do two trips. Assuming that position in the car does not matter, find how many ways in which the group can be taken?

4.   A village has five good dominoes players and seven good darts players. A team of two dominoes players and three darts players is to be picked for a match against a neighbouring village. In how many ways can this be done?

5.   Two aircraft are standing by to take passengers aboard. Each has seats for 98, but owing to 'overbooking' 100 passengers have been booked for each flight. The tour operator has to select two from each group to go on a later flight. In how many different ways could the resulting group of four be chosen?

6.   Simplify as far as possible $^{n+1}C_2 + {^nC_2}$.

7.   If $3 \cdot {^{n+1}C_3} = 7 \cdot {^nC_2}$, find the value of n.

8.   If $2 \cdot {^{2n}C_2} = 9 \cdot {^nC_3}$, find the value of n.

Using combinations to calculate probability

In the examples that follow, symmetry is assumed.

EXAMPLE 13   A storekeeper selects three screwdrivers at random from five large, six medium and four small screwdrivers. What is the probability that they are: (a) all large, (b) all the same size?

The total number of screwdrivers is 15, so the possibility space S consists of all possible selections of 3 screwdrivers out of 15. Hence

$$n(S) = {}^{15}C_3.$$

(a) If A is the event {all 3 screwdrivers are large}, $n(A)$ is the number of ways of selecting 3 out of 5, so that

$$n(A) = {}^5C_3.$$

Hence $P(A) = \dfrac{n(A)}{n(S)} = {}^5C_3 \div {}^{15}C_3 = \dfrac{5 \cdot 4 \cdot 3}{3!} \div \dfrac{15 \cdot 14 \cdot 13}{3!} = \dfrac{2}{91}.$

(b) If B is the event {all the screwdrivers are the same size}, then, by a similar argument,

$$n(B) = {}^5C_3 + {}^6C_3 + {}^4C_3,$$

and $P(B) = ({}^5C_3 + {}^6C_3 + {}^4C_3) \div {}^{15}C_3 = \dfrac{34}{155}.$

EXAMPLE 14   Find the probability that a 'hand' of 13 cards dealt at random from a pack of 52 cards will contain exactly 9 hearts.

The possibility space S consists of all possible hands of 13 cards, chosen from 52, so that

$$n(S) = {}^{52}C_{13}.$$

If a hand is to contain exactly 9 hearts it must consist of 9 hearts and 4 'non-hearts'. Now there are 13 hearts and 39 'non-hearts' in the pack, so that, if A is the event {exactly 9 hearts},

$$n(A) = \text{(the number of ways of selecting 9 from 13)}$$
$$\times \text{(the number of ways of selecting 4 from 39)}$$
$$= {}^{13}C_9 \times {}^{39}C_4.$$

Hence $P(A) = {}^{13}C_9 \times {}^{39}C_4 \div {}^{52}C_{13} \approx 9.26 \times 10^{-5}.$

EXERCISE 4d

1.   In a raffle 60 blue tickets and 6 green tickets are put into a hat and 2 are drawn out. What is the probability that both are green?

2.   Three tickets for an opera performance are sent to a music club. Four men and 13 women would like a ticket. If the 3 people to receive a ticket are chosen at random, what is the probability that they will all be: (a) men, (b) women?

3.   A committee of 4 is chosen at random from 9 civil servants and 11 members of parliament. What is the probability that the committee will consist of 2 civil servants and 2 members of parliament? If a committee of 5 is chosen, what is the probability that members of parliament will be in a majority?

4.   A box at a jumble sale contains an assortment of books. Seven are light romances, 4 are

science fiction and 8 are non-fiction. If 5 books are selected at random, what is the probability that they will be: (a) all non-fiction, (b) 2 light romances and 3 science fiction?

5.   At the bottom of a bag a fisherman has 25 worms of which 5 are lugworms. He picks out 6 worms at random. Find the probability that he gets 3 lugworms and 3 which are not.

6.   In a group of 16 convicts, 11 were smash-and-grab men. Five were chosen at random to go to an open prison for rehabilitation. What is the probability that those chosen included at least 1 smash-and-grab man?

7.   A store room contains 6 TV tubes of type A, 5 of type B and 7 of type C. Three tubes are removed at random. What is the probability that they are: (a) all of type A, (b) all of the same type?

8.   A poker hand of 5 cards is drawn at random from a pack of 52. What is the probability that it will contain exactly 1 ace?

9.   Three cards are drawn from a normal pack of 52. What is the probability that they will: (a) all be hearts, (b) all be from black suits?

EXERCISE 4e (miscellaneous)

1.   The sides of an octahedral die are numbered from 1 to 8. What is the probability of getting a total score of 8 if it is thrown twice? If an octahedral die and an ordinary six-sided die are thrown together, what is the probability of getting a score of '6 or 8'? (You may assume both dice are unbiased.)

2.   From three positive and two negative integers, three are chosen at random. What is the probability that their product will be negative?

3.   Two French, four German and five Italian members of a delegation sit in a line to have their photograph taken. In how many ways can this be done: (a) if they may sit in any order, (b) if nationals must sit together?

4.   In how many different ways can three or more people be selected from seven?

5.   A party of six couples go to a dance together. During the interval they sit down in a single line at a straight counter for a buffet supper. In how many ways can they sit: (a) if each man sits next to his own partner, (b) if at least one man is separated from his partner?   (C)

6.   Find the number of different arrangements that can be formed using the letters of the word MANAGER. How many of these arrangements have a consonant in the 1st, 3rd, 5th and 7th positions?

7.   In a certain game a player invents a 'code' by placing a coloured peg in each one of a row of four holes. Pegs are available in six different colours. How many different 'codes' can he invent:
(a) if he is not allowed to use more than one peg of the same colour;
(b) if he is allowed to use more than one peg of the same colour?
   In the second case, how many of the 'codes' will contain two pegs of the same colour next to each other, with the other two pegs each a different colour?   (MEI)

8.   A party of eight is to be fitted into two four-seater cars. In how many ways can this be done, ignoring the positions inside the cars?

9.   For the final assault on a mountain peak, any or all of six climbers may be taken. The

leader decides to select at least three of them. In how many different ways could he make his selection?

10.   Five sailors and four marines are available for a landing party.
(a) In how many ways can a landing party of four be selected?
(b) If the party must contain two sailors and two marines, how many landing parties can be selected?   (C)

11.   In a bag there are six black balls, six yellow ones and eight red ones. Five are taken at random from the bag. What is the probability that they will be: (a) all yellow, (b) three black and two red, (c) two black, one yellow and two red, (d) not any of them red?

12.   Gift packs of perfume contain 18 small bottles, identical in size. Each pack contains 5 'Hazel', 6 'Dawn Rose' and 7 'Jasmyn'. A pack is opened and 6 bottles are removed at random. What is the probability that: (a) all the 'Hazel' are taken out, (b) 2 of each kind are taken out?

13.   A collection of 10 gramophone records consist of 1 record of an opera, 2 records of symphonies, 3 of concertos, 2 of chamber music, and 2 of solo piano pieces. How many selections of 3 different records can be made?
    If 3 records are selected at random, what is the probability that
(a) the selection consists of 1 record of a symphony, 1 of a concerto and 1 of chamber music;
(b) the selection does not include the opera record?   (MEI)

14.   Five equally competent clarinet players are available for the school senior orchestra. Their names are put into a hat by the director of music and two names are drawn out. What is the probability that the brothers Boehm (John and Carl) will not both be chosen?

15.   (a) A number is chosen at random from the integers from 1 to 30 inclusive. Find the probability that the number chosen is
 (i) a multiple of either 3 or 11,
(ii) a multiple of either 3 or 5 or both.
    (b) A bag contains four white balls and five blue balls. If two balls are drawn out at random, calculate the probability that they are of the same colour.   (C)

16.   A bag of 30 sweets contains 6 each of 5 different colours. If I select 3 sweets at random, what is the probability that they are: (a) all the same colour, (b) all different colours?   (MEI)

17.   A three-digit number between 100 and 999 inclusive is chosen at random. What is the probability that
(a) the sum of the digits is exactly divisible by 3;
(b) the product of the digits is an odd number?   (MEI)

18.★ A bridge hand of 13 cards is dealt from a pack of 52. What is the probability that it contains: (a) all four aces, (b) 3 aces and a king? (You may take $^{52}C_{13}$ as 6.35×10^{11}.)

5 Probability laws

Having introduced the idea of probability in the last chapter, we must now develop formally the laws of probability. Once again we shall constantly be using games of chance as illustrations, not only because of their simplicity, but because, as a matter of history, the laws were derived from such games, using the idea of symmetry.

The first two laws are taken as axiomatic, however probabilities are allocated.

Law 1

$$0 \leqslant P(A) \leqslant 1.$$

The relative frequency with which an event occurs must be less than or equal to one, and so must the corresponding probability.

Law 2

For events $A_1, A_2, \ldots, A_n$, that between them make up a possibility space without overlapping,

$$P(A_1) + P(A_2) + \ldots + P(A_n) = 1.$$

The formal way of describing these events is that they are *mutually exclusive* (they cannot occur simultaneously) and *exhaustive* (taken together, they make up the whole of the possibility space).

As an example, if we take the possibility space for the results of throwing a die as $S = \{1, 2, 3, 4, 5, 6\}$ and the events A_1, A_2 and A_3 as $\{1, 2\}$, $\{3, 4, 5\}$ and $\{6\}$ respectively, then the events do not overlap and do cover S between them.

This law is almost self-evident, but we shall nevertheless show how it may be derived from the relative-frequency definition of probability.

Suppose that in N experiments the number of occurrences of A_1, A_2, A_3 are $n(A_1), n(A_2), n(A_3)$ respectively, then

$$n(A_1) + n(A_2) + n(A_3) = N \Rightarrow n(A_1)/N + n(A_2)/N + n(A_3)/N = 1$$

and since $n(A_1)/N \to P(A_1)$ the sum of the probabilities $P(A_1)$, $P(A_2)$ and $P(A_3)$ must also equal 1.

Corollary: If A' stands for the event 'not A'

$$P(A') = 1 - P(A).$$

5.1   The addition laws

We are now dealing with the probability of *compound* events, that is, combinations of two or more events taken together. The law states that if A and B are *mutually exclusive*, then the probability that either will occur is the sum of $P(A)$ and $P(B)$:

Law 3

If A and B are mutually exclusive events

$$P(A \text{ or } B) = P(A \cup B) = P(A) + P(B).$$

In establishing a result of this kind, the argument is more easily followed by using a Venn diagram. We will look at a die-throwing example again, with $A = \{1,3\}$ and $B = \{4\}$.

Each of the possible outcomes 1,2,3,4,5,6 is represented by a dot and we have placed alongside 1, 3 and 4 the *proportions* of occurrences of these outcomes in 1000 throws.

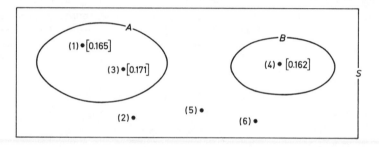

In this case we have supposed that in 1000 throws there have been 165 '1s', 171 '3s' and 162 '4s'.

Now as A and B do not overlap, the *number* of times in which A or B has occurred equals the number of times in which A has occurred plus the number of times in which B has occurred $= (165 + 171) + 162$.

The same is true for the *proportions* and therefore also for the *probabilities*.

The argument in the general case follows the same lines using the notation in Law 2, with $n(A_1)/N$ as the proportion of occurrences of A_1.

This result may be generalised:

If $A, B, C, \ldots$ are mutually exclusive events, then

$$P(A \text{ or } B \text{ or } C \ldots) = P(A \cup B \cup C \cup \ldots) = P(A) + P(B) + P(C) + \ldots.$$

The advantage of using a Venn diagram is that it immediately suggests what the equivalent result must be when A and B are *not* mutually exclusive:

If A and B are *any* two events,

$$P(A \text{ or } B) = P(A \cup B) = P(A) + P(B) - P(A \cap B)$$

where $P(A \cap B)$ means the probability that A *and* B occur.

In the diagram, where A and B are overlapping events, each dot stands for a possible outcome so that A and B have outcomes in common.

The broad argument is that the *number* of occurrences of A or B in N trials equals the number of occurrences of A plus the number of occurrences of B *minus* the number of occurrences of outcomes common to A and B (to avoid counting the common occurrences twice).

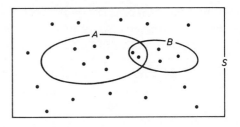

Dividing by N gives the equivalent result for *proportions* and hence for *probabilities*.

Note: 1. When A and B are mutually exclusive, A and B cannot occur together so that $P(A \text{ and } B) = 0$ and $P(A \text{ or } B) = P(A) + P(B)$.
2. The more general addition law will not often be required in the subsequent chapters of this book.
3. The addition laws and later laws may be proved formally by using Boolean algebra (the algebra of sets).

EXAMPLE 1   A pack of cards is numbered from 1 to 30. If a card is picked at random, what is the probability that the number on the card chosen is either divisible by 3 or divisible by 5?

This trivial problem is used merely as an illustration of the addition laws when the events are not mutually exclusive.

$$\text{If}\quad A = \{\text{the number is divisible by 3}\}\quad \text{and}$$
$$B = \{\text{the number is divisible by 5}\},$$
$$A = \{3, 6, 9, 12, 15, 18, 21, 24, 27, 30\},\quad B = \{5, 10, 15, 20, 25, 30\},$$
$$A \cap B = \{15, 30\}.$$

The number of events in the possibility space is 30, so that

$$P(A \cup B) = P(A) + P(B) - P(A \cap B) = \frac{10}{30} + \frac{6}{30} - \frac{2}{30} = \frac{7}{15}.$$

EXAMPLE 2   A committee of three is selected at random from five councillors from the Ratepayers Party and four from the Radical Party. What is the probability that the committee will contain at least two Ratepayers?

If A is the event {2 Ratepayers and 1 Radical} and B is the event {3 Ratepayers}, then, using combinations,

$$P(A) = (^{5}C_{2} \times {}^{4}C_{1}) \div {}^{9}C_{3} = \frac{10}{21}\quad \text{and}\quad P(B) = {}^{5}C_{3} \div {}^{9}C_{3} = \frac{5}{42}.$$

A and B are mutually exclusive, so that

$$P(A \cup B) = P(A) + P(B) = \frac{10}{21} + \frac{5}{42} = \frac{25}{42}.$$

In applying the addition laws there are two useful devices. As we saw above, in using a Venn diagram the argument implied that probabilities, rather than outcomes, may be marked on the appropriate parts of the diagram. In problems involving symmetry, *possibility space diagrams* are often particularly helpful.

The following examples illustrate each of these.

EXAMPLE 3   A survey of the readership of two popular papers shows that the probability that a man chosen at random reads *Blast* is 0.4 and that the probability that he reads *Blast* but not *Look* is 0.28. The probability that he reads neither is 0.05. Find the probability that he reads *Look*, and the probability that he only reads *Look*.

In the diagram we first insert

$$P(B \text{ and not } -L) \text{ or } P(B \cap L') = 0.28$$

and hence

$$P(B \text{ and } L) \text{ or } P(B \cap L) = 0.12.$$

Now $P(B \text{ or } L) = 1 - P(\text{neither})$

$$= 1 - 0.05 = 0.95.$$

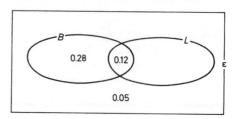

A glance at the figure shows that $P(L) = 0.95 - 0.28 = 0.67$ and that $P(L$ and not $-B)$ or $P(L \cap B') = 0.67 - 0.12 = 0.55.$

Possibility space diagrams

We have been using the words 'possibility space' to describe a set of distinct possible outcomes, and although the use of 'space' does not imply that these outcomes should be shown on a geometrical diagram, such a diagram is sometimes helpful.

Two unbiased dice are thrown.

The possible pairs of scores $(1, 1)$, $(2, 1)$, $(1, 2) \dots$ may be represented by 36 crosses, as shown in the figure opposite.

As the dice are unbiased, each point pair is equally likely and we may read off from the diagram the probability of any event.

Suppose, for example, the event A is {the sum of the scores is less than 6}. $P(A) =$ the number of crosses shown as enclosed, divided by the total number of crosses $= 10/36.$

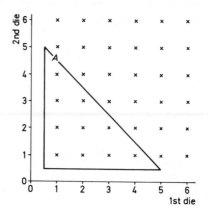

EXAMPLE 4   Two unbiased dice are thrown. What is the probability that either the sum of the scores is 7 or that they differ by at most 1?

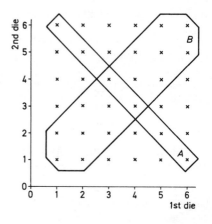

If A is the event {the sum is 7} and B is the event {the difference is at most 1}, then from the figure:

A contains 6 possibilities and

B contains 16 possibilities.

Since A and B are not mutually exclusive, we need the number in $A \cap B$, which is 2. Hence

$$P(A \cup B) = P(A) + P(B) - P(A \cap B) = \frac{6}{36} + \frac{16}{36} - \frac{2}{36} = \frac{5}{9}.$$

(Alternatively the total number of crosses in A or B or both could be counted, giving 20, so that $P(A \cup B) = 20/36 = 5/9$.)

EXERCISE 5a

1.   A card is drawn at random from a pack of 52. What is the probability that it is: (a) a queen, (b) a spade, (c) either a spade or a queen, (d) either a queen or a king?

2.   Twenty cards are numbered 1 to 20. If a card is drawn at random, what is the probability that it is: (a) even, (b) divisible by 5, (c) even or divisible by 5?

3.   Three balls are drawn at random from six white and five red balls. What is the probability that they are all the same colour?

4.   Three cards are taken at random from a normal pack of 52. What is the probability that: (a) exactly two are jacks, (b) at least two are jacks?

5.   Four cards are taken at random from a normal pack of 52. What is the probability that: (a) none of them is a king, (b) at least one is a king?

6.   In a lodging house attached to the University of Ruritania there are 15 nobly-born students. Seven of them have duelling scars on their cheeks and the other eight have closely cropped hair.
   The rector of the University selects five of these students at random to act as noblemen-ushers at the Royal Wedding. What is the probability that:
   (a) the selection will include at least one student with closely cropped hair,
   (b) there will be among the ushers more students with duelling scars than with closely cropped hair.

7.   Use Venn diagrams to illustrate:
(a) $P(A) = P(A \cap B') + P(A \cap B)$,
(b) $P(A \cap B') = P(A) - P(B) + P(A' \cap B)$.

8.   In Ruritania 20% of the population live in the capital city, 25% own a horse, 12% live in the capital *and* own a horse.
   Use a Venn diagram to find the probability that a Ruritanian selected at random:
(a) either lives in the capital or owns a horse,
(b) neither lives in the capital nor owns a horse,
(c) lives in the capital but does not own a horse.

9.   In a smart part of a town, 32% of the houses contain a wall safe but not a burglar alarm; 81% contain either a wall safe or a burglar alarm but not both; 2% of the houses contain neither a wall safe nor a burglar alarm.
   Find the probability that a house picked at random will contain both a wall safe and a burglar alarm.

10. Use a Venn diagram to illustrate:

$$P(A \cup B \cup C) = P(A) + P(B) + P(C) - P(B \cap C) - P(C \cap A) - P(A \cap B) + P(A \cap B \cap C).$$

11. Two unbiased dice are thrown. A is the event {the scores on the two dice are equal}. B is the event {the sum of the scores on the two dice is less than 9}.
Draw a possibility space diagram and use it to find:

(a) $P(A)$, (b) $P(B)$, (c) $P(A \cap B)$, (d) $P(A \cup B)$.

12. Two unbiased dice are thrown. C is the event {the sum of the scores on the two dice is 7}. D is the event {the difference between the two scores is 2 or less}.
Draw a possibility space diagram and use it to find:

(a) $P(C)$, (b) $P(D)$, (c) $P(C \cap D)$, (d) $P(C \cup D)$

(e) $P(D')$, (f) $P(C \cap D')$, (g) $P(C \cup D')$.

5.2 Conditional probability

Suppose A and B are two events, where $P(A) \neq 0$, then $P(B|A)$ means the *conditional* probability of B, *given that A has occurred*.

As a simple illustration of the idea, consider a bowl which contains 12 hard-centred chocolates and 12 soft-centred ones, all wrapped so as to be indistinguishable. A girl picks one out and finds it is hard-centred. There are now 23 left, of which 11 are hard-centred, so that if, after shaking the bowl, she picks another, the probability that this one will be hard-centred is 11/23. In other words the probability that *the second is hard-centred, given that the first is hard-centred* is 11/23.

Before defining conditional probability formally, we shall consider two examples, each of them involving symmetry.

EXAMPLE 5 Two unbiased coins are thrown. If A is {at least one head} and B is {both coins show the same}, what is $P(B|A)$?

The possibility space is $\{HH, HT, TH, TT\}$

A is {at least one head} or $\{HH, HT, TH\}$

B is {both the same}   or $\{HH, TT\}$

These events are shown in a Venn diagram. (The overall rectangular shape is chosen, since A and B cover the *whole* possibility space.)

From the diagram it seems sensible to say that

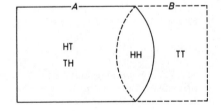

$$P(B|A) = \frac{1}{3}$$

since the only outcomes that matter when we are concerned with the probability of B given that A has occurred are those in A.

It is, in fact, the *number of outcomes in $A \cap B$, divided by the number of outcomes in A alone*.

Now $P(A \cap B) = \frac{1}{4}$ and $P(A) = \frac{3}{4}$, so that what we have done is equivalent to dividing $P(A \cap B)$ by $P(A)$.

Provisionally, therefore, we say that

$$P(B \mid A) = \frac{P(A \cap B)}{P(A)}.$$

Note: In this definition we are using the fact that when considering $P(B \mid A)$, the possibility space is no longer the whole diagram but has been *reduced* to the outcomes in A alone.

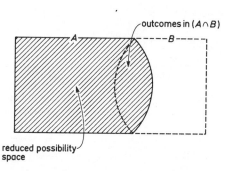

EXAMPLE 6 Two unbiased dice are thrown. Find the probability that the sum of the scores is less than 10, given that at least one of the dice shows a '5'.

Let A be {at least one of the dice shows a '5'}, and let B be {the sum of the scores is less than 10}.

As the diagram shows, A consists of 11 of the total of 36 equally likely outcomes.

Of these 11, 8 also occur in B, so that, as in the previous example it seems right to say that

$$P(B \mid A) = 8/11$$

since the only outcomes that are relevant to $P(B \mid A)$ are those in which A has occurred. Once again, the possibility space is no longer the whole diagram but has been reduced to the outcomes in A alone.

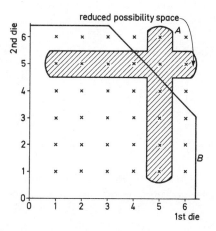

Now $P(A) = 11/36$ and $P(A \cap B) = 8/36$ so that our expression for $P(B \mid A)$ can be written as $P(A \cap B) \div P(A)$.

The *general argument* for defining $P(B \mid A)$ in this way, without assuming symmetry, is as follows.

Suppose that in N experiments A occurs $n(A)$ times, B occurs $n(B)$ times and $A \cap B$ occurs $n(A \cap B)$ times.

Algebraically, the relative frequencies

$$\frac{n(A \cap B)}{N}, \quad \frac{n(A)}{N}, \quad \frac{n(A \cap B)}{n(A)}$$

are related as follows:

$$\frac{n(A \cap B)}{n(A)} = \frac{n(A \cap B)}{N} \div \frac{n(A)}{N}.$$

Now as N increases, $[n(A \cap B)]/N$ tends to $P(A \cap B)$, $n(A)/N$ tends to $P(A)$ and $[n(A \cap B)]/n(A)$, which is the proportion of occurrences of B, given that A has occurred, tends to $P(B|A)$.

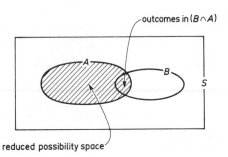

outcomes in $(B \cap A)$

reduced possibility space

Hence $P(B|A) = \dfrac{P(A \cap B)}{P(A)}$.

Note: As in the previous example, this argument implies a reduced possibility space, consisting of the outcomes in A alone.

5.3   The multiplication law and tree diagrams

One consequence of the definition of $P(B|A)$ is that, by rearranging the result we have:

Law 4

$$P(A \cap B) = P(A) \cdot P(B|A)$$

and, since $P(B \cap A) = P(A \cap B)$,

$$P(A \cap B) = P(B) \cdot P(A|B).$$

This is the *multiplication law.*

EXAMPLE 7   From a bag containing 6 black and 4 white balls, 2 balls are selected, one after the other, without replacement.
    Find: (a) the probability that they are both black,
          (b) the probability that they are both the same colour,
          (c) the probability that they are different colours.

We shall use the notation B_1 for {the first ball drawn is black}, W_2 for {the second ball drawn is white}, and so on.

(a) $P(B_1 \cap B_2) = P(B_1) \cdot P(B_2|B_1)$.

$$P(B_1) = \frac{6}{10} \quad \text{and} \quad P(B_2|B_1) = \frac{5}{9},$$

since, when a black ball has been drawn, there are 5 equally likely outcomes out of a reduced possibility space of 9.

Hence $P(B_1 \cap B_2) = \dfrac{6}{10} \cdot \dfrac{5}{9} = \dfrac{1}{3}$.

(b) By a similar argument

$$P(W_1 \cap W_2) = \frac{4}{10} \cdot \frac{3}{9} = \frac{2}{15}.$$

Since $B_1 \cap B_2$ and $W_1 \cap W_2$ are mutually exclusive events,

$$P[(B_1 \cap B_2) \quad \text{or} \quad (W_1 \cap W_2)] = \frac{1}{3} + \frac{2}{15} = \frac{7}{15}.$$

(c) P(the colours are different) $= 1 - P$(the colours are the same)
$$= 1 - 7/15 = 8/15.$$

This method is often best used in conjunction with a 'tree-diagram', which gives the relevant probabilities at a glance.

Starting at O, there are two possible outcomes for the first ball drawn, labelled B_1 and W_1, with the appropriate probabilities 6/10 and 4/10 marked on the branches OB_1 and OW_1. Similarly there are two possible branches from each of B_1 and W_1 to the points representing the possible outcomes for the second draw, B_2 and W_2. The probabilities on these branches take account of the numbers of black 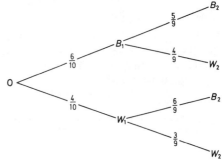 and white balls remaining after the first draw and can be put in by inspection. From the diagram,

P(both black) $=$ the product of the probabilities along the line $OB_1 B_2$

$$= (6/10) \times (5/9) = 1/3.$$

P(same colour) $=$ the sum of the products of the probabilities along the lines $OB_1 B_2$ and $OW_1 W_2$

$$= (6/10) \times (5/9) + (4/10) \times (3/9) = 7/15.$$

P(different colours) $=$ the sum of the products of the probabilities along the lines $OB_1 W_2$ and $OW_1 B_2$

$$= (6/10) \times (4/9) + (4/10) \times (6/9) = 8/15.$$

The problem may also be done, of course, by using combinations.

(a) The total number of possible selections of 2 balls out of 10 is $^{10}C_2$ and the number of possible selections of 2 black balls out of 6 is $^6C_2$, so that the required probability is

$$^6C_2 \div {}^{10}C_2 = \frac{1}{3}.$$

(b) Similarly the probability that the two balls are the same colour is

$$(^6C_2 + {}^4C_2) \div {}^{10}C_2 = \frac{7}{15}.$$

EXERCISE 5b

1.   A bag contains 12 small electric fuses. Through inadvertence, 9 of these are new while the other 3 are 'used' and therefore blown.
   If 2 fuses are taken out at random, what is the probability that: (a) both are new, (b) both are used, (c) one is new and one is used? If 5 fuses are picked out, what is the probability that all are new?

2.   Two cards are taken at random from a normal pack of 52. What is the probability that they are: (a) both queens, (b) the jack of spades and the jack of clubs?

3.   A bag contains five blue and three yellow balls. Balls are taken without being replaced. Show that the probability that the first yellow ball to appear will do so at the fifth draw is 3/56.

4.   A bag contains five apples of which three are sound and two are bruised. Four girls A, B, C and D each take an apple at random, in the order named. For each of A, B, C and D find the probability that she will be the first to pick a bruised apple.
   It is given that A picks a sound apple and B a bruised apple. Show that the probability of C picking a bruised apple is then equal to the probability of D picking a bruised apple.   (L)

5.   A hand of 13 playing cards happens to contain 2 kings. Four cards are taken at random from this hand. What is the probability that the cards selected do not include either of the kings?

6.   The Ruritanian coinage is peculiar in that all coins are the same size, shape and weight. A citizen has in his pocket two zlotyls, three futs and four pnins. If he takes out two coins, one after the other without replacement, what is the probability that he will get a zlotyl and a pnin in either order? If, on the other hand the first coin is replaced before the second is taken out, what will the probability be?

7.   In a raffle, a hat contains 10 red tickets, 15 blue tickets and 5 green tickets. If 3 tickets are taken out at random, what is the probability that they will consist of 2 red tickets and 1 green ticket?

8.   Fifteen cards are numbered 1 to 15. Two cards are taken at random, without replacement. If A is the event {the first card is 7} and B is the event {the second card is even} what are the probabilities $P(A)$, $P(B|A)$, $P(B|A')$, $P(B'|A)$?

9.   An unbiased coin happens to give 20 heads in successive throws. What is the conditional probability that the next throw will also give a head?

10.   In a school, 40% of the boys have fair hair, 25% have blue eyes; 15% have both fair hair and blue eyes. A boy is selected at random.
   (a) Given that he has fair hair, find the probability that he has blue eyes.
   (b) Given that he has blue eyes, find the probability that he does not have fair hair.
   (c) Find the probability that he has neither fair hair nor blue eyes.   (L)

11.   Two events A and B are such that $P(A) = 0.2$, $P(A' \cap B) = 0.22$, $P(A \cap B) = 0.18$.
   Evaluate: (a) $P(A \cap B')$, (b) $P(A|B)$.   (JMB)

12.   A man rolls a die to select one of three boxes. If he rolls a 6, he selects the red box; if he rolls a 5 or a 4, he selects the blue box; and if he rolls a 3 or a 2 or a 1, he selects the yellow box. He opens the box he has chosen and selects a coin at random from it. The red box contains three gold coins, the blue box two gold coins and one silver coin, and the yellow box one gold coin and two bronze coins. Using a tree diagram, or otherwise, find the probability that: (a) he selects a silver coin, (b) he selects a gold coin, (c) having selected and retained a gold coin, if he now selects at random a second coin from the same box, it will also be gold.

(L)

13.   Two unbiased dice are thrown. A is the event {the sum of the scores is 7} and B is the event {the difference between the scores is less than 3}. Use a possibility space diagram to find $P(B \mid A)$.

14.   Two unbiased dice are thrown.
(a) Use a possibility space diagram to find $P(D \mid C)$ where
     C is the event {the sum of the scores is 6},
     D is the event {the difference between the scores is 2}.
(b) Use a possibility space diagram to find $P(F \mid E)$ where
     E is the event {the difference between the scores is 4}.
     F is the event {the sum of the scores is 8}.

15.   A red and a green die are thrown together. Each is unbiased.
(a) Use a possibility space diagram to find $P(H \mid G)$ where
     G is the event {the sum of the scores is less than 8},
     H is the event {the score on the red die is 4}.
(b) Find also $P(K \mid G)$ where
     K is the event {the score on one and only one of the dice is 4}.

16.   Two unbiased dice are thrown, one of which is blue.
A is the event {the score on the blue die is 5},
B is the event {the sum of the two scores is 7},
C is the event {the sum of the two scores is 10}.
   Show that: (a) $P(B \mid A) = P(B)$, (b) $P(A \mid B) = P(A)$, (c) $P(C \mid A) \neq P(C)$.

5.4   Independent events

In ordinary usage, to say that two events are 'independent' implies that the probability that one will occur is unaffected by the occurrence or non-occurrence of the other.

If event A is {I will get to work on time} and event B is {The trains are running normally}, $P(A)$ may well be dependent on $P(B)$, whereas if event C is {The milkman is on time with the delivery of milk}, $P(A)$ is likely to be independent of $P(C)$.

To express this idea mathematically, that is, to define *statistical independence*, it will be helpful to consider an experiment in which an unbiased coin and an unbiased die are thrown simultaneously a large number (N) of times. A is the event {the coin shows its head} and B is the event {the die shows a '5'}. A and B are obviously independent in the ordinary sense of the word.

In N simultaneous throws, a head will turn up about $\frac{1}{2}N$ times, and a '5' will turn up in about $\frac{1}{6}$ of these $\frac{1}{2}N$ times, so that the relative frequency of $A \cap B$ is about

$$\frac{1}{6} \cdot \frac{1}{2}.$$

In terms of probability, then, we can say that

$$P(A \cap B) = P(A) \cdot P(B).$$

We take this as the definition of independence.

Two events A and B are said to be statistically independent if and only if

$$P(A \cap B) = P(A) \cdot P(B).$$

Independent events are of two kinds:

1.   those that occur as a result of independent experiments, as in the coin and die example,

2.   those that occur as a result of repeating an experiment in the same conditions.

As an example of the second kind we could take the experiment in example 7, where 2 balls are taken at random from a bag containing 6 black and 4 white balls, but this time replacing a ball after being selected, so that the results of the two selections are independent. In this case

$$P(B_1 \cap B_2) = \frac{6}{10} \cdot \frac{6}{10} = 0.36.$$

This definition of independence means that for independent events, we use the simplest possible kind of multiplication rule. Moreover, if we look again at the general multiplication law:

$$P(A \cap B) = P(A) \cdot P(B \mid A)$$
$$\text{and} \quad P(A \cap B) = P(B) \cdot P(A \mid B),$$

the definition is equivalent to saying that

$$P(B \mid A) = P(B) \quad \text{and} \quad P(A \mid B) = P(A).$$

This also seems intuitively right as a way of defining independence, as it implies that $P(B)$ is not conditional on whether or not A has occurred.

Now if *either* of the last two conditions is true, then the other will also be. This is to be expected since our original definition is symmetrical in A and B. The proof is as follows.

It is always true that

$$P(B \mid A) \cdot P(A) = P(A \mid B) \cdot P(B) (= P(A \cap B)).$$

If, in addition, $P(B \mid A) = P(B)$, then

$$P(B) \cdot P(A) = P(A \mid B) \cdot P(B) \Rightarrow P(A) = P(A \mid B).$$

We have therefore *three* different necessary and sufficient conditions for A and B to be independent:

$$P(A \cap B) = P(A) \cdot P(B)$$
$$P(B \mid A) = P(B)$$
$$P(A \mid B) = P(A),$$

and any one of them may be used as a test for independence.

Finally, before doing some examples, notice that we have dropped the word 'statistically'. If events are independent in the usual sense of the word, they are statistically independent. (The converse is, as a matter of fact, not necessarily true, but the argument involved in such cases can involve hair-splitting logic of a kind which has no relevance to statistical work.)

EXAMPLE 8   The tea-lady notices that the chairman asks for Earl Grey tea four times out of five, that the vice-chairman does so three times out of four, and the company secretary two times out of three.

What is the probability that on any given occasion she will be asked to supply:
(a) one and only one cup of Earl Grey,
(b) not more than one cup of Earl Grey,
(c) at least one cup of Earl Grey.

We make the assumption that the three people's preferences are independent in the ordinary sense of the word, so that we may use the simple form of multiplication law.

With the obvious notation, $P(C) = 4/5$, $P(V) = 3/4$, $P(S) = 2/3$.
(a) The required probability is

$$P(C \cap V' \cap S') + P(C' \cap V \cap S') + P(C' \cap V' \cap S)$$

$$= \left(\frac{4}{5} \cdot \frac{1}{4} \cdot \frac{1}{3}\right) + \left(\frac{1}{5} \cdot \frac{3}{4} \cdot \frac{1}{3}\right) + \left(\frac{1}{5} \cdot \frac{1}{4} \cdot \frac{2}{3}\right) = \frac{3}{20},$$

since the three compound events are mutually exclusive.
(b) The required probability is

$$P(C' \cap V' \cap S') + \frac{3}{20} = \left(\frac{1}{5} \cdot \frac{1}{4} \cdot \frac{1}{3}\right) + \frac{3}{20} = \frac{1}{6}.$$

(c) $P(\text{no cups of Earl Grey}) = P(C' \cap V' \cap S') = \frac{1}{5} \cdot \frac{1}{4} \cdot \frac{1}{3} = \frac{1}{60}$

$$\Rightarrow P(\text{at least one cup}) = 1 - \frac{1}{60} = \frac{59}{60}.$$

EXAMPLE 9   Two unbiased dice are thrown. A is the event {the first die shows 5} and B is the event {the total score is 7}.

Are A and B independent?

There are two ways of doing this problem.
(a) From the diagram,

$$P(A \cap B) = 1/36,$$

$$P(A) = 6/36, \quad P(B) = 6/36.$$

Hence $P(A \cap B) = P(A) \cdot P(B)$

and A and B are independent.

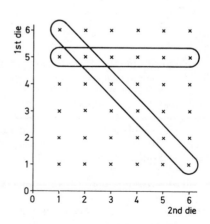

(b) From the diagram

$$P(B\mid A) = \frac{P(B\cap A)}{P(A)} = \frac{1/36}{6/36} = 1/6,$$

$$P(B) = 6/36.$$

Hence $P(B\mid A) = P(B)$ and A and B are therefore independent.

EXAMPLE 10   A pack of cards has all the hearts from 2 to 9 inclusive removed. A card is drawn at random from the reduced pack; A is the event {the card is a king} and B is the event {the card is a spade}.

Are A and B independent?

The pack contains 44 cards, so that $P(A) = 4/44$, $P(B) = 13/44$ and $P(A\cap B) =$ the probability that the card is the king of spades $= 1/44$.

Since $P(A\cap B) \neq P(A)\cdot P(B)$, A and B are not independent.

EXERCISE 5c

1.   A motorist makes regularly a certain journey which involves three traffic lights A,B,C. He reckons that the probabilities of being delayed at these lights are respectively 0.6, 0.75 and 0.8 and that the delays at the points are independent.

Calculate the probabilities of 0, 1, 2, 3 delays on the journey.   (MEI)

2.   The probability of passing a driving test at the first attempt is 1/3. At every subsequent attempt, all previous attempts having resulted in failure, the probability of passing the test is 3/7, independently of any previous record of failure.

Find the probability of passing the test in not more than three attempts.   (L)

3.   A man selects cards, with replacement, from a pack of playing cards recording a score of 1 for an ace, 2 for a two and similarly for all the numbered cards, but zero for all court cards.

Find the probability that:

(a) after three selections his total score is zero,

(b) after two selections his total score is more than five.   (L)

4.   Two guns fire at a target. For each shot by gun A the probability of hitting is 2/5. For each shot by gun B the probability of hitting is $\frac{1}{2}$. Gun A fires 3 shots at the target. Find the probability that there is at least one hit.

This series of shots is in fact unsuccessful. In a second series of shots, gun A fires 3 shots and gun B fires 2 shots. Find the probabilities of: (a) no hits, (b) at least one hit, (c) exactly one hit.   (MEI)

5.   Two chess players A and B find that during the past year A has won 4/9 of the games, B has won 1/3 of the games and the rest were drawn.

Use these probabilities to estimate the probability that in the next three games they play:

(a) B will win all three, (b) there will be one drawn game, (c) A and B will win alternately.

6.   A game is played by two players A and B who have alternate turns. When it is A's turn there is a probability of 1/3 that he will win that turn; otherwise it is a draw. When it is B's turn there is a probability of $\frac{1}{4}$ that he will win; otherwise it is a draw. The game continues until either A or B wins.

(a) If A starts, show that the probability of B winning at his first turn is 1/6.

(b) If B starts, find the probability that he wins at his next turn.

(c) If A starts, find the probability that he wins either at his first or second turn.

(d) If A and B toss an unbiased coin to decide who starts, find the probability that after two turns each neither has yet won.

7.   A card is taken at random from a normal pack of 52.
If A is the event {the card is a heart},
  B is the event {the card is a '3'},
  C is the event {the card shows a red suit},
prove that A and B are independent, B and C are independent, but that A and C are not independent.

8.   All the hearts are removed from a normal pack, and a card is taken at random from this reduced pack.
If C is the event {the card is a king},
  D is the event {the card is a spade},
are C and D independent?

9. A and B are independent events and $P(A) = 0.7$, $P(B) = 0.1$. What are: (a) $P(B\,|\,A)$, (b) $P(A \cap B)$, (c) $P(A \cup B)$?

10.   Two cards are drawn at random from a normal pack. If C is the event {both cards are black} and D is the event {one card is the king of spades}, are C and D independent?

11.   Two unbiased dice are thrown, one of which is coloured red.
E is the event {the score on the red die is 3},
F is the event {the sum of the scores is 7}.
Are E and F independent?

12.   Two unbiased dice are thrown, one of which is coloured blue.
G is the event {the score on the blue die is 4},
H is the event {the sum of the scores is 9}.
Are G and H independent?

5.5   Summary of the probability laws for compound events

$$P(A \cup B) = P(A) + P(B) - P(A \cap B), \qquad \text{in general.}$$

$$P(A \cup B) = P(A) + P(B), \qquad \text{if } A \text{ and } B \text{ are } \textit{mutually exclusive.}$$

$$P(A \cap B) = P(A) \cdot P(B\,|\,A)$$
$$= P(B) \cdot P(A\,|\,B), \qquad \text{in general.}$$

$$P(A \cap B) = P(A) \cdot P(B) \qquad \text{if and only if } A \text{ and } B \text{ are } \textit{independent.}$$

This is equivalent to

$$P(A\,|\,B) = P(A) \quad \text{or} \quad P(B\,|\,A) = P(B).$$

Be careful to avoid confusing the terms 'mutually exclusive' and 'independent'. Mutually exclusive events occur in the context of several possible outcomes of *one* experiment. If A and B are mutually exclusive, A and B cannot occur together and $P(A \cap B) = 0$.

Independent events occur in the context of two (or more) experiments taking place together, or one experiment being repeated one or more times.

5.6   A further application of conditional probability: Bayes' theorem

The multiplication law

$$P(A \mid B) \cdot P(B) = P(A \cap B) = P(B \cap A) = P(B \mid A) \cdot P(A)$$

has the important consequence that if $P(A)$, $P(B)$ and $P(A \mid B)$ are known, $P(B \mid A)$ can be found immediately.

Suppose that Amy and Beryl are two women assembling packets of parts for model railway stations in a small factory. Neither is very efficient, and the probability that a packet assembled by Amy is incomplete is 1/30. The equivalent probability for Beryl is 1/20. However Amy is the slower worker and produces 3 packets to every 5 produced by Beryl. One packet is selected at random from the total output of the two women and is found to be incomplete. What is the probability that it was assembled by Amy?

We will call A the event {a packet is assembled by Amy}
B the event {a packet is assembled by Beryl}
F the event {a packet is incomplete}.

From the data we know that $P(A) = 3/8$ and $P(B) = 5/8$.

Also, since the conditional probability that a packet is incomplete *if* it is one of Amy's is 1/30, so that

$$P(F \mid A) = 1/30.$$

Similarly for Beryl

$$P(F \mid B) = 1/20.$$

What we want to know is the conditional probability that a packet is one of Amy's, given that it is incomplete, or

$$P(A \mid F).$$

This will be calculated using

$$P(A \mid F) = \frac{P(A \cap F)}{P(F)}.$$

The first step is to find $P(F)$.

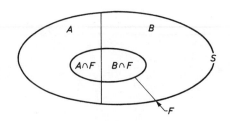

F can occur in either of two ways. Either the packet is faulty and comes from Amy, or it is faulty and comes from Beryl. These two ways are mutually exclusive (see the Venn diagram), so that

$$
\begin{aligned}
P(F) &= P(A \cap F) + P(B \cap F) \\
&= P(F \cap A) + P(F \cap B) \qquad [P(A \cap F) = P(F \cap A)] \\
&= P(F \mid A) \cdot P(A) + P(F \mid B) \cdot P(B) \\
&= (1/30)(3/8) + (1/20)(5/8) = 7/160.
\end{aligned}
$$

Also $P(A \cap F) = P(F \cap A) = P(F \mid A) \cdot P(A) = (1/30)(3/8) = 1/80.$

Hence $P(A \mid F) = \dfrac{1/80}{7/160} = 2/7.$

Looking again at the method of this example, it will be seen that

$$P(A\,|\,F) = \frac{P(F\,|\,A)\cdot P(A)}{P(F\,|\,A)\cdot P(A)+P(F\,|\,B)\cdot P(B)}.$$

This result is an example of *Bayes' theorem*, discovered in the eighteenth century by the Rev. T. Bayes. It may be proved formally by following the steps in this numerical example.

It is sometimes written in the form:

$$P(A\,|\,F) = \frac{P(F\,|\,A)\cdot P(A)}{P(F\,|\,A)\cdot P(A)+P(F\,|\,A')\cdot P(A')},$$

where A and A' are two mutually exclusive events making up the possibility space.

In the last example, $P(A)$, $P(B)$, $P(F\,|\,A)$ and $P(F\,|\,B)$ are called *a priori* probabilities, since they have to be known beforehand in order to calculate $P(A\,|\,F)$, which is called an *a posteriori* probability.

In the next example we shall show how Bayes' theorem may be extended.

EXAMPLE 11   Of three indistinguishable dice, A, B and C, it is known that the last two are unbiased so that the probability of getting a six on a single throw is $1/6$, but A is biased so that the probability of getting a six is $1/5$. One of the dice is selected at random and thrown twice. If the result is two sixes, what is the probability that the die selected was B?

As before we call A the event {die A is thrown}
B the event {die B is thrown}
C the event {die C is thrown}
T the event {two throws produce two '6s'}.

Since a die is chosen at random, $P(A) = P(B) = P(C) = 1/3$.
$P(T\,|\,A) = (1/5)^2 = 1/25$, $P(T\,|\,B) = P(T\,|\,C) = (1/6)^2 = 1/36$, and we want to find

$$P(B\,|\,T) = \frac{P(B\cap T)}{P(T)}.$$

As before, T can occur in three mutually exclusive ways: die A is used and two '6s' result, die B is used and two '6s' result, die C is used and two '6s' result, so that

$$\begin{aligned}
P(T) &= P(A\cap T)+P(B\cap T)+P(C\cap T)\\
&= P(T\cap A)+P(T\cap B)+P(T\cap C)\\
&= P(T\,|\,A)\cdot P(A)+P(T\,|\,B)\cdot P(B)+P(T\,|\,C)\cdot P(C)\\
&= \left(\frac{1}{25}\right)\left(\frac{1}{3}\right)+\left(\frac{1}{36}\right)\left(\frac{1}{3}\right)+\left(\frac{1}{36}\right)\left(\frac{1}{3}\right) = \frac{1}{75}+\frac{1}{54}.
\end{aligned}$$

$$P(B\cap T) = P(T\cap B) = P(T\,|\,B)\cdot P(B) = \left(\frac{1}{36}\right)\left(\frac{1}{3}\right) = \frac{1}{108}.$$

Hence
$$P(B\,|\,T) = \frac{1}{108}\div\left(\frac{1}{75}+\frac{1}{54}\right) = 0.29.$$

As in the previous example, this calculation is equivalent to using the formula

$$P(B \mid T) = \frac{P(T \mid B) \cdot P(B)}{P(T \mid A) \cdot P(A) + P(T \mid B) \cdot P(B) + P(T \mid C) \cdot P(C)},$$

but it is better to avoid using formulae and to think through each problem from first principles.

EXERCISE 5d

1. Suppose that two machines A and B are turning out the same kind of electrical fuse, contributing respectively 3/5 and 2/5 of the total output. It is known by experience that the probability that a fuse from machine A will be faulty is 0.01 and that the equivalent probability for machine B is 0.03. A fuse from the total output, selected at random, is found to be faulty. What is the probability that it came from machine A?

2. A company advertises a professional vacancy in three national daily newspapers, A, B and C which have readerships in the proportions $2:3:1$, respectively. From a survey of the occupations of the readers of these papers, it is thought that the probabilities of an individual reader replying are 0.002, 0.001, 0.005 respectively.
(a) If the company receives one reply, what are the probabilities that the applicant is a reader of paper A, B and C respectively?
(b) If two replies are received, what is the probability that both applicants are readers of paper A?
You may assume that each reader sees only one paper.   (MEI)

3. Four coins are biased so that the probabilities of getting a head are 0.4, 0.48, 0.52 and 0.6. One coin is chosen at random and, on being tossed three times, gives three heads. What is the probability that the coin chosen was the coin with the bias 0.52?

4. On a certain Pacific island, a particular disease is caught by one person in a thousand. 88 % of those who have the disease respond positively to a diagnostic test, but 3 % of those who do not have the disease also respond positively to the same test. If a person selected at random responds positively to the test, what is the probability that he actually has the disease?

EXERCISE 5e (miscellaneous)

1. In a box of counters for playing the game of Ludo there are 5 green, 4 red, 5 blue and 3 yellow counters. I take out from the box three counters at random. What is the probability that I pick out: (a) 3 red counters; (b) 1 green, 1 yellow, 1 blue in that order; (c) 1 green, 1 yellow and 1 blue in any order?   (SMP)

2. The two events A and B are such that $P(A) = 0.6$, $P(B) = 0.2$, $P(A \mid B) = 0.1$. Calculate the probabilities that (a) both of the events occur, (b) at least one of the events occurs, (c) exactly one of the events occurs, (d) B occurs given that A has occurred.   (JMB)

3. When I oversleep there is a probability of 0.8 that I shall miss breakfast. There is a probability of 0.3 that I shall miss breakfast even when I do not oversleep. What is the probability that I am in time for breakfast on a day that I oversleep?
I find that I have a probability of 0.4 of oversleeping. Draw a tree diagram to show the probabilities of my oversleeping or not, with the consequent probabilities of my having breakfast or not.
On a day chosen at random, what is the probability:
(a) that I oversleep and miss breakfast,
(b) that I do not oversleep but do not miss breakfast,

(c) that I do not miss breakfast?

If I am observed to have missed breakfast, what is the probability that I over-slept.   (SMP)

4. (a) A signal consisting of 7 dots and/or dashes is to be given. The probability of a dot in any position is 2/5 and of a dash is 3/5.

Find the probability that, in a signal, no two consecutive characters are the same.

(b) A die is loaded so that the chance of throwing a one is $x/4$, the chance of a two is $1/4$ and the chance of a six is $(1-x)/4$. The chance of a three, four or five is $1/6$. The die is thrown twice. Prove that the chance of throwing a total of 7 is

$$\frac{9x - 9x^2 + 10}{72}.$$

Find the value of x which make this chance a maximum, and find this maximum probability.   (SUJB)

5. For 7 consecutive weeks I decide to have a night out, chosen at random but never Sunday, once in each (6 day) week.

(a) What is the probability: (i) that my first night out is on a Wednesday, (ii) that my second night out is not on a Saturday?

(b) What are the answers to each of the above questions if there is the additional restriction that I never go out on the same night two weeks in succession?   (SMP)

6. The random events A, B and C are defined in a finite sample space S. The events A and B are mutually exclusive and the events A and C are independent. $P(A) = 1/5$, $P(B) = 1/10$, $P(A \cup C) = 7/15$ and $P(B \cup C) = 23/60$. Evaluate $P(A \cup B)$, $P(A \cap B)$, $P(A \cap C)$ and $P(B \cap C)$ and state whether B and C are independent.   (L)

7. State the addition and multiplication rules of probability and explain what is meant by the terms mutually exclusive events and statistical independence.

A bag contains 7 similar balls of which 4 are red and 3 are black. Two balls are drawn at random from the bag, their colours are noted and they are then replaced. Two balls are again drawn at random from the bag and their colours noted.

Determine: (a) the chance that two red balls are obtained on the first drawing and two black balls on the second drawing, (b) the chance that, of the four balls drawn, two are red and two are black, (c) the chance that, of the four balls drawn, three are black and one is red, (d) the chance that all the balls drawn are the same colour.   (MEI)

8. An unbiased die is thrown six times. Calculate the probabilities that the six scores obtained will consist of exactly two sixes and four odd numbers.

9. Two telephone numbers are chosen at random from a directory, and the last digit of each is noted.

(a) Calculate: (i) x, the probability that the first number ends in a 7, (ii) y, the probability that the second ends in a 7, (iii) z, the probability that both numbers end in a 7.

(b) The last digits are now added together, to give a score somewhere between 0 and 18. Calculate: (i) p, the probability that the score is 13, (ii) q, the probability that one or both of the last digits is 7, (iii) r, the probability that the score is 13 and one of the last digits is 7.

(c) Explain why, although $xy = z$, it is not true that $pq = r$.   (SMP)

10. Show that there are 5040 different orders (counting clockwise and anticlockwise as distinct) in which eight people, made up of four married couples, can be seated round a table.

Assuming that each different order is equally likely, find the probability that: (a) Alf is sitting next to his wife; (b) Ben, Chris and Don are sitting together (in any order); (c) each woman is sitting between two men.   (MEI)

11.   The following information is known about the three events A, B and C:

$P(A) = 1/3$, $P(B) = 1/4$, $P(C) = \frac{1}{2}$, $P(A \cup B \cup C) = 1$, $P(B \cap C) = 0$, $P(C \cap A) = 0$.

Show that:
(a) $P(A \cap B) = P(A) \cdot P(B)$,
(b) $P(A \cup B) = P(C)$,
(c) $P(A \cup C) = P(A) + P(C)$.
Write down two of the events A, B, C which are independent.   (L)

12.   A hand of 13 cards is dealt from a standard pack of 52 cards.
(a) Write down, but do not calculate, an expression for the probability that the hand consists of 3 spades, 4 hearts and 6 cards from the other suits.
(b) Calculate, to 3 decimal places, the conditional probability that the hand contains exactly 3 diamonds, given that it contains exactly 3 spades and 4 hearts.
(c) Calculate, again to 3 decimal places, the probability that the hand contains at least 2 diamonds given the same conditions as in part (b).   (O)

13.   Two unbiased dice are thrown. Find the probabilities of each of the following events:
(a) the sum of the numbers showing is odd;
(b) there is at least one 6 showing;
(c) the sum of the numbers is odd *and* there is at least one 6 showing;
(d) the sum of the numbers showing is odd *or* there is at least one 6 showing (where *or* includes *both*);
(e) the sum of the numbers showing is odd, given that there is at least one 6 showing.   (MEI)

14.   Two events A and B are such that $P(A) = 0.4$ and $P(A \cup B) = 0.7$.
(a) Find the value of $P(A' \cap B)$.
(b) Find the value of $P(B)$ if A and B are mutually exclusive.
(c) Find the value of $P(B)$ if A and B are independent.   (JMB)

15.   The events A and B are independent, the events B and C are independent, and the events A and C are mutually exclusive. Given that $P(A) = P(B) = 1/3$ and $P(B \cap C) = 1/8$, find:

(a) $P(C)$,   (b) $P(A' \cap B' \cap C')$,   (c) $P(A \cup B \cup C)$.   (L)

16.   An analysis of the other subjects taken by A-level mathematics candidates in a certain year showed that 20% of them took further mathematics, 50% took physics and 5% took both further mathematics and physics. A candidate is chosen at random from those who took A-level mathematics.
(a) Calculate the probability that the chosen candidate took neither further mathematics nor physics.
(b) Given that the chosen candidate took at least one of further mathematics and physics, calculate the probability that the candidate took further mathematics.   (JMB)

17.   Alec and Bill frequently play each other in a series of games of table tennis. Records of the outcomes of these games show that whenever they play a series of games, Alec has a probability 0.6 of winning the first game and that in every subsequent game in the series, Alec's probability of winning the game is 0.7 if he won the preceding game but only 0.5 if he lost the preceding game. A table-tennis game cannot be drawn. Find the probability that Alec will win the third game in the next series of games played against Bill.   (JMB)

18. A and B are two events. Show that $P(A)$ is between $P(A \mid B)$ and $P(A \mid B')$.

19.   What is the probability of getting a hand of all one suit in bridge?

20.   The Earl of Yarborough used to bet 1000 to 1 against a hand of 13 cards containing no card higher than a 9 turning up in a game of whist or bridge. What is the actual probability of such a hand (now called a Yarborough)? (Ace counts high.)

21.   A pack of 52 cards contains 4 suits each of 13 cards. If 13 cards are taken at random from the pack what is the probability that exactly 10 of them are spades? (You may take $^{52}C_{13} = 6.35 \times 10^{11}$.)

22.   A box contains 9 discs, of which 4 are red, 3 are white and 2 are blue. Three discs are to be drawn at random without replacement from the box. Calculate:
(a) the probability that the discs, in the order drawn, will be coloured red, white and blue, respectively,
(b) the probability that one disc of each colour will be drawn,
(c) the probability that the third disc drawn will be red,
(d) the probability that no red disc will be drawn.   (JMB)

23.★ A pack of cards is dealt to four persons so that each has 13 cards. Calculate, correct to 3 decimal places, the probability p_1 that a specified player has at least two aces, irrespective of the hands of the other players.
   If p_2 is the probability that some one player of the four has at least two aces, irrespective of the hands of the other players, explain briefly without calculating p_2 why $p_2 \neq 4p_1$.   (MEI)

A set of harder examples on probability may be found on page 512.

6 Probability distributions

6.1 Relative frequency distributions

In chapters 2 and 3 we were concerned with frequency distributions, that is, with actual observations derived from a *sample*. We now turn our attention to the distribution of the *population* from which the sample is taken. As we saw in chapter 1, a population may be real, finite, hypothetical or infinite.

Consider an experiment in which three coins are tossed and the number of heads is recorded, the experiment being repeated 200 times. Let us suppose the results come out like this:

	Number of heads	0	1	2	3
Table 1	Frequency (f)	22	78	76	24
	Relative frequency ($f/200$)	0.11	0.39	0.38	0.12

The equivalent population distribution will be the set of relative frequencies for all possible repetitions of the experiment. This is clearly a hypothetical and infinite population, but we can calculate these theoretical relative frequencies provided we make one assumption – that the coins are unbiased. Of course, we do not know whether this is true, but comparison of the actual and theoretical results will soon tell us whether the assumption is a reasonable one.

With $P(\text{head}) = P(\text{tail}) = \frac{1}{2}$, the 'equally likely outcome' definition of probability will give the following table:

Outcome	TTT	TTH THT HTT	HHT HTH THH	HHH
Number of heads	0	1	2	3
Probability	1/8	3/8	3/8	1/8

and since, by definition, the probability in this case is the relative frequency, the population distribution is:

	Number of heads (X)	0	1	2	3
Table 2	Probability (relative frequency)	0.125	0.375	0.375	0.125

Looking at tables 1 and 2 together, the first question to be asked is whether the results are close enough for it to be reasonably certain that table 2 is the right population distribution or, as we shall often call it, the right *model* for the experimental situation. In other words how close is 'close enough' for any particular size of sample? In fact, as we shall see later, there is a way of testing whether sets of experimental and theoretical results *are* close enough for a model to be accepted. Let us suppose that such a test has been carried out and that we are able to accept this model provisionally. In that case we shall be able to *predict* what the relative frequencies might be if the experiment were repeated a further, say, 70 times.

This is typical of the relation between a sample distribution and the population distribution from which it is presumed to come. The model *accounts for the pattern of events* and makes the *prediction* of further events possible. Prediction is, of course, a vital part of much statistical analysis. The supermarket manager, for example, having noted the proportions of French, Italian, German and Spanish wine being bought by a sample of his wine-buying customers, needs to find a model for the proportions bought by wine-buying customers as a whole in order to be able to plan ahead.

We must now define some terms that will be in constant use from now on.

6.2 Random variable

In table 2, the number of heads X is a variable which can take different values with stated probabilities. Such a variable is called a 'random variable' and is always represented by a capital letter. We can now define this as:

A *random variable* is a numerical quantity which can take any of a range of values with specified probabilities.

There are two kinds of random variable, *discrete* and *continuous*. The number of heads X in table 2 can only take distinct values and is therefore discrete. The values that a discrete random variable can take are usually integers and can therefore be counted. Two further examples of discrete random variables are:

1.   the number of elementary particles emitted by a radioactive source in a given time;

2.   the number of defective items when a sample of ten is tested from the output of a machine.

These are obviously distinct integers, and, as we will show later, each value can be associated with specific probabilities. They are also *finite* in number. However, if X is the number of throws of a die until a '6' appears, it can, in theory, have any of the values $1, 2, 3, \ldots$ to infinity. Such a discrete random variable is called 'countably infinite'.

Suppose now that X is the height of a man in metres. It is a continuous random variable since it can, for example, be $1.7286\ldots$ to any number of decimal places. In other words it is part of a continuum and it does not make sense to talk of the probability that X takes one particular distinct value. Nevertheless, we shall find (see chapter 10) that probabilities may be associated with values of X in different ranges, such as 1.7 m to 1.8 m. Further examples of continuous random variables are:

1. the length of time before a rocket burns out;

2. the amount of impurity in samples of a chemical solution.

Here the quantities are measured rather than counted.

6.3   Probability distributions as models

A table such as table 2, giving the values of the random variable X, together with the associated probabilities is called a *probability distribution* and the values $P(X = 0) = 0.125$, $P(X = 1) = 0.375$ and so on define what is called the *probability function*.

Sometimes a probability function can be expressed in a formula rather than as a list of probabilities. For example, if X is the number of throws of an unbiased die until a six appears, the probabilities are:

$$P(X = 1) = 1/6$$
$$P(X = 2) = \text{the probability that the 1st throw does not}$$
$$\text{give a '6' and the 2nd does}$$
$$= (5/6)(1/6)$$
$$P(X = 3) = \text{the probability that the first two throws}$$
$$\text{do not give '6s', and that the third does}$$
$$= (5/6)^2(1/6), \text{ and so on}$$

The probability function here can be given concisely by the formula

$$P(X = x) = (5/6)^{x-1}(1/6), \quad x = 1, 2, 3, \ldots$$

Notation

It is customary to use a small x to represent a particular value of the random variable X, and $P(x)$ is often a convenient short way of referring to $P(X = x)$. We will also occasionally use p_i as an abbreviation for $P(X = x_i)$.

Models

The word 'model' used earlier occurs frequently in the natural and social sciences and needs further explanation.

In science, a model is a way of describing or accounting for experimental results using some general theory. The inverse square law is a mathematical model for the way a planet moves round the Sun. In biology, the Mendelian theory of inheritance is a model for results in genetics. Using such a model establishes a framework to make sense of experimental or observed results, as well as enabling predictions of other results to be made. As long as experimental results agree with results predicted by the model, it serves a useful purpose. When it ceases to do so it has to be modified – or, more drastically, a new kind of model devised.

In biology, it sometimes happens that the number of animals of a given species in a given area fluctuates in a regular way. It is possible to devise a mathematical formula for the animal population, having first made certain assumptions about such things as the presence of predators, length of time for the animal to mature and so on. This formula *is* the model. If observations do not agree with the model it will be modified, taking into account other factors that had been overlooked. For an example of a fundamental change of model there is the case of the movements of natural objects in space. The model based on Newton's laws of motion was accepted until some fresh observations were found not to 'fit'; the outcome was an entirely new type of model, the theory of relativity.

Probability models are similar. In the coin-tossing example, the model is the theoretical distribution (based on the assumption that the coins are unbiased) which predicts the relative frequencies with which, in the long run, the number of heads will be 0, 1, 2 and 3. If it turned out that the observed relative frequencies were not close enough to the predicted ones, a new model would be needed, taking account of bias in one or more of the coins.

In the chapters that follow, we shall meet the most important models (which will sometimes be referred to as 'the underlying population distribution', or 'the distribution of the parent population') and we shall find that very many different types of statistical investigation are covered by comparatively few models. Before we describe these, however, it will be necessary to consider some properties of discrete probability distributions in general.

6.4 Expectation

In our discussion of an experimental frequency distribution and the corresponding theoretical probability distribution, it was repeatedly stressed that *relative frequency* in the one corresponds to *probability* in the other. This leads to a natural way of defining the *mean* of a probability distribution.

We shall take as an example an experiment in which two dice (assumed to be unbiased) are tossed 100 times and the sum of the two scores noted on each occasion. The results are shown in the following frequency distribution:

Total score X	Frequency (f_i)	Relative frequency (f_i/N)
2	3	0.03
3	6	0.06
4	8	0.08
5	10	0.10
6	14	0.14
7	17	0.17
8	13	0.13
9	12	0.12
10	9	0.09
11	6	0.06
12	2	0.02
$N = \Sigma f_i = 100$		

The theoretical probabilities are most easily obtained by counting points on a possibility-space diagram, so that, for example,

$$P(X = 4) = 3/36; \quad P(X = 9) = 4/36.$$

The probability distribution is therefore:

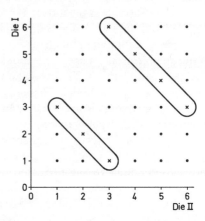

X	Probability (p_i)
2	$1/36 = 0.028$
3	$2/36 = 0.056$
4	$3/36 = 0.083$
5	$4/36 = 0.111$
6	$5/36 = 0.139$
7	$6/36 = 0.167$
8	$5/36 = 0.139$
9	$4/36 = 0.111$
10	$3/36 = 0.083$
11	$2/36 = 0.056$
12	$1/36 = 0.028$

Now the mean, $\bar{x}$, of the frequency distribution $= \Sigma \ (f_i x_i)/N$

$$= \frac{2 \times 3 + 3 \times 6 + 4 \times 8 + 5 \times 10 + 6 \times 14 + 7 \times 17 + 8 \times 13 + 9 \times 12 + 10 \times 9 + 11 \times 6 + 12 \times 2}{100}$$

$$= 7.01.$$

But $\Sigma \ (f_i x_i)/N$ may be written as $\Sigma \ (f_i/N)x_i$, that is, the sum of each value of x multiplied by its *relative frequency*:

$$2 \times 0.03 + 3 \times 0.06 + 4 \times 0.08 + 5 \times 0.10 + 6 \times 0.14 + 7 \times 0.17 + 8 \times 0.13$$
$$+ 9 \times 0.12 + 10 \times 0.09 + 11 \times 0.06 + 12 \times 0.02 = 7.01.$$

The equivalent procedure for the probability distribution is clearly to find the sum of each value of X multiplied by its *probability*.

The mean is therefore

$$\Sigma \ p_i x_i = 2 \times \frac{1}{36} + 3 \times \frac{2}{36} + 4 \times \frac{3}{36} + 5 \times \frac{4}{36} + 6 \times \frac{5}{36} + 7 \times \frac{6}{36} + 8 \times \frac{5}{36}$$

$$+ 9 \times \frac{4}{36} + 10 \times \frac{3}{36} + 11 \times \frac{2}{36} + 12 \times \frac{1}{36} = 7.$$

The mean of a probability distribution is called the *expected value* of the random variable X, or the *expectation* of X. It is written as $E(X)$ for short, its symbol is μ, and it is defined formally as follows:

For the probability distribution

X	x_1	x_2	x_3	$\ldots$	x_n
Probability	p_1	p_2	p_3	$\ldots$	p_n

$$\mu = E(X) = \sum_{i=1}^{n} p_i x_i.$$

This is often written without suffixes as $E(X) = \Sigma\ px$.

It is important to be clear about the distinction between the mean, $\bar{x}$, of the frequency distribution and μ or $E(X)$, which is the mean of the theoretical probability distribution, that is, of the population of all possible repetitions of the experiment.

$$\bar{x} = \frac{\Sigma\ fx}{N}$$

and is a sample *statistic*.

$$\mu = E(X) = \Sigma\ px.$$

A mean of this kind, calculated from a probability distribution in the same way as a statistic is calculated from a frequency distribution is called a *parameter*.

EXAMPLE 1   If X is the score when an unbiased die is thrown, what is $E(X)$?

The probability distribution is

X	1	2	3	4	5	6
Probability	1/6	1/6	1/6	1/6	1/6	1/6

so that $E(X) = (1+2+3+4+5+6) \times \dfrac{1}{6} = 3.5$.

You can see from this example that the words 'expected value' are used in a technical sense. We could not expect to get a score of 3.5 with a die, although for a large number of throws the average would be expected to be 3.5.

EXAMPLE 2   In a game, a player throws two unbiased dice. If he scores 12 he receives £1 from the banker; if he scores 11, 10 or 9 he receives 40p. Find his expected winnings if he has to pay the banker 15p for each throw of the two dice.

X is his winnings and the probability distribution (using the probabilities on page 88) is as follows:

Score	12	11, 10 or 9	any other score
X	$100-15 = 85$	$40-15 = 25$	-15
Probability	1/36	9/36	26/36

$$E(X) = 85 \times \frac{1}{36} + 25 \times \frac{9}{36} + (-15) \times \frac{26}{36} = \frac{-80}{36} = -2\tfrac{2}{9}\text{p}.$$

EXERCISE 6a

1. Calculate $E(X)$ for the following probability distributions:

(a)

X	0	1	2	3	4	5
Probability	1/12	1/6	1/6	5/12	1/12	1/12

(b)

X	-3	-2	-1	0	1	2	3
Probability	0.05	0.2	0.3	0.2	0.1	0.1	0.05

2. Two unbiased dice are thrown. If X is the higher of the two scores showing (or the score, if the same score shows on each), calculate $E(X)$.

3. Two unbiased dice are thrown. If X is the difference between the numbers on the two uppermost faces, find $E(X)$.

4. A die is biased in such a way that the probability of scoring n ($n = 1$ to 6) is proportional to n. Find the mean score.

5. In a game, a player pays 3p each time he throws two unbiased dice. He receives back from the banker a number of pence equal to the difference between his score and 8. Find his expected winnings.

6. A man tosses three unbiased coins one after the other. He pays an entrance fee of 10p and if he gets three heads in turn he receives from the banker 30p. If he gets two heads in turn he receives 10p. In both these cases his entrance fee is also returned. Any other result leads him to lose his entrance money. Will he win or lose in the long run?

7. In a game a man tosses two unbiased coins. He receives 10p for two heads, 2p for two tails and nothing for a head and a tail. In addition, he loses his entrance fee whether he wins or not. If the game is to be 'fair', how much should the entrance fee be?

8. Use the data in question 1 to calculate, in each part of the question, $E(Y)$ where $Y = X + 3$, it being assumed that the probabilities remain unaltered, for example, $P(X = 0) = P(Y = 3)$.

9. Use the data in question 1 to calculate, in each part of the question, $E(Y)$ where $Y = X^2$.

6.5   Expectation of functions of a random variable

If you have completed questions 8 and 9 of exercise 6a, you will have seen that there is a natural way of extending the idea of the expectation of a random variable X to the idea of the *expectation of a function of* X.

EXAMPLE 3   A random variable X has the following probability distribution:

X	2	4	6	8	10
$P(X)$	0.1	0.2	0.4	0.2	0.1

Find $E(X + 3)$.

The problem amounts to finding $E(Y)$ where Y is the new random variable given by $Y = X + 3$. Now X has the value 2 with probability 0.1, so that Y has the value 5 with the same probability. We therefore have the following distribution for Y:

Y	5	7	9	11	13
$P(Y)$	0.1	0.2	0.4	0.2	0.1

and $E(Y) = 0.1 \times 5 + 0.2 \times 7 + 0.4 \times 9 + 0.2 \times 11 + 0.1 \times 13 = 9$.

In fact $E(X) = 6$, so that $E(X + 3)$ turns out to be $E(X) + 3$, as you may well have anticipated. This is a special case of a general result which is mentioned later (see pages 94–5).

EXAMPLE 4   Find $E(X^2)$ if X has the probability distribution:

X	-2	-1	0	1	2
$P(X)$	0.1	0.2	0.3	0.3	0.1

In this case, if $Y = X^2$, it takes the values 4, 1 and 0 with probabilities $0.1 + 0.1$, $0.2 + 0.3$ and 0.3 respectively, and

$$E(Y) = 0.1 \times 4 + 0.2 \times 1 + 0.3 \times 0 + 0.3 \times 1 + 0.1 \times 4 = 1.3.$$

As these examples show $E(X + 3) = \Sigma \ p(x + 3)$

$$\text{and} \quad E(X^2) = \Sigma \ p(x^2).$$

We therefore define the expectation of $g(X)$, any function of X, in the following way:

$$E\{g(X)\} = \Sigma \ p \, g(x).$$

(The use of $g(X)$ instead of the customary $f(X)$ may seem odd. This will become clear when we consider continuous probability distributions later.)

6.6   The variance of a probability distribution

On page 88 we arrived at a definition of the mean of a probability distribution by considering the mean of an experimental frequency distribution

$$\bar{x} = \frac{\Sigma \ f_i x_i}{N}$$

(the summation being from $i = 1$ to $i = n$) and replacing relative frequency f_i/N, by probability, p_i. This led to

$$\mu = E(X) = \Sigma \ p_i x_i.$$

Now for a frequency distribution the usual measure of 'spread', the variance, is given by the formula

$$\text{var}(x) = \frac{\Sigma \ f_i(x_i - \bar{x})^2}{N}$$

so that we must define the variance of a probability distribution by the formula

$$\text{variance} = \Sigma\ p_i(x-\mu)^2.$$

The variance of X is written $\text{Var}(X)$ and is given the symbol σ^2. From the definition, it is the expectation of $(X-\mu)^2$, or the expected value of the square of the deviation from the mean. [In other words, this is a special case of our definition in the previous section. The variance, in fact, is $E\{g(X)\}$ where $g(X) = (X-\mu)^2$.]

To sum up:

$$\sigma^2 = \text{Var}(X) = E(X-\mu)^2 = \sum_{i=1}^{n}\ p_i(x_i-\mu)^2.$$

This is often written without suffixes as $\Sigma\ p(x-\mu)^2$.

When we calculated the variance of a frequency distribution, it was often easier to find

$$\text{var}(x) = \frac{\Sigma\ f_i(x_i-\bar{x})^2}{N}$$

from the alternative formula

$$\text{var}(x) = \frac{\Sigma\ f_i x_i^2}{N} - \bar{x}^2.$$

The corresponding formula for

$$\text{Var}(X) = \Sigma\ p_i(x_i-\mu)^2 \quad \text{is} \quad \Sigma\ p_i x_i^2 - \mu^2.$$

Proof

$$\Sigma\ p_i(x_i-\mu)^2$$
$$= \Sigma\ (p_i x_i^2 - 2p_i x_i \mu + \mu^2 p_i)$$
$$= \Sigma\ p_i x_i^2 - 2\mu\ \Sigma\ p_i x_i + \mu^2\ \Sigma\ p_i \quad \text{(since } \mu \text{ and } \mu^2 \text{ are the same}$$
$$\text{for each } x_i)$$
$$= \Sigma\ p_i x_i^2 - 2\mu \cdot \mu + \mu^2 \quad\quad\quad\quad \text{(since } \Sigma\ p_i x_i = \mu \text{ and } \Sigma\ p_i = 1)$$
$$= \Sigma\ p_i x_i^2 - \mu^2.$$

This may also be written in the form

$$E(X^2) - \mu^2 \quad \text{or} \quad E(X^2) - [E(X)]^2.$$

EXAMPLE 5   Find the mean and variance of the random variable with the following probability distribution:

X	1	2	3	4	5
$P(X)$	0.05	0.1	0.3	0.4	0.15

Using a tabular form:

x_i	p_i	$p_i x_i$	$p_i x_i^2$
1	0.05	0.05	0.05
2	0.1	0.2	0.4
3	0.3	0.9	2.7
4	0.4	1.6	6.4
5	0.15	0.75	3.75
	1.0	3.50	13.30

Note: The 4th column is obtained by multiplying the values of x_i by the values of $p_i x_i$.

Hence $\mu = \Sigma\, p_i x_i = 3.5$ and $\sigma^2 = \Sigma\, p_i x_i^2 - \mu^2 = 13.30 - (3.5)^2 = 1.05.$

EXAMPLE 6$\star$ If the random variable X is the number of throws of an unbiased coin until a head turns up, find $E(X)$.

In this case, X is countably infinite, since it can, in theory, take any of an infinite number of values.

$P(X = 1) = \frac{1}{2}$. $P(X = 2) =$ the probability that the first throw gives a tail and the second a head $= \frac{1}{2} \times \frac{1}{2}$. $P(X = 3) =$ the probability that the first two throws give tails and the third a head $= (\frac{1}{2})^2(\frac{1}{2})$, and so on. Hence the probability distribution is

X	1	2	3	4	...
$P(X)$	$\frac{1}{2}$	$(\frac{1}{2})^2$	$(\frac{1}{2})^3$	$(\frac{1}{2})^4$	...

and $E(X) = \frac{1}{2} \cdot 1 + (\frac{1}{2})^2 \cdot 2 + (\frac{1}{2})^3 \cdot 3 + (\frac{1}{2})^4 \cdot 4 \dots$ to infinity.

We therefore need a formula for the sum to infinity of the series

$$S = x + 2x^2 + 3x^3 + 4x^4 \dots.$$

Multiplying this series by x gives

$$xS = x^2 + 2x^3 + 3x^4 + 4x^5 \dots$$

and subtracting xS from S we get

$$S(1 - x) = x + x^2 + x^3 + x^4 \dots.$$

Now the right-hand side is an infinite geometrical distribution, whose sum, provided $|x| < 1$, is $x/(1-x)$.

Hence $S(1-x) = x/(1-x)$ and $S = x/(1-x)^2.$

Putting $x = \frac{1}{2}$,

$$E(X) = \tfrac{1}{2}/(1 - \tfrac{1}{2})^2 = 2.$$

6.7   Summary of formulae for mean and variance

Frequency distributions	Probability distributions
mean $\bar{x} = \dfrac{\Sigma\, f_i x_i}{N}$	mean $\mu = E(X) = \Sigma\, p_i x_i$
(where $\Sigma\, f_i = N$)	$(\Sigma\, p_i = 1)$
$\text{var}(x) = \dfrac{\Sigma\, f_i (x_i - \bar{x})^2}{N}$	$\text{Var}(X) = \sigma^2 = \Sigma\, p_i (x_i - \mu)^2$
$\qquad = \dfrac{\Sigma\, f_i x_i^2}{N} - \bar{x}^2$	$\qquad = \Sigma\, p_i x_i^2 - \mu^2$

Other ways of writing variance are
$$\text{Var}(X) = E(X^2) - \mu^2$$
$$= E(X^2) - [E(X)]^2$$

EXERCISE 6b

1. For each of the following probability distributions find (a) $E(2X - 1)$, (b) $E(X^2)$.

(i)

X	0	1	2	3	4
$P(X)$	$\frac{1}{8}$	$\frac{1}{4}$	$\frac{1}{4}$	$\frac{1}{4}$	$\frac{1}{8}$

(ii)

X	1	2	3	4	5	6
$P(X)$	0.2	0.2	0.3	0.1	0.1	0.1

(iii)

X	-4	-3	-2	-1	0	1	2	3
$P(X)$	0.04	0.16	0.24	0.16	0.15	0.1	0.1	0.05

2. Calculate the mean and variance of the distributions in question 1.

3. Find the variance for the total score when two unbiased dice are thrown.

4. A die is biased in such a way that the probability of getting a particular score is proportional to that score. Find the variance of the score.

5. Two unbiased dice each have two faces marked 1, two marked 3 and two marked 5. Find the mean and variance of the total score when they are thrown together.

6. If three unbiased coins are tossed find the variance of the number of heads obtained.

6.8   Some important theoretical results

Suppose the random variable X represents masses measured in kilograms and that the distribution has mean 62 kg and variance 5 kg^2.

If now we formed a new random variable giving the masses in pounds, each observation would have to be multiplied by 2.1875. Clearly the mean of the distribution of the new variable will be (62×2.1875)lb.

Similarly, deviations from the mean would also be multiplied by the factor 2.1875, so that the squares of these deviations, and hence the variance, would be multiplied by $(2.1875)^2$. The variance of the new distribution would therefore be $5 \times (2.1875)$lb^2.

Intuitively, the general results must be:

1. $E(kX) = k\,E(X) = k\mu$

2. $\mathrm{Var}(kX) = k^2\,\mathrm{Var}(X) = k^2\sigma^2.$

Now consider the effect of adding 10 kg to each observation. This would obviously result in the mean becoming $(62+10)$kg. The variance, however, would be unaltered since the *deviations* from the mean would remain the same. In other words the spread is unchanged when the 'origin' is changed. In general, then,

3. $E(X+a) = E(X)+a = \mu+a.$

4. $\mathrm{Var}(X+a) = \mathrm{Var}(X) = \sigma^2.$

Putting these results into words:

If the random variable is multiplied by a factor k, the mean will be multiplied by the same factor, and the variance by the factor k^2.

If a constant a is added to the random variable, the mean will be increased by a and the variance will be unaltered.

Formal proofs of these results follow:

1. $E(kX) = \Sigma\, p_i(kx_i) = k\,\Sigma\, p_i x_i = kE(X) = k\mu.$

2. $\mathrm{Var}(kX) = E(kX - k\mu)^2$ (since the mean of kX is $k\mu$)
$$= \Sigma\, p_i(kx_i - k\mu)^2 = k^2\,\Sigma\, p_i(x_i - \mu)^2 = k^2\,\mathrm{Var}(X) = k^2\sigma^2.$$

3. $E(X+a) = \Sigma\, p_i(x_i + a) = \Sigma\, p_i x_i + a\,\Sigma\, p_i$
$$= E(X) + a \quad (\text{since } \Sigma\, p_i = 1)$$
$$= \mu + a.$$

4.   The mean of $X+a$ is $\mu+a$, from (3), and, by definition,
$$\mathrm{Var}(X+a) = E[\{(X+a)-(\mu+a)\}^2] = E(X-\mu)^2$$
$$= \mathrm{Var}(X) = \sigma^2.$$

6.9   The general linear transformation of a random variable

The results we have just proved are special cases of the more general result:

$$E(kX+a) = kE(X)+a = k\mu+a$$
$$\mathrm{Var}(kX+a) = k^2\,\mathrm{Var}(X) = k^2\sigma^2,$$

which can be proved formally using the method of (1) to (4) above.

In other words, if the random variable X has mean μ and variance σ^2, and we form a new variable Y, where

$$Y = kX + a$$

then the mean and variance of Y are:

$$\mu_y = k\mu + a, \qquad \sigma_y^2 = k^2\sigma^2.$$

This relation, $Y = kX + a$ is called a 'linear transformation' of X. We have already met this idea before on page 39, where we showed that it was easier to calculate the mean and variance of a frequency distribution by coding the values of x by means of the relation

$$u = \frac{x - 209.5}{20}$$

which is a linear transformation of the form $u = kx + a$ with $k = 1/20$ and $a = -209.5/20$.

6.10   Standardisation of a random variable

The most important practical application of the idea of a linear transformation is found in what is called 'standardisation'.

Suppose a student takes a public examination and scores 85 % in mathematics and 62 % in economics. He is criticised by his parents for apparently doing so much less well in economics than in mathematics. How might he convince them that this was not necessarily true?

To do this he would need to know how his performances in the two subjects compared with those of all the students who took the examination; in particular he would somehow have to discover the mean and standard deviation of the population distributions of marks in these subjects.

Let us suppose that these turn out to be:

Subject	μ	σ
Mathematics	52	15.0
Economics	41	8.4

It is immediately obvious that the 'raw' marks of 85 and 62 cannot fairly be compared since the mean for economics is lower than for mathematics and the economics marks are much less spread out. What the student should now do is to convert the raw scores (X) for each subject by the linear transformation

$$Z = \frac{X - \mu}{\sigma}$$

For mathematics this gives

$$z = \frac{85 - 52}{15} = 2.2$$

and for economics

$$z = \frac{62 - 41}{8.4} = 2.5.$$

These 'z-scores' show how far each of the original marks is from the mean, *measured by the number of standard deviations.* They therefore provide the only true comparison. In this case, the economics mark is *relatively* better than the mathematics mark.

This process of finding the z-score is called *standardisation*, since it converts the values of X into standard units. The resulting distribution of Z has mean 0 and standard deviation 1, since

$$\mu_z = \frac{1}{\sigma} \times \mu - \frac{\mu}{\sigma} \qquad (Z = kX + a, \text{ where } k = 1/\sigma \text{ and } a = -\mu/\sigma)$$

$$= 0.$$

and $\qquad \sigma_z^2 = \left(\frac{1}{\sigma}\right)^2 \times \sigma^2 = 1.$

We shall be using this idea a great deal in later chapters.

EXAMPLE 7   The discrete random variable X has mean 36 and variance 4. Find the mean and variance of:

$$\text{(a) } X/3, \quad \text{(b) } X-5 \quad \text{(c) } 5X+1 \quad \text{(d) } (X-4)/5.$$

(a) $E(X/3) = \frac{1}{3}E(X) = \frac{1}{3} \times 36 = 12.$

$\qquad \text{Var}(X/3) = \left(\frac{1}{3}\right)^2 \times \text{Var}(X) = \frac{1}{9} \times 4 = \frac{4}{9}.$

(b) $E(X-5) = E(X) - 5 = 36 - 5 = 31.$

$\qquad \text{Var}(X-5) = \text{Var}(X) = 4.$

(c) $E(5X+1) = 5 \times E(X) + 1 = 5 \times 36 + 1 = 181.$

$\qquad \text{Var}(5X+1) = 5^2 \text{Var}(X) = 25 \times 4 = 100.$

(d) $E\left[\dfrac{(X-4)}{5}\right] = \frac{1}{5}E(X-4) = \frac{1}{5}[E(X)-4] = \frac{1}{5}(36-4) = 6.4.$

$\qquad \text{Var}\left[\dfrac{(X-4)}{5}\right] = \left(\frac{1}{5}\right)^2 \text{Var}(X-4) = \frac{1}{25}\text{Var}(X) = \frac{1}{25} \times 4 = 0.16.$

EXERCISE 6c

1.   The distribution of the random variable X has mean 103 and variance 15. Find the mean and variance of the distribution of: (a) $3X/2$, (b) $\frac{1}{2}X - 1$, (c) $(2X+5)/3$.

2.   The discrete random variable X has mean μ and variance σ^2. Find expressions for: (a) $E(4X)$, (b) $E(2X-6)$, (c) $\text{Var}(3X/4)$, (d) $\text{Var}(2X+1)$, (e) $\text{Var}(4-X)$.

3.   It is proposed to convert a set of values of a random variable X, whose mean and standard deviation are 20 and 5 respectively, to a set of values of the variable Y whose mean and standard deviation are 42 and 8 respectively. If the conversion formula is $Y = aX + b$, calculate the values of a and b. (C)

4.   A multiple-choice test is taken by a large number of students. There are only 5 questions and 1 mark is given for the correct answer to a question. The resulting probability distribution is:

Marks (X)	1	2	3	4	5
Probability	0.02	0.08	0.26	0.41	0.23

Find the mean and standard deviation of this distribution.

If the marks are scaled up to go from 10 to 30, so that the new marks Y are given by the linear transformation

$$Y = 5X + 5$$

what will the mean and standard deviation of Y be?

5.   A publisher has a 'back list' of books which are reprinted over a period of years. He finds, on examining the figures, that the average increase in the retail price of reprinted novels over the last ten years is £1.52 with a standard deviation of 20p. The equivalent figures for non-fiction are £2.55 and 34p. He is now reprinting a novel whose original price was £6.95 at a retail price of £8.80 and a history of the Crusades, originally selling at £15.50, is being retailed at £18.60. Which of the two represents the greater relative increase in price?

6.   Prove that, if the discrete random variable X has mean μ and variance σ^2,

$$E(kX + a) = k\mu + a \quad \text{and} \quad \text{Var}(kX + a) = k^2\sigma^2$$

using the method for proving 1 to 4 on page 95.

7.   The mean of the distribution of the random variable X is μ. Use the linear transformation $Y = X - c$, to show that an alternative formula for the variance of X is

$$\text{Var}(X) = E(X - c)^2 - (\mu - c)^2$$

where c is any constant.

(This is another way of stating that the variance is unaffected by subtracting an arbitrary constant. It is sometimes needed in theoretical work.)

6.11   Combinations of random variables

Up to now we have been dealing with the distribution of single random variables, whereas statistical investigations are often concerned with the distribution of two or more variables combined together in some way.

Suppose X and Y are two random variables. It might be important to know the distribution of $X + Y$, that is, of all possible values of X added to all possible values of Y. If, for example, X is the weight of a bottle and Y the weight of its contents, $X + Y$ would be the combined weight. Alternatively, a firm might manufacture a component in two parts, so that if X is the length of one part and Y the length of the other, $X + Y$ would be the combined length when the parts are fitted together.

Similarly for $X - Y$, Y could be the diameter of a bolt designed to fit a hole in a steel plate, and X the diameter of the hole, so that $X - Y$, the difference, is a matter of importance.

In the above examples, X and Y were continuous variables. In this section, we shall be concerned with the distribution of the combinations $X + Y$, $X - Y$ and XY where X and Y are discrete. In particular we shall be looking at the mean and variance of such combinations. We should naturally expect that these will be related to the mean and variance of the separate variables, but the outcome of our discussion will depend on whether or not the random variables X and Y are independent, and we have not yet defined what independence means for random variables. We have already seen that two *events*, A and B, are said to be statistically independent if and only if

$$P(A \cap B) = P(A) \cdot P(B)$$

which is equivalent to saying that $P(A|B) = P(A)$ (see page 74), or that the probability that A occurs is unaffected by the probability of the occurrence of B. We therefore say that:

Two random variables X and Y are independent if and only if the probability that X takes any particular value x is unaffected by Y taking any particular value y, or

$$P(X = x \cap Y = y) = P(X = x) \cdot P(Y = y)$$

for all values of x and y.

An example of a combination of two independent random variables

We shall again take a rather artificial example, for the reason that it does illustrate the idea clearly.

Each of two 'triangular dice' is shaped in the form of a prism with an equilateral triangle as cross-section, so that it has three rectangular faces marked 1, 2 and 3. They are rolled together on a table, and the scores X and Y are the numbers on the faces in contact with the table when they come to rest. We shall assume that the dice are known to be biased so that X and Y have probability distributions

X	$P(X)$
1	0.3
2	0.4
3	0.3

Y	$P(Y)$
1	0.1
2	0.5
3	0.4

From these figures we can construct a *joint probability distribution*, giving the probabilities that $X = 1$ and $Y = 1$, $X = 1$ and $Y = 2$ and so on.

In this case, X and Y are clearly independent, so that

$$P(X = 1 \cap Y = 1) = 0.3 \times 0.1 = 0.03$$
$$P(X = 1 \cap Y = 2) = 0.3 \times 0.5 = 0.15$$

and so on, giving, in tabular form:

		Values of X				
		1	2	3		
	1	0.03	0.04	0.03	0.1	'Marginal'
Values of Y	2	0.15	0.20	0.15	0.5	probabilities of the
	3	0.12	0.16	0.12	0.4	values of Y
		0.3	0.4	0.3		

'Marginal' probabilities of the values of X

$P(X = 1)$, $P(Y = 1)$... are called 'marginal' because they appear in the margin.

It is worth repeating that this table has been constructed on the basis that *because X and Y are independent*, $P(X = 1 \cap Y = 1) = P(X = 1) \cdot P(Y = 1)$, etc.

Conversely, if we are *given* a joint probability distribution like this:

Y \ X	0	1	
0	0.60	0.02	0.62
1	0.04	0.34	0.38
	0.64	0.36	

then X and Y cannot be independent since, for example, $P(X = 0) = 0.64$, $P(Y = 0) = 0.62$, $P(X = 0 \cap Y = 0) = 0.60$ and $0.64 \times 0.62 \neq 0.60$.

This last example, however, is not very illuminating as no indication is given as to where the distribution comes from. For a more realistic case, consider the following figures recorded during a smallpox epidemic in which a large number of people (about 3000) caught the disease. The table gives the proportions (relative frequencies) of those who recovered or died and were vaccinated or not vaccinated. For convenience of reference 'recovered', 'died', 'vaccinated', 'not vaccinated' have been given numerical values 1 or 0.

		X 1 *Recovered*	X 0 *Died*	
Y	*Vaccinated* 1	0.844	0.085	0.929
	Not vaccinated 0	0.034	0.037	0.071
		0.878	0.122	1.000

As the sample of observations is large, the proportions (relative frequencies) may be taken as being close to the population probabilities in the various categories.

Now if the variables X and Y were independent, so that being vaccinated or not did affect survival, then

$$P(X = 1) \cdot P(Y = 1) \text{ should give } P(X = 1 \cap Y = 1) \text{ and so on.}$$

In fact, if we construct the model (theoretical joint probability distribution) on this assumption, the resulting distribution would be:

Y \ X	1	0	
1	0.816	0.113	0.929
0	0.062	0.009	0.071
	0.878	0.122	1.000

On the face of it, these figures do not seem to be particularly close to the actual results and we might assume, provisionally that X and Y are *not* independent.

Nevertheless, as always in this kind of investigation, we would want to test whether actual and theoretical results are sufficiently different for independence to be likely. (For an example of a joint probability distribution for two random variables which are certainly not independent, see the Appendix, page 110.)

Finding the probability distribution of $X + Y$, $X - Y$ and XY

We shall illustrate the method by using the first example in the previous section – the throwing of two 'triangular dice' with faces marked 1, 2 and 3.

Although X and Y are independent here, the method is the same when the random variables concerned are not independent. All we need is the joint distribution table, giving $P(X = 1 \cap Y = 1)$, etc.

			X	
		1	2	3
Y	1	0.03	0.04	0.03
	2	0.15	0.2	0.15
	3	0.12	0.16	0.12

To find the distribution of $X + Y$, we read from the table all possible values of $X + Y$ together with their probabilities:

X	Y	$X+Y$	*Probability*
1	1	2	0.03
1	2	3	0.15
1	3	4	0.12
2	1	3	0.04
2	2	4	0.2
2	3	5	0.16
3	1	4	0.03
3	2	5	0.15
3	3	6	0.12

leading to

$X+Y$	*Probability*
2	0.03
3	0.19
4	0.35
5	0.31
6	0.12

The distribution of $X - Y$ and XY may be found in the same way.
We shall now use these results to calculate the mean and variance of $X + Y$. From the final table

$$E(X + Y) = 2 \times 0.03 + 3 \times 0.19 + 4 \times 0.35 + 5 \times 0.31 + 6 \times 0.12 = 4.3.$$

Using the original table

$$E(X) = 1 \times 1.03 + 2 \times 0.4 + 3 \times 0.3 = 2.0$$

and $E(Y) = 1 \times 0.1 + 2 \times 0.5 + 3 \times 0.4 = 2.3.$

In this case, therefore, $E(X + Y) = E(X) + E(Y)$, a result that might have been anticipated.
Again,

$$\mathrm{Var}(X + Y) = E(X + Y)^2 - [E(X + Y)]^2$$

$$= 4 \times 0.03 + 9 \times 0.19 + 16 \times 0.35 + 25 \times 0.31 + 36 \times 0.12 - (4.3)^2$$

$$= 1.01.$$

$$\mathrm{Var}(X) = 1 \times 0.3 + 4 \times 0.4 + 9 \times 0.3 - 2^2 = 0.6$$

and $\text{Var}(Y) = 1 \times 0.1 + 4 \times 0.5 + 9 \times 0.4 - (2.3)^2 = 0.41$

so that $\text{Var}(X + Y) = \text{Var}(X) + \text{Var}(Y)$.

These are, in fact, special cases of some general results that will shortly be proved. Meanwhile, if you work through the examples in exercise 6d you will begin to discover what these general results might be.

EXERCISE 6d

1. The random variables X and Y have the following distributions:

X	$P(X)$
1	$\frac{3}{4}$
0	$\frac{1}{4}$

Y	$P(Y)$
1	$\frac{5}{8}$
2	$\frac{3}{8}$

Construct the joint probability distribution on the assumption that X and Y are independent.

2. Use the joint probability distribution table in question 1 to find the probability distribution of: (a) $X + Y$, (b) $X - Y$, (c) XY.

3. Use the data of questions 1 and 2 to find:
(a) $E(X)$,   (b) $E(Y)$,   (c) $\text{Var}(X)$,   (d) $\text{Var}(Y)$,
(e) $E(X + Y)$,   (f) $E(X - Y)$,   (g) $\text{Var}(X + Y)$,   (h) $\text{Var}(X - Y)$.

4. X and Y are two independent random variables with probability distributions.

X	$P(X)$
1	0.4
2	0.6

Y	$P(Y)$
1	0.5
2	0.3
3	0.2

Find: (a) $E(X)$, (b) $E(Y)$, (c) $E(X + Y)$, (d) $\text{Var}(X + Y)$.

5. The random variables X and Y have the following joint probability distribution:

		X		
		-1	0	1
Y	0	0.02	0.05	0.03
	1	0.14	0.35	0.21
	2	0.04	0.10	0.06

Find the marginal probabilities and show that X and Y are independent. Calculate: (a) $E(X)$, (b) $E(Y)$, (c) $E(XY)$.

6. The random variables X and Y have the following joint probability distribution.

Y \ X	0	1	2
0	0.1	0.1	0
1	0.1	0.4	0.1
2	0	0.1	0.1

Show that X and Y are not independent. Calculate: (a) $E(X)$, (b) $E(Y)$, (c) $\text{Var}(X)$, (d) $\text{Var}(Y)$, (e) $E(X+Y)$, (f) $\text{Var}(X+Y)$.

7.★ Construct the joint probability distribution for the following experiment.
Three unbiased coins are tossed together, and the number, X, of heads is noted. Any coin showing a head is then removed and the remaining coin or coins (if any) are tossed. The number of heads, Y, on the second throw is noted. (Y is taken as 0 if there is no second throw.)
Verify that the marginal probabilities for Y are:

$$P(Y = 0) = 27/64, P(Y = 1) = 27/64, P(Y = 2) = 9/64, P(Y = 3) = 1/64.$$

Calculate: (a) $E(X)$, (b) $E(Y)$, (c) $E(X+Y)$, (d) $\text{Var}(X)$, (e) $\text{Var}(Y)$, (f) $\text{Var}(X+Y)$.

(*Hint:* see page 110.)

6.12   The mean and variance of $X + Y$, $X - Y$, XY and related combinations

The examples in exercise 6d contained special cases of the following important results:

$$E(X + Y) = E(X) + E(Y).$$

$$E(X - Y) = E(X) - E(Y).$$

$$E(XY) = E(X) \cdot E(Y) \qquad \textit{if X and Y are independent.}$$

$$\text{Var}(X + Y) = \text{Var}(X) + \text{Var}(Y) \qquad \textit{if X and Y are independent.}$$

$$\text{Var}(X - Y) = \text{Var}(X) + \text{Var}(Y) \qquad \textit{if X and Y are independent.}$$

They hold for both discrete and continuous random variables.

Proofs for discrete random variables

Proof 1

$$E(X + Y) = E(X) + E(Y).$$

The general proof is complicated and you may find it easier to follow if we first prove the result when X can take three values x_1, x_2 and x_3 and Y can take two values y_1 and y_2.
In the joint distribution table we will use the notation p_{11} for $P(X = x_1 \cap Y = y_1)$, p_{12} for $P(X = x_1 \cap Y = y_2)$ and so on.
The joint probability distribution table (*whether or not X and Y are independent*) is therefore

		X		
		x_1	x_2	x_3
Y	y_1	p_{11}	p_{21}	p_{31}
	y_2	p_{12}	p_{22}	p_{32}

Using the same procedure as in the numerical example on page 101, the possible values of $X + Y$ are:

$$x_1 + y_1 \quad \text{with probability } p_{11}$$
$$x_1 + y_2 \quad \text{with probability } p_{12} \text{ and so on,}$$

so that

$$E(X + Y) = p_{11}(x_1 + y_1) + p_{12}(x_1 + y_2) + p_{21}(x_2 + y_1) + p_{22}(x_2 + y_2)$$
$$+ p_{31}(x_3 + y_1) + p_{32}(x_3 + y_2)$$
$$= x_1(p_{11} + p_{12}) + x_2(p_{21} + p_{22}) + x_3(p_{31} + p_{32})$$
$$+ y_1(p_{11} + p_{21} + p_{31}) + y_2(p_{12} + p_{22} + p_{32}).$$

Now $p_{11} + p_{12} =$ the sum of the probabilities in the first column

$$= P(X = x_1) = P(x_1).$$

$p_{11} + p_{21} + p_{31} =$ the sum of the probabilities in the first row

$$= P(Y = y_1) = P(y_1),$$

and similarly for the other brackets.

Hence $E(X + Y) = x_1 P(x_1) + x_2 P(x_2) + x_3 P(x_3) + y_1 P(y_1) + y_2 P(y_2)$

$$= E(X) + E(Y).$$

The proof in the general case follows similar lines and is given in the Appendix on page 109.

This result may be extended as follows:

(a) $E(aX + bY) = aE(X) + bE(Y)$

This is proved by letting $U = aX$ and $V = bY$.

$$E(U + V) = E(U) + E(V) = E(aX) + E(bY) = aE(X) + bE(Y)$$

using the result on page 95.

Putting $a = 1$ and $b = -1$ proves that

$$E(X - Y) = E(X) - E(Y),$$

the second of the results given at the beginning of this section.

(b) $E(X + Y + Z + \ldots) = E(X) + E(Y) + E(Z) + \ldots$

To prove this for three random variables, let $U = X + Y$.

Then $E(U + Z) = E(U) + E(Z) = E(X + Y) + E(Z) = E(X) + E(Y) + E(Z).$

Proof 2

$$E(XY) = E(X) \cdot E(Y) \text{ if } X \text{ and } Y \text{ are independent}$$

Once again we shall prove this result for the special case when X can take the values x_1, x_2 and x_3 and Y the values y_1 and y_2.

Using the joint distribution table given in the proof of 1, XY can take the values:

$$x_1 \, y_1 \text{ with probability } p_{11}, x_1 \, y_2 \text{ with probability } p_{12}$$

and so on, so that

$$E(XY) = p_{11} x_1 y_1 + p_{12} x_1 y_2 + p_{21} x_2 y_1 + p_{22} x_2 y_2 + p_{31} x_3 y_1 + p_{32} x_3 y_2.$$

Now if X and Y are independent

$$p_{11} = P(X = x_1) \cdot P(Y = y_1)$$
$$= P(x_1) \cdot P(y_1), \text{ etc.}$$

Hence
$$E(XY) = P(x_1) \, P(y_1) x_1 \, y_1 + P(x_1) \, P(y_2) x_1 \, y_2 + P(x_2) \, P(y_1) x_2 \, y_1$$
$$+ P(x_2) \, P(y_2) x_2 \, y_2 + P(x_3) \, P(y_1) x_3 y_1 + P(x_3) \, P(y_2) x_3 y_2$$
$$= [x_1 \, P(x_1) + x_2 \, P(x_2) + x_3 \, P(x_3)] \times [y_1 \, P(y_1) + y_2 \, P(y_2)]$$
$$= E(X) \cdot E(Y).$$

The general proof is given in the Appendix on page 110.

Proof 3

$$Var(X + Y) = Var(X) + Var(Y) \text{ if } X \text{ and } Y \text{ are independent}$$

Writing $E(X) = \mu_x$ and $E(Y) = \mu_y$, so that $E(X + Y) = \mu_x + \mu_y$,

$$Var(X + Y) = E[(X + Y)^2] - [E(X + Y)]^2$$
$$= E(X^2 + 2XY + Y^2) - (\mu_x + \mu_y)^2$$
$$= E(X^2) + 2E(XY) + E(Y^2) - (\mu_x^2 + 2\mu_x \mu_y + \mu_y^2)$$
$$= E(X^2) - \mu_x^2 + E(Y^2) - \mu_y^2 + 2E(XY) - 2\mu_x \mu_y.$$

Now, since $E(XY) = E(X) \cdot E(Y) = \mu_x \mu_y$ we get

$$Var(X + Y) = Var(X) + Var(Y).$$

This result may also be extended:

(a) $Var(aX + bY) = a^2 \, Var(X) + b^2 \, Var(Y)$

This is proved by putting $U = aX$ and $V = bY$ and using the result on page 95 that $Var(aX) = a^2 \, Var(X)$.

Letting $a = 1$ and $b = -1$, gives $Var(X - Y) = Var(X) + Var(Y)$.

(b) $Var(X + Y + Z + \ldots) = Var(X) + Var(Y) + Var(Z) + \ldots$

All these results will be used extensively in subsequent chapters.

EXERCISE 6e

1.   The random variable X has mean 55 and variance 8.4, and the random variable Y has

mean 48 and variance 6.1. X and Y are independent. Find:
(a) $E(X+Y)$,   (b) $\text{Var}(X+Y)$,
(c) $E(X-Y)$,   (d) $\text{Var}(X-Y)$,
(e) $E(3X+Y)$,   (f) $\text{Var}(3X+Y)$,
(g) $E(2X-3Y)$,   (h) $\text{Var}(2X-3Y)$.
  A third independent random variable Z has mean 28 and variance 4.3. Find:
(i) $E(X+Y+Z)$,   (j) $\text{Var}(X+Y+Z)$,
(k) $E(X-Y+2Z)$,   (l) $\text{Var}(X-Y+2Z)$.

2. The random variable X is the score when an unbiased die is thrown. Find $E(X)$ and $\text{Var}(X)$.
  If Y and Z are the scores when two other unbiased dice are thrown, what are the mean and variance of the sum of the scores when all three dice are thrown together?

3. A public examination in physics consists of two papers, one theoretical and one practical.
  The marks for theoretical physics (X_1) have mean 61 and standard deviation 5; the marks for practical physics (X_2) have mean 52 and standard deviation 3.
  Find the mean and standard deviation of $\bar{X}$, the average of X_1 and X_2.
  If it is decided to give greater weight to theoretical physics, find the mean and standard deviation of the weighted average $(2X_1+X_2)/3$. (X_1 and X_2 are assumed to be independent.)

4. X and Y are two independent random variables.

$$E(X) = 20, \qquad \text{Var}(X) = 9, \qquad E(Y) = 10, \qquad \text{Var}(Y) = 4.$$

  If the mean and variance of the random variable $aX+bY$ are 20 and 145 respectively, where a and b are both integers, find a and b.

5. A random sample of observations x_1, x_2, x_3, x_4, x_5 is taken from a distribution whose mean is μ and whose variance is σ^2.
  Find the mean and variance of $\bar{X} = (x_1+x_2+x_3+x_4+x_5)/5$.

Note: The mean and variance of each x_i are μ and σ^2, and the x_i are independent.
  Generalise this result for a sample of size n:

$$x_1, x_2, x_3, \ldots, x_n$$

and $\bar{X} = \left(\sum_{i=1}^{n} x_i \right) \bigg/ n.$

EXERCISE 6f *(miscellaneous)*

1. A cubical die has its faces marked with the numbers 1, 3, 5, 7, 9, 11. It is biased in such a way that the probability of getting any particular score when the die is thrown is proportional to that score. If X is the score, find the probability distribution of X.
(a) Calculate $E(X)$ and $\text{Var}(X)$;
(b) If $E(X) = \mu$ and $\text{Var}(X) = \sigma^2$, find $P(\mu-\sigma < X < \mu+\sigma)$.

2. Two unbiased dice are thrown and the random variable X is the smaller of the two scores obtained, unless these are equal, in which case $X = 0$.
(a) Draw a sample space diagram and use it to find the probability distribution of X.
(b) Calculate $E(X)$ and $\text{Var}(X)$.

3. A box contains eight fuses, three of which are 'blown' and five good. Fuses are taken from the box, one at a time without replacement until a good fuse is found. If X is the number of fuses taken out, find the probability distribution of X and calculate the mean and variance of X.

4. A discrete random variable X takes the values 0, 1 and 2 only, with probabilities p_0, p_1

and p_2 respectively.

Given that $E(X) = 4/3$ and $\mathrm{Var}(X) = 5/9$, find the values of p_0, p_1 and p_2. (C)

5.   The random variable X has the probability distribution:

X	1	2	4	5
$P(X)$	0.1	0.4	0.3	0.2

If Z is the sum of two independent observations from this distribution, find $E(Z)$ and $\mathrm{Var}(Z)$.

6. X and Y are random variables which take six possible pairs of values, with the probabilities shown.

	$Y = -1$	$Y = 0$	$Y = 1$
$X = 0$	1/56	3/56	3/56
$X = 1$	1/4	1/8	1/2

Find the probability that $X = 1$, and the probability that $Y = 1$.

Calculate $E(X)$, $E(Y)$ and $E(XY)$. Are X and Y independent random variables? Give your reason. (SMP)

7.   The discrete random variable X has the distribution

X	1	2	3	4
$P(X)$	0.1	0.2	0.3	0.4

Another random variable Y, independent of X has the same distribution. Find:
(a) the mean and variance of $X + Y$,
(b) the mean and variance of $X - Y$.

Find the probability distribution of XY and calculate $E(XY)$.

8.   The discrete random variable, X, has the following probability distribution.

X	-1	0	1
$P(X)$	$\frac{1}{4}$	$\frac{1}{2}$	$\frac{1}{4}$

Find the mean and variance of X.

If two independent random variables X_1 and X_2, have the same distribution as X, find the distribution of $X_1 - X_2$ and give its mean and variance.

What is the distribution of $X_1 + X_2$? Comment on the relationship between this distribution and that of $X_1 - X_2$. (O)

9.   One turn of a game is as follows. Two coins are tossed. If the exposed faces of the two coins are the same as each other, then both are tossed for a second time and the turn then ends. Otherwise, the turn ends after the first toss of the coins. The score, X, obtained in the turn, is equal to the total number of heads exposed during that turn.
(a) Show that $P(X = 3) = 1/8$, and that $P(X = 1) = 5/8$.
(b) Find the expectation and variance of X.

The scores obtained on two randomly chosen turns are X_1 and X_2. State the value of $E(X_1 - X_2)$. Find $P(X_1 = X_2)$. (C)

10.   Two fair six-sided dice, one coloured red and one blue, are tossed simultaneously. Let the value on the red die be i and the value shown on the blue die be j.

(a) The random variable X is defined as the sum of the two face values, that is $X = i+j$. What values may X take? Determine the probability distribution of X.

(b) The random variable Y is defined as the product of the two face values, that is $Y = ij$. What values may Y take? Determine the probability distribution of Y.

(c) What is the probability that $Y = 12$ given that $X = 7$?

(d) Are the random variables X and Y independent? Justify your reply briefly.

(AEB 1983)

11.   A cubical die has three faces marked with a '1', two faces marked with a '2' and one face marked with a '3'. Calculate the expectation and variance of the score obtained when this die is thrown once.

Deduce, or find otherwise, the expectation and variance of the score obtained in one throw of a second cubical die, which has one face marked '1', two faces marked '2' and three faces marked '3'.

Two of the first type of die and one of the second type are thrown together, and X denotes the total score obtained. Denoting the expectation and variance of X by μ and σ^2 respectively, show that: (a) $\sigma^2 = 5/3$; (b) $P(|X - \mu| > 2\sigma) = 1/18$.   (C)

12.   The random variable X has probability distribution
$$P(X = x) = k(1/4)^x, \qquad x = 0, 1, 2, \ldots.$$

(a) Find the value of k.

(b) Find $P(X > 5)$.

13.★ When N children are vaccinated it is known that each child may experience an adverse reaction. Let X be the number of children who react adversely. Given that the probability distribution of X is
$$P(X = r) = A/2^r, \qquad r = 0, 1, 2, \ldots, N,$$
where A is a positive constant, show that
$$A = 2^N/(2^{N+1} - 1).$$

Find, in terms of A and M, the probability that at least M of the children react adversely.

Show that, when $N = 4$, the probability of there being at least one adverse reaction is 15/31.   (L)

14.★ X and Y are independent random variables with probability distributions

X	$P(X)$		Y	$P(Y)$
9	$\frac{1}{2}$		8	$\frac{1}{2}$
11	$\frac{1}{2}$		12	$\frac{1}{2}$

If $Z = aX + bY$ and $E(Z) = 10$, find the values of a and b to make $\mathrm{Var}(Z)$ as small as possible.

15.★ The random variable X has probability distribution given by
$$P(x) = kp(1-p)^{x-1} \qquad \text{for } x = 1, 2, 3, \ldots$$
$(0 < p < 1)$.

(a) Show that $k = 1$.

(b) Derive the mean of X. $\left(\text{You may assume that } \sum_{r=1}^{\infty} r\theta^{r-1} = (1-\theta)^{-2} \text{ if } -1 < \theta < +1. \right)$

The Square Table organisation is holding a summer fête in order to raise funds for local charities. At one of the sideshows customers are invited to take up to five shots with an ancient crossbow at a small circular target. The entry fee is £2 with the customer receiving £4 for a hit after only one shot, £2 if two or three shots are needed, £1 if four or five are needed and zero otherwise. Firing ceases after a hit, or after 5 shots if no hits are recorded. If a customer, Mr. Robin Tell, has a constant probability of a hit equal to 0.25 and successive shots are independent, what is his expected gain from one trial?

It is decided to reduce the maximum number of shots to three in the above game, with the customer receiving £5 for a hit after only one shot, £3 if two shots are needed, £1 if three are needed and zero otherwise. What entry fee should be charged for Mr. Tell's expected gain from one trial to be zero?   (AEB 1982)

6.13   Appendix

$E(X+Y) = E(X)+E(Y)$: Proof for the general case

The procedure is similar to the method used on page 101, where we took just three values of X and two values of Y. However it is better, in the general case, to avoid a complicated system of suffixes by using the Σ notation.

The diagrams show just one typical element in the joint probability distribution table: the one indicating that $P(X = x \cap Y = y)$ is p_{xy}

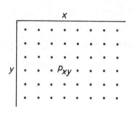

By definition,

$E(X+Y) =$ the sum of all possible
values of $(x+y)$

$\times P(X = x \cap Y = y)$,

which may be written in the form

$$\sum_{\text{all }x}\sum_{\text{all }y}(x+y)\cdot p_{xy},$$

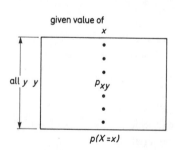

given value of
x

where x is *any* of the values that X can take, y is *any* value that Y can take and

$$p_{xy} = P(X = x \cap Y = y).$$

all y y

Now for any *given* value of x,

$$\sum_{\text{all }y}(x\cdot p_{xy})$$

$p(X=x)$

$= x \times$ (the sum of the 'x-column')

$$= x\times\sum_{\text{all }y}(p_{xy}) = x\cdot P(X=x) \qquad (1)$$

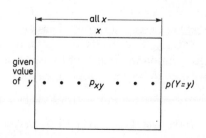

all x
x

Similarly for any given value of y,

$$\sum_{\text{all }x}(y\cdot p_{xy})$$

given
value
of y

$p(Y=y)$

$= y \times$ (the sum of the 'y-row')

$$= y\times\sum_{\text{all }x}(p_{xy}) = y\cdot P(Y=y). \qquad (2)$$

We now use these results in the expression for $E(X+Y)$.

$$E(X+Y) = \sum_{\text{all }x} \sum_{\text{all }y} (x+y) \cdot p_{xy}$$

$$= \sum_{\text{all }x} \sum_{\text{all }y} (x \cdot p_{xy}) + \sum_{\text{all }x} \sum_{\text{all }y} (y \cdot p_{xy})$$

$$= \sum_{\text{all }x} \left(\sum_{\text{all }y} x \cdot p_{xy} \right) + \sum_{\text{all }y} \left(\sum_{\text{all }x} y \cdot p_{xy} \right)$$

$$= \sum_{\text{all }x} [x \cdot P(X = x)] + \sum_{\text{all }y} [y \cdot P(Y = y)]$$

using relations (1) and (2), which gives us

$$E(X+Y) = E(X) + E(Y).$$

Proof that $E(XY) = E(X) \cdot E(Y)$ if X and Y are independent

The proof follows lines similar to the proof for $E(X+Y)$.

By definition, $E(XY) = \sum_{\text{all }x} \sum_{\text{all }y} (xy)p_{xy}.$

Now, since X and Y are independent,

$$p_{xy} = P(X = x \cap Y = y) = P(X = x) \cdot P(Y = y)$$

so that

$$E(XY) = \sum_{\text{all }x} \sum_{\text{all }y} xy \cdot P(X = x) \cdot P(Y = y)$$

$$= \sum_{\text{all }x} \sum_{\text{all }y} [x \cdot P(X = x)] [y \cdot P(Y = y)]$$

$$= \left[\sum_{\text{all }x} x \cdot P(X = x) \right] \left[\sum_{\text{all }y} y \cdot P(Y = y) \right]$$

$$= E(X) \cdot E(Y).$$

An example of a joint probability distribution of two random variables that are not independent

Two coins are tossed and the number of heads, X, is noted. Any coin or coins showing a head is then removed, and the remaining coin or coins tossed again. Y is the number of heads showing after the second throw (which is taken to be zero if there is no second throw). The coins are unbiased.

Clearly Y depends on X, and in order to construct the joint probability distribution we shall have to use the idea of conditional probability. In other words we need to use relations of the type

$$P(Y = 0 \cap X = 1) = P(Y = 0 \,|\, X = 1) \cdot P(X = 1).$$

If the first throw gives TT, then $X = 0$ and $P(X = 0) = \frac{1}{4}$.

For the second throw both coins are available and

$$P(Y = 0 \mid X = 0) = \tfrac{1}{4}, \qquad P(Y = 1 \mid X = 0) = \tfrac{1}{2}, \qquad P(Y = 2 \mid X = 0) = \tfrac{1}{4}.$$

Hence $\quad P(Y = 0 \cap X = 0) = \tfrac{1}{4} \cdot \tfrac{1}{4} = 1/16$

$$P(Y = 1 \cap X = 0) = \tfrac{1}{2} \cdot \tfrac{1}{4} = 1/8$$

$$P(Y = 2 \cap X = 0) = \tfrac{1}{4} \cdot \tfrac{1}{4} = 1/16.$$

In a similar way

$$P(Y = 0 \cap X = 1) = 1/4$$

$$P(Y = 1 \cap X = 1) = 1/4$$

$$P(Y = 0 \cap X = 2) = 1/4$$

and the remaining joint probabilities are 0. The distribution is thus:

		X			
		0	1	2	
	0	1/16	1/4	1/4	9/16
Y	1	1/8	1/4	0	3/8
	2	1/16	0	0	1/16
		1/4	1/2	1/4	1

From the table, the marginal probabilities $P(Y = 0)$, $P(Y = 1)$ and $P(Y = 2)$ are obtained by addition of the rows; the marginal probabilities for X are, of course, given.

The table shows, as it must since X and Y are not independent, that $P(X = 0 \cap Y = 0) \neq 9/16 \times 1/4$ and so on.

7 Probability generating functions

7.1 Definition

As we saw in chapter 6, in a simple numerical case, calculating the mean and variance of a probability distribution is easy using the formulae

$$E(X) = \Sigma\, Px \quad \text{and} \quad \text{Var}(X) = \Sigma\, px^2 - \mu^2.$$

In chapter 8, however, we shall be describing some important probability models, and applying these formulae can lead to heavy, unattractive algebra. Fortunately, when the random variable takes *integral* values only, there is a way of avoiding this complexity by using 'probability generating functions', and you will find it very helpful to master the idea at this stage.

We shall start with a simple example. If the random variable X takes integral values (0 to 4) with the following distribution

X	0	1	2	3	4
$P(X)$	$\frac{1}{8}$	$\frac{1}{4}$	$\frac{1}{4}$	$\frac{1}{4}$	$\frac{1}{8}$

the probability generating function is defined as the function of t:

$$\tfrac{1}{8}t^0 + \tfrac{1}{4}t^1 + \tfrac{1}{4}t^2 + \tfrac{1}{4}t^3 + \tfrac{1}{8}t^4.$$

It is an 'invented' function in which 't' has no significance in itself but is a sort of dummy variable, and the powers of t, t^0, t^1, t^2, ... just act as a framework on which to hang the probabilities $P(X = 0)$, $P(X = 1)$, $P(X = 2)$, ...

Students who have met functions like $\sin x$ and e^x (which are associated with the natural phenomena of oscillation and growth) are inclined, when they meet the idea of an artificial function like the probability generating function for the first time, to ask immediately: 'What's the point?' Its justification is simply that it is a very useful device.

The symbol for a probability generating function (p.g.f. for short) is $G(t)$ and is defined as follows.

If the discrete random variable X can only take integral values and has the probability distribution

X	0	1	2	...	n
$P(X)$	p_0	p_1	p_2	...	p_n

$$G(t) = p_0 t^0 + p_1 t^1 + p_2 t^2 + \ldots + p_n t^n$$

$$= \sum_{i=0}^{n} p_i t^i = E(t^X).$$

It can be seen that the value of $G(t)$ when $t = 1$, $G(1) = 1$.

Further examples of a p.g.f.

1. Suppose that X is the score when an unbiased die is thrown. In this case X can only take values from 1 to 6 and

$$G(t) = \frac{1}{6}(t^1 + t^2 + t^3 + t^4 + t^5 + t^6).$$

2. If X is the number of points on a playing card drawn at random from a normal pack, where an ace scores 1 and a court card 10,

$$G(t) = \frac{1}{13}(t^1 + t^2 + t^3 + t^4 + t^5 + t^6 + t^7 + t^8 + t^9) + \frac{4}{13}t^{10}.$$

A p.g.f. when the random variable is countably infinite

If X is countably infinite, the p.g.f. is similarly defined:

$$G(t) = p_0 t^0 + p_1 t^1 + p_2 t^2 + \ldots \text{ to } \infty$$

$$= \sum_{i=0}^{\infty} p_i t^i.$$

As an example, if X is the number of throws of an unbiased die needed to get a score of six,

$$G(t) = \left(\frac{1}{6}\right)t^1 + \left(\frac{5}{6}\right)\left(\frac{1}{6}\right)t^2 + \left(\frac{5}{6}\right)^2\left(\frac{1}{6}\right)t^3 + \ldots \text{ to } \infty$$

(see page 86).

7.2 Using a p.g.f. to find mean and variance

First of all we note the forms that the mean and variance take when X takes integral values only.

$$E(X) = 0 \cdot p_0 + 1 \cdot p_1 + 2 \cdot p_2 + \ldots + n \cdot p_n$$

$$= \sum_{i=0}^{n} i \cdot p_i \quad (i = 0 \text{ gives a zero term}).$$

$$E(X^2) = 0^2 \cdot p_0 + 1^2 \cdot p_1 + 2^2 \cdot p_2 + \ldots + n^2 \cdot p_n = \sum_{i=0}^{n} i^2 \cdot p_i$$

so that $\mathrm{Var}(X) = \sum\limits_{i=0}^{n} i^2 \cdot p_i - \mu^2.$

Since $G(t) = p_0 t^0 + p_1 t^1 + p_2 t^2 + p_3 t^3 + p_4 t^4 + \ldots + p_n t^n,$

$$\frac{dG(t)}{dt} = G'(t) = p_1 + 2p_2 t + 3p_3 t^2 + 4p_4 t^3 + \ldots + np_n t^{n-1}.$$

Putting $t = 1$, we get

$$G'(1) = p_1 + 2p_2 + 3p_3 + 4p_4 + \ldots + np_n = \sum_{i=0}^{n} i \cdot p_i$$

$$\Rightarrow G'(1) = \mu.$$

Differentiating the expression for $G'(t)$,

$$G''(t) = 2p_2 + 3 \cdot 2p_3 t + 4 \cdot 3p_4 t^2 + \ldots + n(n-1)t^{n-2}.$$

Putting $t = 1$ again,

$$G''(1) = 2 \cdot 1p_2 + 3 \cdot 2p_3 + 4 \cdot 3p_4 + \ldots + n(n-1)p_n$$

$$= \sum_{i=0}^{n} i(i-1)p_i \quad (i = 0 \text{ and } 1 \text{ give zero terms})$$

$$= \sum_{i=0}^{n} i^2 \cdot p_i - \sum_{i=0}^{n} i \cdot p_i.$$

Now $\sigma^2 = \sum\limits_{i=0}^{n} i^2 \cdot p_i - \mu^2$

$$= \left[G''(1) + \sum_{i=0}^{n} i \cdot p_i \right] - \mu^2$$

$$= G''(1) + \mu - \mu^2.$$

Hence $G''(1) = \sigma^2 + \mu^2 - \mu.$

The three properties of $G(t)$ that we have arrived at,

$$G(1) = 1, \qquad G'(1) = \mu \quad \text{and} \quad G''(1) = \sigma^2 + \mu^2 - \mu,$$

hold also for a countably infinite random variable and the proofs are very similar. It is the last two results that lead to neat ways of finding the mean and variance of the theoretical probability distributions that we shall meet in the next chapter. They do not have any advantage over the direct method in simple numerical cases (see example 1). Nevertheless, you should work through some numerical examples before tackling the theoretical cases in the next chapter.

EXAMPLE 1   An unbiased die is thrown. For a face value that is odd, the random variable X is that face value. For an even face value, X is 0. Find the mean and variance of X.

The probability distribution is

X	0	1	3	5
$P(X)$	$\frac{1}{2}$	$\frac{1}{6}$	$\frac{1}{6}$	$\frac{1}{6}$

and $G(t) = \dfrac{1}{2} + \dfrac{1}{6}t + \dfrac{1}{6}t^3 + \dfrac{1}{6}t^5$

$$G'(t) = \dfrac{1}{6} + \dfrac{1}{2}t^2 + \dfrac{5}{6}t^4 \qquad \Rightarrow G'(1) = \dfrac{3}{2} \Rightarrow \mu = 1\tfrac{1}{2}.$$

$$G''(t) = t + \dfrac{10}{3}t^3 \qquad\qquad \Rightarrow G''(1) = \dfrac{13}{3}.$$

Hence $\sigma^2 + \mu^2 - \mu = \dfrac{13}{3} \Rightarrow \sigma^2 = \dfrac{13}{3} - \left(\dfrac{3}{2}\right)^2 + \dfrac{3}{2} = \dfrac{43}{12}.$

EXAMPLE 2   If X is the number of throws of an unbiased die needed to get a six, find $E(X)$ and Var(X).

$$P(X = 1) = \tfrac{1}{6}, \; P(X = 2) = \tfrac{5}{6}\cdot\tfrac{1}{6}, \; P(X = 3) = (\tfrac{5}{6})^2 \cdot \tfrac{1}{6} + \ldots$$

so that

$$G(t) = (\tfrac{1}{6})t^1 + (\tfrac{5}{6})(\tfrac{1}{6})t^2 + (\tfrac{5}{6})^2 \cdot \tfrac{1}{6}t^3 + \ldots \text{ to } \infty.$$

Here $G(t)$ is an infinite geometrical progression and can be expressed in compact form, using the formula $a/(1 - r)$, where a (the first term) $= \tfrac{1}{6}t$ and r (the common ratio) $= \tfrac{5}{6}t$.

Hence $G(t) = \dfrac{t/6}{1 - 5t/6} = \dfrac{t}{6 - 5t}$

and $G'(t) = \dfrac{(6 - 5t)\cdot 1 - t(-5)}{(6 - 5t)^2} = \dfrac{6}{(6 - 5t)^2}.$

Thus $G'(1) = 6 \Rightarrow E(X) = 6.$

$$G'(t) = 6(6 - 5t)^{-2} \Rightarrow G''(t) = 60(6 - 5t)^{-3} \Rightarrow G''(1) = 60.$$

Hence $\sigma^2 + \mu^2 - \mu = 60 \Rightarrow \sigma^2 = 60 + 6 - 6^2.$

Thus   Var$(X) = 30.$

EXERCISE 7a

1.   The random variable X has probability distribution:

X	0	1	2
$P(X)$	$\tfrac{1}{2}$	$\tfrac{3}{8}$	$\tfrac{1}{8}$

Write down the p.g.f. and use it to find $E(X)$ and Var(X).

2.   Use the p.g.f. to find the mean and variance of the following probability distribution:

X	1	2	3	4	5	6
$P(X)$	0.2	0.2	0.3	0.1	0.1	0.1

3.   Use the p.g.f. for the score when an unbiased die is thrown to calculate the mean and variance of the score.

4.   Three unbiased coins are tossed and the number of tails (X) is noted. Show that the p.g.f. is

$$\tfrac{1}{8}(1+3t+3t^2+t^3).$$

Hence calculate the mean and variance of X.

5.   A box contains three white balls and two blue balls. They are taken out one at a time, without replacement, and X is the number of withdrawals until a white ball is chosen. Find the p.g.f. and use it to find $E(X)$ and Var(X).
   What will the p.g.f. be if X is the number of withdrawals *with* replacement until a white ball is found?

6.   The random variable X is the number of throws of an unbiased coin until a head is obtained, and is therefore countably infinite.
   Write down the p.g.f. and use the formula for the sum of an infinite geometrical progression to show that

$$G(t) = \frac{t}{2-t}.$$

Hence show that $E(X) = 2$ and Var$(X) = 2$.

7.3   The p.g.f. for the sum of two or more random variables

Suppose a coin-tossing experiment is carried out in which six unbiased coins are tossed and we want to know the probability distribution of U, the sum of the number of heads obtained. If X is the number of heads obtained from one of the coins, X can be either 0 or 1, with a probability of $\tfrac{1}{2}$ for each. U can therefore have any value from 0 to 6, but finding the probability associated with each of these values would be rather laborious. Now, it happens that the distribution of U is a standard distribution (see chapter 8). In fact, we do not need to know this, as there is an easy method of wide general application which will help us find the probability distribution. This method involves finding the p.g.f. of the *sum* of a number of independent random variables whose p.g.f.s we know. It turns out that all we have to do is to *multiply* together the p.g.f.s of the individual random variables.
   We shall first prove this theorem and then apply it to the coin-tossing example. The theorem (for two variables) states that:

If X and Y are two independent random variables with p.g.f.s $G_x(t)$ and $G_y(t)$ respectively, the p.g.f. of $X + Y$ is $G_x(t) \cdot G_y(t)$.

In other words, the p.g.f. of the sum of X and Y is the product of their p.g.f.s.

Proof

Let $U = X + Y$ with p.g.f. $G_u(t)$.
   We shall suppose, first, that X and Y can take the same number of values, $0, 1, 2, \ldots, n$, with probabilities $p_0, p_1, p_2, \ldots, p_n$ and $q_0, q_1, q_2, \ldots, q_n$ respectively.

   Then $G_x(t) = p_0 + p_1 t + p_2 t^2 + \ldots + p_n t^n$

   and $G_y(t) = q_0 + q_1 t + q_2 t^2 + \ldots + q_n t^n$

so that $G_x(t) \cdot G_y(t) = (p_0 + p_1 t + p_2 t^2 + p_3 t^3 + \ldots + p_n t^n)$

$$\times (q_0 + q_1 t + q_2 t^2 + q_3 t^3 + \ldots + q_n t^n).$$

It can be seen that the coefficient of t^3, for example, in this product is

$$p_0 q_3 + p_1 q_2 + p_2 q_1 + p_3 q_0$$

which is

$$P(X = 0) \cdot P(Y = 3) + P(X = 1) \cdot P(Y = 2) + P(X = 2) \cdot P(Y = 1) + P(X = 3) \cdot P(Y = 0)$$

$$= P(X + Y = 3).$$

In general

$$P(X + Y = r) = P(X = 0) \cdot P(Y = r) + P(X = 1) \cdot P(Y = r - 1)$$
$$+ \ldots + P(X = r) \cdot P(Y = 0)$$

$$= p_0 q_r + p_1 q_{r-1} + \ldots + p_r q_0$$

and this is the coefficient of t^r in the product.

Hence $G_u(t) = G_x(t) \cdot G_y(t)$.

We may now remove the restriction that X and Y take the same integral values, by assigning zero probabilities where necessary.

This theorem may be extended:

(a) The p.g.f. of $X + Y + Z = G_x(t) \cdot G_y(t) \cdot G_z(t)$
    and similarly for any number of variables.

(b) If n variables all have the *same* p.g.f., $G(t)$, then the p.g.f. for their sum $= [G(t)]^n$.
    Let us now have another look at the coin-tossing example.

EXAMPLE 3   Find the probability generating function for the total number of heads when six unbiased coins are tossed. Deduce the mean and variance.

For each of the six coins, if X is the number of heads, $P(X = 0) = \frac{1}{2}$ and $P(X = 1) = \frac{1}{2}$, so that the p.g.f. of X,

$$G_x(t) = \tfrac{1}{2}t^0 + \tfrac{1}{2}t^1 = \tfrac{1}{2} + \tfrac{1}{2}t.$$

It follows that the p.g.f. of the sum,

$$G(t) = (\tfrac{1}{2} + \tfrac{1}{2}t)^6 = \frac{1}{64}(1 + t)^6.$$

Expanding this expression by the binomial theorem

$$G(t) = \frac{1}{64}(1 + 6t + 15t^2 + 20t^3 + 15t^4 + 6t^5 + t^6)$$

and the coefficients give the probability distribution.

To find the mean and variance of the sum,

$$G(t) = \frac{1}{64}(1 + t)^6 \Rightarrow G'(t) = \frac{3}{32}(1 + t)^5 \Rightarrow G''(t) = \frac{15}{32}(1 + t)^4.$$

Hence $\mu = G'(1) = 3$
$$G''(1) = 7.5 = \sigma^2 + \mu^2 - \mu \Rightarrow \sigma^2 = 7.5 - 9 + 3 = 1.5.$$

The mean and variance could alternatively have been found by adding the means and variances of 6 independent random variables, each with mean $\frac{1}{2}$ and variance $\frac{1}{4}$.

The numerical examples in exercise 7b that follow are, like this one, of no particular importance in themselves, but they should help you to get used to the general idea of multiplying together a number of p.g.f.s, which you will find useful in later work.

EXERCISE 7b

1.   A coin is biased so that the probability of getting a head when it is tossed is 0.4. Find the p.g.f. of the number of heads when it is tossed four times and hence the probability distribution. Find also the mean and variance of this distribution.

2.   Two unbiased dice are thrown. Find the p.g.f. for the total score and use it to find the mean.

3.   Two unbiased dice are marked so that one of the dice has three faces marked '1' and the other three faces marked '0', and the second die has four faces marked '2' and the other two marked '0'. Find the p.g.f. for the total score when the dice are thrown together, and use it to find the mean and variance of the total score.

4.   An unbiased die is thrown three times. The score for each throw is the face value if an odd number shows and is zero if an even number shows. Find the p.g.f. of X, the sum of the three scores, and hence $E(X)$. What will $E(X)$ be if the die is thrown n times?

5.   The probability distributions of two independent random variablés X and Y are

X	1	2	3
$P(X)$	$\frac{1}{4}$	$\frac{1}{4}$	$\frac{1}{2}$

Y	1	2
$P(Y)$	$\frac{1}{2}$	$\frac{1}{2}$

Prove that the p.g.f. of the random variable U where
$$U = X + Y + 1$$
is $(t^3/8)(1 + 2t + 3t^2 + 2t^3)$. Find $E(U)$ and $\text{Var}(U)$.

6.★ Prove that if $E(X) = \mu_x$ and $E(Y) = \mu_y$, where X and Y are independent, then
$$E(X + Y) = \mu_x + \mu_y.$$
(This is the result proved for *any* two random variables on page 103. Let $Z = X + Y$, so that $G_z(t) = G_x(t) \cdot G_y(t)$. Find $G'_z(1)$.)

7.★ Use the method of question 6 to prove that, if X and Y are independent random variables with variances σ_x^2 and σ_y^2 respectively, then
$$\text{Var}(X + Y) = \sigma_x^2 + \sigma_y^2.$$

7.4★ The p.g.f. for a random variable that can take negative values

In introducing probability generating functions we confined our examples to cases where the random variable X could take non-negative values $0, 1, 2, \ldots$. It can be shown, without much difficulty, that all the properties we proved apply when some or all of the values of X can be negative integers, with $G(t)$ defined as $E(t^X)$ as before.

As a simple example, suppose an unbiased coin is tossed and the random variable X is $+1$ if a head turns up and -1 if a tail turns up. The probability distribution is then

X	$+1$	-1
$P(X)$	$\frac{1}{2}$	$\frac{1}{2}$

and the p.g.f. $G(t) = \frac{1}{2}t^1 + \frac{1}{2}t^{-1}$.

EXAMPLE 4   An unbiased coin is tossed n times. On a single throw the score for a head is $+1$ and for a tail the score is -1. Find the p.g.f. of the total score X for n throws and deduce the mean and variance.

Since the p.g.f. for a single throw is $\frac{1}{2}t + \frac{1}{2}t^{-1} = \frac{1}{2}(t + 1/t)$, the p.g.f. of X is

$$G(t) = [\tfrac{1}{2}(t + 1/t)]^n.$$

Hence $G'(t) = (1/2^n)n(t + 1/t)^{n-1}(1 - 1/t^2)$

and $\mu = G'(1) = 0.$

Differentiating again,

$$G''(t) = (1/2^n)n[(n-1)(t+1/t)^{n-2}(1-1/t^2)^2 + (t+1/t)^{n-1}(2/t^3)]$$
$$\Rightarrow G''(1) = n/2^n[2^{n-1} \cdot 2] = n \Rightarrow \sigma^2 = n \quad (\text{since } \mu = 0).$$

(The mean and variance could, of course, have been derived by adding n values of the mean and variance of the score for a single throw, which are 0 and 1 respectively.)

EXAMPLE 5   If the probability distributions of the independent random variables X and Y are

X	1	2	...	n
$P(X)$	p_1	p_2	...	p_n

Y	1	2	...	n
$P(Y)$	q_1	q_2	...	q_n

find: (a) the probability generating function of $-Y$;
   (b) the probability generating function of $X - Y$.

(a) The p.g.f. of Y,

$$G_y(t) = q_1 t^1 + q_2 t^2 + \ldots + q_n t^n.$$

If $Z = -Y$, the p.g.f. of Z,

$$G_z(t) = q_1 t^{-1} + q_2 t^{-2} + \ldots + q_n t^{-n}$$
$$= G_y(1/t).$$

(b) The p.g.f. of $X - Y$,

$$G_{x-y}(t) = G_{x+z}(t)$$

$$= G_x(t) \cdot G_z(t) \quad \text{(assuming the rule for multiplying p.g.f.s)}$$

$$= G_x(t) \cdot G_y(1/t)$$

$$= (p_1 t^1 + p_2 t^2 + \ldots + p_n t^n)(q_1 t^{-1} + q_2 t^{-2} + \ldots + q_n t^{-n}).$$

As a check, the coefficient of t^2 in the product

$$= p_3 q_1 + p_4 q_2 + \ldots + p_n q_{n-2}$$

$$= P(X = 3 \cap Y = 1) + P(X = 4 \cap Y = 2) + \ldots + P(X = n \cap Y = n-2)$$

$$= P(X - Y = 2), \quad \text{which is as it should be.}$$

EXERCISE 7c (miscellaneous)

1.   A random variable X is the score when a card is drawn at random from a normal pack. The score is the face value, with aces counting as 1 and court cards as 10. Write down the p.g.f. and deduce $E(X)$.

2.   The random variable X can take integral values from 1 to n, and its p.g.f. is $G_x(t)$.
If $Z = x + a$, prove that its p.g.f. $G_z(t) = t^a G_x(t)$.

3.   The probability generating function of a discrete random variable X is

$$G(z) = (az + 1 - a)^n,$$

where a is a constant $(0 \leqslant a \leqslant 1)$ and n is a positive integer. Prove that $E(X) = an$ and find the variance of X.   (C)

4.   The random variable X can take values $1, 2, \ldots$ to ∞, and its p.g.f. is $G(t)$.
Show that the probability that X is even is given by

$$\tfrac{1}{2}[1 + G(-1)].$$

5.   A biased die has probability $1/5$ of showing a '6' in a single throw. The die is thrown repeatedly until a '6' shows. Write down the probability that the first '6' shows at the rth throw.
   The random variable R is the number of the throw on which the first '6' occurs. Obtain:
(a) the probability generating function of R,
(b) the mean, r_1, of R,
(c) the variance of R,
(d) the probability that R is greater than r_1,
(e) the largest value, r_2, of R for which $P(R < r_2)$ is less than $\tfrac{1}{2}$.   (C)

6.   Three balls a, b, c are placed at random in three boxes A, B, C (with one ball in each box). The variable x is defined by

$$\begin{cases} x = 1 & \text{if ball } a \text{ is in box } A, \\ x = 0 & \text{otherwise.} \end{cases}$$

Find the probability generator for x.

The variable y is defined in the table

	Box			Value of y
	A	B	C	
Contents	a	b	c	1
	c	a	b	0
	b	c	a	0
	a	c	b	0
	b	a	c	1
	c	b	a	1

Deduce the probability generator of y.

The variable s is defined by $s = x+y+1$. Given that x and y are independent, show that s has generator

$$\tfrac{1}{6}(t^3 + 3t^2 + 2t)$$

and deduce the mean and variance of s. (SMP)

7. Two people, A and B, fire alternately at a target, the winner of the game being the first to hit the target. The probability that A hits the target with any particular shot is $1/3$ and the probability that B hits the target with any particular shot is $1/4$.

Given that A fires first, find:
(a) the probability that B wins the game with his first shot;
(b) the probability that A wins the game with his second shot;
(c) the probability that A wins the game.

If R is the total number of shots fired by A and B show that the probability generating function of R is given by

$$G(z) = \frac{2z + z^2}{3(2 - z^2)}.$$

Find $E(R)$. (C)

8.★ Urn A contains four balls numbered 1 to 4, urn B six balls numbered 1 to 6 and urn C eight balls numbered 1 to 8. One ball is selected at random from each urn. The random variable X is the sum of the numbers on the three selected balls. Find $E(X)$ and $Var(X)$.

In either order:
(a) find $P(X \leqslant 6)$;
(b) show that the probability generating function of X is

$$G(z) = \frac{z^3(1 - z^4)(1 - z^6)(1 - z^8)}{192(1 - z)^3}.$$

(C)

9.★ An ordinary die is thrown r times. Show that the probability generating function for the total score is

$$\left[\frac{t(1 - t^6)}{6(1 - t)}\right]^r.$$

Hence show that the probability of the total score being $r + 6$ is

$$\left[\binom{r+5}{6} - r\right]\frac{1}{6^r}.$$

(O & C)

10.★★ An unbiased coin is tossed eight times. On a single throw the score for a head is 1 and for a tail is -1. If X is the sum of the scores for eight throws, find the probability generating function.

Deduce $P(-4 \leqslant X \leqslant 4)$.

11.★★ An unbiased coin is tossed n times. If a head appears $+2$ is scored; if a tail appears, -1 is scored. Find the probability generating function of X, the total score for n throws.

Deduce $E(X)$ and $\text{Var}(X)$.

12.★★ Two dice, one red and the other blue, are thrown together. X is the score on the red die and Y the score on the blue die. If $Z = X - Y$, find its probability generating function.

8 The binomial and geometric distributions

In chapter 6 we discussed probability distributions in general and made the point that they are theoretical distributions, used as 'models' for experimental or observed results. (Look back to pages 84–7 if you need to remind yourself of the connection between observations and a model.)

Suppose, for example, a researcher obtained the following figures for the number of boys born in 1000 families of three children. The figures might give the following frequency distribution:

Number of boys (X)	0	1	2	3
Number of families (f)	106	370	383	141
Relative frequency ($f/1000$)	0.106	0.370	0.383	0.141

A suitable model for these results would be a probability distribution in which the *probabilities* of 0, 1, 2, 3 boys in a three-child family would be close to the *relative frequencies* in this table. To set up such a model it is obvious that some reasonable assumption must be made about the probability that in any given birth a boy will be born. Suppose the researcher had assumed a figure of 0.515 for this probability (this figure was quoted on page 51), what would the resulting theoretical probability distribution be? The answer to this question is that it is an example of what is called the *binomial distribution*, one of the most important distributions in statistical theory.

We shall first look at a simpler discrete probability distribution, to which it is related.

8.1 The Bernouilli distribution

This refers to a 'trial' in which there are only *two* possible outcomes, one of which it is convenient to call a 'success' and the other a 'failure'.

Suppose a coin is biased so that the probability of a 'head' is 0.6. If X is the number of heads ('successes') after it has been tossed, then after a single trial X can only be 0 or 1. The resulting probability distribution is:

Number of successes (X)	0	1
Probability	0.4	0.6

Similarly, if an unbiased die is thrown and 'success' is 'throwing a four', then, in a single trial, there are only two possible outcomes 'throwing a four' and 'not throwing a four'.

The resulting probability model will be:

Number of successes (X)	0	1
Probability	5/6	1/6

If these experiments were repeated a number of times they would give rise to the subject of the next section, a binomial distribution.

Although it may seem odd that something as apparently trivial as an 'either-or' trial should be given a special name, the Bernouilli family were a very distinguished group of mathematicians in the 17th and 18th centuries and it was Jacob Bernouilli (1654–1703) who first put forward the binomial distribution.

8.2   The binomial distribution

We shall take as our example the second of the Bernouilli trials mentioned in the last paragraph and consider what may happen when it is repeated *five times*. If this is done and if we take as random variable X the *total* number of successes ('fours') in five trials, then the possible values of X are 0, 1, 2, 3, 4, 5.

Consider $P(X = 2)$. The probability of getting two successes followed by three failures is

$$(\tfrac{1}{6})(\tfrac{1}{6})(\tfrac{5}{6})(\tfrac{5}{6})(\tfrac{5}{6}) = (\tfrac{1}{6})^2(\tfrac{5}{6})^3.$$

If, however, the pattern is not SSFFF (where S = success and F = failure), but SFFSF or any other arrangement of two successes and three failures, the probability of each of these arrangements will still be

$$(\tfrac{1}{6})^2(\tfrac{5}{6})^3.$$

It follows that the probability of getting two successes and three failures *in any order* is $(1/6)^2(5/6)^3$ multiplied by the number of ways of selecting the positions for the two S's out of the five possible positions, which is $^5C_2$, or, as we shall now write it more compactly, $\binom{5}{2}$.   Hence we have $P(X = 2) = \binom{5}{2}(\tfrac{1}{6})^2(\tfrac{5}{6})^3.$

By a similar argument the probabilities that $X = 0, 1, 2, 3, 4$ and 5 are as shown in the following table:

X	0	1	2	3	4	5
Probability	$\left(\dfrac{5}{6}\right)^5$	$\binom{5}{1}\left(\dfrac{5}{6}\right)^4\left(\dfrac{1}{6}\right)$	$\binom{5}{2}\left(\dfrac{5}{6}\right)^3\left(\dfrac{1}{6}\right)^2$	$\binom{5}{3}\left(\dfrac{5}{6}\right)^2\left(\dfrac{1}{6}\right)^3$	$\binom{5}{4}\left(\dfrac{5}{6}\right)\left(\dfrac{1}{6}\right)^4$	$\left(\dfrac{1}{6}\right)^5$

As these probabilities are clearly the successive terms of the binomial expansion of

$$\left(\frac{5}{6}+\frac{1}{6}\right)^5$$

the probability distribution is called a binomial distribution.

In general, if a Bernouilli trial, in which the probability of success is p, is repeated n times, the probabilities of the number of successes $X = 0, 1, 2, 3, \ldots, n$ are given by the successive terms of the binomial expansion of

$$(q+p)^n,$$

where $q = 1 - p$, so that, for example,

$$P(X = 3) = \binom{n}{3}q^{n-3}p^3.$$

We shall now define the binomial distribution formally.

A discrete random variable X is said to follow the binomial distribution if

$$P(X = r) = \binom{n}{r}q^{n-r}p^r, \quad r = 0, 1, 2, \ldots, n, \text{ where } 0 < p < 1 \text{ and } q = 1 - p.$$

A short way of referring to the binomial distribution where the 'parameters' are n and p is $B(n, p)$.

The quantities n and p are called parameters because when they are given particular values, the distribution is determined. (We shall meet a further use of this word 'parameter' later.)

It is important to notice that for a binomial distribution to arise from a fixed number n of a Bernouilli trial, the probability of success p in each trial must be *constant* and the trials must be *independent*.

The binomial distribution as a model

Examples of situations where the binomial distribution is the appropriate model are:

Nature of the 'trial'	'Success'
1. Picking out a sample from the results of an industrial process (e.g. a machine producing steel rods) and testing (e.g. measuring the lengths of the rods).	The length of a rod is within certain prescribed limits.
2. Selecting a sample of cattle that have been injected with a certain serum.	The animal has been affected by the injection in some prescribed way.
3. Inspecting a sample of tins of shandy from a factory's output and measuring the beer content.	The tin contains a certain minimum quantity of beer.

In each of these examples, n is the size of the sample inspected or tested.

The shape of the binomial distribution's histogram

The diagrams show the shape for: (a) $n = 5$, $p = 1/6$ (the die experiment); (b) $n = 10$, $p = \frac{1}{2}$ (tossing a coin 10 times, where 'success' is throwing a 'head').

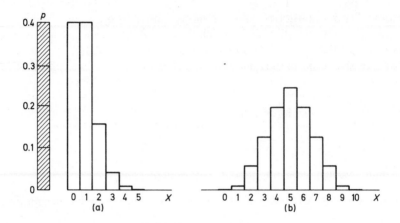

(a)   (b)

We shall now illustrate the distribution with two numerical examples, the first being the one with which this chapter started.

EXAMPLE 1   Assuming that the probability of a new-born baby being a boy is 0.515, calculate the probabilities that in a family of three children there will be 0, 1, 2, 3 boys.

The model here is $B(3, 0.515)$, that is, $n = 3$, $p = 0.515$ and $q = 1 - 0.515 = 0.485$, so that the probabilities are:

$$P(X = 0) = (0.485)^3 \qquad\qquad = 0.114$$

$$P(X = 1) = \binom{3}{1}(0.485)^2(0.515) = 0.363$$

$$P(X = 2) = \binom{3}{2}(0.485)(0.515)^2 = 0.386$$

$$P(X = 3) = (0.515)^3 \qquad\qquad = 0.137.$$

These probabilities seem reasonably close to the relative frequencies 0.106, 0.370, 0.383, 0.141 quoted on page 123, which were calculated from the researcher's observations. Whether or not they are close enough to justify taking $B(3, 0.515)$ as a model is a question that we are not yet in a position to answer.

EXAMPLE 2   A machine produces metal rods. It is known that the probability that the length of a rod will be outside certain specified limits is $\frac{1}{10}$. Such a rod is unacceptable. Find:
(a) the probability that exactly 3 are unacceptable,
(b) the probability that 2 or more are unacceptable,
when a sample of 12 is examined.

If 5 samples of 12 are examined, what is the probability that none of them contains 2 or more unacceptable rods?

In (a) and (b) the distribution $B(12, 0.1)$ is involved. If X is the number of unacceptable rods,

(a) $P(X = 3) = \binom{12}{3}(0.9)^9(0.1)^3 = 0.085$ to 3 dec. places.

(b) $P(X = 0 \text{ or } 1) = (0.9)^{12} + \binom{12}{1}(0.9)^{11}(0.1) = 0.659$

$P(X = 2 \text{ or more}) = 1 - P(X = 0 \text{ or } 1) = 0.341$ to 3 dec. places.

In the last part of the problem we are concerned with a trial where 'success' means 'a sample contains 2 or more unacceptable rods' and the number of trials is 5. The distribution is therefore $B(5, 0.341)$.

Hence $P(X = 0) = (0.659)^5 = 0.124$ to 3 dec. places.

EXERCISE 8a

1.   It is known that only three-quarters of a large batch of seeds will germinate. If six seeds are sown, find the probability that:
(a) two seeds will germinate,
(b) four seeds will germinate,
(c) at least two seeds will germinate.

2.   A machine produces large quantities of components, 15% of which are defective. If three components are selected at random, what is the probability that they will include:
(a) no defectives,
(b) at least one defective?

3.   An unbiased die is thrown seven times. Find the probability that: (a) three '6's are thrown, (b) at least three '6's are thrown.

4.   In a large consignment of light bulbs, it is known that one-tenth are of foreign manufacture. If eight bulbs are bought, which is more probable: (a) that none are foreign, or (b) that at least one of the bulbs is foreign?

5.   Two dice, each unbiased, are thrown. Find the probability that the total score is two-figured.
   If five people throw two dice, find the probability that exactly four of them get a two-figured score.

6.   If the chance of a male child is 0.515, find the probability that, in a three-child family chosen at random: (a) all will be girls, (b) there will be at least one boy.

7.   The probability that a test rocket will fail is 0.2. In eight firings show that the probability of exactly two failures is twice the probability of exactly three failures.

8.   A die is biased so that the probability of getting a '6' is 0.18. If the die is thrown ten times, what is the probability that the number of '6's that turn up is odd?

9.   The probability that a certain type of missile hits its target is 0.6.
(a) Find the number of missiles that must be fired in order that the probability of hitting the target at least once is greater than 0.99.

(b) If five missiles are fired, calculate the probability that at least two of them are on target.   (C)

10.   (a) In a certain manufacturing process, it is known that approximately 10% of the items produced are defective. A quality control scheme is set up, by selecting 20 items out of a large batch, and rejecting the whole batch if three or more are defective. Find the probability that the batch is rejected.

(b) Two boys, John and David, play a game with an unbiased die. The die will be thrown four times. David will give John £x if there is an odd number of '6's, otherwise John will give David £1. If the game is to be a fair one to both John and David, find x.   (SUJB)

11.   The probability that an apple tree of one particular type will produce apples in the first year after planting is 0.8. The owner of a garden centre sells these to customers five trees at a time with a guarantee that if fewer than three trees produce apples in the first year he will replace all the ones that fail.
(a) Calculate the probability that no replacements will be required.
(b) If he sells five trees to each of 3000 customers, how many replacement trees can he expect to supply?   (C)

12.   Four unbiased coins are thrown. Find the probability of two or more heads. If six people each throw four coins, find the probability that:
(a) exactly two of them get two or more heads,
(b) at least one of them gets two or more heads.

13.   Among the population of a large city, the proportion of people with brown eyes is 0.15. If a random group of eight people is selected, find the probability that it will contain fewer than three brown-eyed people.
     If a second group of eight people is chosen, find the probability that the total number with brown eyes in the two groups is exactly two.   (C)

14.   In a quality-control procedure, the probability that a mass-produced article is found to be defective is p. Find expressions in terms of p for the probabilities of (a) exactly two, (b) two or more defective articles in a random sample of ten articles. Find an expression in terms of p for the probability that of ten such samples exactly one contains two or more defective articles.   (L)

8.3   The mean and variance of $B(n, p)$

Suppose that, using the data of example (2) in which a sample of 12 rods is examined from the output of a machine producing rods of which 1 in 10 are unacceptable, we wished to estimate the *average* number of unacceptable rods per sample for a large number of samples. This means calculating the mean or $E(X)$. In this case, you might guess immediately that since $n = 12$ and $p = 1/10$, the mean must be $12 \times 1/10 = 1.2$. Such a guess would be correct. The formula for the mean of $B(n, p)$ is

$$\mu = E(X) = np$$

In order to prove this result there are two methods, the first of which involves using the definition of $E(X)$ for a probability distribution:

$$\mu = E(X) = \sum_{i=1}^{n} p_i x_i,$$

where $p_i = P(X = x_i)$ (see page 88).

Let us first see how this method works out in a special case, when $n = 4$, for example. The probability distribution is then

X	0	1	2	3	4
P	q^4	$4q^3p$	$6q^2p^2$	$4qp^3$	p^4

and

$$E(X) = q^4 \cdot 0 + 4q^3p \cdot 1 + 6q^2p^2 \cdot 2 + 4qp^3 \cdot 3 + p^4 \cdot 4$$
$$= 4q^3p + 12q^2p^2 + 12qp^3 + 4p^4$$
$$= 4p(q^3 + 3q^2p + 3qp^2 + p^3)$$
$$= 4p(q + p)^3.$$

Now $q + p = 1$, so that

$$\mu = E(X) = 4p,$$

which confirms the correctness of the formula for $n = 4$.

We shall now investigate the variance for this special case, using the formula

$$\sigma^2 = \text{Var}(X) = E(X^2) - \mu^2.$$

$$E(X^2) = 4q^3p \cdot 1^2 + 6q^2p^2 \cdot 2^2 + 4qp^3 \cdot 3^2 + p^4 \cdot 4^2$$
$$= 4p(q^3 + 6q^2p + 9qp^2 + 4p^3).$$

Rearranging gives

$$E(X^2) = 4p[(q^3 + 3q^2p + 3qp^2 + p^3) + (3q^2p + 6qp^2 + 3p^3)]$$
$$= 4p[(q + p)^3 + 3p(q + p)^2]$$
$$= 4p(1 + 3p),$$

since $q + p = 1$.

Hence

$$\sigma^2 = 4p(1 + 3p) - (4p)^2$$
$$= 4p + 12p^2 - 16p^2$$
$$= 4p - 4p^2$$
$$= 4p(1 - p)$$
$$= 4pq.$$

Since $n = 4$, an intelligent guess would be that the variance of $B(n, p)$ is given by the formula

$$\sigma^2 = npq = np(1 - p)$$

which is, in fact, correct.

Unfortunately, as you will probably have felt by now, the application of this direct method to the general case, that is, for *any* value of n leads to some very complicated algebra (see the Appendix on pages 141–3).

There is, however, an alternative and much neater method which is available for those who have mastered probability generating functions (pages 113–14). In the next section we shall use this method. It will be the first of several examples in which the use of a p.g.f. makes for simplicity.

Mean and variance of $B(n, p)$, using the p.g.f.

The probability distribution for $B(n, p)$ is

X	0	1	2	...	r	...	n
P	q^n	$\binom{n}{1}q^{n-1}p$	$\binom{n}{2}q^{n-2}p^2$		$\binom{n}{r}q^{n-r}p^r$		p^n

so that the p.g.f.,

$$G(t) = q^n t^0 + \binom{n}{1}q^{n-1}pt^1 + \binom{n}{2}q^{n-2}p^2 t^2 + \ldots + \binom{n}{r}q^{n-r}p^r t^r + \ldots + p^n t^n$$

$$= (q + pt)^n.$$

$$G'(t) = np(q + pt)^{n-1} \quad \text{and} \quad G''(t) = n(n-1)p^2(q + pt)^{n-2}.$$

Now
$$\mu = E(X) = G'(1) \qquad \text{(see page 114)}$$

$$= np(q + p)^{n-1} = np,$$

since $q + p = 1$.

To find the variance σ^2 we use the result on page 114

$$G''(1) = \sigma^2 + \mu^2 - \mu.$$

In this case $G''(1) = n(n-1)p^2(q+p)^{n-2} = n(n-1)p^2$, since $q + p = 1$.

Hence
$$\sigma^2 = G''(1) - \mu^2 + \mu$$

$$= n(n-1)p^2 - n^2 p^2 + np \quad (\mu = np)$$

$$= np - np^2 = np(1 - p) = npq.$$

Thus:

$$\text{For } B(n, p) \quad \text{the mean } \mu = E(X) = np,$$

$$\text{the variance } \sigma^2 = \text{Var}(X) = npq = np(1 - p).$$

For a third method of proving these results, see exercise 8b, question 12.

8.4   Fitting a binomial distribution to given data

Suppose an experiment produces some sample data which the experimenter has reason to believe would be accounted for by assuming that the appropriate model is binomial. If he does not know the parameter p for the distribution, he may nevertheless estimate p from the sample and use it to calculate what the theoretical distribution would be using this estimated value of p. This theoretical distribution is then compared with the actual results. The method is demonstrated in the next example.

EXAMPLE 3   In a genetic experiment with a plant having a particular consti-
tution, offspring are raised from four of its seeds. The experiment is repeated 200
times with plants of the same kind and the number of offspring which are white are
counted. The results are:

Number of white offspring in a sample of four (X)	0	1	2	3	4
Number of plants (f)	51	80	50	16	3

Assuming that a binomial distribution is the appropriate model, calculate the
expected frequencies.

In order to calculate the expected frequencies we need to know p, the probability
of a plant being white in the *population* from which this sample of 200 has been
taken. The best we can do in the circumstances is to take the mean of the sample as
an estimate of the population mean, and calculate an estimate of p using the fact
that the mean of a binomial distribution is np. (The problem of estimating the mean
of a population when only a sample is available is discussed in chapter 12.)

$$\text{The mean of this sample} = \frac{51 \times 0 + 80 \times 1 + 50 \times 2 + 16 \times 3 + 3 \times 4}{200} = 1.2.$$

Here $n = 4$ and the mean of $B(4, p)$ is $4p$, so that we estimate p to be $1.2/4 = 0.3$.
The probabilities for $B(4, 0.3)$ are given by

$$P(X = r) = \binom{4}{3}(0.7)^{4-r}(0.3)^r$$

$$\Rightarrow P(0) = 0.240, \; P(1) = 0.412, \; P(2) = 0.265, \; P(3) = 0.076, \; P(4) = 0.008.$$

Multiplying by 200 to obtain the expected theoretical distribution gives the
following table:

X	0	1	2	3	4
f	48	82	53	15	2 (to the nearest integer)

The 'fit' between the observed and expected frequencies seems to be quite close.

Sampling with and without replacement

For a practical example of a binomial distribution, consider the following. Suppose
a machine is known to produce 5000 articles daily of which 96% are satisfactory
and 4% (or 200) are unsatisfactory. (It is possible, of course, that a machine will not
perform as consistently as this, but the example will illustrate an important point.)
If a daily inspection is made by picking out a random sample of 20 articles, each of
which is *replaced* before another article is selected, then the probability on each
occasion of an article being unsatisfactory will be 0.04 and the distribution of the
number of unsatisfactory articles in the sample will be $B(20, 0.04)$.

If, however, as is more likely, the first article is *not replaced* after it has been
selected, the probability of the second being unsatisfactory is no longer 0.04. If the
first was in fact unsatisfactory, the probability of the next being so is $199/4999$; if it

was satisfactory the probability of the next being unsatisfactory is 200/4999. The distribution will, therefore, no longer be binomial.

In practice, however, if the sample size is small in comparison with the total number from which the sample is taken, as in the example we have given, the distribution *is* approximately binomial. Indeed the approximation to a binomial is quite good so long as the ratio of sample size to total number is less than about 0.1.

EXERCISE 8b

1.   For the binomial distribution $B(5, \frac{1}{4})$, draw up a table giving the probabilities that the number of 'successes' X has values 0, 1, 2, 3, 4, 5.
   Use this table to calculate the mean of X and the variance of X, and verify that your answers agree with the results of using the formulae

$$\text{mean} = np, \qquad \text{variance} = np(1-p).$$

2.   The binomial distribution $B(n, p)$ has mean 15 and variance 10.5. Find the values of n and p.

3.   An unbiased coin is tossed 10 times. Calculate the mean and variance of the number of heads. What is the probability that the number of heads will exceed the mean number?

4.   Forty per cent of students at a large university are known to have some defect of vision. If samples of 20 students were selected at random, find the mean μ and the standard deviation σ of the number of students in a sample with defective vision.
   Find also the probability that the number with defective vision in a sample is: (a) less than $\mu - \sigma$, (b) less than $\mu - 2\sigma$.

5.   A wholesaler receives from a manufacturer regular consignments of chocolates, a fixed proportion of which are meant to have hard centres. Samples of four are taken from each consignment and the number of hard centres is noted. The results after 200 samples are as follows:

Number of hard-centred chocolates	0	1	2	3	4
Number of samples	64	85	40	9	2

   Find the mean number of hard-centred chocolates per sample and use it to calculate the theoretical binomial distribution with this mean.

6.   Razor blades are made up into packets of five. After a consignment of 100 packets had been sent to a retailer, it was discovered that the blades from which these packets had been made up contained a proportion that were sub-standard. The consignment was therefore recalled and each of the 100 packets was examined to see how many of the razor blades were sub-standard. The following were the results:

Number of sub-standard blades in a packet	0	1	2	3	4	5
Number of packets	3	6	21	36	26	8

   Calculate the theoretical frequencies of 0, 1, 2, 3, 4, 5 sub-standard blades, using the associated theoretical binomial distribution.

7.   One hundred samples, with ten items in each sample, were examined to find the number of flawed items. The results are summarised in the following table:

Number of flawed items in a sample	0	1	2	3	4	5	6 and over
Number of samples	7	30	31	21	10	1	0

Assuming that there is a constant chance p that an item is flawed, estimate p from these data.

Use this estimated value of p to calculate the theoretical frequencies of $0, 1, \ldots, 10$ flawed items in 100 samples with ten items in each sample.

Write down the probability that a sample of ten items chosen at random will contain three or more defective items.

(MEI)

8.   Testing the output of packets of detergent from a packing machine involves selecting samples of six packets and checking that they are not more than 2 grams underweight. The numbers of underweight packets in 140 successive samples are shown in the following table:

Number of underweight packets per sample	0	1	2	3	4	5	6
Number of samples	27	36	39	22	10	4	2

Find the mean number of underweight packets per sample and calculate the theoretical frequencies for a binomial distribution with this mean.

9.★ For the binomial distribution $B(n, p)$ prove that $P(X = r+1) > P(X = r)$ if $r+1 < (n+1)p$.

Hence show that if the distribution of the random discrete variable X is $B(12, 1/3)$, $P(X = r)$ is greatest when $r = 4$.

10.★ Find the most probable number of 'successes' in 16 independent trials of a random experiment, in each of which the probability of a success is 0.2.

11.   Write down the probability distribution of $B(5, 1/3)$ and *hence* show that its probability generating function is

$$G(t) = \left(\frac{2}{3} + \frac{1}{3}t\right)^5.$$

Use this expression for $G(t)$ to calculate $G'(1)$ and $G''(1)$ and hence derive the mean and variance of the distribution.

Verify that your results agree with the mean and variance obtained by using the formulae

$$\text{mean} = np, \qquad \text{variance} = np(1-p).$$

12.★ Obtain the mean and variance of $B(n, p)$ by the following method.

If the outcome of the first trial is X_1, which can take values 1 (success) and 0 (failure) with probabilities p and q, find $E(X_1)$ and $\text{Var}(X_1)$.

The outcomes of the 2nd, 3rd, ... and nth trials are $X_2, X_3, \ldots, X_n$ and these trials are independent, with the same distribution as X_1. If X is the total number of successes, use the fact that

$$X = X_1 + X_2 + \ldots + X_n$$

and the method of finding the mean and variance of the sum of a number of independent variables to show that $E(X) = np$ and $\text{Var}(X) = npq$.

Approximations to the binomial distribution

Clearly, the calculation of binomial probabilities will be unwieldy when n is large. To find, for example, $P(X \leqslant 30)$ for $B(200, 0.4)$ it would be necessary to calculate

$$P(X = 30) = \binom{400}{30} (0.4)^{30}(0.6)^{370} \text{ and 30 other terms.}$$

Under certain circumstances approximate methods are available (see page 207).

8.5   The geometric distribution

The binomial distribution arises when we are concerned with a fixed number of trials and the random variable X is the number of *successes*. The geometric distribution is needed when a Bernouilli trial is repeated until the first success is achieved, and the random variable is the *number of trials until this first success.*

We shall take as an example a pack of cards which is cut, the card revealed and placed back in the pack which is then shuffled. 'Success' is defined as 'drawing an ace'.

If X is the number of cuts needed, up to and including the drawing of the first ace, the possible values of X are, theoretically $1, 2, 3, \ldots$ to infinity. Clearly, for X to be 3, the first two cards must *not* be aces and the third must be an ace, so that

$$P(X = 3) = (12/13)^2(1/13).$$

Similarly $P(X = 5) = (12/13)^4(1/13)$, and so on.

As a second example we shall pursue the theme with which this chapter started (the births of boys and girls in a family), but we shall now imagine that a family is determined to go on having children until a girl is born. As before, we shall take the probability of a girl as 0.485. The probability distribution for X, the number of births until the first girl, will be:

X	1	2	3	$\ldots$
Probability	0.485	$(0.515)(0.485) = 0.250$	$(0.515)^2(0.485) = 0.129$	$\ldots$

In general, then:

If the probability of success in any one trial is p $(0 < p < 1)$ and $q = 1-p$, then

$$P(X = r) = q^{r-1}p, \quad r = 1, 2, 3, \ldots.$$

Any random variable with this distribution is said to have a *geometric distribution*, which is so called since the successive probabilities

$$p, qp, q^2p, q^3p, \ldots$$

form a geometric series. It can be seen that the sum of these probabilities is

$$p + qp + q^2p + q^3p + \ldots = p(1 + q + q^2 + q^3 + \ldots)$$

$$= p \times \frac{1}{1-q} \quad \text{(using the formula for the 'sum to infinity' of an infinite geometric series)}$$

$$= 1$$

since $1 - q = p$.

8.6   The mean and variance of a geometric distribution

In the example of cutting a pack of cards until an ace is drawn, what would be the *average* number of cuts made until an ace appears if the experiment were repeated a large number of times? As in the case of the binomial distribution it is possible to guess the result. The probability of getting an ace in a single trial is 1/13, so that it seems likely that the mean number of trials for an ace to appear should be 13. The formula for the mean of a geometric distribution with parameter p is, in fact

$$\mu = 1/p$$

so that, as a further example, the mean number of births needed for a daughter to be born is $1/0.485 = 2.06$, or just over two.

A direct proof of this result, and of the equivalent formula for the variance

$$\sigma^2 = q/p^2$$

is given in the Appendix on page 143, but as with the binomial distribution, neat proofs using the probability generating function are better and simpler.

The table of probabilities is:

X	1	2	3	...	r	...
Probability	p	qp	q^2p	...	$q^{r-1}p$	...

and the p.g.f.,

$$G(t) = pt^1 + qpt^2 + q^2pt^3 + \ldots + q^{r-1}pt^r \ldots \text{to } \infty$$

$$= pt(1 + qt + q^2t^2 + \ldots \text{to } \infty)$$

$$= pt/(1 - qt)$$

$$G'(t) = \frac{(1-qt)p - pt(-q)}{(1-qt)^2} = \frac{p}{(1-qt)^2}$$

Hence
$$\mu = G'(1) = p/(1-q)^2 = p/p^2 = 1/p.$$

$$G''(t) = 2pq/(1-qt)^3 \quad \text{and} \quad G''(1) = 2pq/(1-q)^3 = 2q/p^2.$$

$$\Rightarrow \sigma^2 = G''(1) - \mu^2 + \mu = 2q/p^2 - 1/p^2 + 1/p = \frac{2q - 1 + p}{p^2}$$

$$= q/p^2.$$

Thus:

For the geometric distribution with parameter p

$$\mu = E(X) = 1/p, \qquad \sigma^2 = \text{Var}(X) = \frac{q}{p^2} = \frac{1-p}{p^2}.$$

EXAMPLE 4   Single items from the output of a machine are extracted and tested. It is known that 1 in 20 is faulty. Find: (a) the probability that the first faulty item found is the 8th selected, (b) the probability that the first faulty item found comes before the 8th, (c) the expected number of items extracted before a faulty one is found.

$$p = 0.05 \text{ and } q = 0.95$$

(a) $P(X = 8) = (0.95)^7(0.05) = 0.035$ to 3 dec. places.

(b) $P(X < 8) = 1 - P(X \geqslant 8)$

$$= 1 - (q^7 p + q^8 p + \ldots) = 1 - q^7 p/(1 - q) = 1 - q^7$$

$$= 0.302 \quad \text{to 3 dec. places}$$

(c) $E(X) = 1/p = 1/0.05 = 20.$

8.7★ The negative binomial (or Pascal) distribution

Looking once more at the example on page 134 of cutting a pack of cards, a more complicated question could be asked. What is the probability that the 7th trial will give the 3rd ace to appear?

This means that the 7th trial must produce an ace, for which the probability is $1/13$ *and* that in the previous 6 trials there must have been exactly 2 aces (no more and no less) for which the probability, using the binomial distribution $B(6, 1/13)$, is

$$\binom{6}{2}(12/13)^4(1/13)^2.$$

Thus the required probability is

$$(1/13) \times \binom{6}{2}(12/13)^4(1/13)^2 \approx 0.005.$$

For the general case, suppose the probability of 'success' is p and the random variable X is the *number of trials up to and including the kth success.*

Clearly the rth trial will give the kth success if

(a) the rth trial is a success *and*

(b) the previous $(r-1)$ trials gave exactly $(k-1)$ successes.

Hence

$$P(X = r) = p \times \binom{r-1}{k-1} q^{(r-1)-(k-1)} \times p^{k-1}$$

$$= \binom{r-1}{k-1} q^{r-k} p^k, \quad r = k, k+1, \ldots.$$

This defines the *negative binomial distribution.*

To find the mean and variance of this distribution it should be noted that if

$X_1 = $ the number of trials up to and including the first success,

$X_2 = $ the number of trials after the first success up to and including the second success,

$\vdots$

$X_k = $ the number of trials after the $(k-1)$th success up to and including the kth success, and $X = $ the number of trials up to and including the kth success,

then $X = X_1 + X_2 + \ldots + X_k.$

Now each of $X_1, X_2, \ldots, X_k$ is an independent random variable with a geometric distribution. The mean of each is $1/p$ and the variance of each is q/p^2.

Hence $E(X) = E(X_1) + E(X_2) + \ldots + E(X_k) = k/p,$

and $\text{Var}(X) = \text{Var}(X_1) + \text{Var}(X_2) + \ldots + \text{Var}(X_k) = kq/p^2.$

EXAMPLE 5   An unbiased die is thrown until three '6's have appeared. Find: (a) the probability that 5 throws are needed, (b) the expected number of throws.

(a) $p = 1/6, q = 5/6, k = 3$

$$P(X = 5) = \binom{5-1}{3-1}(5/6)^{5-3}(1/6)^3 = \binom{4}{2}(5/6)^2(1/6)^3$$

$$= 0.019 \quad \text{to 3 dec. places.}$$

(b) $E(X) = k/p = 3/(1/6) = 18.$

EXERCISE 8c

1.   Ten per cent of a large batch of model piston-engined aircraft have inadvertently been assembled without a propellor.
   If the manufacturer takes the aircraft out one at a time, what is the probability that the fourth model is the first to be without a propellor?

2.   A man throws an unbiased die until he gets a '6'. What is the probability that the first '6' appears on: (a) the 3rd throw, (b) the 9th throw?
   Calculate the average number of throws he should expect to make in order to get a '6', if he repeats the experiment many times.

3.   A large quantity of mixed pink and black snooker balls are stored in a warehouse. The warehouse keeper's young son, with nothing better to do, performs the experiment of picking a ball out until he gets a pink one. After doing this many times he finds that the average number of balls he has to pick in order to get a pink one is 9.
   Estimate the proportion of pink balls in the store.
   What is the probability that the first black ball picked will be the fifth?

4.   In an inspection process, the output of a machine is tested until the first faulty article is found. If 4% of the output is faulty, find the expected number of articles tested until a faulty one is found. Find also the probability that the first faulty article is found before this expected number is reached.

5.   Find the least value of r such that, when a die is thrown repeatedly, there is more than a 50% chance of obtaining a '6' on the rth throw or earlier.

6.   Show that, for a geometric distribution with parameter p, $P(X > k)$ is $(1 - p)^k$.

7.★ A die is thrown until two '6's appear. Find the probability that: (a) five throws are needed, (b) eight throws are needed.

8.★ The output of the machine in question 4 is tested until two faulty articles are found. What will be the average number of articles tested? What is the probability that exactly this number of articles will be needed for two faulty ones to be found?

EXERCISE 8d (miscellaneous)

1. Taking the probability of a male child as 0.52, find the probability that a family of four children will be: (a) all boys, (b) all girls, (c) 3 boys and a girl.

2. Nine per cent of articles in a large consignment of pyjama coats have buttons missing. Calculate the probability that a sample of 30 pyjama coats will contain more than 2 with buttons missing.

3. Ten unbiased dice are thrown. Find the probabilities of obtaining: (a) no 'successes', (b) at least 8 failures, (c) at least twice as many successes as failures, if 'success' means throwing a '5' or a '6'.

4. The probability that a marksman will hit a target is 4/5. He fires 10 shots.
     Calculate, correct to three decimal places, the probability that he will hit the target: (a) at least 8 times, (b) no more than 7 times.
     If he hits the target exactly 7 times, calculate the probability that the 3 misses are with successive shots.   (C)

5. The random variable X has distribution $B(n, p)$. Its mean is 18 and its variance is 7.2.
     Calculate: (a) $P(X < 2)$, (b) $P(X \geqslant 3)$.

6. A die is biased so that the probability of throwing a '6' is 0.2.
(a) What is the mean number of throws needed to obtain a '6' in the long run?
(b) What is the probability that the number of throws needed to get a '6' is equal to this mean number?

7. A man and his wife decide to go on having children until a girl is born. If the probability of a child being a girl is 0.48 find the probability that the first girl born will be: (a) the 3rd child, (b) the 5th child?
     What is the probability that they will have to have more than two children in order to have a girl?

8. Given the following data, fit a theoretical binomial distribution:

X	0	1	2	3	4	5	6
f	251	361	244	83	21	3	1

working to one decimal place.

9. Groups of six people are chosen at random and the number, x, of people in each group who normally wear glasses is recorded. The results obtained from 200 groups of six are shown in the table.

Number of people in group wearing glasses (x)	0	1	2	3	4	5	6
Number of occurrences	17	53	65	45	18	2	0

     Calculate, from the above data, the mean value of x.
     Assuming that the situation can be modelled by a binomial distribution having the same mean as the one calculated above, state the appropriate values for the binomial parameters n and p. Calculate the theoretical frequencies corresponding to those in the table.   (C)

10. A manufacturing company sends out invoices to its customers allowing 5% discount to those who settle their accounts within 15 days. In the past, 40% of the customers took advantage of the discount terms. On a particular day the company sent out 10 invoices.
     Assuming a binomial model, calculate the probability that, of the 10 invoices, less than 2

settled within 15 days. For days when 10 invoices are sent out, calculate the mean and variance of the distribution of the number of invoices settled within 15 days. (L)

11. A nurseryman is raising plants from seed and selling them in batches of 5 plants. He expects that one quarter of the seeds will produce white-flowered plants.
(a) If he makes up 1024 batches of plants, how many of these will he expect to contain 0, 1, 2, 3, 4 and 5 white-flowered plants?
(b) What will be the expected mean and standard deviation of the number of white-flowered plants per batch?
(c) What is the probability that a batch chosen at random will contain 4 or more white-flowered plants?
(d) If, instead, he makes up batches of 10 plants, what will be the mean number of white-flowered plants per batch? Show what will then be the most commonly occurring number of white-flowered plants per batch. (O)

12. If $p(k)$ is the probability of obtaining exactly k successes in n independent trials, where the probability of success at each trial is p, find an expression for $p(k)/p(k-1)$. Hence find the condition that r must satisfy if no other value of $p(k)$ is greater than $p(r)$.
Prove also that the mode of the distribution of successes differs from the mean by less than one. (MEI)

13. What is the probability of at least one double '6' in three throws of two dice? What is the probability in n throws?
Show that with 25 throws of two dice, one would feel that a wager on a double '6' occurring would be worthwhile. (SUJB)

14. Five fair dice are thrown. Find the probabilities that there are:
(a) exactly three '6's, (b) at least three '6's,
(c) three '6's and two '5's, (d) two '5's, *given* that there are exactly three '6's showing.
(MEI)

15. A man canvasses people to join an organisation. For each new recruit he receives 25p. The probability that a person he canvasses will join is 0.2. Calculate to three decimal places the probability that he will obtain three or more recruits from ten people canvassed. State the amount of money that he would be expected to obtain on average from ten canvassings. State how many people he would need to canvas each evening to average ten new recruits per evening.
Calculate the number of people he must arrange to canvass to be 99 % certain of obtaining at least one recruit. (JMB)

16.★ Thatcher's Pottery produces large batches of coffee mugs decorated with the faces of famous politicians. They are considering adopting one of the following sampling plans for batch inspection.

Method A (single sample plan). Select 10 mugs from the batch at random and accept the batch if there are 2 or less defectives, otherwise reject the batch.

Method B (double sample plan). Select 5 mugs from the batch at random and accept the batch if there are no defectives, reject the batch if there are 2 or more defectives, otherwise select another 5 mugs at random. When the second sample is drawn count the number of defectives in the combined sample of 10 and accept the batch if the number of defectives is 2 or less, otherwise reject the batch.

(a) If the proportion of defectives in a batch is p, find in terms of p, for each method in turn, the probability that the batch will be accepted.
(b) Evaluate *both* the above probabilities for $p = 0.2$ and $p = 0.5$.
(c) Hence, or otherwise, decide which of these two plans is more appropriate, and why. (AEB 1981)

17. The random variable X is the number of success in n independent trials of an experiment in which the probability of success at any one trial is p. Show that

$$\frac{P(X = k+1)}{P(X = k)} = \frac{(n-k)p}{(k+1)(1-p)}, \quad k = 0, 1, 2, \ldots, (n-1).$$

Find the most probable number of successes when $n = 10$ and $p = \frac{1}{4}$. \hfill (C)

18.★ In n independent trials with constant probability p of success at each trial, show that the ratio of the probabilities of $r+1$ and r successes is

$$\frac{(n-r)p}{(r+1)(1-p)} \quad (0 \leqslant r \leqslant n-1).$$

A manufacturer produces bracelets consisting of six sections linked together. After this linking-up process, each section independently has a probability q of having been scratched. The bracelets are inspected and those which are unscratched are sold for £2 each. Those with a single scratched section are sold as substandard for £1 each. Any with more than one scratched section are scrapped. It is found that the ratio of the number of bracelets sold as perfect to those sold as substandard is $4:1$. Calculate q and hence find to three decimal places the proportion of the bracelets that are scrapped.

Each bracelet costs £1 to produce. Find to the nearest £1 the expected profit on a batch which consists of 1000 bracelets before inspection. \hfill (JMB)

19.★ The probability that a biased coin comes up heads when it is tossed is p. For an experiment it is tossed n times, where n is chosen at random from a preliminary experiment, and can take one of two values, 2 and 3, with

$$P(n = 2) = \frac{1}{3}, \qquad P(n = 3) = \frac{2}{3}.$$

Obtain the probability distribution of the number of heads in the combined experiment, and find the mean of this distribution. \hfill (O)

20.★ In a binomial distribution let X be the number of successes scored in n independent trials when the probability of success in each trial is p. Write down the values of $E(1)$ and $E(X)$ and show (using standard formulae) that

$$E(X^2) = n(n-1)p^2 + np.$$

In a simple model of a diffusion process, a particle undergoes a sequence of independent jumps, each of unit length, along a straight line. The probability of any particular jump being to the right is p. Show that the probability that the particle is at a distance $2r-n$ to the right of its starting point after n jumps is

$$\binom{n}{r} p^r (1-p)^{n-r}.$$

Calculate the expected value of the square of the particle's distance from its starting point after n jumps. \hfill (SMP)

21.★ In each batch of manufactured articles, the proportion of defective articles is p. From each batch a random sample of nine is taken and each of the nine articles is examined. If two or more of the nine articles are found to be defective the batch is rejected; otherwise it is accepted. Prove that the probability that a batch is accepted is

$$(1-p)^8(1+8p).$$

It is decided to modify the sampling scheme so that when one defective is found in the sample, a second sample of nine is taken and the batch rejected if this contains any defectives. With this exception, the original scheme is continued. Find an expression in terms of p for the probability that a batch is accepted.

For this modified scheme evaluate the average number sampled per manufactured batch over a large number of batches when p has the value 0.1.   (C)

22.★★ A series of independent trials is being carried out, in each of which the probability of success is p and of failure $q = 1 - p$.

(a) Suppose n trials are carried out. *Derive* the probability that there are exactly r successes among these n trials $(r = 0, 1, 2, \ldots, n)$.

(b) Suppose, instead, that the series of trials is to continue until the kth success has been achieved. Show that the probability that $k + m$ trials are needed $(m = 0, 1, 2, \ldots)$ is

$$\binom{k+m-1}{m} p^k q^m.$$

By considering in two different ways the probability of at least h successes being achieved in $h + j$ trials, prove the identity

$$\sum_{i=0}^{j} \binom{h+i-1}{i} p^h q^i = \sum_{s=0}^{j} \binom{h+j}{s} p^{h+j-s} q^s. \qquad \text{(MEI)}$$

23.★★ A coin, which falls heads with probability p and tails with probability $q \, (= 1 - p)$, is tossed repeatedly until a total of $N + 1$ heads occur. Show that the probability that K tails occur in this period is

$$\binom{N+K}{N} p^{N+1} q^K.$$

A smoker initially has N matches in each of his two coat pockets. When he wants a match, he selects his left pocket with probability p and his right pocket with probability $q \, (= 1 - p)$. He will discover that a pocket is empty only when he selects that pocket for the $(N+1)$th time. Using the above result, show that, at the moment he first discovers one of his pockets is empty, the probability that the other contains exactly r matches is

$$\binom{2N-r}{N} (pq)^{N-r} (p^{r+1} + q^{r+1})$$

for $r = 0, 1, \ldots, N$.   (O)

24.★★ The random variable R is distributed binomially, n being the number of trials and p the probability of 'success'. It is however truncated in the sense that $R = n$ is not observable. Prove that the expected value of R is

$$np(1 - p^{n-1})/(1 - p^n).$$

8.8   Appendix

The mean and variance of $B(n, p)$

The probabilities (here called P_r) for $X = r, r = 0, 1, 2, 3, \ldots, n$ are set out in tabular form:

X	P_r	
0	q^n	$= P_0$
1	$\binom{n}{1} q^{n-1} p = nq^{n-1}p$	$= P_1$
2	$\binom{n}{2} q^{n-2} p^2 = \dfrac{n(n-1)}{2!} q^{n-2} p^2$	$= P_2$
3	$\binom{n}{3} q^{n-3} p^3 = \dfrac{n(n-1)(n-2)}{3!} q^{n-3} p^3 = P_3$	
$\vdots$		
n	p^n	$= P_n$

$$\mu = E(X) = \sum_{r=0}^{n} P_r x_r = P_0 \cdot 0 + P_1 \cdot 1 + P_2 \cdot 2 + \ldots + P_n \cdot n$$

$$= nq^{n-1}p \cdot 1 + \frac{n(n-1)}{2!}q^{n-2}p^2 \cdot 2 + \frac{n(n-1)(n-2)}{3!}q^{n-3}p^3 \cdot 3$$

$$+ \ldots + p^n \cdot n$$

$$= n\left[q^{n-1}p + \frac{(n-1)}{1!}q^{n-2}p^2 + \frac{(n-1)(n-2)}{2!}q^{n-3}p^3 + \ldots + p^n \right]$$

(since $2/2! = 1/1!$, $3/3! = 1/2!, \ldots$)

$$= np\left[q^{n-1} + \binom{n-1}{1}q^{n-2}p + \binom{n-1}{2}q^{n-3}p^2 + \ldots + p^{n-1} \right]$$

$$= np(q+p)^{n-1}$$

$$= np$$

since $q + p = 1$.

The variance $\sigma^2 = E(X^2) - \mu^2 = \sum_{r=0}^{n} P_r \cdot r^2 - \mu^2.$

It is convenient to write

$$\sum_{r=0}^{n} P_r \cdot r^2 = \sum_{r=1}^{n} P_r \cdot r^2$$

in the form

$$\sum_{r=1}^{n} P_r[r(r-1)+r] = \sum_{r=1}^{n} P_r \cdot r(r-1) + \sum_{r=1}^{n} P_r \cdot r$$

$$= \sum_{r=2}^{n} P_r \cdot r(r-1) + \mu$$

(since $r = 1$ gives a zero term)

so that

$$\sigma^2 = \sum_{r=2}^{n} P_r \cdot r(r-1) + \mu - \mu^2.$$

Now $\displaystyle\sum_{r=2}^{n} P_r \cdot r(r-1)$

$$= \frac{n(n-1)}{2!}q^{n-2}p^2 \cdot 2 \cdot 1 + \frac{n(n-1)(n-2)}{3!}q^{n-3}p^3 \cdot 3 \cdot 2 + \ldots + p^n \cdot n(n-1)$$

$$= n(n-1)p^2\left[\frac{2 \cdot 1}{2!}q^{n-2} + \frac{(n-2)3 \cdot 2}{3!}q^{n-3}p + \frac{(n-2)(n-3) \cdot 4 \cdot 3}{4!}q^{n-4}p^2 \right.$$

$$\left. + \ldots + p^{n-2} \right]$$

$$= n(n-1)p^2\left[q^{n-2} + \frac{(n-2)}{1!}q^{n-3}p + \frac{(n-2)(n-3)}{2!}q^{n-4}p^2 + \ldots + p^{n-2} \right]$$

$$\left(\text{since } \frac{3\cdot 2}{3!} = 1/1!, \frac{4\cdot 3}{4!} = 1/2!, \dots\right)$$

$$= n(n-1)p^2(q+p)^{n-2} = n(n-1)p^2.$$

Hence
$$\sigma^2 = \sum_{r=2}^{n} P_r \cdot r(r-1) + \mu - \mu^2$$

$$= n(n-1)p^2 + \mu - \mu^2$$

$$= (n^2-n)p^2 + np - n^2p^2 \quad (\text{since } \mu = np)$$

$$= np - np^2$$

$$= np(1-p)$$

$$= npq.$$

The mean and variance of the geometric distribution

The following results are needed:

$$1+2x+3x^2+\dots+rx^{r-1}+\dots \text{ to } \infty = 1/(1-x)^2 \tag{1}$$
$$1^2+2^2x+3^2x^2+\dots+r^2x^{r-1}+\dots \text{ to } \infty = (1+x)/(1-x)^3 \tag{2}$$

when $|x| < 1$.

Proof of (1)

$$1+2x+3x^2+\dots+rx^{r-1}+\dots$$

$$= \frac{d}{dx}(x+x^2+x^3+\dots+x^r+\dots) = \frac{d}{dx}\left(\frac{x}{1-x}\right) = (1-x)^{-2}.$$

(It is assumed here that it is valid to differentiate the sum of an infinite convergent series by differentiating each term separately.)

Proof of (2)

$$1^2+2^2x+3^2x^2+\dots+r^2x^{r-1}+\dots$$

$$= \frac{d}{dx}(x+2x^2+3x^3+\dots+rx^r+\dots)$$

$$= \frac{d}{dx}[x(1+2x+3x^2+\dots+rx^{r-1}+\dots)] = \frac{d}{dx}[x(1-x)^{-2}]$$

using (1)

$$= 2x(1-x)^{-3}+(1-x)^{-2} = (1+x)/(1-x)^3.$$

These results are now applied to find the mean and variance.

From the table of probabilities:

X	1	2	3	$\ldots$	r	$\ldots$
Probability	p	qp	q^2p	$\ldots$	$q^{r-1}p$	$\ldots$

$$\mu = E(X) = p\cdot 1 + qp\cdot 2 + q^2p\cdot 3 + \ldots + q^{r-1}p\cdot r + \ldots$$

$$= p(1 + 2q + 3q^2 + \ldots) = p/(1-q)^2 \quad \text{[using (1)]}$$

$$= p/p^2 = 1/p.$$

$$E(X^2) = p\cdot 1^2 + qp\cdot 2^2 + q^2p\cdot 3^2 + \ldots$$

$$= p(1^2 + 2^2q + 3^2q^2 + \ldots) = p(1+q)/(1-q)^3 \quad \text{[using (2)]}$$

$$= (1+q)/p^2$$

Hence $\quad \sigma^2 = E(X^2) - \mu^2 = (1+q)/p^2 - 1/p^2 \quad$ (since $\mu = 1/p$)

$$= q/p^2.$$

9* The Poisson distribution

9.1 The Poisson distribution as the limit of a binomial distribution

The binomial distribution described in chapter 8 is one of the most important models when we are dealing with a discrete random variable. In this chapter we shall be concerned with another discrete distribution with a very wide range of applications.

Consider the following situation. In a certain country, the proportion of children under four years of age who suffer from rickets is 1/100. What are the probabilities that in a sample of 500 children chosen at random from this age group there will be $0, 1, 2, 3, \ldots$ children with rickets?

Clearly we could take $B(500, 0.01)$ as the appropriate model and use the binomial probabilities, so that, for example, if X is the number of children in the sample who have rickets:

$$P(X = 1) = \binom{500}{1}(0.99)^{499}(0.01)$$

$$P(X = 4) = \binom{500}{4}(0.99)^{496}(0.01)^4.$$

However, calculations of this kind can be laborious. There is a much more convenient alternative for situations such as this, using the *Poisson distribution*.

Poisson, a 19th century French mathematician, set out to find a distribution which would give a good approximation to $B(n, p)$ when n *is large* and p *is small* and np is of moderate size. The result was the Poisson distribution. (We shall be more specific about the words 'large' and 'small' later.) Moreover, as we shall see, it turns out that this distribution has a much wider range of application than this approach would suggest.

The development of the Poisson distribution is complicated and it may help you to understand the steps that are taken if you know in advance what we are working towards. We shall therefore give the conclusion first, check that it gives good approximations in our particular example, and then go on to establish the results formally.

The Poisson probabilities for $B(n, p)$ where the mean $np = a$, are

X	0	1	2	3	4	$\ldots$
Probability	e^{-a}	ae^{-a}	$\dfrac{a^2 e^{-a}}{2!}$	$\dfrac{a^3 e^{-a}}{3!}$	$\dfrac{a^4 e^{-a}}{4!}$	$\ldots$

(e^x is the exponential function).

Applying this to the problem of children with rickets, where the distribution is $B(500, 0.01)$, so that $a = np = 500 \times 0.01 = 5$, the first five Poisson probabilities, to 4 decimal places are:

$$P(X = 0) = 0.0067$$
$$P(X = 1) = 0.0337$$
$$P(X = 2) = 0.0842$$
$$P(X = 3) = 0.1404$$
$$P(X = 4) = 0.1755.$$

Now let us compare these with the binomial probabilities:

$$P(X = 0) = (0.99)^{500} \qquad\qquad = 0.0066$$

$$P(X = 1) = 500(0.99)^{499}(0.01) \qquad\qquad = 0.0332$$

$$P(X = 2) = \frac{500 \times 499}{2!}(0.99)^{498}(0.01)^2 \qquad = 0.0836$$

$$P(X = 3) = \frac{500 \times 499 \times 498}{3!}(0.99)^{497}(0.01)^3 \qquad = 0.1402$$

$$P(X = 4) = \frac{500 \times 499 \times 498 \times 497}{4!}(0.99)^{496}(0.01)^4 = 0.1760.$$

The two sets of results are close to 4 decimal places and very close to 3 decimal places.

Proof

The method is to find the limit of the binomial probabilities as $n \to \infty$ and $p \to 0$ in such a way that np, the mean of $B(n, p)$, remains constant and equal to a.

To do this the following important limit is needed:

$$\lim_{n \to \infty} (1 + x/n)^n = e^x.$$

Consider the expansion of $(1 + x/n)^n$ by the binomial theorem.

$$(1 + x/n)^n = 1 + \frac{n}{1!} \times \frac{x}{n} + \frac{n(n-1)}{2!} \times \frac{x^2}{n^2} + \frac{n(n-1)(n-2)}{3!} \times \frac{x^3}{n^3} + \dots$$

$$+ \frac{n(n-1)(n-2)\dots(n-r+1)}{r!} \times \frac{x^r}{n^r} \dots \quad \text{to } n \text{ terms}$$

$$= 1 + x + \frac{[1 - (1/n)]}{2!}x^2 + \frac{[1 - (1/n)][1 - (2/n)]}{3!}x^3 + \dots$$

$$+ \frac{[1 - (1/n)][1 - (2/n)]\dots[1 - (r-1)/n]}{r!}x^r \dots \quad \text{to } n \text{ terms.}$$

Since each of the brackets $[1 - (1/n)]$, $[1 - (2/n)], \dots \to 1$ as $n \to \infty$, it seems reasonable to say that the limit of this (finite) series as $n \to \infty$ is $1 + x/1! + x^2/2! + \dots + x^r/r! + \dots$ to infinity $= e^x$.

Note: This is not a rigorous proof, as we have not considered the limit of the product of n factors, each of which tends to 1, as n tends to ∞.

We shall now see what happens to the probabilities $P(X = 0)$, $P(X = 1)$ and $P(X = 2)$ for the binomial $B(n, p)$ when $n \to \infty$, $p \to 0$ and $np = a$.

$$P(X = 0) = q^n$$

$$= (1 - p)^n$$

$$= \left(1 - \frac{a}{n}\right)^n \quad (\text{since} \quad np = a)$$

As $n \to \infty$, $\left(1 - \dfrac{a}{n}\right)^n \to e^{-a}$ (using the above result).

Hence
$$P(X = 0) \to e^{-a}.$$

$$P(X = 1) = nq^{n-1}p$$

$$= n(1 - p)^{n-1}\frac{a}{n}$$

$$= a\frac{(1 - p)^n}{1 - p}.$$

Now as $p \to 0$, $1 - p \to 1$.

And as $n \to \infty$, $(1 - p)^n = \left(1 - \dfrac{a}{n}\right)^n \to e^{-a}$

Hence
$$P(X = 1) \to ae^{-a}$$

$$P(X = 2) = \frac{n(n-1)}{2!}q^{n-2}p^2$$

$$= \frac{n(n-1)}{2!}\left(\frac{a}{n}\right)^2 (1 - p)^{n-2}$$

Rearranging gives

$$P(X = 2) = \frac{a^2}{2!}\frac{n(n-1)}{n^2}(1 - p)^{n-2}$$

$$= \frac{a^2}{2!} \cdot 1 \cdot (1 - 1/n)\frac{(1 - p)^n}{(1 - p)^2}.$$

As $p \to 0$, $(1 - p)^2 \to 1$.

As $n \to \infty$, $(1 - p)^n = \left(1 - \dfrac{a}{n}\right)^n \to e^{-a}$ as before.

As $n \to \infty$, $(1 - 1/n) \to 1$.

Hence
$$P(X = 2) \to \frac{a^2}{2!}e^{-a}.$$

The proof for $P(X = r)$ follows similar lines, and is given in the Appendix on page 163.

The resulting distribution is defined formally as follows:

The discrete random variable X which can assume values $0, 1, 2, \ldots$ with probabilities

$$P(X = r) = \frac{e^{-a}a^r}{r!}, \quad r = 0, 1, 2, \ldots$$

(where $a > 0$) is the *Poisson distribution*.

It can be seen that while $B(n, p)$ has two parameters, the Poisson distribution has only one, a.

Later in this chapter, and in chapter 22, it will be found that the Poisson distribution is a suitable model for a wide range of situations. For the moment, however, we shall regard it merely as a suitable approximation for the binomial distribution whenever n is large and p is small.

As a rough guide, the approximation is quite good even for moderate values of n, say 50 or more, provided p is less than about 0.1.

The table of probabilities, quoted earlier, is:

X	0	1	2	3	$\ldots$	r	$\ldots$
Probability	e^{-a}	$\dfrac{e^{-a} \cdot a}{1!}$	$\dfrac{e^{-a} \cdot a^2}{2!}$	$\dfrac{e^{-a} \cdot a^3}{3!}$	$\ldots$	$\dfrac{e^{-a} \cdot a^r}{r!}$	$\ldots$

where the parameter $a = np$.

The sum of these probabilities $= e^{-a}(1 + a/1! + a^2/2! + \ldots) = e^{-a}e^{a}$

$$= 1,$$

a necessary condition for any probability distribution.

Using the Poisson distribution

EXAMPLE 1   The number X of radioactive particles emitted in a minute from a substance is registered on a Geiger counter. The distribution of X is Poisson with parameter 4. Find:
(a) $P(X = 0)$,   (b) $P(X = 3)$,   (c) $P(X > 2)$.

The parameter $a = 4$, so that
(a) $P(X = 0) = e^{-4} = 0.018$ to 3 dec. places.

(b) $P(X = 3) = \dfrac{4^3}{3!}e^{-4} = 0.195.$

(c) $P(X > 2) = 1 - P(X = 0 \text{ or } 1)$

$$= 1 - (e^{-4} + 4e^{-4}) = 1 - 5e^{-4} = 0.908.$$

EXAMPLE 2   The probability that a bottle of chemicals produced by a machine will contain some impurity is 1/20. If a sample of 60 bottles is tested, find the probability that: (a) exactly two bottles will contain some impurity, (b) that more than two bottles will do so.

Here $n = 60$, $p = 0.05$ and $a = np = 3$.

(a) $P(X = 2) = \dfrac{e^{-3}}{2!} \times 3^2 = 0.2240$ to 4 dec. places.

(b) $P(X > 2) = 1 - P(X = 0 \text{ or } 1 \text{ or } 2)$

$$= 1 - e^{-a}(1 + a + a^2/2), \quad \text{where } a = 3$$

$$= 1 - e^{-3}(1 + 3 + 4.5) = 0.5768 \quad \text{to 4 dec. places.}$$

Note: Using $n = 60$, $p = 0.05$ the binomial probability that $X = 2$ is

$$\binom{60}{2}(0.05)^2(0.95)^{58} = 0.2259 \quad \text{to 4 dec. places.}$$

If $n = 120$ and $p = 0.025$ (so that $a = np$ is still 3), the binomial probability that $X = 2$ is 0.2250.

If $n = 240$ and $p = 0.0125$ (so that $a = np$ is still 3), the binomial probability that $X = 2$ is 0.2245.

In other words, the Poisson approximation improves, the larger the value of n and the smaller the value of p.

EXAMPLE 3   With the data of example (2), suppose five samples of 60 bottles are taken. What is the probability that three of these samples will contain exactly two bottles with some impurity?

From example (2), the probability that a *single* sample will contain exactly two impure bottles is 0.224. Let us call this probability p.

If Y is the number of *occurrences of 'exactly two impure bottles'* in a sample, and the number of samples is five, the distribution of Y is binomial, $B(n, p)$, where $n = 5$ and $p = 0.224$.

For $B(5, 0.224)$, $P(Y = 3) = \dbinom{5}{3}(0.224)^3(0.776)^2$

$$= 0.068 \text{ to 3 dec. places.}$$

EXERCISES 9a

1.   If the distribution of a random variable X is Poisson with parameter $a = 2.5$ find:
(a) $P(X = 0)$,   (b) $P(X = 1)$,   (c) $P(X = 2)$,
(d) $P(X \leqslant 2)$,   (e) $P(X \geqslant 5)$.

2.   Show that for a Poisson distribution with mean 0.5, the probability of 0, 1, 2 'successes' is in close agreement with the probability of 0, 1, 2 'successes' in the binomial distribution $B(50, 0.01)$.

3.   The probability that an item produced by a machine is faulty is 0.03. A sample of 60 items is selected. Using the Poisson distribution, calculate the probability that three faulty

items will be found.

Compare this with the probability obtained by using the binomial distribution.

4.   The probability of an employee having an accident in a factory during a year is $1/1500$. If there are 400 employees, find the probability of there being: (a) one accident, (b) two accidents, (c) more than two accidents during a year.

5.   A piano emporium keeps three pianos which are available for hire for an evening. The number of applications for a piano on each day has a Poisson distribution with mean 1.4.

What is the probability that, on any given day, the emporium will have to turn down requests for a piano?

6.   The proportion of defective hearing aids in a large batch of aids produced by a manufacturer is 0.005. What is the probability that there will be 0, 1, 2, 3, 4 defective items in a sample of 400?

7.   A random variable X has a Poisson distribution with parameter a. If $P(X = 0) = 0.2$, find: (a) the value of a, (b) $P(X = 4)$.

8.   The distribution of X is Poisson. If $P(X = 2) = \frac{2}{3} P(X = 1)$, what is the parameter of the distribution?

9.   If p_k is $P(X = k)$ for a Poisson distribution with parameter 3, show that

$$p_{k+1}/p_k = 3/(k+1).$$

For what value of k is $P(X = k)$ largest?

10.   Bacteria are distributed independently of one another in a solution, and it is known that the number of bacteria per millilitre follows a Poisson distribution with mean 2.
(a) Show that the probability of a sample of 1 ml of solution containing three or more bacteria is 0.32 approximately.
(b) Five samples, each of 1 ml of solution, are taken. Find the probability that less than two of these samples contain three or more bacteria.   (L)

11.   A large consignment of oranges for export from Spain is inspected by having samples of 400 examined for flaws. If one orange in a hundred is expected to be imperfect, estimate the probability that two samples of 400 will contain:
(a) no imperfect oranges between them,
(b) two imperfect oranges between them.   (C)

12.   It is known that when a patient is injected with a certain drug, the probability of a reaction taking place is 0.002. Find the probability that, out of 3000 patients injected:
(a) exactly three will suffer a reaction,
(b) more than two will suffer a reaction.

Find also the probability that, in each of three groups of 1000 patients, exactly one would suffer a reaction.   (L)

9.2   A further application of the Poisson distribution

We shall now consider a situation where, at first sight, the binomial model in its limiting Poisson form does not seem to apply.

Over a long period of time, the average number of calls put through to a small telephone exchange during the hour 2 pm to 3 pm on a weekday is 2. The problem is to find: (a) the probability that, on any given weekday between these times, the exchange will receive $0, 1, 2, 3, \ldots$ calls, (b) the probability that, on any given weekday, the exchange will receive more than 3 calls between 2.15 pm and 2.45 pm.

(a) This can be converted into a binomial situation in the following way. Divide the hour up into 60 intervals of one minute, that is, a period short enough for it to be a reasonable assumption that more than one call will not come in. The hour can then be regarded as being made up of 60 independent periods (or 'trials') in each of which a call will either be made or not made.

   If the probability that a call will be made in a minute is p, the appropriate binomial model for the situation is $B(60, p)$. Now the mean number of calls for 60 'trials' is $60p$. But the mean is given to be 2. Hence $60p = 2$ and $p = 1/30$.

   We take as random variable, X, the number of calls received during the hour. As n is large and p is small we can use the Poisson approximation with $a = 2$, giving

$$P(X = 0) = e^{-2} = 0.135 \quad \text{to 3 dec. places.}$$

$$P(X = 1) = 2e^{-2} = 0.271$$

$$P(X = 2) = \frac{2^2}{2}e^{-2} = 0.271 \quad \text{and so on.}$$

   In arguing in this way it is necessary to imagine sub-divisions of the hour small enough for the likelihood of there being more than one call during each interval to be negligible. In this example, if there were any doubt, we could divide the hour up into, say, 240 periods of 15 seconds, in each of which it would be assumed that a call is either made or not made. The binomial model would now be $B(240, p)$ where the mean $240p = 2$, so that $p = 1/120$. As before, since n is large and p is small, the use of the Poisson approximation with $a = 2$ is justified.

(b) In this part of the problem we are concerned with a half-hour period, so that the mean number of calls is 1. Applying the Poisson approximation with $a = 1$,

$$P(X > 3) = 1 - P(X = 0 \text{ or } 1 \text{ or } 2 \text{ or } 3)$$

$$= 1 - e^{-1}(1 + 1 + 1^2/2 + 1^3/6) = 0.019 \quad \text{to 3 dec. places.}$$

   It is important to notice that the argument used in this example depends on the assumptions that:

1. the events (arrivals of calls) occur *randomly*,

2. they occur *uniformly* in the sense that the expected number (the mean) arriving in a given time is proportional to the time,

3. they are *independent*. (This matter is dealt with in some detail in chapter 22.)

   Now this example concerned events that could be represented along a *time* axis, but the Poisson distribution may also be used when the events can be represented along a *distance* axis.

   We might, for example be interested in a process in which a loom is weaving a length of Harris Tweed, knowing from past experience that, on average, there will be 1.5 flaws in a length of 15 m. The probability that there will be $0, 1, 2, \ldots$ flaws in a given length of 15 m may be calculated by using a Poisson distribution with parameter 1.5. The argument is the same as for the time problem. We imagine the 15 m length to be divided into lengths small enough for the assumption that such a length interval will contain one flaw at most.

In the same way, the Poisson distribution may be used for a random variable which is the number of events occurring in a given *area* or a given *volume*. The next example illustrates this.

9.3   Fitting a Poisson distribution to given data

EXAMPLE 4   The number of bacteria in a number of 0.1 litre samples of liquid are given in the following table:

Number of bacteria per sample	0	1	2	3	4	5 or more
Number of samples	23	14	9	3	1	0

On the assumption that the model is Poisson, estimate the probabilities that a 0.1 litre sample will contain 0, 1, 2, 3, 4 bacteria and deduce the theoretical frequency distribution.

In order to use the Poisson distribution we need to know the mean number of bacteria in a 0.1 litre sample, and we have to make the assumption that the mean estimated from the data we are given may be taken as the population mean.

$$\text{From the table, the mean} = \frac{0 \times 23 + 1 \times 14 + 2 \times 9 + 3 \times 3 + 4 \times 1}{50}$$

$$= 0.9.$$

Using $a = 0.9$ the Poisson probabilities and the theoretical frequencies (obtained by multiplying the probabilities by 50) are:

Number of bacteria in a sample	0	1	2	3	4	5 or more
Probability	0.4066	0.3659	0.1647	0.0494	0.0111	0.0023
Frequency (nearest integer)	20	18	8	2	1	0

(The un-rounded frequencies give a total of 50.)

EXERCISE 9b

1.   In a large factory, accidents requiring the summoning of an ambulance occur with a frequency on average of 1.5 per week. Use the Poisson distribution to calculate the probability that: (a) there will not be an accident in a given week, (b) there will be fewer than three accidents in a given week, (c) there will be no accidents in a given fortnight, (d) there will be fewer than three accidents in a given fortnight.

2.   Customers enter a post office at an average of 30 per hour between 1.00 and 2.00 pm. Calculate the probability that: (a) between 1.30 and 1.35 no customers enter, (b) between 1.40 and 1.45 two customers enter, (c) between 1.05 and 1.15 two customers enter.

3.   In 11 successive years, the number of road accidents involving pedestrians at a pedestrian crossing were:

$$8 \quad 9 \quad 9 \quad 8 \quad 8 \quad 7 \quad 8 \quad 11 \quad 4 \quad 2 \quad 3$$

On the assumption that a Poisson distribution is appropriate, calculate the probability of there being 4 or more accidents in a year.

4.   Breakdowns in an electricity supply system occur on average once in every 50 days. The probability distribution of the number of breakdowns occurring is given by a Poisson distribution. Show that the probability that a period of 200 days will pass without any breakdowns is approximately 0.018. Show also that the probability of one and only one breakdown in a year of 365 days is approximately 0.005.
   Find the probability that two or more breakdowns occur in a month of 30 days.   (L)

5.   Samples of 40 articles at a time are taken periodically from the continuous production of a machine and the number of samples containing $0, 1, 2, \ldots$ defective articles are recorded in the following table:

Defectives per sample	0	1	2	3	4	5	6
Number of samples	30	23	27	14	4	2	0

Find the mean number of defectives per sample. Assuming this is the mean of the population and that the Poisson distribution applies find the probability that:
(a) a sample contains 4 or more defectives,
(b) two successive samples contain between them 4 or more defectives.   (O & C)

6.   The following table gives a record of the number of goals scored by a certain football team in each of 450 matches:

Goals per match	0	1	2	3	4	5	6 or more
Frequency	201	163	65	18	2	1	0

Calculate the expected frequencies for a Poisson distribution having the same mean number of goals per match.   (C)

7.   A rather bizarre application of the Poisson distribution occurred in the 19th century.
   L. von Bortkiewicz examined the records of 10 Prussian army corps over 20 years and obtained the following records showing the number of deaths caused by the kick of a horse.

Number of deaths per army corps in a year	0	1	2	3	4
Frequency	109	65	22	3	1

Using the Poisson distribution as a model, calculate the theoretical distribution.

8.   A shopkeeper's sales of refrigerators are four per week on average. Assuming that the weekly sales fit a Poisson distribution, find to what number he should make up his stock at the beginning of each week so that his chance of running out of machines during the week will be less than 4%.

9.4   The mean and variance of a Poisson distribution

In deriving the Poisson distribution we made use of the fact that the mean of $B(n, p)$

is np $(= a)$, so it is clear that the mean of a Poisson distribution with parameter a must be a. Nevertheless, as we shall see later, the distribution can be derived by another method. We will therefore obtain the mean and variance from scratch, starting with the formal definition of the distribution already given on page 148.

As with the binomial distribution, there are two methods, the easier of which uses the probability generating function.

$$P(X = r) = \frac{a^r}{r!}e^{-a}$$

Hence the p.g.f. $\qquad G(t) = e^{-a}\left(t^0 + \frac{at^1}{1!} + \frac{a^2}{2!}t^2 + \ldots\right)$

$$= e^{-a}e^{at}$$

$$\Rightarrow G'(t) = ae^{-a}e^{at} \quad \text{and} \quad G''(t) = a^2e^{-a}e^{at}.$$

Hence $\quad \mu = E(X) = G'(1) = a.$

$G''(1) = a^2$, and, since $G''(1) = \sigma^2 + \mu^2 - \mu$, $\text{Var}(X) = \sigma^2 = a.$

Thus:

For a Poisson distribution with parameter a

the mean = the variance = a

The fact that the mean and variance of a Poisson distribution are equal may be useful in deciding whether or not a given distribution is likely to be modelled by a Poisson distribution. (In example 4 on page 152, the mean of the given distribution is 0.9 and its variance is 1.02, so that the choice of a Poisson model is not unreasonable, seeing that it is only a sample.)

Note: As we derived the Poisson distribution with parameter a as the limit of $B(n, p)$, whose variance is $npq = np(1-p)$, we would expect the variance of the Poisson distribution to be the limit of $np(1-p)$ as $p \to 0$ and $n \to \infty$, while $np = a$, that is, $a(1-0) = a$.

The mean and variance are obtained without using the probability generating function in the Appendix to this chapter on page 162.

9.5★ The sum of two independent Poisson variables

It sometimes happens that we want to calculate the probabilities for the *sum* of two independent random variables, each with a different Poisson distribution.

Suppose, for example, that a librarian in a public library receives requests for information from 'Who's Who', either by personal application or by telephone. The daily average for personal applications is 2, and for telephoned applications is 5, and we want to know the probability that on any given day he will receive exactly 4 requests.

This problem *could* be dealt with in the following way.

If X is the number of personal applications in a day, and Y is the number of telephoned applications in a day, then, assuming a Poisson distribution:

$$P(X = 0 \text{ and } Y = 4) = (e^{-2})\left(\frac{5^4}{4!} \times e^{-5}\right)$$

$$P(X = 1 \text{ and } Y = 3) = (2 \times e^{-2})\left(\frac{5^3}{3!} \times e^{-5}\right)$$

$$P(X = 2 \text{ and } Y = 2) = \left(\frac{2^2}{2!} \times e^{-2}\right)\left(\frac{5^2}{2!} \times e^{-5}\right)$$

$$P(X = 3 \text{ and } Y = 1) = \left(\frac{2^3}{3!} \times e^{-2}\right)(5 \times e^{-5})$$

$$P(X = 4 \text{ and } Y = 0) = \left(\frac{2^4}{4!} \times e^{-2}\right)(e^{-5}).$$

Adding these gives: $P(X + Y = 4) = 0.091$ to 3 dec. places.

This is laborious and there is a much easier method using the following important result.

If a random variable X has a Poisson distribution with mean a and another random variable Y has a Poisson distribution with mean b, the two distributions being independent, then the distribution of $X + Y$ is Poisson with mean $a + b$.

Proof (using probability generating functions)

The p.g.f. for X is $G_x(t) = e^{a(t-1)}$ (see page 154) and the p.g.f. for Y is $G_y(t) = e^{b(t-1)}$.

It follows that the p.g.f. for $X + Y = G_x(t) \cdot G_y(t)$ (see page 116)

$$= e^{a(t-1)} \times e^{b(t-1)}$$
$$= e^{a(t-1)+b(t-1)}$$
$$= e^{(a+b)(t-1)},$$

which is the p.g.f. for a Poisson distribution with mean $a + b$.

(A proof avoiding the use of the probability generating function is given in the Appendix on page 162.)

We shall now apply this result to our example of a librarian receiving requests for information from a reference book.

If X is the number of personal applications received daily, the distribution of X is Poisson with mean 2. Similarly the distribution of Y, the number of telephoned applications received daily is Poisson with mean 5.

The distribution of the total number of daily applications, $X + Y$, is therefore Poisson with mean $2 + 5 = 7$, so that

$$P(X + Y = 4) = \frac{7^4}{4!}e^{-7} = 0.091 \quad \text{to 3 dec. places.}$$

EXERCISE 9c

1.   A series of fixed interval tests were carried out on a car component. The results were as follows:

Number of failures per interval	0	1	2	3	4	5	6
Frequency	92	142	96	46	18	6	0

Show that the distribution may reasonably be taken as Poisson, by calculating the mean and variance to 1 decimal place.

2.   A wholesaler supplies boxes of fireworks to each of two retailers when asked. The numbers of boxes asked for per week have Poisson distributions with mean 1 and 1.2 respectively. Calculate the probability that the number of boxes requested in a given week by both retailers together is: (a) exactly 3, (b) more than 2.

3.   The number of vehicles that pass a motorway service station from the direction of Birmingham is 15 per minute on average. The average number passing the same point in the opposite direction is 18 per minute. Assuming a Poisson distribution, calculate the probability that in a given period of 10 seconds:
(a)   a total of 2 cars pass,
(b)   more than 4 cars pass.
   What is the probability that no cars pass in a given period of 20 seconds?

4.   Two teams A and B play a game of football. The number of goals scored by A has a Poisson distribution with mean 2 and the number of goals scored by B has independent Poisson distribution with mean 1. Find, to three decimal places, the probabilities that: (a) A wins by 2 goals to 1; (b) B wins by 2 goals to 1; (c) the total number of goals scored is 4.
   Determine the most likely results, and comment on the assumptions made in this model.   (MEI)

9.6   Approximations to Poisson probabilities

As for the binomial distribution, the calculation of some Poisson probabilities can be unwieldy. Consider, for example, the problem of calculating $P(X \leqslant 30)$ for a Poisson distribution with parameter 25.
   Under certain circumstances approximate methods are available (see page 209).

9.7★   Truncated Poisson distributions

A truncated Poisson distribution arises when certain values of the variable cannot exist. The following example illustrates the method.

EXAMPLE 5   A Geiger counter records the number of atomic particles arriving at the counter, *provided that the number is less than 4*. The mean number arriving in a given time is μ. Find the probability that k arrivals are recorded ($k = 0, 1, 2, 3$) during this time.
   If $X =$ the number of arrivals recorded

$$P(X = k \,|\, k < 4) = \frac{P(X = k \text{ and } k < 4)}{P(k < 4)}$$

(using conditional probability)

$$= \frac{e^{-\mu}\mu^k/k!}{e^{-\mu}[(1+\mu+(\mu^2/2!)+(\mu^3/3!)]},$$

$$= \frac{\mu^k/k!}{(1+\mu+(\mu^2/2!)+(\mu^3/3!))}, \quad k = 0, 1, 2, 3.$$

EXERCISE 9d (*miscellaneous*)

1.   Records kept at a lost-property office over a period of 200 days showed the following results:

Number of articles handed in	0	1	2	3	4	5	6 or more
Number of days in which stated number of articles were handed in	48	71	48	22	8	3	0

Show that the mean is 1.4 and, assuming a Poisson distribution with this mean, estimate the probability that fewer than three articles are handed in on a given day.   (C)

2.   Define the Poisson distribution and derive its mean and variance.
   In the first years of the life of a certain type of machine, the number of times a maintenance engineer is required has a Poisson distribution with mean four. Find the probability that more than four calls are necessary.
   The first call is free of charge and subsequent calls cost £20 each. Find the mean cost of maintenance in the first year.   (JMB)

3.   (a) The number of accidents notified in a factory per day over a period of 200 days gave rise to the following table:

Number of accidents	0	1	2	3	4	5
Number of days	127	54	14	3	1	1

Calculate the mean number of accidents per day and, assuming that this situation can be represented by a suitable Poisson distribution, calculate the corresponding frequencies.
   (b) Of items produced by a machine, approximately 3% are defective, and these occur at random. What is the probability that, in a sample of 144 items, there will be at least two which are defective?   (SUJB)

4.   The number of telephone calls received at a switchboard in any time interval of length T minutes has a Poisson distribution with mean $\frac{1}{2}T$. The operator leaves the switchboard unattended for five minutes. Calculate to three decimal places that there are: (a) no calls, (b) four or more calls in her absence.   (JMB)

5.   Show that, for the Poisson distribution in which the probabilities of $0, 1, 2, \ldots$ successes are

$$e^{-m}, \, me^{-m}, \, \frac{m^2 e^{-m}}{2!}, \ldots$$

the mean number of successes is equal to m. State the variance.
   A sales manager receives six telephone calls on average between 9.30 am and 10.30 am on a weekday. Find the probability that:
(a) he will receive two or more calls between 9.30 and 10.30 on a certain weekday;
(b) he will receive exactly two calls between 9.30 and 9.40;

(c) during a normal five-day working week, there will be exactly three days on which he will receive no calls between 9.30 and 9.40.   (SUJB)

6.   During a week of air attacks on a city, the authorities kept a record of the number of bombs that fell in 250 areas of the city. The results were as follows:

Number of bombs	7 or more	6	5	4	3	2	1	0
Number of areas	0	1	3	3	4	12	56	171

Calculate the mean number of hits per area and use the Poisson distribution with this mean to calculate the theoretical chance of an area getting fewer than two bombs.   (C)

7.   State the conditions under which it is permissible to use the Poisson distribution as an approximation to the binomial distribution.
It is known that 0.6% of the components produced by a factory are defective. Each day a random sample of 200 components is inspected. Find the probability that there are no defectives in this daily sample: (a) using the binomial distribution, (b) using the Poisson distribution.
Find, to two decimal places, the probability that there is at least one defective on each of three successive days. Taking the components inspected on the three successive days as a single sample, use the Poisson distribution to calculate, to two decimal places, the probability of three or more defectives in the three days. State briefly why you would expect the second of the last two probabilities to be greater than the first.   (JMB)

8.   A prospector for gold examines exactly 800 pans of material every month. The contents of each pan may be assumed to be independent of the others with the probability of a pan containing gold being 0.005 for each pan. The prospector has a good month when 4 or more of the pans that he examines contains gold. Show that, to three significant figures, the probability that a randomly chosen month is a good month is 0.567.
Determine the probability that:
(a) a randomly chosen period of four months contains more than two good months,
(b) a randomly chosen period of 24 months contains more than two good months.   (C)

9.   Records are kept daily of the number of times a certain fire brigade is called out. For a sample of 200 days the frequency distribution of the number of times per day that the brigade was called out is shown in the following table.

Number of calls	0	1	2	3	4	5	6	7
Number of days	31	53	51	33	20	8	2	2

(a) Prepare a table for the given data showing all the terms in the summations that have to be calculated to obtain the mean and variance of this sample. Show that the mean is 2 and calculate the variance.
(b) It has been suggested that the number of times the fire brigade will be called out in a day has a Poisson distribution for which the probability of the brigade being called out k times in a day is

$$p(k) = e^{-m}m^k/k!, \quad k = 0, 1, 2, \ldots$$

where m is the mean number of calls per day. List the sample frequencies and, taking $m = 2$, the corresponding Poisson probabilities. State the Poisson variance. Give reasons for believing that the Poisson distribution is appropriate.
Use the above Poisson distribution to obtain an estimate of the probability that on two randomly chosen days the fire brigade will be called out four times in all.   (JMB)

10.   Derive the Poisson distribution as the limiting form of the binomial distribution when n becomes very large and p becomes very small in such a way that np remains constant.

The mean number of bacteria per millilitre of a liquid is known to be 3. Ten samples of the liquid, chosen at random and each of volume 1 ml, are examined. Assuming the Poisson distribution is applicable, obtain expressions for the probabilities: (a) that each of the ten samples contains at least one bacterium, (b) that exactly eight of the samples contain at least one bacterium.

If 3 ml of the liquid is examined, show that it is rather improbable that it will contain fewer than 3 or more than 15 bacteria.   (MEI)

11.   Every year a wildlife photographer takes a large number of colour photographs. Every photograph has a probability 1/18 of being judged excellent. Using a Poisson approximation, show that the probability for a collection of 36 randomly selected photographs to include 3 or more photographs judged excellent is approximately 1/3.

When this occurs, the photographer calls the collection a 'star' collection. Using the value 1/3, determine the probability that 5 randomly selected collections include more than 3 'star' collections.   (C)

12.   A random variable X has a Poisson distribution given by

$$P(X = r) = p_r = e^{-\lambda}\lambda^r/r!, \quad r = 0, 1, 2, \ldots.$$

Prove that the mean of X is λ. Give two examples (other than that suggested below) of situations where you would expect a Poisson distribution to occur.

The number of white corpuscles on a slide has a Poisson distribution with mean 3.2. By considering the values of r for which $p_{r+1}/p_r > 1$ find the most likely number of white corpuscles on a slide. Calculate correct to three decimal places the probability of obtaining this number. If two such slides are prepared what is the probability, correct to three decimal places, of obtaining at least two white corpuscles in total on the two slides?   (SUJB)

13.   The number of goals X scored by football team A in home matches against team B has a Poisson distribution with mean 2. Verify that

$$rP(X = r) = 2P(X = r-1), \quad r = 1, 2, \ldots.$$

The number of goals Y scored by team B in away matches against A has a Poisson distribution with mean 1. Write down the relationship connecting $P(Y = r)$ and $P(Y = r-1)$, for $r \geqslant 1$. Assuming that X and Y are independent, find the probabilities, correct to two decimal places, that in a match between A and B, on A's home ground,
(a) neither A nor B scores,
(b) A and B score an equal, non-zero number of goals,
(c) A scores more goals than B.   (JMB)

14.   Given that X and Y are independent and have Poisson distributions with means a and b respectively, show that the sum of these quantities, $X + Y$, has probabilities of values $0, 1, 2$ which are in accordance with those of a Poisson distribution with mean $a+b$.

A shopkeeper has two shops which are supplied from a central store. For a particular product each shop asks the store for a complete box when required. The number of boxes per week requested by the two shops are independent and have Poisson distributions with means 4/5 and 1/5 respectively. Find the probabilities that two or more boxes are requested from the store in a week:
(a) by the first shop, (b) by the second shop, (c) together.

Calculate the lowest level of stock at the central store for which there is a probability greater than 90 % that all demands from the shops in the next week can be met.   (JMB)

15.   The number of flaws per roll of manufactured material has the Poisson distribution with mean 0.5.
(a) Find the probability that two randomly chosen rolls will have no flaw.
(b) Five rolls are chosen at random one after the other. Find the probability that the fifth roll chosen is the *only* one which contains one or more flaws.
(c) Find the largest number of rolls that can be chosen at random for there to be a

probability at least 0.1 that none of the rolls contain a flaw.   (JMB)

16. Independent random variables X and Y follow Poisson distributions with parameters μ_1 and μ_2 respectively.
  Derive the distribution of the random variable $X + Y$.
  Lorry drivers call at random at a roadside transport cafe on a north–south main road; the arrival rates are constant at 6 drivers per hour northbound and 2 drivers per hour southbound over the whole time that the cafe is open. Find the probabilities that, in a half-hour period,
(a) 1 or more northbound and 1 or more southbound drivers call,
(b) a total of at least 2 drivers call.
  The cafe manager has found that southbound drivers spend more than northbound ones, and that the cafe will run profitably in a half-hour period provided that a total of at least 3 drivers or at least 2 southbound drivers call. What is the probability of making a profit in a half-hour period?   (MEI)

17.★ X and Y are random variables from Poisson distributions with parameters μ_x and μ_y respectively. Prove that $Z = X + Y$ is distributed in a Poisson distribution.
  Buns are made in batches of 2500 from dough preparations of 100 kg. Currants are dispersed randomly in large volumes of dough, suitable for several batches of buns. If $10c$ currants and $10s$ sultanas per kilogram of dough are used, find the probability that there are a total of t currants and sultanas in one bun.
  On average there are three times as many currants as sultanas. How many currants (to the nearest whole number) per kilogram of dough must be used in order that there is at least one currant and one sultana per bun with probability 0.99.
  (Take the root of $q^4 - q^3 - q + 0.01 = 0$ to be 0.01.)   (MEI)

18.★ Gnat larvae are distributed at random in pond water so that the number of larvae contained in a random sample of $10\,\text{cm}^3$ of pond water may be regarded as a random variable having a Poisson distribution with mean 0.2. Ten independent random samples, each of $10\,\text{cm}^3$ of pond water, are taken by a zoologist. Determine (correct to three significant figures):
(a) the probability that none of the samples contain larvae,
(b) the probability that one sample contains a single larva and the remainder contain no larvae,
(c) the probability that one sample contains two or more larvae and the remainder contain no larvae,
(d) the expectation of the total number of larvae contained in the ten samples,
(e) the expectation of the number of samples containing no larvae.   (C)

19.★ The number X of eggs laid by an insect of a certain species has a Poisson distribution with mean μ. Independently, for each egg laid, there is a probability p that it will hatch. Let Y denote the number of eggs laid by an insect that will hatch.
(a) Write down an expression in terms of r, k, p and μ for the joint probability that $X = r + k$ and $Y = r$, where r and k are non-negative integers. Hence show that Y has a Poisson distribution.
(b) Given that $\mu = 6$, $p = 0.5$ and that six of the eggs laid by an insect hatched, calculate the probability that the insect actually laid exactly six eggs.
(c) Evaluate $P(Y < 6)$ when $p = 0.02$ and $X = 150$.   (JMB)

20. The independent random variables X and Y have Poisson distributions with parameters θ_1 and θ_2 respectively.
  Write down expressions for $P(X = x)$ for $x = 0, 1, 2, \ldots$ and $P(Y = y)$ for $y = 0, 1, 2, \ldots$ and *hence* derive an expression for $P(X + Y = z)$ for $z = 0, 1, 2, \ldots$.
  Obtain the probability generating function (p.g.f.) of the random variable X; deduce the p.g.f. of Y, and hence obtain the p.g.f. of $X + Y$. *From this p.g.f.*, determine $P(X + Y = z)$ for $z = 0, 1, 2, \ldots$ and find also the mean and variance of $X + Y$. (*Hint:* see page 162.)   (MEI)

21.★ The random variables X and Y are independent and have Poisson distributions with means α and β respectively. Show that Z, where $Z = X + Y$ has a Poisson distribution.

Given that $Z = n$, show that the conditional distribution of X is binomial.

A certain species of plant may be attacked by two distinct types of caterpillar, A and B. The numbers of type A and of type B on a plant are independent and have Poisson distributions with means 2 and 3 respectively.
(a) Calculate the probability that a plant is free of both types of caterpillar.
(b) Given that there is at least one caterpillar on a plant, calculate the conditional probability that all the caterpillars on the plant are of the same type.
(c) Given that there are a total of 10 caterpillars on a plant, calculate the conditional probability that there are more of type A than of type B. (JMB)

22.★ The random variable X is said to have a Poisson distribution with parameter λ if its probability distribution is given by

$$P(x) = \begin{cases} e^{-\lambda} \cdot \lambda^x / x! & \text{for } x = 0, 1, 2, \ldots \quad \lambda > 0 \\ 0 & \text{otherwise.} \end{cases}$$

(a) Derive the mean and variance of X.
(b) The mode m of a discrete distribution satisfies the relationships $P(m-1) \leqslant P(m)$ and $P(m) \geqslant P(m+1)$. Show that for the above Poisson distribution these lead to the relation $\lambda - 1 \leqslant m \leqslant \lambda$. Hence or otherwise, for values of $\lambda \leqslant 2$, sketch four distinct shapes that may be taken by this distribution. (AEB 1981)

23.★ A certain blood disease is caused by the presence of two types of deformed corpuscles, X and Y. The probability that there are r type X corpuscles in 0.001 cm^3 of blood is known to be $\lambda^r e^{-\lambda} / r!$ and the corresponding probability for type Y corpuscles is $\mu^r e^{-\mu} / r!$, where λ and μ are constants and $r = 0, 1, 2, 3, \ldots$.

Suppose that a particular 0.001 cm^3 of blood is known to contain a total of n deformed corpuscles. If the random variable S denotes the number of these n corpuscles which are of type X, show that S has a binomial distribution with parameters n and $\lambda/(\lambda + \mu)$. (C)

24.★ The leaves of a certain type of fruit tree are often infested by a pest, and an agricultural research station is carrying out an investigation into this phenomenon. The pest reveals its presence by eating small holes in the leaves. A large number of leaves having such holes have been brought to the research station, and it has been found that the random variable X number of holes per leaf follows a Poisson distribution with parameter λ, *except* that the *only* values X can take are $1, 2, 3, \ldots$ since cases having $X = 0$ will not be found among the leaves brought in. Show that

$$P(X = x) = \frac{e^{-\lambda} \lambda^x}{x!(1 - e^{-\lambda})} \quad x = 1, 2, 3, \ldots.$$

Suggest another situation which may be expected to be modelled in this way.

Write down the corresponding expression for $P(X = x)$ for a case where *neither* $X = 0$ *nor* $X = 1$ will be found. (MEI)

25.★★ The random variable X has a Poisson distribution (for $\mu > 0$) given by

$$(X = r) = e^{-\mu} \mu^r / r!, \quad r = 0, 1, 2, \ldots.$$

Show that the mean and variance of X are both equal to μ.

Show that $\phi = \mu^{-\frac{1}{2}}(X - \mu)$ has an expected value of zero and variance of 1. By expanding $X^{\frac{1}{2}}$ as a power series in ϕ when μ is very large, show that $X^{\frac{1}{2}}$ has mean and variance approximately given by $\mu^{\frac{1}{2}}$ and $\frac{1}{4}$ respectively. (MEI)

9.8 Appendix

The mean and variance of the Poisson distribution

Writing P_r for $P(X = r)$,

$$\mu = E(X) = \sum_{r=0}^{\infty} P_r \cdot r = \sum_{r=1}^{\infty} P_r \cdot r \quad (r = 0 \text{ gives a zero term})$$

$$= \frac{ae^{-a}}{1!} \cdot 1 + \frac{a^2 e^{-a}}{2!} \cdot 2 + \ldots + \frac{a^r e^{-a}}{r!} \cdot r + \ldots$$

$$= ae^{-a}[1 + a/1! + a^2/2! + \ldots + a^{r-1}/(r-1)! + \ldots]$$

$$= ae^{-a} \cdot e^a$$

$$= a.$$

$$E(X^2) = \sum_{r=1}^{\infty} P_r \cdot r^2 = \sum_{r=1}^{\infty} P_r \cdot r(r-1) + \sum_{r=1}^{\infty} P_r \cdot r$$

$$\text{(since } r^2 = r(r-1) + r)$$

$$= \sum_{r=2}^{\infty} P_r \cdot r(r-1) + \mu \quad \text{(since } r = 1 \text{ gives a zero term).}$$

Now $\displaystyle\sum_{r=2}^{\infty} P_r \cdot r(r-1) = \frac{a^2}{2!} e^{-a} \cdot 2 \cdot 1 + \frac{a^3}{3!} e^{-a} \cdot 3 \cdot 2 + \ldots + \frac{a^r}{r!} e^{-a} \cdot r(r-1) + \ldots$

$$= a^2 e^{-a}[1 + a/1! + a^2/2! + \ldots + a^{r-2}/(r-2)! + \ldots]$$

$$\text{(since } r(r-1)/r! = 1/(r-2)!)$$

$$= a^2 e^{-a} \cdot e^a = a^2.$$

$$\Rightarrow E(X^2) = a^2 + \mu = a^2 + a.$$

Hence $\qquad \sigma^2 = E(X^2) - \mu^2 = a^2 + a - a^2 = a.$

The sum of two independent Poisson variables

The distribution of X_1 is Poisson with parameter a_1, so that the probabilities that $X_1 = 0, 1, 2, \ldots$ are:

$P(X_1 = r)$	0	1	2	3	...	r	...
	e^{-a_1},	$\dfrac{a_1 e^{-a_1}}{1!}$,	$\dfrac{a_1^2 e^{-a_1}}{2!}$,	$\dfrac{a_1^3 e^{-a_1}}{3!}$,	...,	$\dfrac{a_1^r e^{-a_1}}{r!}$,	...

and similarly for X_2 with parameter a_2.

$$P(X_1+X_2=0) = P(X_1=0 \text{ and } X_2=0) = e^{-a_1} \cdot e^{-a_2} = e^{-(a_1+a_2)}$$

$$P(X_1+X_2=1) = P(X_1=0 \text{ and } X_2=1)+P(X_1=1 \text{ and } X_2=0)$$

$$= e^{-a_1} \cdot a_2 \frac{e^{-a_2}}{1!} + e^{-a_2} \cdot a_1 \frac{e^{-a_1}}{1!} = (a_1+a_2) \frac{e^{-(a_1+a_2)}}{1!}.$$

Similarly,

$$P(X_1+X_2=2) = P(X_1=0 \text{ and } X_2=2)+P(X_1=1 \text{ and } X_2=1)$$
$$+ P(X_1=2 \text{ and } X_2=0)$$
$$= e^{-(a_1+a_2)} (a_1^2/2! + a_1 a_2 + a_2^2/2!)$$
$$= (a_1+a_2)^2 \frac{e^{-(a_1+a_2)}}{2!}.$$

In general

$$P(X_1+X_2=r) = P(X_1=0 \text{ and } X_2=r)+P(X_1=1 \text{ and } X_2=r-1)$$
$$+ \ldots + P(X_1=r \text{ and } X_2=0)$$
$$= e^{-(a_1+a_2)}[a_1^r/r! + a_1^{r-1}a_2/(r-1)! + a_1^{r-2}a_2^2/(r-2)!2!$$
$$+ \ldots + a_1^{r-k}a_2^k/(r-k)!k! + \ldots + a_2^r/r!]$$

Now $1/(r-k)!k! = \binom{r}{k}/r!$

so that the series inside the square bracket

$$= \frac{1}{r!}\left[a_1^r + \binom{r}{1}a_1^{r-1}a_2 + \binom{r}{2}a_1^{r-2}a_2^2 + \ldots + \binom{r}{k}a_1^{r-k}a_2^k + \ldots + a_2^r\right]$$
$$= \frac{1}{r!}(a_1+a_2)^r$$

and $P(X_1+X_2=r) = (a_1+a_2)^r \dfrac{e^{-(a_1+a_2)}}{r!}$

which is the Poisson probability for parameter a_1+a_2.

Proof that $P(X = r)$ for $B(n,p) \to a^r e^{-a}/r!$, $(a = np)$, as $n \to \infty$

For $B(n,p)$ $P(X = r) = \binom{n}{r}p^r(1-p)^{n-r}$

$$= \frac{n(n-1)(n-2)\ldots(n-r+1)}{r!}p^r(1-p)^{n-r}$$

Let $np = a$ and replace p^r by $(a/n)^r$.

$$P(X = r) = \frac{n(n-1)(n-2)\ldots(n-r+1)}{r!}\frac{a^r}{n^r}(1-p)^{n-r}$$

$$= \frac{a^r}{r!} \frac{n(n-1)(n-2)\ldots(n-r+1)}{n^r}(1-p)^{n-r} \quad \text{(re-arranging)}$$

$$= \frac{a^r}{r!} 1 \cdot [1-(1/n)][1-(2/n)]\ldots\{1-[(r-1)/n]\} \cdot (1-p)^{n-r}$$

$$= \frac{a^r}{r!} [1-(1/n)][1-(2/n)]\ldots\{1-[(r-1)/n]\} \cdot (1-p)^n(1-p)^{-r}.$$

As $n \to \infty$ and $p \to 0$, each of the terms $[1-(1/n)]$, $[1-(2/n)]\ldots \to 1$ and $(1-p)^{-r} \to 1$, since r is fixed.

However $\quad (1-p)^n = [1-(a/n)]^n \to e^{-a}$ as $n \to \infty$.

Hence $\quad P(X = r) \to \dfrac{a^r}{r!} e^{-a}$.

10 Continuous probability distributions

The probability distributions described so far have all been models for the distribution of discrete random variables – the number of people in a population with brown eyes, the number of articles tested in an inspection process until a faulty one is found, the number of breakdowns in an electricity supply system. In distributions like these, the random variable X can only take integral values, arrived at by counting, and it is possible to associate a probability with each distinct value of X.

By contrast, a continuous random variable, such as a man's height, cannot be stated precisely. In practice, of course, we state that his height is, say, 1.73 m to the nearest centimetre, but it could be given as 1.7286 m if it were possible to measure heights to that degree of accuracy. In fact whatever the degree of accuracy of the measuring instrument, a finer measurement can always be imagined. As a consequence the only probability we can associate with $X = 1.73$ is zero. A way out of this difficulty is to use the concept of *probability density*.

10.1 Probability density functions

Suppose the masses of 1000 men, measured in kilograms to the nearest kilogram are given by the following frequency distribution:

Mass (X)	50–	55–	60–	65–	70–	75–	80–	85–	90–	95–100
f	18	42	150	250	220	155	95	47	13	10

The end points of the classes are 49.5, 54.5,... and the class width is 5; the figure shows part of the histogram from $X = 49.5$ to $X = 74.5$.

In the usual way, since the area of each rectangle represents the frequency, the vertical axis is labelled *frequency density*, the frequency per unit value of X. The height of the rectangle representing $f = 250$ is therefore 50, since the class width is 5.

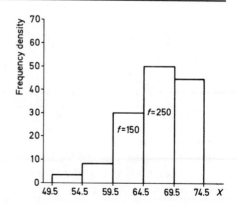

In this figure, the distribution has been converted into a *relative frequency* distribution, showing the proportion in each class (or the *probability* that the mass of a man chosen at random will be in a given class).

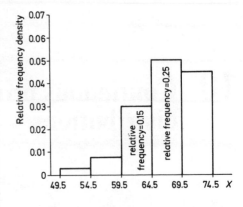

The vertical axis is therefore now labelled *relative frequency density*, which is frequency density divided by 1000.

Suppose now that the data available give the masses to the nearest 0.1 kg and that a new frequency distribution is compiled with the class width narrowed to 1. The part of the resulting distribution from $X = 60$ to $X = 70$ is:

Mass (X)	60–	61–	62–	63–	64–	65–	66–	67–	68–	69–70
f	17	21	28	38	46	48	52	56	51	43

and the relevant part of the histogram is shown in the figure below, with the vertical axis now labelled *probability density* rather than relative frequency density.

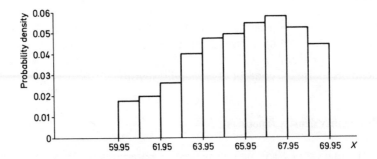

Assuming that the accuracy of measurement permitted it, a further reduction in the class width would lead to a histogram such as the figure below.

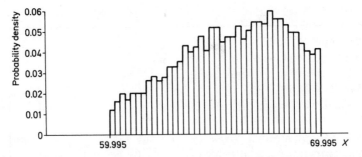

Finally, imagining a progressive reduction in the class width in this way, it seems natural to take as a model for the distribution a *curve*, as in the following figure, chosen in such a way that the *probability that the mass of a man selected at random lies in a given interval* is given by the *area under the curve* over this interval.

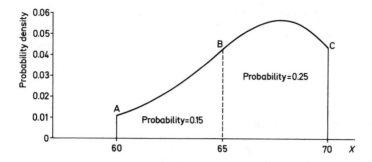

Such a curve is an idealised model, designed to fit the given distribution as closely as possible in the way just described, so that, for example, the area under the curve from A to B is 0.15 and the area from B to C is 0.25. Moreover, if this large sample 1000 were a random one we would expect the curve to be a reasonable model for the distribution of the masses in the population. Obviously a model like this would not be of much value unless the curve were the graph of some known function of *x*. In practice it turns out that well-known functions of *x* are appropriate models for a very wide range of distributions.

Such a function of *x* is called the *probability density function* and integration (where possible) is used to find the area under any part of it.

Definition of a probability density function

Using the idea that probabilities are obtained by integration, we define the probability density function of a random variable *X* formally in the following way:

If *X* is a random variable and there is a function $f(x)$ such that

$$P(a \leqslant X \leqslant b) = \int_a^b f(x)\,dx, \quad (b > a)$$

then $f(x)$ is the probability density function of *X*.

We shall often shorten 'probability density function' to 'density function' or p.d.f., and as usual we write *x* for any particular value that the random variable *X* can take. The range of possible values of *X* is generally taken as $-\infty$ to ∞, with the understanding that $f(x)$ may be zero for values of *x* that are not reached. Here, for example, are two probability density functions defined in this way:

1. $f(x) = \begin{cases} (3/32)(4-x^2), & -2 \leqslant x \leqslant 2 \\ 0 & , \quad \text{otherwise.} \end{cases}$

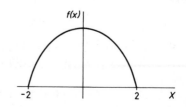

where 'otherwise' is a short way of writing

$$-\infty < x < -2 \quad \text{and} \quad 2 < x < \infty.$$

2. $f(x) = \begin{cases} 1/x^2, & 1 \leqslant x < \infty \\ 0, & x < 1. \end{cases}$

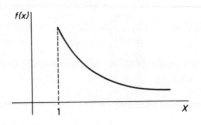

Conditions for a probability density function

Clearly if a function $f(x)$ is to be a possible density function it must be positive or zero for all values of X, and the area under the graph over the whole range of X must be 1. Hence $f(x)$ must satisfy the following conditions:

(a) $f(x) \geqslant 0$ for all values of x

(b) $\displaystyle\int_{-\infty}^{\infty} f(x)\,dx = 1.$

It can be seen that both functions 1 and 2 above satisfy these two conditions. Each is clearly never negative and

$$\int_{-\infty}^{\infty} \frac{3}{32}(4-x^2)\,dx = \int_{-2}^{2} \frac{3}{32}(4-x^2)\,dx = \frac{3}{32}\left[4x-x^3/3\right]_{-2}^{2} = \frac{3}{32}\times\frac{32}{3} = 1$$

$$\int_{-\infty}^{\infty} (1/x^2)\,dx = \int_{1}^{\infty} x^{-2}\,dx = \left[-1/x\right]_{1}^{\infty} = 0-(-1) = 1.$$

It is important to realise that for a continuous random variable X, $P(X=a)=0$, since

$$P(a \leqslant X \leqslant a) = \int_{a}^{a} f(x)\,dx = 0.$$

We can, however, define the probability that X lies within a small interval near $X = x$ as follows:

In the figure, a small rectangle with base of width δx, containing the point $X = x$, is shown. Taking the area of this rectangle as a close approximation to the area under the curve for the interval δx, we say that the probability that X lies within an interval of width δx containing $X = x \approx f(x)\,\delta x$.

In practice, this is often stated in the form

$$P(x \leqslant X \leqslant x+\delta x) \approx f(x)\,\delta x.$$

(We shall need this form when we come to define the mean of a continuous distribution.)

It should also be noted that we can-

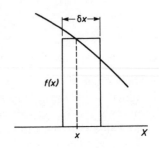

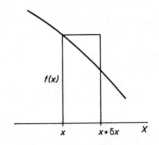

not distinguish between $P(a \leqslant X \leqslant b)$, $P(a \leqslant X < b)$, $P(a < X \leqslant b)$ and $P(a < X < b)$ since $P(X = a)$ and $P(X = b)$ are both zero.

EXAMPLE 1   The continuous random variable X has density function defined as follows:

$$f(x) = \begin{cases} k(9 - x^2), & -3 \leqslant x \leqslant 3 \\ 0 & , \quad \text{otherwise.} \end{cases}$$

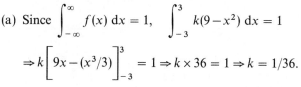

Find: (a) the value of k,
   (b) $P(-2 \leqslant X \leqslant 1)$.

(a) Since $\displaystyle\int_{-\infty}^{\infty} f(x)\, \mathrm{d}x = 1$, $\quad \displaystyle\int_{-3}^{3} k(9 - x^2)\, \mathrm{d}x = 1$

$$\Rightarrow k \left[9x - (x^3/3) \right]_{-3}^{3} = 1 \Rightarrow k \times 36 = 1 \Rightarrow k = 1/36.$$

(b) $P(-2 \leqslant X \leqslant 1) = \displaystyle\int_{-2}^{1} f(x)\, \mathrm{d}x = \dfrac{1}{36} \displaystyle\int_{-2}^{1} (9 - x^2)\, \mathrm{d}x$

$$= \frac{1}{36} \left[9x - (x^3/3) \right]_{-2}^{1} = \frac{24}{36} = 2/3.$$

EXAMPLE 2   Trains are known to arrive at a station every 20 minutes. A man arrives at the station at a random time. Find the density function of X, the time he will have to wait for the arrival of a train.

Find also the probability that he will have to wait less than 9 minutes.

If X is the time he will have to wait, it can have any value from 0 to 20 minutes and the probability that he will have to wait between, say, 4 and 5 minutes is the same as the probability that he will have to wait between 12 and 13 minutes. In other words the probability density (the probability per unit time) is constant.

The appropriate model will therefore be:

$$f(x) = \begin{cases} k, & 0 \leqslant x \leqslant 20 \\ 0, & \text{otherwise.} \end{cases}$$

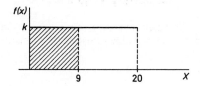

By inspection of the graph $k = 1/20$.

Hence $P(X \leqslant 9) = \displaystyle\int_{0}^{9} f(x)\, \mathrm{d}x = \displaystyle\int_{0}^{9} (1/20)\, \mathrm{d}x = 9/20.$

It can be seen that integration is not really needed here. Indeed it is intuitively obvious that $P(X \leqslant 9) = 9/20$.

This is an example of a *uniform distribution*, which will be discussed later (see page 182).

EXAMPLE 3   For the continuous random variable with density function defined in example 1, find $P(X \leqslant 1)$, *given that* $X \geqslant 0$.

Conditional probability is defined in the same way as for discrete variables, so that

$$P(X \leqslant 1 \mid X \geqslant 0) = \frac{P(0 \leqslant X \leqslant 1)}{P(X \geqslant 0)}$$

$$= \frac{\displaystyle\int_0^1 f(x)\,dx}{\displaystyle\int_0^3 f(x)\,dx} = \frac{26}{54} = \frac{13}{27}.$$

EXERCISE 10a

1. If X is a random variable show that $f(x)$ defined as

$$f(x) = \begin{cases} x/8, & 0 \leqslant x \leqslant 4 \\ 0, & \text{elsewhere} \end{cases}$$

satisfies the conditions for a probability density function.
    Find: (a) $P(X < 2)$,   (b) $P(1 < X < 3)$.

2. The p.d.f. for X, the time at which an event happens, is given by

$$f(x) = \begin{cases} kx^3, & 0 \leqslant x \leqslant 4 \\ 0, & \text{otherwise.} \end{cases}$$

Find k and the probability that the event happens before $x = 3$.

3. If the probability density function for the random variable X is

$$f(x) = \begin{cases} kx^{-2}, & x \geqslant 1 \\ 0, & x < 1 \end{cases}$$

find k and $P(X < 3)$.

4. The p.d.f. of the random variable Z is

$$f(z) = \begin{cases} kz(100 - z), & 0 \leqslant z \leqslant 100 \\ 0, & \text{otherwise.} \end{cases}$$

Find the probability that $Z \leqslant 25$.

5. The length of time T (hours) that an insect survives under certain laboratory conditions is a random variable whose density function is

$$f(t) = \begin{cases} kt^3(10 - t), & 0 \leqslant t \leqslant 10 \\ 0, & \text{otherwise.} \end{cases}$$

Calculate: (a) the value of k, (b) the probability that an insect will live for more than 8 hours, (c) the probability that it will live for between 6 and 7 hours.

6. X is the amount of a catalyst needed for a chemical reaction to be started. Its p.d.f. is

$$f(x) = \begin{cases} k(x^3 - 3x + 4), & 0 \leqslant x \leqslant 2 \\ 0, & \text{otherwise.} \end{cases}$$

Find k and the probability that the reaction will take place when X is between 1 and 2.

7.   If the probability density function of the random variable X is

$$f(x) = \begin{cases} k \sin \pi x, & 0 \leqslant x \leqslant 1 \\ 0 & , \quad \text{otherwise} \end{cases}$$

find the value of k and $P(X > 1/3)$.

8.   I arrive at a London underground platform just as a train draws out, and have to wait x minutes for the next train. You are given that x has probability density which increases linearly from zero in the interval $0 \leqslant x \leqslant 10$, then decreases linearly to zero over the interval $10 \leqslant x \leqslant 20$, and is zero for $x = 20$. Sketch carefully the graph of the probability density function.

Calculate the probability that I have to wait more than eight minutes for the next train.   (SMP)

9.   A girl hits a tennis ball against a wall, aiming at a fixed vertical line marked on the wall. The horizontal distance, X metres, from the vertical line of the point where the ball strikes the wall is a random variable whose density function is

$$f(x) = \begin{cases} k(1-x^2), & -1 \leqslant x \leqslant 1 \\ 0 & , \quad \text{otherwise.} \end{cases}$$

Find: (a) $P(-\tfrac{1}{2} < X < \tfrac{1}{2})$,   (b) $P(X < \tfrac{3}{4} \mid X > \tfrac{1}{2})$.

10.   The weekly demand for petrol, in thousands of gallons, at a garage is X, a random variable whose density function is

$$f(x) = \begin{cases} kx(a-x), & 0 \leqslant x \leqslant 2 \\ 0 & , \quad \text{otherwise.} \end{cases}$$

Given that there is a probability of 7/20 that the demand in a given week will be less than 1000 gallons, find the constants k and a.

11.★   A probability density function has the form

$$f(x) = \begin{cases} a+bx(x-3), & 0 \leqslant x \leqslant 3 \\ 0 & , \quad \text{otherwise} \end{cases}$$

where a and b are constants. The conditional probability $P(1 \leqslant X \leqslant 2 \mid X \geqslant 1) = 0.2$. Find the constants a and b.   (MEI)

10.2   The mean and variance of a continuous distribution

Mean

The mean of the distribution of a discrete random variable X

X	x_0	x_1	x_2	$\cdots$	x_n
Probability	p_0	p_1	p_2	$\cdots$	p_n

is $E(X) = \sum\limits_{i=0}^{n} p_i x_i$ or $\sum\limits_{\text{all } x} px.$

Finding an equivalent definition for a continuous variable involves calculus, and you may find it useful to refresh your memory about the usual abbreviated argument in calculus for finding the area between a curve $y = f(x)$

and part of the *x*-axis. We shall then use a similar approach to find the mean of *X* whose density function is $f(x)$.

In finding the area under $y = f(x)$ the range concerned is divided into equal small intervals δx, giving an element between *x* and $x + \delta x$ whose area is approximately $f(x)\,\delta x$.

The short argument then goes like this.

The sum of all such elements of area

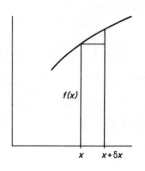

$$\approx \sum_{\text{all } x} f(x)\,\delta x.$$

As the number of intervals $\rightarrow \infty$, so that $\delta x \rightarrow 0$, this sum $\rightarrow \int f(x)\,dx$.

Hence the area $= \displaystyle\int_{a}^{b} f(x)\,dx$, where the range is $a \leqslant x \leqslant b$.

A similar process, proceeding from summation Σ to integration $\int$, is involved in defining the mean of a continuous distribution.

We imagine the range of values of *X*, the random variable, to be divided into small intervals of width δx, and consider one such interval containing $X = x$.

As we saw on page 168, the probability that *X* lies within this interval $\approx f(x)\,\delta x$. In other words, the probability associated with $(X = x) \approx f(x)\,\delta x$, and we can say that

$$E(X) \approx \sum_{\text{all } x} [f(x)\,\delta x]x \quad \text{or} \quad \sum_{\text{all } x} xf(x)\,\delta x \quad \text{(rearranging the order).}$$

As $\delta x \rightarrow 0$, this sum $\rightarrow \int xf(x)\,dx$, integrating over the whole range of *X*, and since this range is taken to be $-\infty$ to ∞, we have arrived at the definition we needed:

The mean of the distribution of a continuous random variable *X*, whose probability density function is $f(x)$, $-\infty < x < \infty$, is

$$\mu = E(X) = \int_{-\infty}^{\infty} xf(x)\,dx.$$

Comparing this with the formula $E(X) = \Sigma px$ for a discrete variable, it will be seen that we have exchanged Σ for $\int$ and replaced *p* by '$f(x)\,dx$'. If you bear this in mind, you will find the definitions that follow easier to grasp.

(The preliminary discussion about finding an area was chosen for simplicity. There is, in fact, an exact analogy between finding the mean and finding the *moment* of an area about the *y*-axis, used in finding centre of gravity of the area.)

EXAMPLE 4   Find the mean of the continuous random variable whose density function is

$$f(x) = \begin{cases} \frac{1}{2}x, & 0 \leqslant x \leqslant 2 \\ 0, & \text{otherwise.} \end{cases}$$

$$\mu = E(X) = \int_0^2 x \cdot \tfrac{1}{2}x \, dx = \left[x^3/6 \right]_0^2 = 4/3.$$

Variance

For a discrete random variable X, the variance is defined as

$$\sigma^2 = \text{Var}(X) = E(X - \mu)^2 = \sum_{i=0}^{n} p_i(x_i - \mu)^2$$

with the alternative form

$$\sum_{i=0}^{n} p_i x_i^2 - \mu^2 = E(X^2) - \mu^2.$$

The analogous definition for a continuous random variable is found, as with the mean, by exchanging Σ for $\int$ and replacing p by '$f(x) \, dx$'. Thus:

The variance of the distribution of the random variable X, whose p.d.f. is $f(x)$, $-\infty < x < \infty$, is

$$\sigma^2 = \text{Var}(X) = E(X - \mu)^2 = \int_{-\infty}^{\infty} (x - \mu)^2 f(x) \, dx$$

$$= \int_{-\infty}^{\infty} x^2 f(x) \, dx - \mu^2 = E(X^2) - \mu^2.$$

This definition may be justified by an argument similar to the one we used for the mean. The equivalence of the two forms for $\text{Var}(X)$ is proved formally in the Appendix on page 194.

EXAMPLE 5   Find the mean and variance of the continuous random variable whose density function is

$$f(x) = \begin{cases} 4x(1 - x^2), & 0 \leqslant x \leqslant 1 \\ 0, & \text{otherwise.} \end{cases}$$

$$\mu = E(X) = \int_0^1 xf(x) \, dx = \int_0^1 x(4x - 4x^3) \, dx = \left[(4x^3/3) - (4x^5/5) \right]_0^1 = 8/15.$$

$$E(X^2) = \int_0^1 x^2 f(x) \, dx = \int_0^1 x^2(4x - 4x^3) \, dx = \left[x^4 - (2x^6/3) \right]_0^1 = 1/3.$$

Hence $\sigma^2 = E(X^2) - \mu^2 = 1/3 - (8/15)^2 = 11/225.$

10.3   The expectation of a function of a random variable

The formula for variance is a special case of a general result. Just as we defined the expectation of the function $g(X)$, where X is a discrete random variable, as

$$E[g(x)] = \sum_{i=0}^{n} p_i g(x_i) \quad \text{(page 91)},$$

so, for a continuous random variable X with density function $f(x)$, $-\infty < x < \infty$,

$$E[g(x)] = \int_{-\infty}^{\infty} g(x)f(x)\, dx.$$

EXAMPLE 6★ Find the mean and the mean deviation from the mean of X, a random variable whose probability density function is

$$f(x) = \begin{cases} \frac{1}{4}x^3, & 0 \leqslant x \leqslant 2 \\ 0, & \text{otherwise.} \end{cases}$$

$$\mu = E(X) = \int_0^2 xf(x)\, dx = \int_0^2 \tfrac{1}{4}x^4\, dx = \left[x^5/20 \right]_0^2 = 1.6.$$

Now the mean deviation from the mean $= E[|x - \mu|]$,   (see page 26)

$$= \int_0^2 |x - \mu| f(x)\, dx$$

and since, when $x < \mu$, $|x - \mu| = \mu - x$ and when $x > \mu$, $|x - \mu| = x - \mu$, we have to split the integral into two parts:

$$\int_0^\mu (\mu - x)f(x)\, dx + \int_\mu^2 (x - \mu)f(x)\, dx$$

$$= \int_0^{1.6} (1.6 - x)\tfrac{1}{4}x^3\, dx + \int_{1.6}^2 (x - 1.6)\tfrac{1}{4}x^3\, dx,$$

which, after integration, gives the value 0.26 to 2 dec. places.

(When the mean deviation was first defined we pointed out that it is awkward to handle. This example is a good illustration.)

10.4   The mode and median of a continuous distribution

The definitions of mode and median for a continuous distribution follow naturally from the equivalent definitions for the discrete case:

A *mode* is a value of X where its density function has a maximum.

The *median* is the value m such that $\displaystyle\int_{-\infty}^{m} f(x)\, dx = \tfrac{1}{2}.$

Quartiles, deciles and percentiles are similarly defined.

EXAMPLE 7 Find the mode, the median and the 30th percentile for the distribution of the random variable X whose density function is

$$f(x) = \begin{cases} 4x(1 - x^2), & 0 \leqslant x \leqslant 1, \\ 0, & \text{otherwise.} \end{cases}$$

Since $f(x)$ is a cubic curve meeting the x-axis at $x = 0$ and $x = 1$, the only possible maximum will be between these two points and is found by differentiation.

$$\frac{df(x)}{dx} = 4 - 12x^2,$$

which is zero when $x^2 = 1/3$. The mode is therefore $1/\sqrt{3}$.

If m is the median, $\displaystyle\int_0^m (4x - 4x^3)\, dx = \tfrac{1}{2} \Rightarrow \left[2x^2 - x^4\right]_0^m = \tfrac{1}{2}$

$$\Rightarrow 2m^2 - m^4 = \tfrac{1}{2} \Rightarrow 2m^4 - 4m^2 + 1 = 0,$$

an equation whose only solution in the range $0 \leqslant x \leqslant 1$ is 0.54.

Similarly if the 30th percentile is a,

$$\int_0^a (4x - 4x^3)\, dx = 0.3 \Rightarrow 2a^2 - a^4 = 0.3 \Rightarrow a^4 - 2a^2 + 0.3 = 0$$

giving $a = 0.40$.

Hence the median is 0.54 and the 30th percentile is 0.40 to 2 dec. places.

In the next example the probability density function is not a smooth curve and a mixture of calculus and geometry is needed.

EXAMPLE 8 Find the mean, median, 10th and 90th percentiles for the distribution of the random variable X whose density function is

$$f(x) = \begin{cases} k & , \ 0 \leqslant x \leqslant 2 \\ k(3 - x), & 2 < x \leqslant 3 \\ 0 & , \ \text{otherwise.} \end{cases}$$

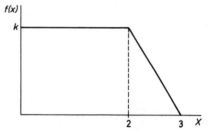

From the sketch the total area $= 2k + \tfrac{1}{2}k$ so that $k = 0.4$.

$$\mu = \int_0^2 x \cdot k\, dx + \int_2^3 xk(3 - x)\, dx$$

which, after integration gives $\mu = 19/15$.

The median divides the total area into two halves. Since the area of the rectangle is 0.8 and of the triangle is 0.2, the median will be m, where $m \times 0.4 = 0.5$

$$\Rightarrow m = 1.25.$$

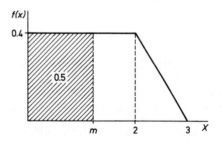

The 10th percentile will be b where $b \times 0.4 = 0.1$

$$\Rightarrow b = 0.25.$$

The 90th percentile will be c, where the area of the small triangle is 0.1.

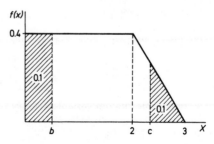

By similar triangles, $\left(\dfrac{3-c}{1}\right)^2 = \dfrac{0.1}{0.2}$.

Hence $c = 3 - 1/\sqrt{2}$.

EXERCISE 10b

1. A random variable X has p.d.f.

$$f(x) = \begin{cases} \frac{3}{4}x(2-x), & 0 \leqslant x \leqslant 2 \\ 0 & , \quad \text{otherwise.} \end{cases}$$

Find the mean, variance, mode and median.

2. A random variable X has density function

$$f(x) = \begin{cases} kx(x-1)^2, & 0 \leqslant x \leqslant 1 \\ 0 & , \quad \text{otherwise.} \end{cases}$$

Find k, the mean, the variance and the mode.

3. The random variable, X, has probability density function kx^3 for $0 \leqslant x \leqslant 2$ and zero outside this range. Find the value of k, and the mean, median and variance of the distribution.
    Find the probability that an observation lies within one standard deviation of the mean.

(O)

4. If the random variable X has p.d.f.

$$f(x) = \begin{cases} k/x^2, & 2 \leqslant x \leqslant 3 \\ 0 & , \quad \text{otherwise,} \end{cases}$$

find k and the median.

5. The random variable X has probability density function

$$f(x) = \begin{cases} kx^2(3-x), & 0 \leqslant x \leqslant 3 \\ 0 & , \quad \text{elsewhere.} \end{cases}$$

Find k and the mean μ. Find also $P(X \leqslant \mu)$.

6. The continuous random variable X has probability density function f given by

$$f(x) = \begin{cases} kx^2(1-x), & \text{for } 0 \leqslant x \leqslant 1, \\ 0 & , \quad \text{otherwise,} \end{cases}$$

where k is a constant.
(a) Determine the value of k, and sketch the graph of the density function.
(b) Find $E(X)$ and $\text{Var}(X)$.
(c) Find $P[X < E(X)]$.

(C)

7. Find the mean and variance of the uniform distribution whose p.d.f. is

$$f(x) = \begin{cases} k, & a \leqslant x \leqslant b \\ 0, & \text{elsewhere.} \end{cases}$$

8. A continuous random variable X takes values in the interval $1 \leqslant x \leqslant 3$. The probability density function of X is given by $f(x) = k/x^2$.
(a) Determine the value of k.

(b) Find $E(X)$ and Var(X).
(c) Find $P(X \geq 5/3)$.
(d) Calculate the lower and upper quartile values of X. (C)

9.   Two species of insect, type A and type B, are distinguished by measuring a particular characteristic X. The probability density function of X for insects of type A is

$$f(x) = \begin{cases} C_1(1-x)^2, & 0 \leq x \leq 1 \\ 0 & , \quad \text{elsewhere} \end{cases}$$

and the probability density function of X for insects of type B is

$$f(x) = \begin{cases} C_2 x^2(1-x), & 0 \leq x \leq 1 \\ 0 & , \quad \text{elsewhere.} \end{cases}$$

Find the values of the constants C_1 and C_2 and calculate the mean and the standard deviation of each distribution.

It is decided that, when X has been measured for a particular insect, then if $X \leq \frac{1}{2}$ the insect will be classified as type A and if $X > \frac{1}{2}$ the insect will be classified as type B. Find:
(a) the probability that an insect of type A is wrongly classified as type B,
(b) the probability that an insect of type B is wrongly classified as type A. (MEI)

10.   The continuous random variable X has a distribution whose probability density function is given by

$$f(x) = \begin{cases} kx(4-x), & 2 \leq x \leq 4, \\ 0 & , \quad \text{otherwise,} \end{cases}$$

where k is a constant.
(a) Determine the value of k.
(b) Sketch the graph of $f(x)$.
(c) State the mode of the distribution.
(d) Show that the distribution has mean equal to $11/4$ and verify that $P(X < 11/4) > \frac{1}{2}$.
(e) Use the results of (c) and (d) to deduce that the median of this distribution lies between the mode and the mean. (C)

11.   The probability density function for values x of a continuous variate X is given by $f(x) = k(n-x)$ for $0 \leq x \leq n$ and $f(x) = 0$ elsewhere. k is a constant. Find k in terms of n. Find also the mean and variance of X.

If b is such that $P(X > b) = 0.05$, show that $[1-(b/n)]^2 = 0.05$ and hence find b when $n = 10$. (O & C)

12.   The lifetime X in *tens of hours* of a torch battery is a random variable with probability density function

$$f(x) = \begin{cases} \frac{3}{4}\{1-(x-2)^2\}, & 1 \leq x \leq 3, \\ 0 & , \quad \text{otherwise.} \end{cases}$$

Calculate the mean of X.
A torch runs on two batteries, both of which have to be working for the torch to function. If two new batteries are put in the torch, what is the probability that the torch will function for at least 22 hours, on the assumption that the lifetimes of the batteries are independent? (O & C)

13.   A continuous random variable X has the probability density function defined by

$$f(x) = \begin{cases} cx/3, & 0 \leq x < 3, \\ c & , \quad 3 \leq x \leq 4, \\ 0 & , \quad \text{otherwise,} \end{cases}$$

where c is a positive constant.

Find: (a) the value of c, (b) the mean of X, (c) the value, a, for there to be a probability of 0.85 that a randomly observed value of X will exceed a.   (JMB)

14.   A continuous random variable X has probability density function $f(x)$ defined by

$$f(x) = \begin{cases} 12(x^2 - x^3), & 0 \leqslant x \leqslant 1, \\ 0 & , \quad \text{otherwise.} \end{cases}$$

Find the mean and standard deviation of X; find also its mean deviation about the mean.   (O & C)

15.   A continuous variable X is distributed at random between two values, $x = 0$ and $\dot{x} = 2$, and has a probability density function of $ax^2 + bx$. The mean is 1.25.
(a) Show that $b = \frac{3}{4}$, and find the value of a.
(b) Find the variance of X.
(c) Verify that the median value of X is approximately 1.3.
(d) Find the mode.   (SUJB)

10.5   The cumulative distribution function

One important difference between discrete and continuous probability distribution lies in the way probability itself can be handled.

If, for example, X is the number of weekly accidents in a factory needing medical treatment and has a Poisson distribution (discrete) with mean 1.5, then we can say that

$$P(X = 2) = \frac{(1.5)^2}{2!} e^{-1.5} = 0.25.$$

On the other hand, if X is the life in hours of a torch battery with a continuous distribution whose density function is known, $P(X = 2) = 0$, $P(X = 3.5) = 0 \ldots$, as we pointed out on page 168, and the only kind of probabilities we can calculate are

$$P(X \leqslant 2), \qquad P(2 \leqslant X \leqslant 3.5) \quad \text{and so on.}$$

For this reason it is useful, in discussing continuous distributions to give a special name to $P(X \leqslant x)$ and call it the *cumulative distribution function* (often shortened to 'distribution function').

If a continuous random variable X has probability density function $f(x)$ for $-\infty < x < \infty$, the cumulative distribution function is

$$F(x) = P(X \leqslant x) = \int_{-\infty}^{x} f(t) \, dt$$

which is the area under the graph of $f(x)$ to the left of $X = x$.

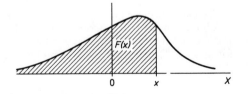

(The variable t is a 'dummy' variable, which does not affect the value of this definite integral.)

We can discover the special features of a cumulative distribution function by taking a simple example.

Suppose X has density function

$$f(x) = \begin{cases} \frac{1}{2}x, & 0 \leqslant x \leqslant 2 \\ 0, & \text{otherwise.} \end{cases}$$

If $0 \leqslant x \leqslant 2$, $F(x) = \int_0^x \frac{1}{2}t\, dt = \frac{1}{4}x^2$,

and $F(2) = 1$ since the total area is 1.

If $x > 2$, $F(x) = \int_0^2 \frac{1}{2}t\, dt$

$$+ \int_2^x 0 \cdot dt = 1.$$

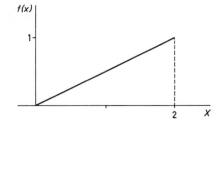

If $x < 0$, $F(x) = 0$.

The resulting graph of $F(x)$ is typical in many ways. To the left of the lower end of the range for which X is not zero, $F(x)$ is always 0. At the upper end of the range for which X is not zero, $F(x)$ is always 1 and remains 1 thereafter. Between these points $F(x)$ increases steadily.

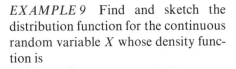

EXAMPLE 9 Find and sketch the distribution function for the continuous random variable X whose density function is

$$f(x) = \begin{cases} (1/36)(9-x^2), & -3 \leqslant x \leqslant 3 \\ 0, & \text{otherwise.} \end{cases}$$

If $-3 \leqslant x \leqslant 3$,

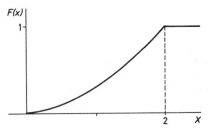

$$F(x) = \int_{-3}^x \frac{1}{36}(9-t^2)\, dt = \frac{1}{36}\left[9t - (t^3/3) \right]_{-3}^x$$

$$= \frac{1}{36}[9x - (x^3/3) + 18] = \frac{1}{2} + x/4 - x^3/108.$$

If $x > 3$, $F(x) = 1$
If $x < -3$, $F(x) = 0$.

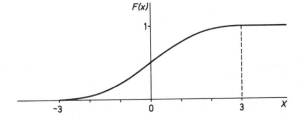

$F(x)$ is sketched by plotting a few points and noting that the gradient at $x = -3$ and $x = 3$ is zero.

The relation between $f(x)$ and $F(x)$

Since

$$F(x) = \int_{-\infty}^{x} f(t)\, dt,$$

it follows that $f(x)$ is the derivative of $F(x)$:

$$\frac{dF(x)}{dx} = f(x).$$

This is justified formally as being an example of what is known as the fundamental theorem of calculus, which states that if

$$F(x) = \int_{c}^{x} f(t)\, dt,$$

where the lower limit of integration is any constant, then $F'(x) = f(x)$.

[If you are unfamiliar with this theorem, which only states formally that integration is the inverse operation of differentiation, you may be convinced by taking a simple example like this:

If $f(x) = x^3$, then $F(x) = \int_{c}^{x} t^3\, dt = \frac{1}{4}x^4 - \frac{1}{4}c^4$, and $F'(x) = f(x)$.]

The importance of this relation lies in the fact that when we are trying to establish the nature of a distribution it is sometimes easier to find the distribution function $F(x)$ first, and deduce the density function $f(x)$ by differentiation (see chapter 22).

EXAMPLE 10   The distribution function of a continuous random variable X is

$$F(x) = \begin{cases} x^3/2 + \frac{1}{2}, & -1 \leqslant x \leqslant 1 \\ 0 & , & x < -1 \\ 1 & , & x > 1. \end{cases}$$

Find and sketch the density function $f(x)$.

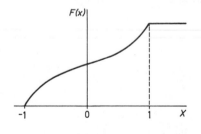

$$f(x) = dF(x)/dx = \begin{cases} 3x^2/2, & -1 \leqslant x \leqslant 1 \\ 0 & , & x < -1 \\ 0 & , & x > 1. \end{cases}$$

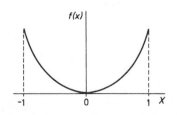

EXERCISE 10c

1.   Find and sketch the distribution function for the following uniform distributions of a random variable X:
(a) $f(x) = 1/20, 0 \leqslant x \leqslant 20; f(x) = 0$ otherwise.
(b) $f(x) = \frac{1}{4}, -3 \leqslant x \leqslant 1; f(x) = 0$ otherwise.

2.   Find and sketch the distribution function for the random variable X whose density function is

$$f(x) = \begin{cases} kx(2-x), & 0 \leqslant x \leqslant 2 \\ 0 & , \quad \text{otherwise.} \end{cases}$$

3.   In each of the following examples, where $f(x)$ is the density function of a random variable X, find k and the distribution function $F(x)$. Sketch $f(x)$ and $F(x)$ in each case.

(a) $f(x) = \begin{cases} kx^{1/2}, & 1 \leqslant x \leqslant 9 \\ 0 & , \quad \text{elsewhere.} \end{cases}$ (b) $f(x) = \begin{cases} k \cos x, & -\pi/2 \leqslant x \leqslant \pi/2 \\ 0 & , \quad \text{elsewhere.} \end{cases}$

(c) $f(x) = \begin{cases} ke^{-2x}, & x > 0 \\ 0 & , \quad \text{elsewhere.} \end{cases}$

4.   Find the distribution function for the random variable X whose density function is

$$f(x) = \begin{cases} kx & , \quad 0 \leqslant x \leqslant 1 \\ k(2-x), & 1 < x \leqslant 2 \\ 0 & , \quad \text{otherwise.} \end{cases}$$

Sketch both functions.

5.   The probability density function of a random variable X is

$$f(x) = \begin{cases} \frac{3}{8}(x^2+1), & -1 \leqslant x \leqslant 1 \\ 0 & , \quad \text{otherwise.} \end{cases}$$

Find and sketch its cumulative distribution function.

6.★ Show that the following function satisfies the conditions for a probability density function:

$$f(x) = \begin{cases} 0 & , \quad x \leqslant 0 \\ 3x(1-x), & 0 < x < 1 \\ 1/2x^2 & , \quad 1 \leqslant x < \infty. \end{cases}$$

Determine the cumulative distribution function and the mode of the distribution, and find $P(X \geqslant 4)$.   (MEI)

7.   In each of the following examples, where the distribution function $F(x)$ is given, find and sketch the density function $f(x)$.

(a) $F(x) = \begin{cases} x/4, & 0 \leqslant x \leqslant 4 \\ 0 & , \quad x < 0 \\ 1 & , \quad x > 4 \end{cases}$ (b) $F(x) = \begin{cases} e^{2x}, & -\infty < x \leqslant 0 \\ 1 & , \quad x > 0 \end{cases}$

(c) $F(x) = \begin{cases} x^3/8, & 0 \leqslant x \leqslant 2 \\ 0 & , \quad x < 0 \\ 1 & , \quad x > 2 \end{cases}$ (d) $F(x) = \begin{cases} (2/\pi) \sin^{-1}(x^{1/2}), & 0 \leqslant x \leqslant 1 \\ 0 & , \quad x < 0 \\ 1 & , \quad x > 1. \end{cases}$

10.6   Some particular continuous distributions

Up to now we have been discussing continuous distributions in general. We shall now describe some particular distributions.

The most important continuous distribution, the *normal* distribution is so fundamental that we shall postpone it to the next chapter so that we can discuss it in depth. Next in importance for the number of situations for which it is the appropriate model is the *exponential* distribution. We shall start, however, with two simpler distributions.

The uniform distribution

In example 2 on page 169 we mentioned briefly an instance of a uniform distribution. A man arrives at a station to catch a train. All he knows is that a train stops every 20 minutes, but he does not know when. If X is the time he will have to wait until a train arrives, $0 \leqslant X \leqslant 20$. Remembering that for a continuous distribution we cannot usefully talk of the probability of X being exactly 3.5 minutes (since this is zero), we might fix our attention on, say, $P(2 \leqslant X \leqslant 6)$ and $P(5 \leqslant X \leqslant 9)$. It seems reasonable to say that these probabilities are equal since the interval length is 4 minutes in both cases. We therefore take the probability density (probability per unit time) to be constant.

It is the distinguishing characteristic of a uniform distribution that $P(a \leqslant X \leqslant b)$ is the same for *any* values a and b ($a < b$) within the total range which are the same distance apart, and does not depend on where the interval (a to b) is.

Further instances where the uniform distribution is the appropriate model follow.

1.   You will often come across phrases like 'A point P is chosen at random on a line AB'. This is to be understood as meaning that P is equally likely to be anywhere on AB, so that if $AP = X$, the distribution of X is uniform.

2.   Suppose the masses of the 1000 men on page 165 were measured in kilograms to one decimal place (54.6, 53.3,...) but by some mischance the recording apparatus recorded only the decimal part of the mass (0.6, 0.3,...), then it is likely that these values will be evenly distributed between 0 and 1, forming a uniform distribution. Although this is a somewhat unlikely situation, it will serve as an introduction to our third example, which is important and which we will examine in detail.

3.   Straight lines drawn at random are measured and recorded in centimetres to the nearest millimetre, so that 10.4 cm could be anything between 10.35 cm and 10.45 cm. The *error* (actual length − recorded length) lies between −0.05 cm and 0.05 cm, and since the lines are drawn at random, no one error should be expected to be more probable than another.

If X is the random error, then, the density function is

$$f(x) = \begin{cases} k, & -0.05 \leqslant x \leqslant 0.05 \\ 0, & \text{otherwise.} \end{cases}$$

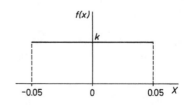

By inspection $k = 10$ and the mean $= 0$, so that $\mathrm{Var}(X) = E(X^2)$.

$$E(X^2) = \int_{-0.05}^{0.05} x^2 \cdot 10 \, dx = \left[10x^3/3 \right]_{-0.05}^{0.05}. \text{ Hence } \mathrm{Var}(X) = 8.3 \times 10^{-4}.$$

We will now find the cumulative distribution function.
If $-0.05 \leqslant x \leqslant 0.05$,

$$F(x) = \int_{-0.05}^{x} 10 \, dt = 10x + 0.5.$$

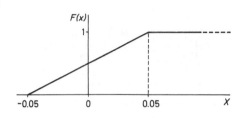

If $\quad x < -0.05, \quad F(x) = 0.$

If $\quad x > 0.05, \quad F(x) = 1.$

A triangular distribution

It can be shown that if X and Y are two independent continuous random variables, each of which is uniformly distributed between 0 and 1, then the distribution of the sum, $X + Y$, has a triangular distribution between 0 and 2, as in the sketch. (For a proof see page 412.)

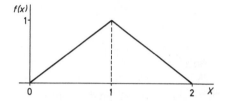

Similarly if X and Y are the errors in measuring pairs of lines in centimetres to the nearest millimetre as in the last example, so that each is uniformly distributed from -0.05 to 0.05, the distribution of the sum of these errors has the triangular shape shown on the right.

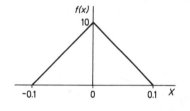

Investigations of these distributions are left as examples in exercise 10d questions 5 and 6.

EXERCISE 10d

1. The continuous random variable X is uniformly distributed over the interval $x = -1$ to $x = 4$. Find its mean and variance. Calculate: (a) $P(X < 3)$, (b) the probability that X is within one standard deviation of the mean.

2. Prove that, for a uniform distribution over the interval $x = a$ to $x = b$, the mean is $\frac{1}{2}(a+b)$ and the variance is $(b-a)^2/12$.

3. A point is chosen at random on a line of length L. Find the probability that the ratio of the shorter to the longer parts of the line is less than $1/3$.

4. The figure shows the graph of the density function of a random variable X whose distribution is triangular.
    Find: (a) $P(X > 0)$,
    (b) $P(-1 < X < 1)$.

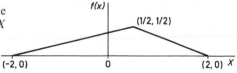

5. Write down the probability density function for the random variable Z whose distribution is shown in the sketch.

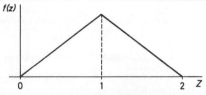

   Find: (a) Var(Z),
         (b) the interquartile range.

6. The random variable Z is the sum of the errors when pairs of lines are measured in centimetres to the nearest millimetre. The probability density function of Z has the triangular shape shown on page 183, $-0.1 \leqslant z \leqslant 0.1$.
(a) Find the variance of this distribution.
(b) Find and sketch the cumulative distribution function.
(c) Calculate $P(0.05 \leqslant Z \leqslant 0.06)$.

7. The continuous random variable X has probability density function

$$f(x) = \begin{cases} (x+a)/a^2, & -a \leqslant x \leqslant 0 \\ (a-x)/a^2, & 0 < x \leqslant a \\ 0 & , \text{ otherwise.} \end{cases}$$

Sketch this function. Find also the cumulative distribution function and sketch it.
Calculate: $P(\tfrac{1}{4}a \leqslant Z \leqslant \tfrac{1}{2}a)$.

10.7★ The exponential distribution

The exponential distribution is of particular importance because of the wide-ranging nature of the practical situations in which it is used. Examples of random variables for whose distributions the exponential distribution is a possible model include:

1. the length of time until an electronic device fails;

2. the time required to wait for the first emission of a particle from a radioactive source;

3. the length of time between successive accidents in a large factory.

Many of these applications involve some quite complicated mathematics and will be examined in detail in chapter 22. At this stage we will introduce the distribution and apply it in a simple situation.

A continuous random variable X has an exponential distribution if, for some constant $\lambda(>0)$, it has the probability density function

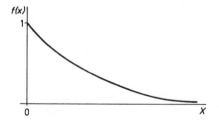

$$f(x) = \begin{cases} \lambda e^{-\lambda x}, & x \geqslant 0 \\ 0 & , \text{ otherwise.} \end{cases}$$

The graph of $f(x)$ is shown in the figure.
The parameter of this distribution is λ.
The function $f(x)$ is clearly positive for all values of x and the area under the curve

$$= \int_0^\infty f(x)\,\mathrm{d}x = \int_0^\infty \lambda\,e^{-\lambda x}\,\mathrm{d}x = \left[-e^{-\lambda x} \right]_0^\infty = 1,$$

so that $f(x)$ satisfies the necessary conditions for a density function.

The mean of the distribution $E(X) = \displaystyle\int_0^\infty x\lambda e^{-\lambda x}\, dx.$

Integrating by parts, this integral $= \displaystyle\int_0^\infty x\,d(-e^{-\lambda x})$

$$= \left[-xe^{-\lambda x}\right]_0^\infty - \int_0^\infty -e^{-\lambda x}\, dx.$$

Now $\displaystyle\lim_{x\to\infty} xe^{-\lambda x} = 0$ (proved in the Appendix on page 193),

so that $E(X) = 0 + \displaystyle\int_0^\infty e^{-\lambda x}\, dx = \left[-e^{-\lambda x}/\lambda\right]_0^\infty = 1/\lambda.$

Hence the mean $= 1/\lambda$.

The variance $= \text{Var}(X) = E(X^2) - [E(X)]^2$

where $E(X^2) = \displaystyle\int_0^\infty x^2\lambda e^{-\lambda x}\, dx.$

Integrating by parts, this integral $= \displaystyle\int_0^\infty x^2\,d(-e^{-\lambda x})$

$$= \left[-x^2 e^{-\lambda x}\right]_0^\infty - \int_0^\infty -e^{-\lambda x}2x\, dx.$$

But $\displaystyle\lim_{x\to\infty} x^2 e^{-\lambda x}$ is also 0   (see Appendix page 193)

so that $E(X^2) = 0 + 2\displaystyle\int_0^\infty xe^{-\lambda x}\, dx = (2/\lambda)\int_0^\infty x\lambda e^{-\lambda x}\, dx$

$$= (2/\lambda)\,E(X) = (2/\lambda)\,(1/\lambda) = 2/\lambda^2.$$

Hence $\text{Var}(X) = 2/\lambda^2 - (1/\lambda)^2 = 1/\lambda^2.$

Thus we have   the mean $= 1/\lambda$ and the variance $= 1/\lambda^2$.

The *cumulative distribution function* is found more easily.

For $x \geqslant 0$, $F(x) = \displaystyle\int_0^x \lambda e^{-\lambda x}\, dx$

$$= \left[-e^{-\lambda x}/\lambda\right]_0^x$$

$$= 1 - e^{-\lambda x}$$

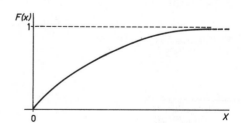

and $F(x) = 0$ for $x < 0$.

It can be seen that $F(x) \to 1$ as $x \to \infty.$

EXAMPLE 11   The mean life of an electrical component is 100 hours and its life has an exponential distribution.

Find:
(a) the probability that it will last less than 60 hours,
(b) the probability that it will last more than 90 hours,
(c) the probability that it will last less than 60 hours, given that it has already lasted 50 hours.
(d) the median life.

The mean of an exponential distribution with parameter λ is $1/\lambda$ so that in this case, if X is the life in hours, the parameter is $1/100$ and the density function is

$$f(x) = \frac{1}{100} e^{-x/100}, \quad x \geq 0 \quad \text{and} \quad 0 \text{ elsewhere.}$$

(a) To find $P(X < 60)$ we either integrate:

$$P(X < 60) = \int_0^{60} \frac{1}{100} e^{-x/100} \, dx = \left[-e^{-x/100} \right]_0^{60}$$

$$= 1 - e^{-0.6} = 0.45 \quad \text{to 2 dec. places}$$

or, quoting the cumulative distribution function, $F(x) = 1 - e^{-x/100}$,

$$P(X < 60) = F(60) = 1 - e^{-0.6} = 0.45.$$

(b) $P(X > 90) = 1 - P(X \leq 90) = 1 - F(90) = 1 - [1 - e^{-0.9}]$
$$= e^{-0.9} = 0.41 \quad \text{to 2 dec. places.}$$

(c) Using conditional probability,

$$P(X < 60 \,|\, X > 50) = \frac{P(50 < X < 60)}{P(X > 50)} = \frac{F(60) - F(50)}{1 - F(50)} = 0.095.$$

(d) If the median is m,

$$\int_0^m \frac{1}{100} e^{-x/100} \, dx = 0.5 \Rightarrow \left[-e^{-x/100} \right]_0^m = 0.5$$

$$\Rightarrow 1 - e^{-m/100} = 0.5 \quad \text{(i.e. } F(m) = 0.5)$$

$$\Rightarrow m = 100 \ln 2 = 69.3 \text{ hours.}$$

EXERCISE 10e

1. The life X (in thousands of miles) of a certain brand of tyre has an exponential distribution with density function

$$f(x) = \frac{1}{4} e^{-x/4}, \quad x \geq 0.$$

(a) Write down the mean and variance.
(b) Find $P(X < 2)$ and $P(5 < X < 6)$.
(c) Find the lower quartile.

2. If the distribution of the random variable X has density function
$$f(x) = k e^{-x/5}, \quad x \geq 0 \quad \text{and} \quad f(x) = 0, \quad \text{elsewhere.}$$
Find k and the probability that X lies between the median and the mean.

3. The distribution of X, the life of a certain make of TV tube, is exponential with a mean of 1000 days. Find: (a) the probability that a tube will last more than 2000 days, (b) the probability that it will last between 1200 and 1500 days, (c) the probability that it will last more than 900 days, given that it has already lasted 800 days.

4. The random variable X comes from an exponential distribution with p.d.f.

$$f(x) = \begin{cases} \lambda e^{-\lambda x}, & x \geq 0, \\ 0 & , & x < 0 \end{cases}$$

where $\lambda > 0$.
   Show that $P(X \geq 3.7/\lambda) = 0.025$ and find L so that $P(X \leq L/\lambda) = 0.025$.   (MEI)

5. The life of a particular kind of insect under laboratory conditions is T hours, whose density function is

$$f(t) = (1/a)\, e^{-t/a}, \quad t \geq 0, \quad \text{where } a > 0.$$

   Find, in terms of k, the probability that an insect will survive beyond $t = ka$. Find also the smallest value of k for which this probability is less than $\frac{1}{2}$.

6. The time to failure, in hundreds of hours, of component A in a machine has density function

$$f(t) = \frac{1}{3} e^{-t/3}, \quad t \geq 0,$$

and the time to failure of another component, B, in the same machine has density function

$$\phi(t) = \frac{1}{6} e^{-t/6}, \quad t \geq 0.$$

   The machine will cease functioning if either A or B fails. Assuming independence find the probability that the machine is still operating after 300 hours.

10.8   Some important theoretical results

A number of general theorems about a discrete random variable were proved on pages 94–5 and 103–5. These results also hold true for a continuous random variable and will often be needed in subsequent chapters. For convenience we repeat them here. If you are interested, proofs of the first three results may be found in the Appendix on pages 194–5. Proofs of the others are omitted as they involve 'double integration'.

1. $E(kX) = kE(X) = k\mu$.

2. $E(kX + a) = kE(X) + a = k\mu + a$.

3. $\text{Var}(kX + a) = k^2\text{Var}(X) = k^2\sigma^2$.

4. $E(X \pm Y) = E(X) \pm (E(Y))$, whether or not X and Y are independent.

5. $E(XY) = E(X) \cdot E(Y)$ if X and Y are independent.

6. $\text{Var}(X \pm Y) = \text{Var}(X) + \text{Var}(Y)$ if X and Y are independent.

   (For the equivalent result when X and Y are not independent see page 379.)

EXERCISE 10f (*miscellaneous*)

1.   The continuous random variable X has probability density function f given by

$$f(x) = \begin{cases} k & , & \text{if } 0 \leqslant x \leqslant 2, \\ k(3-x), & \text{if } 2 < x \leqslant 3, \\ 0 & , & \text{otherwise.} \end{cases}$$

(a) Find the value of k.
(b) Hence evaluate $E(X)$ and the median of X.
(c) Prove that σ, the standard deviation of X, is 0.75 correct to 2 decimal places.
(d) Denoting $E(X)$ by μ, find $P[X < (\mu - \sigma)]$. (C)

2.   The number of kilograms of metal extracted from 10 kg of ore from a certain mine is a continuous random variable X with probability density function $f(x)$, where $f(x) = cx(2-x)^2$ if $0 \leqslant x \leqslant 2$ and $f(x) = 0$ otherwise, where c is a constant.
   Show that c is $\frac{3}{4}$, and find the mean and variance of X.
   The cost of extracting the metal from 10 kg of ore is £10x. Find the expected cost of extracting the metal from 10 kg of ore. (MEI)

3.   A random variable X has probability density function

$$f(x) = \begin{cases} kx^2(1-x), & 0 \leqslant x \leqslant 1 \\ 0 & , & \text{otherwise.} \end{cases}$$

   Find k and sketch the curve $y = f(x)$.
   Find μ, the mean of X; find also the mode of X. Obtain the mean deviation of X, defined by $E(|X - \mu|)$. (O & C)

4.   The continuous random variable X has probability density function

$$f(x) = \begin{cases} kx, & 5 < x < 10, \\ 0, & \text{otherwise.} \end{cases}$$

(a) Find the value of k.
(b) Find the expected value of X.
(c) Find the probability that $X > 8$.

   The annual income from money invested in a unit trust fund is $X\%$ of the amount invested, where X has the above distribution. Suppose that you have a sum of money to invest and that you are prepared to leave the money invested over a period of several years.
   State, with your reasons, whether you would invest in the unit trust fund or in a money bond offering a guaranteed annual income of 8 % on money invested. (JMB)

5.   A continuous random variable X has probability density function $\lambda \sin x$ $(0 \leqslant x \leqslant \pi)$, and zero outside this range. Find the value of the constant λ, the mean, the variance, the median and the quartiles. What is the probability that a random observation lies within one standard deviation of the mean? (O)

6.   A random variable X has cumulative distribution function

$$F(x) = \begin{cases} cx^3, & 0 \leqslant x \leqslant 2, \\ 0 & , & x \leqslant 0, \\ 1 & , & x \geqslant 2, \end{cases}$$

where c is a constant.
   Find c, the mean, variance and median of the distribution.

7.　The random variable X is the distance, in metres, that an inexperienced tightrope walker has moved along a given tightrope before falling off. It is given that

$$P(X > x) = 1 - x^3/64, \quad 0 \leqslant x \leqslant 4.$$

(a) Show that $E(X) = 3$.
(b) Find the standard deviation, σ, of X.

(c) Show that $P(|X - 3| < \sigma) = \dfrac{69}{80}\sqrt{\dfrac{3}{5}}$.

　　　　(C)

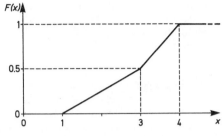

8.　A continuous random variable X takes values in the interval $1 \leqslant x \leqslant 4$. The cumulative distribution function, $F(x)$, of X is a different linear function in each of the intervals $1 \leqslant x \leqslant 3$ and $3 \leqslant x \leqslant 4$, and takes the values $0, 0.5$ and 1 corresponding to $x = 1, 3$ and 4, as shown in the diagram.
(a) Sketch the graph of the probability density function of X.
(b) Show that the expectation, μ, of X, is equal to $11/4$.
(c) Determine the standard deviation of X, leaving your answer in surd form.
(d) Determine the probability that $|X - \mu|$ exceeds 1.　　　　(C)

9.　The probability density function of x is given by

$$f(x) = \begin{cases} k(ax - x^2), & 0 \leqslant x \leqslant 2 \\ 0 & , \quad x < 0, \quad x > 2, \end{cases}$$

where k and a are positive constants.
　Show that $a \geqslant 2$ and that $k = 3/(6a - 8)$.
　Given that the mean value of x is 1, calculate the values of a and k.
　For these values of a and k sketch the graph of the probability density function and find the variance of x.　　　　(JMB)

10.　The probability density function of x is

$$f(x) = \begin{cases} kx(1 - ax^2), & 0 \leqslant x \leqslant 1 \\ 0 & , \quad x < 0, \quad x > 1, \end{cases}$$

where k and a are positive constants. Show that $a \leqslant 1$ and that $k = 4/(2 - a)$.
　Sketch the probability density function for the case when $a = 1$ and in this case find the mean, variance and 75th percentile of the distribution.　　　　(JMB)

11.　The random variable X has probability density function f given by

$$f(x) = \frac{c}{(x + 1)^2 + 1} \quad (-\infty < x < \infty),$$

where c is a constant. By considering the derivative of $\arctan(x + 1)$, or otherwise, show that $c = 1/\pi$.
　Write down the median value of X, and determine the value X_0 of X such that there is a probability 0.1 that $X > X_0$.
　Calculate the probabilities that: (a) $X > 0$, (b) $X > -2$.
　Three independent measurements of X are taken. What is the probability that at least one of them is greater than 0?　　　　(MEI)

12.   The weekly demand for petrol, in thousands of gallons, at a certain garage is a continuous random variable X distributed with probability density function f given by

$$f(x) = ax(b-2x), \quad 1/2 \leqslant x \leqslant 3/2,$$
$$f(x) = 0 \qquad , \quad \text{otherwise.}$$

(a) Given that the mean weekly demand is 900 gallons, determine each of the constants a and b.
(b) Show that there is a probability of 0.35 that the demand in a given week will exceed 1000 gallons.   (JMB)

13.   In the study of atmospheric pollution, the concentration X of the pollutant is a random variable. Units of concentration can be chosen so that the value of X always lies between 1 and 10. In certain circumstances, an appropriate mathematical model of the probability that $X \leqslant x$ is

$$c\{\sqrt{(\ln 10)} - \sqrt{(\ln 10 - \ln x)}\}, \quad (1 \leqslant x \leqslant 10),$$

where c is a constant. Explain how the value of c can be determined by considering the probability that $X \leqslant 10$. Hence show that $c = 0.659$ correct to 3 decimal places.
   Find, correct to 2 decimal places, the probabilities that: (a) $X \leqslant 8$, (b) $2 \leqslant X \leqslant 8$.
   Show that the median concentration is about 5.6.   (MEI)

14.   The random variable X can take all values between 0 and a inclusive, where $a > 0$. Its probability density function $f(x)$ is zero for $x < 0$ and $x > a$, and, for $0 \leqslant x \leqslant a$, satisfies

$$f(x) = (A/a)\,e^{-x/a},$$

where A is a positive constant. Show by integration that $A = 1.582$ to 3 decimal places.
   Also use integration to find to 2 decimal places:
(a) the probability that X is less than $\frac{1}{2}a$;
(b) the number λ for which there is a probability $\frac{1}{2}$ that X is less than λa.   (MEI)

15.   The probability density function for the life X of a motor car sparking plug is given by

$$f(x) = \theta e^{-\theta x} \quad (x \geqslant 0),$$

where x is measured in units of 1000 miles. What is the probability that a single sparking plug will last more than 10 000 miles if the average life is 5000 miles?
   A car has four plugs in use and they are all of the same age. What is the probability that three of them will need replacing before 15 000 miles after they were installed?   (MEI)

16.   The time t seconds between the arrival of successive vehicles at a point on a country road has a probability density function $ke^{-\lambda t}$, $t \geqslant 0$, where k and λ are positive constants. Find k in terms of λ and sketch the graph of the probability density function.
   Given that $\lambda = 0.01$, find the mean and variance of t.

$$\left(\text{You may use the result } \int_0^\infty t^r e^{-t}\, dt = \frac{r!}{\lambda^{r+1}}, \quad r = 0, 1, 2, \ldots.\right)$$

   An elderly pedestrian takes 50 seconds to cross the road at this point. With $\lambda = 0.01$, calculate the probability that, if he sets off as one vehicle passes, he will complete the crossing before the next vehicle arrives. Calculate also the probability that, if he adopts the same procedure on the return journey, he completes each crossing without a vehicle arriving while he is doing so.   (JMB)

17.   In 1972 the distribution of annual incomes (in pounds sterling) of people employed in a certain occupation might be represented by the continuous probability density function,

$$f(x) = \begin{cases} kx^{-3.5}, & x \geqslant x_0 \\ 0 & , \quad x < x_0 \end{cases}$$

where k and x_0 are constants. Show that $k = 2.5x_0^{2.5}$ and sketch the graph of $f(x)$. What is the interpretation of the constant x_0?

Given that $x_0 = 900$, find the mean and variance of the annual incomes. Determine the proportion of people employed in this occupation who earned between £1600 and £2500 per year.   (JMB)

18.   The lifetime in years of a television tube of a certain make is a random variable T and its probability density function $f(t)$ is given by

$$f(t) = Ae^{-kt}, \quad \text{for } 0 \leqslant t < \infty \quad (k > 0)$$
$$f(t) = 0 \quad , \quad \text{elsewhere.}$$

Obtain A in terms of k.
(a) If the manufacturer, after some research, finds that out of 1000 such tubes 371 failed within the first two years of use, estimate the value of k.
(b) Using this value of k correct to 3 significant figures, calculate the mean and variance of T, giving answers correct to 2 significant figures ($t^r e^{-kt} = 0$ when $t = \infty$ for finite r).
(c) If two such tubes are bought, what is the probability that one fails within its first year and the other lasts longer than six years?   (SUJB)

19.   A random variable X has exponential distribution with density function

$$f(x) = \begin{cases} \alpha e^{-\alpha x}, & x > 0 \\ 0 & , \quad \text{elsewhere,} \end{cases}$$

where α is a positive constant.

Sketch the distribution and find its mean and standard deviation.

The lifetime X of a certain article may be considered as a continuous random variable with the above distribution. There are two processes by which the article may be manufactured: process A costs Cp per article and produces articles with an expected lifetime of 1000 hours, while process B costs $2C$p per article and produces articles with an expected lifetime of 1500 hours. If an article lasts less than 500 hours then the manufacturer refunds kp to the user. Find a criterion by which the manufacturer should decide which process to employ.   (MEI)

20.★ For a certain type of bacterium, the time X from birth to death is a random variable with density function $(1+x)^{-2}$, $(x \geqslant 0)$. A culture of such bacteria is routinely inspected at regular intervals of time.

One bacterium, inspected at time t after birth, is found to be dead already. Find the probability that this bacterium has been dead for at least time kt ($0 < k < 1$).

Another bacterium of the same type is alive at time t after birth but is found to be dead by the end of a further time t. Find the probability that this bacterium has then been dead for at least time kt ($0 < k < 1$).   (MEI)

21.★ The random continuous variable X takes all values x in the range $0 \leqslant x \leqslant 1$, and has a continuous probability density function $f(x)$ defined by

$$f(x) = kx^{\theta-1}(1-x)^2 \quad (\theta \geqslant 1).$$

(a) Show that $k = \frac{1}{2}\theta(\theta+1)(\theta+2)$.
(b) Find $E(X)$ and $E(X^2)$.
(c) Deduce the variance of X.
(d) For $\theta = 3/2$, find the location of the mode and sketch $f(x)$.   (O & C)

22.★ The quality of a product is measured by the continuous random variable X, whose probability density function f is given by

$$f(x) = \begin{cases} k[1-(x-\mu)^2], & \text{if } |x-\mu| < 1 \\ 0 & , \quad \text{otherwise.} \end{cases}$$

Prove that $k = \frac{3}{4}$.

Items for which $X < 1$ can be sold at a profit of £P each; those for which $1 \leqslant X \leqslant 2$ can be sold at a profit of £$3P$ each, and those for which $X > 2$ can be sold at a profit of £$2P$ each. Assuming that $1 < \mu < 2$, show that the expected profit per item is

$$£\tfrac{3}{4}P[2+\mu+\tfrac{2}{3}(1-\mu)^3-\tfrac{1}{3}(2-\mu)^3]. \tag{C}$$

23.★ A continuous random variable X has probability density function

$$x/\lambda^2, \quad (0 \leqslant x \leqslant \lambda)$$
$$(2\lambda-x)/\lambda^2, \quad (\lambda \leqslant x \leqslant 2\lambda).$$

Find a value t such that $P[t\lambda \leqslant X \leqslant (2-t)\lambda] = 0.95$.

A single observation x is taken from such a distribution for which the value of λ is unknown. For what values of λ does x lie between $t\lambda$ and $(2-t)\lambda$ with probability 0.95?   (O)

24.★ An overnight mail train calls at a certain station. Instructions are that, in order to transact its business, the train must stand at the station for not less than two minutes and will then leave as soon as the business has been completed. The total time for which the train stands at the station is modelled by the random variable X whose cumulative distribution function is

$$F(x) = P(X \leqslant x) = \begin{cases} 0 & , \quad x < 2 \\ 1-\theta e^{-\lambda(x-2)}, & x \geqslant 2. \end{cases}$$

Show that $P(X = 2) = 1-\theta$; obtain the p.d.f. of X for $X > 2$, and *hence* verify that $P(X > 2) = \theta$.

Obtain the mean and variance of X.

Suggest another situation that may be modelled by a random variable having the same general form.   (MEI)

25.★ A continuous random variable X is such that $0 \leqslant X \leqslant 1$ and the probability that $X \leqslant x$ is

$$k(mx^n - nx^m).$$

$(k, m, n$ are positive constants, $m \neq n)$.

Find k in terms of m and n and show that m is greater than n.

Find, in terms of m and n, the probability density function of X and the mean of X.

When $m = 2n$, calculate in terms of n the median value of X and the value of X for which the probability density function is a maximum. Find the value of n for which these two quantities are equal.   (JMB)

26.★ The probability density function of the random variable X is

$$f(x) = \begin{cases} kxe^{-\lambda x}, & x \geqslant 0 \\ 0 & , \quad \text{elsewhere,} \end{cases}$$

where k, λ are constants $(\lambda > 0)$.

Determine k in terms of λ, and show that the mean and variance of X are $2/\lambda$ and $2/\lambda^2$ respectively.

For the case $\lambda = 1$, determine the probability that the value of X lies within two standard deviations of its mean (you may leave your answer in terms of e).   (MEI)

27.★ The probability density function of x is

$$f(x) = \begin{cases} kx(1-x)^\beta, & 0 \leqslant x \leqslant 1, \\ 0 & , \quad x < 0, \quad x > 1, \end{cases}$$

where $\beta \geqslant 0$. Calculate k in terms of β and find the mean and variance of x.

The probability that $X \leqslant x$ is denoted by $F(x)$; find $F(x)$. Sketch $f(x)$ and $F(x)$ when $\beta = 1$. (JMB)

28.★ A random variable X has the probability density function

$$f(x) = \begin{cases} a+cx^2, & |x| \leqslant k, \\ 0 & , \quad |x| > k, \quad 0 < k \leqslant 10. \end{cases}$$

Show that the conditions for a p.d.f. imply that $a \geqslant 0$. Find the values of k giving the maximum variance of X and the values of the maximum when: (a) $a > 0$, (b) when $a = 0$. (MEI)

29.★ Show that

$$f(x) = \begin{cases} 0 & , \quad x < 0 \\ k^2 x e^{-kx} & , \quad 0 \leqslant x \leqslant L, \quad k > 0 \\ k(1+kL)e^{-kx}, & x > L, \quad k > 0 \end{cases}$$

is a probability density function.
    Find the cumulative distribution function.
    What is the mode of the distribution: (a) when $1 < kL$, (b) when $1 > kL$?
    Sketch carefully the p.d.f. for these two cases. (MEI)

10.9 Appendix

Proof that $\lim_{x \to \infty} xe^{-\lambda x} = \lim_{x \to \infty} x^2 e^{-\lambda x} = 0$

These two limits are needed in finding the mean and variance of the exponential distribution (page 185).

$$xe^{-\lambda x} = \frac{x}{1+\lambda x+\lambda^2 x^2/2!+\lambda^3 x^3/3!+\dots}$$

$$= \frac{1}{1/x+\lambda+\lambda^2 x/2!+\lambda^3 x^2/3!+\dots}$$

by dividing numerator and denominator by x.
    As $x \to \infty$ the denominator $\to \infty$ and $xe^{-\lambda x} \to 0$.
Similarly

$$x^2 e^{-\lambda x} = \frac{x^2}{1+\lambda x+\lambda^2 x^2/2!+\lambda^3 x^3/3!+\dots}$$

$$= \frac{1}{1/x^2+\lambda/x+\lambda^2/2!+\lambda^3 x/3!+\dots}$$

by dividing numerator and denominator by x^2.
    As $x \to \infty$ the denominator $\to \infty$ as before and $x^2 e^{-\lambda x} \to 0$.

Expectation and variance of a continuous random variable

On page 187 various theoretical results were listed, without proof. The formal proofs of 1, 2 and 3 are given here. They all depend on the definition of expectation for a continuous random variable X whose p.d.f. is $f(x)$, $-\infty < x < \infty$:

$$E[g(X)] = \int_{-\infty}^{\infty} g(x)f(x)\,dx.$$

1. $E(kX) = kE(X) = k\mu$

Proof

$$E(kX) = \int_{-\infty}^{\infty} kxf(x)\,dx = k\int_{-\infty}^{\infty} xf(x)\,dx = kE(X) = k\mu.$$

2. $E(kX+a) = kE(X)+a = k\mu+a$

Proof

$$E(kX+a) = \int_{-\infty}^{\infty} (kx+a)f(x)\,dx = k\int_{-\infty}^{\infty} xf(x)\,dx + a\int_{-\infty}^{\infty} f(x)\,dx$$

$$= kE(X)+a \quad \left(\text{since } \int_{-\infty}^{\infty} f(x)\,dx = 1 \right)$$

$$= k\mu+a.$$

3. $\mathrm{Var}(kX+a) = k^2\mathrm{Var}(X) = k^2\sigma^2$

Proof

If $Y = kX+a$, its mean is $k\mu+a$ (by 2).

Hence $\mathrm{Var}(Y) = E[(kX+a)-(k\mu+a)^2]$ (definition of variance)

$$= E[(kX-k\mu)^2]$$

$$= \int_{-\infty}^{\infty} (kx-k\mu)^2 f(x)\,dx = k^2\int_{-\infty}^{\infty} (x-\mu)^2 f(x)\,dx$$

$$= k^2\mathrm{Var}(X) = k^2\sigma^2.$$

The alternative forms for variance

A formal proof for a continuous random variable X that

$$E(X-\mu)^2 \quad \text{and} \quad E(X^2)-\mu^2$$

are equivalent forms for $\mathrm{Var}(X)$.

$$\mathrm{Var}(X) = E(X-\mu)^2 = \int_{-\infty}^{\infty} (x-\mu)^2 f(x)\,dx = \int_{-\infty}^{\infty} (x^2 - 2\mu x + \mu^2)f(x)\,dx$$

$$= \int_{-\infty}^{\infty} x^2 f(x) \, dx - 2\mu \int_{-\infty}^{\infty} x f(x) \, dx + \mu^2 \int_{-\infty}^{\infty} f(x) \, dx$$

$$= E(X^2) - 2\mu E(X) + \mu^2 \quad \left(\text{since} \int_{-\infty}^{\infty} f(x) \, dx = 1 \right)$$

$$= E(X^2) - 2\mu \times \mu + \mu^2 = E(X^2) - \mu^2.$$

In some later theoretical work it will be useful to have a more general form:

$$\mathrm{Var}(X) = E(X-a)^2 - (\mu-a)^2, \text{where } a \text{ is any constant.}$$

Proof

If $Y = X - a$,

$$\mathrm{Var}(X) = \mathrm{Var}(Y) \quad \text{(by 3 with } k = 1)$$

$$= E(Y^2) - [E(Y)]^2$$

$$= E(X-a)^2 - (\mu-a)^2 \quad \text{(since } E(Y) = \mu - a, \text{ by 2).}$$

11 The normal distribution

The normal distribution, which is a suitable model for a very large number of distributions of data, is the most important distribution in statistical theory.

It was first derived from the binomial distribution; Gauss then used it as a model for errors in astronomical observations, and it remains the appropriate distribution for modelling errors of many kinds, due to random causes, in scientific and economic measurements. Later it was found that biological data such as the heights of humans of the same age and sex are described by the normal model. The normal distribution is also a useful approximation, under certain circumstances, to other probability distributions such as the binomial and Poisson.

Finally, when we come to discuss what is called the central limit theorem, it will be found that the normal distribution takes on an even wider significance. Indeed the central limit theorem sheds light on the reasons *why* data of the kind mentioned above are so often accounted for by the normal model.

11.1 Normal curves

The common feature of the distribution models for all the examples just mentioned is that the curves of their graphs all have the same basic shape:

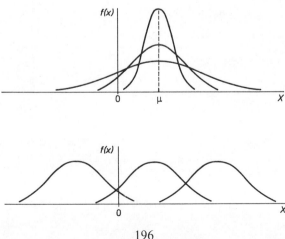

1.   they are symmetrical,

2.   they die away rapidly at each 'tail',

3.   they are shaped like a bell.

The actual shape of any particular normal curve is determined completely by its mean μ and its variance σ^2, the mean giving the position of the axis of symmetry and the variance the spread. In other words μ and σ^2 are the *parameters* of a normal distribution.

The first figure on page 196 shows normal curves with the same mean and different variances. Since the area under each of the curves must be 1, the curve flattens out as the variance (spread) increases.

The second figure shows normal curves with the same variance and different means.

The normal distribution with mean μ and variance σ^2 is called $N(\mu, \sigma^2)$ for short.

The way any normal curve spreads out from its mean is indicated by the following facts (which will be discussed more fully later):

1.   About 68 % of the area under the curve is between $\mu - \sigma$ and $\mu + \sigma$.

2.   About 95 % of the area under the curve is between $\mu - 2\sigma$ and $\mu + 2\sigma$.

3.   About 99.7 % of the area under the curve is between $\mu - 3\sigma$ and $\mu + 3\sigma$.

The area in the 'tails' beyond $\mu \pm 3\sigma$ is therefore insignificant.

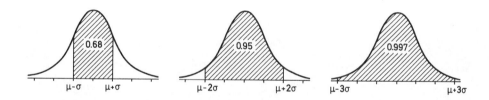

11.2   The general and standard normal distributions

The mathematics of the normal distribution is quite complicated, so we shall start by discussing it informally, without, at this stage, bringing in the equations of the curves.

The simplest and basic normal distribution is called the *standard normal distribution* and is written as $N(0, 1)$, which is a short way of saying that it is *the normal distribution with mean 0 and variance 1*.

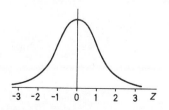

For reasons which become clear later, we will use z for the variable, instead of x, and will assume that the density function is $f(z)$ for some unspecified range of values of z (which are in fact $-\infty$ to $+\infty$).

If, now, we change the scale,

putting $x' = \sigma z$, where σ is any positive constant, we will have a new variable x' and a new curve.

The effect will be to stretch the curve out (assuming $\sigma > 1$) and to flatten it, since the area must remain 1. At the same time the mean is unaltered.

But for any distribution, the effect of multiplying the variable by σ is to increase its variance by a factor σ^2. The variance of the new distribution will therefore be $\sigma^2 \times 1 = \sigma^2$.

The effect of changing the scale like this is illustrated here:

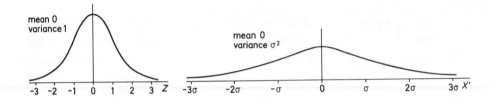

If we now shift the whole curve along the axis by putting $x = x' + \mu$ (where μ is any constant) the mean of the distribution with x as variable becomes μ, while the variance remains unchanged:

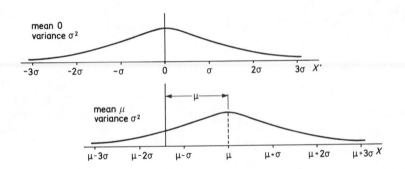

To sum up, by changing the variable z in two stages, $x' = \sigma z$ and $x = x' + \mu$, we have transformed the standard normal distribution with mean 0 and variance 1, to a general normal distribution with mean μ and variance σ^2.

Note: The change of scale followed by change of origin is equivalent to the transformation from z to x:

$$x = \sigma z + \mu$$

What we have, then, are two curves:

1. the general normal distribution

    $N(\mu, \sigma^2)$ with mean μ and variance σ^2.

    This is found to be the appropriate
    model for a wide variety of experimental
    data or observations. We shall continue

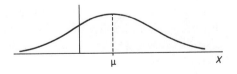

to use X for the random variable associated with this distribution.

2.   the standard normal distribution

$N(0, 1)$ with mean 0 and variance 1.

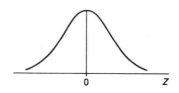

In order to distinguish it from the
general normal it is customary to use Z
for the associated random variable.

We have also said that $N(0, 1)$ may be transformed into $N(\mu, \sigma^2)$ by changing the
scale and shifting the axis, which amounts to using the relation

$$X = \sigma Z + \mu.$$

This relation may also be written in the form

$$Z = \frac{X - \mu}{\sigma}$$

and used to carry out the reverse process, namely, to transform $N(\mu, \sigma^2)$ to $N(0, 1)$.

If you are meeting the normal distribution for the first time, you will find that
these basic ideas are all you need to know for most practical purposes. More
advanced students will find a more detailed and formal approach to the distribution in the Appendix on page 222, but this strictly mathematical treatment is best
avoided on a first reading.

The standard normal distribution

We are now almost ready to use these ideas in a practical problem, but as the
procedure always involves the standard normal distribution some further information is needed about this special distribution.

Its density function is usually written $\phi(z)$, rather than $f(z)$, and is given by

$$\phi(z) = \frac{1}{\sqrt{(2\pi)}} e^{-\frac{1}{2}z^2}, \quad -\infty < z < \infty.$$

(This rather formidable expression is not strictly needed at this stage, but you
should be aware that it exists.)

As usual for continuous distributions the corresponding *distribution function* is
written $\Phi(z)$ [Φ is the capital ϕ in Greek], so that

$$\Phi(z) = \frac{1}{\sqrt{(2\pi)}} \int_{-\infty}^{z} e^{-\frac{1}{2}z^2} \, dz$$

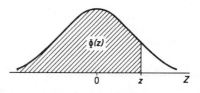

and gives the area to the left of $Z = z$ or
$P(Z \leqslant z)$.

Now knowledge of the density function does not enable us to calculate
$\Phi(z)$ for any particular value of z by integration, as we did with simpler continuous
distributions, since the integration involved cannot be carried out by elementary
methods. However a table of values (calculated by what is called 'numerical
integration') is given on page 518 and from this table the probabilities (or areas)
may be read off in the following way.

From the tables:

$$\Phi(0) = 0.5, \qquad \Phi(1) = 0.8413, \qquad \Phi(2) = 0.9771, \qquad \Phi(3) = 0.9987$$

which means that the proportions of the distributions to the left of $z = 0, 1, 2$ and 3 respectively are 0.5, 0.8413, 0.9771 and 0.9987.

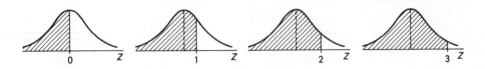

To find $\Phi(z)$ when z is negative, we use the fact that, by symmetry

$$\Phi(-1) = 1 - \Phi(1)$$
$$= 1 - 0.8413$$
$$= 0.1587.$$

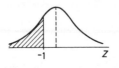
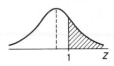

EXAMPLE 1   If the distribution of Z is the standard normal distribution, $N(0, 1)$, find:

(a) $P(Z \leqslant 0.7)$,   (b) $P(0.7 \leqslant Z \leqslant 1.4)$,   (c) $P(-1.4 \leqslant Z \leqslant 1)$.

(a) $P(Z \leqslant 0.7) = \Phi(0.7) = 0.7580.$
(b) $P(0.7 \leqslant Z \leqslant 1.4) = \Phi(1.4) - \Phi(0.7)$
$$= 0.9192 - 0.7580 = 0.1612.$$
(c) $P(-1.4 \leqslant Z \leqslant 1) = \Phi(1) - \Phi(-1.4)$
$$= \Phi(1) - [1 - \Phi(1.4)] = 0.7605.$$

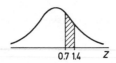

EXAMPLE 2   Find a and b if $P(Z \geqslant a) = 0.05$ and $P(Z \leqslant b) = 0.10$.

Note: In using the table in reverse, the following notation may be used:

If $\Phi(z) = 0.5793, \qquad z = 0.2,$

and we write $\Phi^{-1}(0.5793) = 0.2.$

In this example $\Phi(a) = 1 - 0.05 = 0.95, \quad a = \Phi^{-1}(0.95) = 1.64$ to 2 dec. places
$$\Phi(b) = 0.10 \Rightarrow b = \Phi^{-1}(0.10) = -\Phi^{-1}(0.90) = -1.28.$$

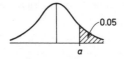
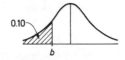

EXERCISE 11a

1.   If Z has distribution $N(0, 1)$ find:
(a) $P(Z \leqslant 1.62)$,   (b) $P(Z \geqslant -0.94)$,   (c) $P(Z \leqslant -1.47)$,
(d) $P(0.3 \leqslant Z \leqslant 2.0)$,   (e) $P(-1.82 \leqslant Z \leqslant 0)$,   (f) $P(-2.0 \leqslant Z \leqslant -1.0)$,
(g) $P(-1.5 \leqslant Z \leqslant 2.0)$,   (h) $P(|Z| \leqslant 2.2)$,   (i) $P(|Z| \geqslant 0.45)$.

2.   Find z if: $\Phi(z) =$ (a) 0.82, (b) 0.63, (c) 0.48, (d) 0.14.

3.   Find a if: (a) $P(Z \leqslant a) = 0.94$,   (b) $P(Z \geqslant a) = 0.02$,   (c) $P(Z \leqslant a) = 0.45$,
(d) $P(Z \leqslant a) = 0.05$, (e) $P(Z \geqslant a) = 0.85$.

4.   Prove that the central 95% of the distribution $N(0, 1)$ lies between $z = -1.96$ and
$z = 1.96$. Between what values of z does 90% of the distribution lie?

11.3   Standardising a normal distribution

Now we are at last in a position to deal with a problem involving any normal
distribution. The procedure is this. If the distribution model (with random variable
X) is $N(\mu, \sigma^2)$ we transform it into the standard normal $N(0, 1)$ (with random
variable Z), using the relation

$$Z = \frac{X - \mu}{\sigma}$$

and then use the standard normal tables in the way we have just outlined. (The
transformation process is known as '*standardising*'.)

Suppose a large number of students take an aptitude test and the resulting scores
are normally distributed with mean 200 and variance 400. The model is, therefore,

$$N(200, 20^2).$$

If we want to know the proportion of
students scoring less than 250, this will
be given by the area to the left of
$x = 250$.

Standardising this X score means
subtracting μ (200) and dividing by
σ (20), so that the standardised score

$$z = \frac{250 - 200}{20} = 2.5.$$

Since the distribution of Z is $N(0, 1)$
we use the tables, which show that

$$\Phi(2.5) = 0.994.$$

In other words $P(Z \leqslant 2.5) = P(X \leqslant 250) = 0.994$ to 3 dec. places.

(The common practice is to use Z for the random variable and z for a particular
value of the random variable.)

EXAMPLE 3   A lathe turns out metal rods whose diameters are normally distributed with mean 1.5 cm and variance 10^{-4} cm^2. Five per cent of the rods are
rejected as being oversize. Find: (a) the critical upper diameter, (b) the percentage
of rods with diameter less than 1.48 cm.

(a) With $\mu = 1.5$ and $\sigma^2 = 10^{-4}$ ($\sigma = 0.01$), the standardised value of x, the critical
upper limit, is

$$z = \frac{x - 1.5}{0.01},$$

where $\Phi(z) = 1 - 0.05$.
Hence $z = \Phi^{-1}(0.95) = 1.64$

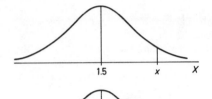

so that

$$x - 1.5 = 0.01 \times 1.64 \Rightarrow x = 1.516.$$

(b) The standardised value of 1.48,

$$z = \frac{1.48 - 1.5}{0.01} = -2$$

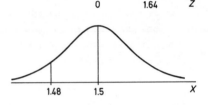

and

$$\Phi(-2) = 1 - \Phi(2) = 1 - 0.9772$$

$$= 0.023 \text{ or } 2.3\% \quad \text{to 2 sig. figs.}$$

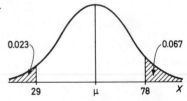

EXAMPLE 4   An examination was taken by 1000 candidates; 67 gained more
than 78% and 23 gained less than 29%. Assuming the distribution to be normal,
estimate the mean mark and the standard deviation.

If the mean is μ and the standard deviation σ

$$\frac{78 - \mu}{\sigma} = \Phi^{-1}(1 - 0.067) = 1.5$$

and $\quad \dfrac{\mu - 29}{\sigma} = \Phi^{-1}(1 - 0.023) = 2.0.$

Solving these equations $\mu = 57$ and $\sigma = 14$.
(Notice that it is often easier to avoid using negative values of the standardised
variable.)

EXERCISE 11b

1.   If X has distribution $N(3, 0.09)$, that is, its mean is 3 and standard deviation is 0.3, find:
(a) $P(X \leqslant 3.5)$,   (b) $P(X \geqslant 3.7)$,   (c) $P(X \geqslant 2.5)$,   (d) $P(2.6 \leqslant X \leqslant 3.2)$.

2.   If X is normally distributed with mean -3 and standard deviation 2, find: (a) $P(X > 0)$,
(b) $P(X < 0.9)$.

3.   The heights of men measured in metres are distributed normally with mean 1.765 and
standard deviation 0.1. What is the probability that a man chosen at random will have
height: (a) more than 1.83 m, (b) between 1.765 and 1.80 m, (c) between 1.70 and 175 m?

4.   The marks of candidates in an examination are normally distributed with mean 56 and standard deviation 22. Calculate the percentage of candidates who score: (a) between 65 and 90, (b) less than 35.

5.   The resistance in ohms of resistors produced by a certain machine is distributed normally with a mean of 80.2 and standard deviation of 0.4. If resistors with resistances outside the range 79 to 81 are to be rejected, estimate the number rejected in each batch of 1000. Show that, if the standard deviation can be reduced to 0.3, the mean remaining constant, then on average less than 4 resistors in 1000 will be rejected.   (L)

6.   The weights of the contents of packets of breakfast cereal are normally distributed with mean 250 g and standard deviation 6 g. If 10 % of the packets have contents weighing less than X g, find X.

7.   Sacks of potatoes packed automatically have an average weight of 114 lb. Ten per cent of the sacks are over 116 lb. Find the standard deviation, assuming a normal distribution.

8.   A machine produces components to any required length specification with a standard deviation of 1.40 mm. At a certain setting it produces to a mean length of 102.30 mm. Assuming the distribution of lengths to be normal, calculate: (a) what percentage would be rejected as less than 100 mm long; (b) to what value, to the nearest 0.01 mm, the mean should be adjusted if this rejection rate is to be 1 %.   (MEI)

9.   In an examination, 30 % of the candidates fail and 10 % achieve distinction. Last year the pass mark (out of 200) was 84 and the minimum mark required for a distinction was 154. Assuming that the marks of the candidates were normally distributed, estimate the mean mark and the standard deviation.   (O & C)

10.   Over a long period of time, scientists using a species of lizard for experimental observations find that the mean weight of lizards bred in captivity is 57.5 g and that 3 % of them weigh more than 58.45 g. Given that the weight distribution is normal, calculate the standard deviation.
   Lizards below a certain weight are not suitable for the experiments. Calculate this lower weight if 1 % are found to be unsuitable.

11.   The manufacturers of 'Whiter-than-white' found that, on average, their machines deliver 3700 kg of the powder in filling 5000 packets. They wish to aim at getting not more than 2 % of their packets underweight, and intend to print on each packet 'Contents not less than X kg'. The standard deviation of the weights in the individual packets was found to be 0.015 kg. What should be the value of X, correct to 3 s.f.?   (MEI)

12.   Each day an aeroplane flies from Liverpool to London and it is scheduled to arrive at London by 0800. During the first year of operation it leaves Liverpool at 0700 and is found to be late arriving at London on an average of once every 4 days. In the second year it leaves Liverpool at 0650 and is late once every 20 days. Assuming the time of flight has a normal distribution find, to the nearest minute, the mean time of flight and the standard deviation.
   Use these values to estimate the latest time of departure of the aeroplane (to the nearest minute) which ensures that its probability of arriving late does not exceed 0.02.   (L)

13.   A petrol station finds that its daily sales in May exceed 1000 gallons of four star petrol 10 % of the time and are greater than 500 gallons for 95 % of the time. If the daily sales are distributed normally, find the average daily sales of four star petrol.
   The daily sales of two star petrol in May are distributed normally with mean 700 gallons and standard deviation 125 gallons. Find the probability that the daily sales of two star petrol exceed 600 gallons.
   What is the probability that for a given period of five working days the daily sales of four star petrol exceed 750 gallons on all of the days and on at least one day the sales of two star petrol are less than 600 gallons.   (MEI)

14. In a particular experiment, the length of a metal bar is measured many times. The measured values are distributed normally with mean 1.340 m and standard deviation 0.021 m. Find the probabilities that any one measured value: (a) exceeds 1.370 m; (b) lies between 1.310 m and 1.370 m; (c) lies between 1.330 m and 1.390 m.

Find the length L for which the probability that any one measured value is less than L is 0.1.   (MEI)

11.4   The sum or difference of two independent random variables with normal distributions

Consider a machine which is producing metal plates whose weights are distributed normally, $N(3.2, 0.004)$ and suppose that another machine is making studs whose weights are distributed normally $N(0.04, 0.001)$, the units being grams in each case. If the studs are slotted into the plates what is the distribution of the combined weights of plates plus studs?

We already know (see page 187) that the distribution of the combination will have mean $3.2 + 0.04$ and variance $0.004 + 0.001$, but what sort of distribution will it be? The answer is that it will also be *normal* – in this case $N(3.24, 0.005)$.

This result will not seem surprising, though it cannot be proved by an elementary method. The theorem stating this property is as follows:

If X and Y are independent random variables with distributions $N(\mu_1, \sigma_1^2)$ and $N(\mu_2, \sigma_2^2)$ respectively, and $Z = X + Y$, then the distribution of Z is $N(\mu_1 + \mu_2, \sigma_1^2 + \sigma_2^2)$.

It is known as the *reproductive* property of the normal distribution. (Those of you who go on to study moment generating functions will find a proof on page 412.)

Now if Y has distribution $N(\mu_2, \sigma_2^2)$ the distribution of $-Y$ will be $N(-\mu_2, \sigma_2^2)$, so that the distribution of $X - Y$ will be $N(\mu_1 - \mu_2, \sigma_1^2 + \sigma_2^2)$.

This result can be extended to any linear combination of a number of normally distributed random variables as follows:

If $X_1, X_2, \ldots, X_n$ are independent random variables with distributions $N(\mu_1, \sigma_1^2)$, $N(\mu_2, \sigma_2^2), \ldots, N(\mu_n, \sigma_n^2)$ respectively and $Y = a_1 X_1 + a_2 X_2 + \ldots + a_n X_n$, where $a_1, a_2, \ldots, a_n$ are constants, then the distribution of Y is normal with mean $a_1 \mu_1 + a_2 \mu_2 + \ldots + a_n \mu_n$ and variance $a_1^2 \sigma_1^2 + a_2^2 \sigma_2^2 + \ldots + a_n^2 \sigma_n^2$.

The distribution of a multiple of a random variable with normal distribution

Some practical applications of the normal distribution make use of the following property:

If X is a random variable with distribution $N(\mu, \sigma^2)$ and $Y = aX$, then the distribution of Y is $N(a\mu, a^2\sigma^2)$.

EXAMPLE 5   The heights of 21-year-old men are normally distributed with mean 170 cm and standard deviation 6 cm. The heights of women of the same age are normally distributed with mean 165 cm and standard deviation 4.5 cm. If a man and a woman are selected at random, what is the probability that: (a) the man is 8 cm taller than the woman, (b) the woman is taller than the man?

X (height of men) has distribution $N(170, 36)$,

Y (height of women) has distribution $N(165, 20.25)$.

Hence, if $U = X - Y$, the distribution of U is normal with mean 170–165 and variance $36 + 20.25$ and is, therefore $N(5, 56.25)$.

(a) $P(X - Y > 8) = P(U > 8)$

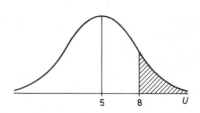

$$= 1 - \Phi\left(\frac{8-5}{\sqrt{56.25}}\right)$$

$$= 1 - \Phi(0.4)$$

$$= 0.345 \quad \text{to 3 dec. places.}$$

(b) $P(Y > X) = P(U < 0)$

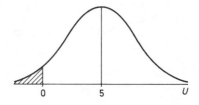

$$= 1 - \Phi\left(\frac{5-0}{\sqrt{56.25}}\right)$$

$$= 1 - \Phi(0.67)$$

$$= 0.250 \quad \text{to 3 dec. places.}$$

This method may also be used to find the probability that two *men* selected at random differ in height by more than 8 cm, say. In theory the two variables, picked from the same population, are not strictly independent, because it is not possible to pick the same man more than once. Nevertheless, this point may be disregarded in practice.

We are now concerned with the distribution $N(0, 36 + 36) = N(0, 72)$ and if D is the difference between the heights of two men,

$$P(D > 8) = 2\left[1 - \Phi\left(\frac{8-0}{\sqrt{72}}\right)\right] = 2[1 - \Phi(0.94)] = 0.35 \quad \text{to 2 dec. places.}$$

EXERCISE 11c

1.   If the distribution of X is $N(15, 2)$ and of Y is $N(8, 7)$, find: (a) $P(X + Y < 27.5)$, (b) $P(X - Y < 10.6)$.

2.   The distribution of the heights of men of a certain age is $N(170, 8)$ and of women of the same age $N(162, 5)$, the measurements being in centimetres. If a man and a woman are selected at random, find:
(a) the probability that the man is more than 10 cm taller than the woman,
(b) the probability that the woman is taller than the man.
If two men are selected at random, what is the probability that the difference in height is less than 5 cm?

3.   A manufacturer produces resistances normally distributed about a mean value of 50 ohms with a standard deviation of 1 ohm. Find the probability that if two resistances are chosen at random, their combined resistance in series will lie between 99 ohms and 101 ohms.

4.   Bolts are manufactured which are to fit holes in steel plates. The diameter of the bolts is normally distributed with mean 3.2 cm and standard deviation 0.03 cm; the diameter of the holes is distributed normally with mean 3.31 cm and standard deviation 0.04 cm.

   If the bolts and holes are selected at random show that the proportion of bolts that will not fit the holes is 1.39 %. Find also the probability that a batch of 100 bolts will all fit their holes.   (O & C)

5.   In the production of an item of furniture, part A fits into part B. For part A, the relevant outer dimension is x, and for B the corresponding inner dimension is y. Both x and y are normally distributed, having means $\mu_x = 2.05$ cm and $\mu_y = 2.10$ cm. Their standard deviations are $\sigma_x = 0.03$ cm and $\sigma_y = 0.04$ cm.

   In assembly, a part A is selected at random and an attempt is made to fit it into a part B, also selected at random. Find the percentage of pairs so selected which must be rejected because part A is too large to fit into part B.   (JMB)

6.   The time of arrival of a bus at a bus stop varies in a normal distribution with a mean of 09.00 am and a standard deviation of 2 minutes. Independently a second bus departs from this stop at a time which varies in a normal distribution with a mean of 09.01 am and a standard deviation of 1 minute. Find the probability that:
(a) the first bus arrives before the second bus leaves,
(b) this happens on 5 given consecutive days.   (O & C)

7.   If the distributions of X, Y, Z are independent normal distributions $N(40, 10)$, $N(35, 8)$ and $N(120, 18)$ respectively, what is the distribution of: (a) $X + Y + Z$, (b) $X - Y + Z$?

8.   Four shrubs are selected from a large distribution of shrubs whose heights are normally distributed with mean 30 cm and standard deviation 4.2 cm. Find the probability that the total height of the shrubs is less than 142 cm.

9.   A firm mass-produced certain articles which in course of manufacture were subjected to three processes, A, B and C. A record of the times articles took for each process and of the times articles were waiting between processes was kept over a long period. The following table gives the mean and standard deviation of these times.

Operation	Mean time (hours)	Standard deviation (hours)
Process A	3.6	0.34
Waiting between A and B	1.4	0.12
Process B	9.4	0.65
Waiting between B and C	2.5	0.18
Process C	6.3	0.45

   Calculate the mean and standard deviation of the distribution of the time taken to produce an article assuming independence of the times in the table. Assuming normal variation calculate: (a) the proportion of articles taking less than 22 hours to produce, (b) the proportion taking more than 24 hours.   (MEI)

10.   If the distribution of X is $N(8, 9)$ and of Y is $(15, 4)$, what is the distribution of: (a) $3X$, (b) $4Y$, (c) $2X + 3Y$, (d) $4X - 2Y$, (e) $(X + Y)/2$?

11.   Resistances type A are normally distributed with mean 100 ohms and standard deviation 2 ohms. Resistances type B are normally distributed with mean 50 ohms and standard deviation 1.3 ohms. If two resistances type A and one type B are selected at random and connected together in series, what is the probability that the resulting resistance will be less than 254 ohms?

11.5   Normal approximation to the binomial distribution

On page 146, the Poisson distribution was developed from the binomial distribution $B(n, p)$ by letting $n \to \infty$ and $p \to 0$, while np remained finite. In this way the Poisson distribution was seen to be a useful approximation to the binomial when n *is large and p is small* (and np is of moderate size).

There is still a need for an approximation to the binomial distribution when n *is large and p is any size* (subject to certain conditions).

If we consider the histograms of $B(10, 1/6)$, $B(10, \frac{1}{2})$ and $B(15, 0.4)$, it can be seen that the last two, where p is $\frac{1}{2}$ or near $\frac{1}{2}$ are, in the first case, symmetrical and in the second nearly so.

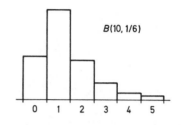

It will be found that if n is large and p is not too small (this point will be discussed later) the polygonal outline of the histogram tends towards a bell-shaped continuous curve which can be shown to approximate to a normal curve.

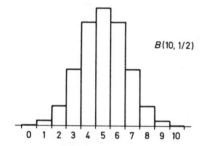

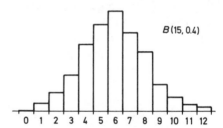

In other words:

If the (discrete) variable X has distribution $B(n, p)$ with large n, and p not too small, the distribution of X is approximately normal, with mean np and variance npq, i.e. its distribution is approximately $N(np, npq)$ or $N[(np, np(1-p)]$. (This is known as de Moivre's approximation.)

In order to justify this approximation, it must be proved that $B(n, p) \to N(np, npq)$ as $n \to \infty$. (A proof appears later in chapter 21, pages 416–17.)

The normal approximation is a good alternative to $B(n, p)$ when the latter is approximately symmetrical, which is true when n is large and p is close to $\frac{1}{2}$. (As a working rule, if $n > 10$ and p is close to $\frac{1}{2}$ the approximation is good.)

If *either* p or q is small, the approximation will still be good provided n is large enough. (*Both* np and nq must exceed 5 is a working rule.)

In using a *continuous* distribution as an approximation to a *discrete* one, care must be taken over the boundaries of the intervals concerned. The following examples illustrate the method.

EXAMPLE 6   The probability of a head when a biased coin is tossed is 0.48. Find the probability of getting 8 heads if the coin is tossed 15 times.

In this case $n = 15$, $p = 0.48$, $q = 0.52$, so we use the normal distribution with mean $np = 7.2$ and variance $npq = 3.744$.

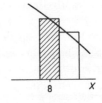

$P(X = 8)$ is the area of the rectangle whose upper and lower bounds are 8.5 and 9.5 (the class width is 1).

The corresponding area under the normal curve $N(7.2, 3.744)$

$$= \Phi\left(\frac{8.5 - 7.2}{\sqrt{3.744}}\right) - \Phi\left(\frac{7.5 - 7.2}{\sqrt{3.744}}\right) = 0.189 \quad \text{to 3 dec. places.}$$

Using $B(15, 0.48)$ the probability of 8 heads is

$$\binom{15}{8}(0.48)^8(0.52)^7 = 0.185 \quad \text{to 3 dec. places.}$$

EXAMPLE 7   If the distribution of X is $B(200, 0.2)$, find: (a) $P(X \geqslant 48)$, (b) $P(X > 48)$, (c) $P(46 \leqslant X \leqslant 50)$.

Here $n = 200$, $p = 0.2$, $q = 0.8$ (and np and nq are both greater than 5). We therefore use the normal distribution with mean $np = 40$ and variance $npq = 32$.

(a) $P(X \geqslant 48)$ is the sum of the areas of the rectangles to the right of, and *including*, the one centred on 48.

The corresponding area under the normal curve is therefore

$$1 - \Phi\left(\frac{47.5 - 40}{\sqrt{32}}\right)$$

$$= 0.093 \quad \text{to 3 dec. places.}$$

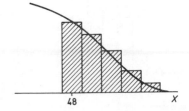

(b) $P(X > 48)$ is the sum of the areas of the rectangles to the right of, and *excluding*, the one centred on 48.

The corresponding area under the normal curve is thus

$$1 - \Phi\left(\frac{48.5 - 40}{\sqrt{32}}\right)$$

$$= 0.067 \quad \text{to 3 dec. places.}$$

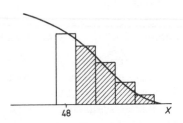

(c) By a similar argument,

$$P(46 \leqslant X \leqslant 50) = \Phi\left(\frac{50.5 - 40}{\sqrt{32}}\right) - \Phi\left(\frac{45.5 - 40}{\sqrt{32}}\right) = 0.135 \quad \text{to 3 dec. places.}$$

In general, for the approximation to $P(X = r)$ we use the area under the normal curve from $r - \frac{1}{2}$ to $r + \frac{1}{2}$.

For the approximation to $P(r_1 \leqslant X \leqslant r_2)$ we use the area under the normal curve from $r_1 - \frac{1}{2}$ to $r_2 + \frac{1}{2}$.

This is called applying a *continuity correction*.

11.6 Normal approximation to the Poisson distribution

It was mentioned on page 156 that calculations of Poisson probabilities can sometimes be laborious. Consider, for example, the problem of finding $P(X \leqslant 20)$ when X has a Poisson distribution with parameter 30. This would involve evaluating 21 Poisson probabilities of the form $30^r e^{-30}/r!$. Although tables have been compiled to deal with this situation, there is an alternative method using the normal distribution.

If the (discrete) random variable X has a Poisson distribution with parameter μ, then, for *large* values of μ, the distribution of X is approximately normal with mean μ and variance μ i.e. its distribution is approximately $N(\mu, \mu)$.

This also is proved on page 416. The approximation will be a good one for μ greater than about 20.

As with the use of the normal as an approximation to the binomial, a continuity correction must be applied.

EXAMPLE 8   The mean number of calls to a small telephone exchange is 2 per hour. Find the probability that fewer than 30 calls arrive in a 12-hour period, assuming the mean remains constant over this interval of time.

We are concerned with approximating to a Poisson distribution with mean 24, so that the normal approximation is $N(24, 24)$.

$$P(X < 30) = \Phi\left(\frac{29.5 - 24}{\sqrt{24}}\right) = 0.821 \quad \text{to 3 dec. places.}$$

11.7   Summary

1.   Normal approximation to $B(n, p)$ is $N(np, npq)$, where $q = 1 - p$ and n large and p close to $\frac{1}{2}$ ($n > 10$ in practice) *or*, n very large and p (or q) small (*np and nq > 5 in practice*).

2.   Normal approximation to Poisson with parameter μ is $N(\mu, \mu)$, μ large ($\mu > 20$ in practice).

3.   Poisson approximation to $B(n, p)$ is Poisson with parameter np, n large and p small. (See page 207.)

EXERCISE 11d

1.   If the distribution of X is binomial with $n = 100$ and $p = \frac{1}{2}$, find: (a) $P(X \geqslant 60)$, (b) $P(X > 60)$, (c) $P(X = 60)$.

2.   An unbiased die is to be thrown 500 times and the number of 6s noted. If the number of 6s is X, find:
(a) $P(70 \leqslant X \leqslant 100)$,   (b) $P(70 < X < 100)$.

3.   A stock of ball-point pens imported from Bulgaria is offered to the public at a clearance sale. One quarter of the stock is defective. If a dealer buys 192 pens, find the probability that: (a) 56 or fewer will be defective, (b) exactly 56 will be defective.

4.   Calculate the normal approximation to the probability of obtaining exactly 30 successes in 90 independent trials in each of which the probability of success is 1/3.

5.   The probability that an insect will be killed when it is sprayed with a commercial insect spray is 0.8. Use the normal distribution to estimate the probability that, when 100 insects are sprayed: (a) at least 75 will be killed, (b) more than 75 will be killed.   (JMB)

6.   It is known that 30% of an apple crop has been attacked by insects. A random sample of 150 is selected from the crop. Assuming that the distribution of the number of damaged apples in random samples of 150 may be approximated to by a normal distribution, estimate the probability that:
(a) more than half the sample is damaged,
(b) less than 10% is damaged,
(c) the number of damaged apples lies between 35 and 50, inclusive.   (L)

7.   The following table gives, correct to 3 decimal places, the values of some probabilities for the binomial distribution when $n = 20$ and $p = \frac{1}{2}$.

Number of 'successes'	1	3	5	7	9
Probability	0.000	0.001	0.015	0.074	0.160

Find the corresponding probabilities from the appropriate normal distribution, and comment on your answers.
     Would the comparison between the exact and approximate distributions be as good for the case $n = 20$, $p = 1/20$?   (MEI)

8.   A pack of cards consists of thirty-six plain cards marked with the numbers $1, 2, 3, \ldots, 36$. If a card is drawn at random what is the probability that the outcome is divisible by 4 or 5 (or both)? If this experiment is repeated 100 times what is the probability of getting more than 45 outcomes which are divisible by 4 or 5 (or both)?

9.   If the probability that a car has some defect in its lighting system is $\frac{1}{4}$ and a random sample of 120 cars is tested, find X such that the probability that fewer than X cars have defective lights is 0.1.

10.   A telephone receives calls at random at an average rate of 30 calls every quarter of an hour. Use the normal approximation to the Poisson distribution to find the probability that, in a quarter of an hour: (a) fewer than 35, (b) between 25 and 35 calls are received.

11.   The distribution of X is Poisson with mean 36. If $P(X > a)$ is less than 0.05, find a.

12.   Five thousand supporters travel from Liverpool to Milan for a football match. Two airlines A and B having identical aircraft intend to offer identical terms for the return journey. Each supporter acts independently and chooses to travel by A or B with equal probability. Use the normal approximation to the binomial distribution to decide how many seats each airline should put on if each is prepared to run a 1% risk that it will have too few.
     Suppose that A changes its mind and reduces its price so that in fact there is a probability 3/5 that an individual supporter will travel with A and a probability 2/5 that he will travel with B. If A puts on 3080 seats and B puts on 2050 seats, what are the probabilities that: (a) A will have too few seats, and (b) B will have too few seats?   (MEI)

11.8   Fitting a normal distribution to given data

On page 131 we gave an example of the method of fitting a discrete probability distribution (binomial) to a set of given data. We shall now do the same for data for which the appropriate model is a normal distribution. The procedure is broadly similar, but as the variable is continuous there are points of detail which need care.

We shall take as our example the following frequency distribution:

Class	Observed frequency
$\leqslant 10$	1
10–	3
15–	7
20–	14
25–	20
30–	9
35–40	4
	58

The problem is to calculate the theoretical frequencies for a normal distribution with the same mean and standard deviation as the given distribution.

There is no information about the degree of accuracy to which these observations have been measured, so we take the class centres as $7.5, 12.5, \ldots$, assuming a class width of 5 for the first class. Calculating the mean and variance in the usual way

$$\bar{x} = 25.4 \quad \text{and} \quad \text{var}(x) = 43.14 = 6.57^2.$$

As the sample is a large one we assume that these are close to the population mean and variance, so that the model is $N(25.4, 6.57^2)$. (Whether or not this is a reasonable assumption will be discussed later, in the next chapter.)

In order to make the process clear we will focus on the area under the normal curve between $x = 30$ and $x = 35$ as an example, setting out the reasoning in full. In practice, however, the working is best done in the form of a table, shown below. The successive steps in the procedure are labelled 1 to 5, corresponding to the columns 1 to 5 in the table.

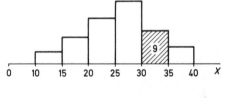

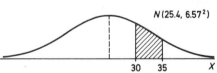

1. Write down the *upper bounds* (x) of the classes $(10, 15, \ldots)$.

2. We wish to find the area under the curve $N(25.4, 6.57^2)$ between $x = 35$ and $x = 30$.
    Standardising $x = 35$, using $z = (x - 25.4)/6.57$ gives $z = 1.46$.
    Similarly, standardising $x = 30$, gives $z = 0.70$.

3. From the normal tables $\Phi(1.46) = 0.928$ and $\Phi(0.70) = 0.758$.

4. The required area is $0.928 - 0.758 = 0.170$.

5. Finally, to find the expected *frequency*, we multiply 0.170 by 58, giving 9.86.

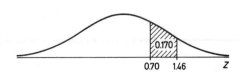

In the model, the upper bound of the last class has to be ∞ even though there are no recorded values over 40. Similarly the lower bound of the first class must be $-\infty$.

(Notice that the upper bounds in column (1) are placed *between* the entries in the frequency column so as to ensure that the final expected frequency will appear opposite the corresponding observed frequency.)

		1	2	3	4	5
Class	f	x (upper bound)	$z = \dfrac{x - \mu}{\sigma}$	$\Phi(z)$	p	*Expected frequency* (p × 58)
$-\infty-$	1				0.010	0.58
		10	-2.34	0.010		
10–	3				0.047	2.73
		15	-1.58	0.057		
15–	7				0.149	8.64
		20	-0.82	0.206		
20–	14				0.270	15.66
		25	-0.06	0.476		
25–	20				0.282	16.36
		30	0.70	0.758		
30–	9				0.170	9.86
		35	1.46	0.928		
35–	4				0.072	4.18
		∞	∞	1.000		

The observed and theoretical frequencies seem reasonably close, but to test whether in fact the assumption of a normal distribution with the stated mean and variance is justified on these results, a special test is needed. This will be described later in chapter 23.

Note: If the class divisions were, say, 0–9, 10–14, 15–19,..., arising perhaps from measurements to the nearest integer, the upper bounds would be 9.5, 14.5, 19.5,...,

A graphical method

A method of testing the hypothesis that a distribution is likely to be normal is to use a graph.

This depends on the relationship $z = (x - \mu)/\sigma$ between x and its standardised value, z. If, then, z is plotted against x the result should be approximately a straight line; μ and σ may then be estimated from the graph. We shall use the above distribution to show the procedure.

1. Write down the upper bounds (x) as before.

2. Observed frequencies (f).

3. Calculate the cumulative frequencies (cf).

4. Convert cumulative frequency to cumulative percentage, which is $\Phi(z)$.

$$\Phi(z) = cf \times (100/58).$$

5.   Find z from the normal table.

1	2	3	4		5
x	f	cf	Cum. %	$= \Phi(z)$	z
10	1	1	1.72%	0.0172	−2.12
15	3	4	6.90%	0.0690	−1.48
20	7	11	18.97%	0.1897	−0.88
25	14	25	43.10%	0.4310	−0.17
30	20	45	77.59%	0.7759	0.76
35	9	54	93.10%	0.9310	1.48
∞	4	58	100.00%	1.0000	∞

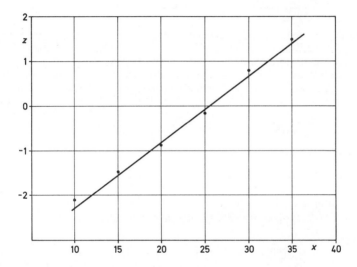

Since the equation of the straight line is $z = (x - \mu)/\sigma$ its gradient is $(1/\sigma)$. In this case the gradient is 0.152, giving an estimate of σ equal to 6.6. The value of x when $z = 0$ gives an estimate of μ, 25.5.

EXERCISE 11e

1.   The table gives the distribution of the breaking strengths of 100 samples of rope (in newtons):

Breaking strength	15–	25–	35–	45–	55–	65–	75–	85–	95–100
Frequency	1	3	7	20	32	25	10	1	1

Show that the mean is 60.45 and the standard deviation 13.5, and use these values to find the theoretical frequencies on the assumption that the distribution is normal.

2.   With the figures in question 1, use a graphical method to show that the distribution is approximately normal, and use the graph to estimate the mean and standard deviation.

3. The number of marks gained by 120 pupils in a public examination are shown in the table:

Marks	10–19	20–29	30–39	40–49	50–59	60–69	70–79	80–89
Frequency	5	10	22	24	28	20	8	3

Show that the mean is 48.4 and the standard deviation 16.25.

Find the theoretical frequencies for a normal distribution with this mean and standard deviation, using the same class intervals.

4. The shape of the human head is the subject of an international project financed by the World Council for Health and Welfare. Observations were taken in many countries and the nose lengths, to the nearest millimetre, of 150 Italians are summarised below.

Nose lengths x (mm)	$-\infty < x \leqslant 44$	$45 \leqslant x \leqslant 47$	$48 \leqslant x \leqslant 50$	$51 \leqslant x \leqslant 53$	$54 \leqslant x \leqslant 56$	$57 \leqslant x < \infty$
Frequency f	4	12	63	59	10	2

Calculate the mean and standard deviation of this sample. (For these calculations you should assume that the lower and upper classes have the same range as the other classes.)

Taking these values as estimates of the mean and standard deviation of the population from which the observations were taken, find the theoretical frequencies, assuming a normal distribution.   (AEB 1982)

11.9   The central limit theorem

We have stated on page 204 that the random variable equal to the sum of a finite number of independent normally distributed random variables is itself normally distributed. This is typical of many results in statistical theory in that it is not really surprising.

The *central limit theorem* is an exception. It is quite remarkably unexpected. What it says is that even if the individual variables are *not* normally distributed their sum *is*, provided the number of variables is large. To put it precisely:

If $X_1, X_2, \ldots, X_n$ are independent random variables and $Y = X_1 + X_2 + \ldots + X_n$, then the distribution of Y will be approximately normal for large values of n, *whatever the nature of the distributions* of $X_1, X_2, \ldots, X_n$.

Apart from the practical consequences, with which we shall be mainly concerned, the general significance of the result is considerable.

Suppose we are interested in the distribution of the diameters of metal rods produced by a machine. The diameters of the rods will vary, perhaps not by much and the variations must be made up of a large number of random variations, no one of which would by itself have much effect, due to a variety of causes. What the central limit theorem is saying is that if these random small variations (whose distributions are quite unknown) are added together, the result is a total variation whose distribution is approximately normal. This being so, the fact that so many of the sets of observations in this chapter have been stated to be normally distributed is not a happy accident, but is to be expected. The same sort of reasoning presumably applies in the case of human measurements like height and mass, although it is perhaps less easy to suggest what the particular factors (biological or otherwise) might be that contribute to the variations.

As for the practical consequences, these stem from the *special case* of the central limit theorem when each of the individual variables, $X_1, X_2, \ldots, X_n$ have the *same* distribution:

If $X_1, X_2, \ldots, X_n$ are independent random variables, all having the *same* distribution, and $Y = X_1 + X_2 + \ldots + X_n$, then the distribution of Y will be approximately normal for large values of n, *whatever the nature of the distribution of the X_i.*

Now if the mean and variance of the distribution of the X_i are μ and σ^2 respectively,

$$E(Y) = n \sum_{i=1}^{n} E(X_i) = n\mu \quad \text{and} \quad \text{Var}(Y) = n \sum_{i=1}^{n} \text{Var}(X_i) = n\sigma^2,$$

so that the distribution of Y will be approximately $N(n\mu, n\sigma^2)$.

It is this special form of the central limit theorem which has such important consequences.

The size of n for which the approximation will be a good one depends on the nature of the distribution of the X_i. If this distribution is very skew, n will have to be much greater than if it is fairly symmetrical. The size of n for a reasonable approximation will also depend on the 'peakedness' (see page 406) of the distribution. We shall, however, be able to be more precise than this when it comes to applying the theorem to sampling.

The proof of the general central limit theorem is difficult but if you are interested, there is an indication of the way the *special* form of the theorem is proved on page 417. It should be added that there are some circumstances in which the statement of the theorem needs to be modified, but these will not concern us.

We conclude this section by showing how the normal approximation to the binomial distribution may be derived from the central limit theorem.

The normal approximation to the binomial theorem

If $X =$ the number of successes in n trials, it can be regarded as the sum of the independent random variables $X_1, X_2, \ldots, X_n$ each of which has the same distribution, defined as follows:

$$X_i = \begin{cases} 1 & \text{if the } i\text{th trial is a success} \\ 0 & \text{if the } i\text{th trial is a failure,} \end{cases}$$

where $P(X_i = 1) = p$ and $P(X_i = 0) = q \, (= 1-p)$.

Clearly X can take any value from 0 to n and has distribution $B(n, p)$.

Now $E(X_i) = p \cdot 1 + q \cdot 0 = p$ and $\text{Var}(X_i) = (p \cdot 1^2 + q \cdot 0^2) - p^2$
$$= p(1-p) = pq.$$

Applying the central limit theorem, the distribution of X will be approximately normal, with mean np and variance npq.

A truncated normal distribution

Suppose a machine produces metal rods whose diameters are normally distributed with mean 1.2 cm and standard deviation 0.02 cm, but rods whose diameters exceed 1.24 cm are discarded as being unsuitable. If X is the length of rods *not* discarded, the distribution of X is a truncated normal distribution.

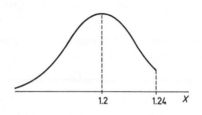

The procedure for finding the density function of X is given in the Appendix on page 226.

EXERCISE 11f (*miscellaneous*)

1. The heights of men, measured in centimetres, are found to be normally distributed with mean 176.5 and standard deviation 10. Find the percentage of men who are:
(a) over 183 cm in height,
(b) over average height but below 180 cm in height,
(c) between 170 cm and 175 cm in height.   (L)

2. A machine makes metal rods, the lengths of which are normally distributed with standard deviation 0.5 cm. The manufacturer is satisfied if at least 80 % of the rods turned out are over 8.98 cm long. He is able to adjust the mean length so that this performance is achieved. Calculate the mean length.
   If a rod is picked out at random, calculate the probability that it will be between 9.1 cm and 10.2 cm long.   (C)

3. The concentration by volume of methane at a point on the centre line of a jet of natural gas mixing with air is distributed approximately normally with mean 20 % and standard deviation 7 %. Find the probabilities that the concentration: (a) exceeds 30 %, (b) is between 5 % and 15 %.
   In another similar jet, the mean concentration is 18 % and the standard deviation is 5 %. Find the probability that in at least one of the jets the concentration is between 5 % and 15 %
(MEI)

4. (a) X is a normal variable with mean μ and standard deviation σ. It is given that $P(X > 128) = 0.15$ and that $P(X > 97) = 0.875$. Calculate μ and σ.
   (b) The quantity of articles manufactured in a certain process is classified as 'poor', 'fair' or 'good' on the basis of a measured quantity Y which may be assumed to have a normal distribution with mean 50 units and standard deviation 5 units. Articles are 'poor' if $Y < 44$, and the proportions of fair and good are equal. Calculate the boundary value of Y separating 'fair' from 'good'.   (C)

5. The time in minutes required to complete a particular task may be assumed to be normally distributed. Given that there is a probability of 0.01 that the task will be completed in less than 30 minutes and a probability of 0.95 that it will be completed in less than one hour, calculate, correct to two decimal places, the probability that the task will be completed in less than 45 minutes.   (JMB)

6. In an examination there are 4000 candidates. In Paper I the marks may be taken as a continuous and normal distribution with mean 47.5 and standard deviation 12.5.
(a) Estimate the numbers of candidates who will obtain:
   (i) less than 40 marks,
   (ii) marks between 36 and 44.

(b) Estimate the mark which will be exceeded by 10% of the candidates.

(c) In Paper II the mean mark is 50.5 and the standard deviation 15 and the total mark for the examination is the average for the two papers. How many candidates will have a total mark less than 40?   (O & C)

7. A machine produces components in batches of 20 000, the lengths of which may be considered to be normally distributed.

At the beginning of production, the machine is set to produce the required mean length of components at 15 mm and it can then be set to give any one of three standard deviations: 0.06 mm, 0.075 mm, 0.09 mm. It costs £850, £550 and £100 respectively to set these deviations. Any length produced must lie in the range 14.82 mm to 15.18 mm, otherwise it is classified as defective and costs the company £1.

Which standard deviation should be used, if the decision is to be made purely on the cost of setting the machine and of the defectives?   (SUJB)

8. The number of meals sold in a refectory for a large factory varies from day to day, it is assumed according to a normal distribution. In a period of 100 days there were 91 occasions when the number of meals was between 960 and 1130, and the number of occasions when between 1075 and 1100 meals were sold was equal to the number of occasions when between 1000 and 1025 meals were sold. Show that the standard deviation is 50 and find the number of meals which will only be exceeded on 1% of occasions, and the average number of meals.   (MEI)

9. A transport company finds that the amount of fuel that its lorries use during a week may be taken as a normal variable with mean 5000 litres and standard deviation 400 litres.

(a) Calculate the probability that in a particular week less than 6000 litres will be used.

(b) The probability that more than L litres will be used in any week is 0.75. Find L.

(c) Calculate the probability that in a particular four-week period more than 21 000 litres will be used, assuming that fuel consumption in any week is independent of that in other weeks.   (C)

10. Books on the top shelf at a library are directly accessible only to a person having a reachable height of at least 250 cm. It may be assumed that the reachable heights of adult male readers at the library are normally distributed with a mean of 264 cm and a standard deviation of 8 cm, and that those of adult female readers are normally distributed with a mean of 254 cm and a standard deviation of 5 cm.

(a) Find, correct to three significant figures, the proportion of adult male readers and the proportion of adult female readers who are able to reach books on the top shelf.

(b) Given that 40% of all adult readers at the library are male, find the proportion, correct to three significant figures, of all adult readers who are able to reach books on the top shelf.

(c) The library decides to lower the top shelf so that 95% of all adult female readers will be able to reach books there. Find the corresponding percentage, correct to the nearest integer, of all adult male readers who will then be able to reach books on the top shelf.   (JMB)

11. The length X of a certain component made by a machine is specified by the manufacturer to be 10 cm. X may be considered to be a random variable distributed normally with mean 10 cm and standard deviation 0.05 cm. All components are tested and are acceptable if they lie between 9.95 cm and 10.03 cm. Those less than 9.95 cm are rejected at a loss of 40p each to the manufacturer; those between 10.03 and 10.05 cm can be shortened at a loss of 20p and those greater than 10.05 cm can be shortened resulting in a loss of 25p.

Calculate the probabilities that if a component is tested the loss $L = 0, 20, 25, 40$ pence and hence calculate the expected value of L.   (SUJB)

12. A population of adult men has a mean mass of 82.5 kg with a standard deviation of 7.5 kg. Assuming a normal distribution:

(a) what is the probability that a sample of 100 men taken at random from the population will have a mean mass equal to or greater than 84 kg;

(b) what is the probability that two men taken at random from the population will have a combined mass greater than 170 kg?

13.   In a certain country, the men have masses which are normally distributed with mean 70 kg and a standard deviation 8 kg. Three men are chosen at random from this country. Find the probability that:
(a) none of them will have a mass greater than 66 kg,
(b) their mean mass will be greater than 72 kg.
   The women in the same country have masses which are normally distributed with mean 60 kg and standard deviation 6 kg. Find the probability that a man and a woman, selected at random, will have masses differing by less than 5 kg.   (C)

14.   At one stage in the manufacture of an article, a cylindrical rod with a circular cross-section has to fit into a circular socket. Quality control measurements show that the distribution of rod diameters is normal with mean 5.01 cm and standard deviation 0.03 cm, while that of socket diameters is independently normal with mean 5.11 and standard deviation 0.04 cm. If components are selected at random for assembly, what proportion of rods will not fit?
   Rods and sockets are randomly paired for delivery to customers. Batches for delivery are made up of n such pairs. What is the largest value of n for which the probability that all the rods in the batch fit into their respective sockets is greater than 0.9? Given that $n = 30$, find the probability that not more than one rod will fail to fit into its socket.   (MEI)

15.   The thickness, P cm, of a randomly chosen paperback may be regarded as an observation from a normal distribution with mean 2.0 and variance 0.730. The thickness, H cm, of a randomly chosen hardback book may be regarded as an observation from a normal distribution with mean 4.9 and variance 1.920.
(a) Determine the probability that the combined thickness of four randomly chosen paperbacks is greater than the combined thickness of two randomly chosen hardbacks.
(b) By considering $X = 2P - H$, or otherwise, determine the probability that a randomly chosen paperback is less than half as thick as a randomly chosen hardback.
(c) Determine the probability that a randomly chosen collection of sixteen paperbacks and eight hardbacks will have a combined thickness of less than 70 cm.
      (Give 3 decimal places in your answers.)   (C)

16.   A fair coin is tossed 10 times. Evaluate the probability that exactly half of the tosses result in heads.
   The same coin is tossed 100 times. Use the normal approximation to the binomial to estimate the probability that exactly half the tosses result in heads. Also estimate the probability that more than 60 of the tosses result in heads.
   Explain briefly the meaning of the term 'continuity correction' in using the normal approximation to the binomial, and the reason for the adoption of this correction.   (MEI)

17.   An old car is never garaged at night. On the morning following a wet night, the probability that the car does not start is 1/3.
(a) There are 6 consecutive wet nights. Determine the probability that the car does not start on at least 2 of the 6 mornings.
(b) During a wet autumn there are 32 wet nights. Using a suitable approximation, determine the probability that the car does not start on less than 16 of the 32 mornings.   (C)

18. During an advertising campaign, the manufacturers of Wolfitt (a dog food) claimed that 60 % of dog owners preferred to buy Wolfitt. Assuming that the manufacturer's claim is correct for the population of dog owners, calculate: (a) using the binomial distribution, and (b) using a normal approximation to the binomial, the probability that at least 6 of a random sample of 8 dog owners prefer Wolfitt. Comment on the agreement, or disagreement, between your two values. Would the agreement be better or worse if the proportion had been 80 % instead of 60 %?

Continuing to assume that the manufacturer's figure of 60 % is correct, use the normal approximation to the binomial to estimate the probability that, of a random sample of 100 dog owners, the number preferring Wolfitt is between 60 and 70 inclusive.   (MEI)

19. In the long run, an insecticide kills 60 % of a particular type of insect. Write down expressions for the probability that exactly one insect survives in a sample of size (a) 5, (b) 100.

Use the normal distribution to find an approximation to the probability that more than 49 % survive out of a sample of size 96.   (L)

20. State the conditions under which it is permissible to use the normal distribution as an approximation to the binomial distribution.

In a multiple-choice examination paper there are 100 questions, each with a choice of three answers of which only one is correct. To pass the examination it is necessary to answer 40 or more questions correctly. Use the normal distribution as an approximation to the binomial distribution to estimate the probability that a candidate who chooses the answer to each question randomly will pass the examination.

Estimate the probability of passing in this way if the choice of answers to each question is increased to four.

Assuming that the choice of answers to each question remains at three, and the proportion of correct choices required to pass remains at 40 % estimate the least number of questions that the paper should contain if the probability of a pass by random choice is not to exceed 1 %.   (JMB)

21. The number of counts on a Geiger counter is 35 per minute on average. Assuming a Poisson distribution, find an approximation to the probability that there will be more than 40 counts in a given minute.

22. The data below relate to the daily coke yield (per cent) of a coke oven plant over a period of 260 days.

Daily coke yield (% to the nearest unit)	67	68	69	70	71	72	73
Number of days	7	23	61	87	56	19	7

Calculate the mean daily coke yield (per cent) for the period and also the standard deviation of the yield.

Calculate the expected frequency in the interval 68.5–70.5 (per cent) for a normal distribution having the same mean and standard deviation.

Estimate the frequency of observations in the above table within one standard deviation of the mean value and find also the expected frequency in the same interval for the normal distribution.   (MEI)

23. An experiment was set up to estimate the potency of a disinfectant. Various concentrations (expressed as the percentage by volume of pure disinfectant in the solution) of disinfectant were used, and the table shows the numbers of insects killed in the trial of each concentration:

Concentration (x)	40	50	60	65	75
Total number of insects at the start of trial	160	200	200	200	300
Number of insects killed	2	20	66	108	258

(a) For each concentration, determine the proportion of insects killed and find the number y such that the probability of obtaining an observation less than y from a standard normal distribution is equal to that proportion.
(b) Plot a graph of y against x and fit by eye a straight line relating y and x.
(c) Assuming that y is given by $y = (x - \mu)/\sigma$, hence estimate the values of μ and σ. What interpretation does μ have?
(d) Hence estimate also the concentration required to kill 90 % of insects.   (MEI)

24.★ A firm makes short bricks and long bricks. The distribution of the lengths of short bricks is normal with mean 25 cm and standard deviation 1.5 cm. The distribution of the lengths of long bricks is normal with mean k cm and standard deviation 2 cm. The random variable S is the combined length (in centimetres) of a random sample of 9 short bricks and the random variable L is the combined length of a random sample of 4 long bricks.
(a) Determine $P(S > 220)$.
(b) Obtain the mean and variance of $S - L$.
(c) Given that $P(S > L) = 0.95$, show that, to two significant figures, $k = 54$.
(d) Taking k to be 54, determine the probability that a randomly chosen long brick is more than twice as long as a randomly chosen short brick.   (C)

25.★ The random variable X has a normal distribution with parameters μ and σ^2. Derive the mean and variance of X.

$$\left(\text{You may assume that } \frac{1}{\sqrt{2\pi}} \int_{-\infty}^{\infty} e^{-\frac{1}{2}t^2} \, dt = 1. \right)$$

Ben Wedgewood and Sons in cooperation with the National Enterprise Commission have just developed a sophisticated new microwave oven. The 'in-use' lifetimes of two vital components may be considered to be random variables, such that the lifetime of the quality sensitiser, X, is normal with mean 60 hours and standard deviation 5 hours and the lifetime of the overheat warning mechanism, Y, is normal with mean 70 hours and standard deviation 4 hours.
(a) What value of x should be quoted such that $P(X > x) = 0.99$?
(b) The intensive inspection period for the overheat warning mechanism begins at 60 hours and ends at 75 hours. What is the probability of the mechanism failing in this period?
(c) Assuming that X and Y are independent and that $W = Y - X$, what are $E(W)$ and $\text{Var}(W)$? Further, what is the probability that the overheat warning mechanism lasts longer than the quality sensitiser?   (AEB 1981)

26★ The school bus leaves the stop near Ben's home at X minutes past 8.00 am, where X is an $N(20, 3^2)$ random variable. Ben reaches the bus stop at Y minutes after 8.00 am, where Y is an $N(15, 2^2)$ random variable, X and Y being independent. Find the probability (to three decimal places) that Ben misses the bus.

The ride from this stop to the school lasts Z minutes, where Z is an $N[30, (\sqrt{7})^2]$ random variable, independent of X. The random variable W is the number of minutes before 9.00 am at which the bus arrives at the school. Express W in terms of X and Z. Calculate the mean and variance of W and find, to three decimal places, the probability that the bus arrives at the school after 9.00 am. (C)

27.★ The random variable X is distributed normally with mean 2 and standard deviation 4. Find the probabilities that (a) $X > 6$, (b) $|X| > 6$.

A second random variable Y, independent of X, is distributed normally with probability density function f, where

$$f(y) = \frac{1}{(18\pi)^{\frac{1}{2}}} e^{-(y-1)^2/18} \quad \text{(see Appendix page 224).}$$

Write down the values of the mean and standard deviation of Y, and hence determine the mean and standard deviation of $(X + Y)$. Find the probabilities that: (c) $X + Y > 6$, (d) $|X + Y| > 6$. (MEI)

28.★ Four sprinters A, B, C and D each run 100 m. The times (in seconds) that they take may be regarded as independent observations from a normal distribution with mean 14 and standard deviation 0.2. The athlete E runs 400 m. The time (in seconds) that this athlete takes may be regarded as an observation from a normal distribution with mean 58 and standard deviation 1.0 and is independent of the sprinters.

Determine the probability that:
(a) the time taken by E is less than 3 seconds greater than the sum of the times taken by the four sprinters,
(b) the time taken by E is less than 4 times as great as the time taken by A.

For the four sprinters A, B, C and D, find the probability that A is the fastest sprinter of the four.

Find also the probability that C and D (in either order) are the two slowest sprinters. (C)

29.★ The following data were collected in a research investigation into the run times of programs in a computer:

Run time (minutes)	Number of programs
0–	108
2–	112
4–	71
6–	50
8–	37
10–	25
12–	16
14–	11
16–	10
18–	8
20–	13
30–	8
40–	2
50–	0
Total	471

Why is the normal distribution unsuitable for these data?

Use a graphical procedure to estimate the average log run time and its standard deviation.
(MEI)

30.   The following data were collected in a firm responsible for transporting its own goods. They relate to the daily work load of the drivers. Define t as the number of hours worked and x_T as the proportion of observations such that $t \leqslant T$. In order to see whether the distribution is fitted well by a normal distribution, plot $\Phi^{-1}(x_T)$ against T where

$$\Phi(T) = (2\pi)^{-\frac{1}{2}} \int_{-\infty}^{T} \exp(-\tfrac{1}{2}y^2) \, \mathrm{d}y.$$

Use the plot to estimate the mean and standard deviation and to find the upper and lower quartiles. What proportion of drivers exceed the legal maximum of 10 hours?

Hours worked	Frequency	% frequency
6.00– 6.50	5	1.06
6.51– 7.00	11	2.33
7.01– 7.50	24	5.08
7.51– 8.00	73	15.47
8.01– 8.50	116	24.58
8.51– 9.00	89	18.86
9.01– 9.50	76	16.10
9.51–10.00	26	5.51
10.01–10.50	24	5.08
10.51–11.00	28	5.93
	472	100.00

(MEI)

31.★ For the standard distribution with p.d.f.
$$\phi(x) = (2\pi)^{-\frac{1}{2}}\exp(-x^2/2), \quad -\infty \leqslant x \leqslant \infty,$$
prove that $E(X) = 0$, $V(X) = 1$, and find $E(X^3)$ and $E(X^4)$.
   Hence find the mean and variance of the distribution with p.d.f.
$$f(x) = K(1+2x)^2\exp(-x^2/2), \quad -\infty \leqslant x \leqslant \infty.$$

(K should be determined from the properties of p.d.f.'s and you may assume that $x^n\exp(-x^2/2) = 0$ at $x = \pm\infty$ for $n = 1, 2, 3, \ldots$.)
(MEI)

32.★ The length of rods produced by a machine are normally distributed with mean 3 units and standard deviation 1.5. Find the density function for X, the length of rods actually produced, that is disregarding negative values of X (see Appendix page 226).

11.10   Appendix

The density function of a normal distribution

Although our discussion on pages 197–9 described in broad terms the relation between the standard normal distribution and the general normal distribution, it obscured at least one important point: how the change of scale affects the *area* and therefore the density function.

We shall now develop the theory of the distribution, starting with the standard normal distribution, as before, but this time it will be *defined by its density function* (making no assumption about its variance). From there we will derive the general normal distribution, changing the scale and shifting the origin as we did before. This will give us the density function of the general distribution. Finally we shall *derive* the mean and standard deviation of the distribution *from the density function.*

Consider the function

$$f(z) = ke^{-\frac{1}{2}z^2}, \quad -\infty < z < \infty.$$

It has the following properties:

1. $f(z) = f(-z)$, so that it is symmetrical about the origin,

2. $f(z) \to 0$ as $z \to \infty$ or $-\infty$,

3. $f(z)$ is positive for all values of z.

If $f(z)$ is to be a possible probability density function for the random variable Z it is necessary that

$$\int_{-\infty}^{\infty} f(z)\, dz = 1.$$

Now it can be proved that $\displaystyle\int_{-\infty}^{\infty} e^{-\frac{1}{2}z^2}\, dz = \sqrt{(2\pi)}.$

(The proof is well beyond the scope of this book.)

We can therefore define as a probability density function the distribution of the random variable Z, whose density function is

$$f(z) = \frac{1}{\sqrt{(2\pi)}}\, e^{-\frac{1}{2}z^2}, \quad -\infty < z < \infty.$$

Z is then said to have a *standard normal distribution.*

The graph of $f(z)$ is shown in the figure.

The mean of the distribution is 0, by symmetry, and the fact that there is a point of inflexion at $z = 1$ may be verified by equating $f''(z)$ to 0.

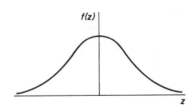

The general normal distribution

Suppose, now, that we change the scale, putting $y = \sigma z$, so that $z = y/\sigma$, where σ is any positive constant.

In order that the function $Ke^{-\frac{1}{2}(y/\sigma)^2}$ may be a possible density function for the random variable Y ($-\infty < y < \infty$), K has to be chosen so that

$$\int_{-\infty}^{\infty} Ke^{-\frac{1}{2}(y/\sigma)^2}\, dy = 1.$$

To find the value of K, put $y = \sigma z$, so that

$$\int_{-\infty}^{\infty} e^{-\frac{1}{2}(y/\sigma)^2}\, dy = \sigma \int_{-\infty}^{\infty} e^{-\frac{1}{2}z^2}\, dz = \sigma\sqrt{(2\pi)} \quad \left[\text{since} \int_{-\infty}^{\infty} e^{-\frac{1}{2}z^2}\, dz = \sqrt{(2\pi)}\right].$$

Hence $K = 1/(\sigma\sqrt{2\pi})$, and the density function for Y is

$$\frac{1}{\sigma\sqrt{(2\pi)}}e^{-\frac{1}{2}(y/\sigma)^2}, \quad -\infty < y < \infty.$$

The graphs of the density functions for Z and Y are shown together in the figure (in which σ is assumed to be greater than 1). Their fundamental shapes are the same, the only difference being their scales.

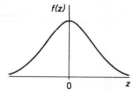
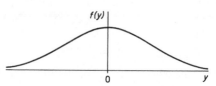

If, finally, we change the origin by putting $x = y+\mu$, so that $y = x-\mu$, where μ is another arbitrary constant, this will have the effect of shifting the origin so that the mean will now be at $x = \mu$. The area under the curve will remain equal to 1, and the density function of X will be

$$\frac{1}{\sigma\sqrt{(2\pi)}}e^{-\frac{1}{2}[(x-\mu)/\sigma]^2}, \quad -\infty < x < \infty.$$

The graphs of the density functions for Y and X are shown together in the next figure (in which μ is assumed to be positive).

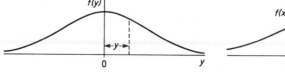

The normal distribution is therefore defined formally as follows:

The distribution of the random variable X, with density function

$$\frac{1}{\sigma\sqrt{(2\pi)}}e^{-\frac{1}{2}[(x-\mu)/\sigma]^2}, \quad -\infty < x < \infty,$$

where μ and σ are two parameters ($\sigma > 0$), is called the normal distribution.

It is referred to shortly as $N(\mu, \sigma^2)$.

Putting $\mu = 0$ and $\sigma = 1$, gives the standard normal distribution, $N(0, 1)$ with density function

$$\frac{1}{\sqrt{(2\pi)}}e^{-\frac{1}{2}z^2}, \quad -\infty < z < \infty \quad \text{(using } z \text{ as variable)}.$$

Note: The transformation of the random variable Z to the random variable X was effected by putting $Y = \sigma Z$, followed by $X = Y+\mu$. Together these are equivalent to

$$X = \sigma Z + \mu \quad \text{or} \quad Z = \frac{X-\mu}{\sigma}.$$

The mean and standard deviation of the normal distribution

In this formal treatment, the standard normal distribution has been defined without reference to its mean and variance, except for noting that the mean must be 0 by symmetry.

As with any continuous random variable whose density function is known, the mean and variance may be found by integration.

Similarly, in the definition of $N(\mu, \sigma^2)$, μ and σ^2 were stated to be *any* constants, so that it remains to be proved that they are in fact the mean and variance.

We start with the standard normal distribution, the distribution of the random variable Z, whose density function is

$$\frac{1}{\sqrt{(2\pi)}}e^{-\frac{1}{2}z^2}, \quad -\infty < z < \infty.$$

$$E(Z) = \frac{1}{\sqrt{(2\pi)}} \int_{-\infty}^{\infty} z e^{-\frac{1}{2}z^2} \, dz = \frac{1}{\sqrt{(2\pi)}} \left[-e^{-\frac{1}{2}z^2} \right]_{-\infty}^{\infty} = 0.$$

$$E(Z^2) = \frac{1}{\sqrt{(2\pi)}} \int_{-\infty}^{\infty} z^2 e^{-\frac{1}{2}z^2} \, dz$$

$$= 2 \times \frac{1}{\sqrt{(2\pi)}} \int_{0}^{\infty} z^2 e^{-\frac{1}{2}z^2} \, dz \quad \begin{array}{l}\text{(since the integrand is an even} \\ \text{function, that is it is} \\ \text{unchanged if } z \text{ is replaced} \\ \text{by } -z).\end{array}$$

Integrating by parts:

Let $u = z$ and $dv = z e^{-\frac{1}{2}z^2} \, dz \Rightarrow v = -e^{-\frac{1}{2}z^2}$.

Hence
$$E(Z^2) = \frac{2}{\sqrt{(2\pi)}} \int_{0}^{\infty} z \cdot z e^{-\frac{1}{2}z^2} \, dz$$

$$= \frac{2}{\sqrt{(2\pi)}} \left\{ \left[-z e^{-\frac{1}{2}z^2} \right]_{0}^{\infty} + \int_{0}^{\infty} e^{-\frac{1}{2}z^2} \, dz \right\}.$$

Now $\lim_{z \to \infty} z e^{-\frac{1}{2}z^2} = 0$ (this may be proved by the method of page 193)

and
$$\int_{0}^{\infty} e^{-\frac{1}{2}z^2} \, dz = \frac{1}{2} \int_{-\infty}^{\infty} e^{-\frac{1}{2}z^2} \, dz = \frac{1}{2} \times \sqrt{(2\pi)},$$

so that
$$E(Z^2) = \frac{2}{\sqrt{(2\pi)}} \times \frac{1}{2}\sqrt{(2\pi)} = 1$$

and
$$\text{Var}(Z) = E(Z^2) - (0)^2 = 1.$$

The general normal distribution $N(\mu, \sigma^2)$ was also defined without reference to mean and variance. To find these we use the results that if $X = aZ + b$,

$$E(X) = aE(Z) + b \quad \text{and} \quad \text{Var}(X) = a^2 \text{Var}(Z).$$

In this case $X = \sigma Z + \mu$, so that $E(X) = \sigma E(Z) + \mu = 0 + \mu = \mu$,

and $\text{Var}(X) = \sigma^2 \times \text{Var}(Z) = \sigma^2 \times 1 = \sigma^2$.

A truncated normal distribution

It sometimes happens that a random variable can be normally distributed with certain values being inadmissible, giving what is called a truncated normal distribution. The following example will illustrate the procedure.

A machine produces metal rods whose diameters are normally distributed with mean 1.2 cm and standard deviation 0.02 cm. Rods whose diameters exceed 1.24 are discarded. Find the density function for X, the lengths of rods *not* discarded.

We first find the distribution function:

If $x > 1.24$, $F(x) = P(X \leqslant x) = 1$, since X is always less than 1.24.

If $x \leqslant 1.24$, $F(x) = P(X \leqslant x \mid x \leqslant 1.24)$

$$= \frac{P(X \leqslant x)}{P(x \leqslant 1.24)}, \quad \text{using conditional probability.}$$

Now
$$P(X \leqslant x) = \Phi\left(\frac{x - 1.2}{0.02}\right)$$

and
$$P(X \leqslant 1.24) = \Phi\left(\frac{1.24 - 1.2}{0.02}\right) = \Phi(2).$$

Hence, for $x \leqslant 1.24$,

$$F(x) = \frac{1}{\Phi(2)} \Phi\left(\frac{x - 1.2}{0.02}\right).$$

Differentiating w.r.t. x,

$$f(x) = F'(x) = \frac{1}{\Phi(2)} \times \Phi'\left(\frac{x - 1.2}{0.02}\right).$$

But $\Phi'(z) = \phi(z)$, where $\phi(z) = \frac{1}{\sqrt{(2\pi)}} e^{-\frac{1}{2}z^2}$,

so that
$$f(x) = \frac{1}{\Phi(2)} \times \frac{1}{0.02} \times \Phi\left(\frac{x - 1.2}{0.02}\right)$$

$$= \frac{1}{\Phi(2)} \times \frac{1}{0.02} \times \frac{1}{\sqrt{(2\pi)}} e^{-\frac{1}{2}[(x - 1.2)/0.02]^2},$$

giving the density function of X as

$$f(x) = \begin{cases} \dfrac{1}{\Phi(2)} \times \dfrac{1}{\sqrt{(2\pi)} \times (0.02)} \times e^{-\frac{1}{2}[(x - 1.2)/0.02]^2}, & x \leqslant 1.24 \\ 0 & , \quad x > 1.24. \end{cases}$$

12 Sampling distributions and estimation

12.1 From sample statistic to population parameter

In chapter 1, two typical problems were described, illustrating one of the central problems of statistical investigation, which is to discover what we can about the nature of a population by examining a sample drawn from that population. In the first example, conclusions about voting intentions were inferred from information gathered from a sample of voters; in the second a conclusion about the likely mean weight of a population of students was inferred from the mean weight of a sample of students.

Most of you will have been made aware that sampling lies at the heart of statistical investigation by references to polls of opinion on matters of current interest that appear regularly in the press. These references often state specifically the size of the sample from which the state of public opinion as a whole is estimated. What is less well known is that the use of samples in this way is comparatively modern. One of the most important social surveys ever undertaken in Great Britain was Seebohm Rowntree's account of the extent and effect of poverty in the city of York during the last decade of the 19th century, a survey which led to the first National Insurance Act and subsequently to the provisions of the Welfare State. Yet Rowntree was obliged to collect his findings by having every single working-class household in the city visited. Fifty years later, when the investigation was repeated it was only necessary to use a sample of households.

In this, and succeeding, chapters we shall show what is involved in this process of inference from sample to population, or more specifically from *sample statistic* to *population parameter*. (Remember that populations are characterised by certain parameters: the parameters of a normal distribution are its mean μ and variance σ^2, for example. The corresponding quantities for a single sample are statistics.)

The distinction between parameters and statistics will be kept clear if we use Greek letters for parameters. For example:

Quantity	Population	Sample
Mean	μ	$\bar{x}$
Variance	σ^2	$\text{var}(x)$
Proportion with a certain characteristic	π	p

Note:

1.   There is no generally accepted symbol for sample variance, as there are, in fact, two possible ways of *defining* this variance (see later on page 234).

2.   Up to now we have used '*p*' for a population proportion. From now on we will use π.

Suppose the Cumberland county agricultural committee wants to estimate the mean length μ of full-grown sheep in their area. It would clearly be impracticable to measure all sheep, so a random sample of 60, say, is measured and the mean of the sample, $\bar{x}$, is taken as an estimate of the population mean. The question then arises: how close is this sample value to the actual population mean? Clearly, no hard and fast answer can be given as the value of $\bar{x}$ will depend on how the sample is made up. Nevertheless, we ought to be able to draw some conclusion, but, as we saw in chapter 1, this conclusion will always be expressed in terms of probability. Diagrammatically, then, the problem is like this:

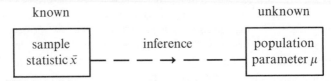

where the 'inference' line is broken to reflect the fact that the conclusion cannot be certain.

In order to deal with this problem we must first look at the *reverse* problem. Suppose we *knew* what the distribution of lengths was, that, for example, they were normally distributed with a known mean and variance. What then will be the nature of the distribution of all possible means of samples of size 60? This will turn out to be a matter of deduction and will lead us, in due course, to the answer to our original problem:

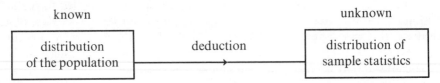

12.2   Sampling distributions

In the last paragraph we referred to a distribution of sample means, so we must now explain this term precisely.

If all possible samples of a given size are taken from a population with known mean and variance, then the *means of the samples* themselves will form a distribution. Similarly the *variances of the samples* will form a distribution. These distributions are called, respectively, the *sampling distribution of the mean* and the *sampling distribution of the variance*.

The same thing applies to other statistics. If, for example, it is known that the population consisting of the output of a machine has a *proportion*, 1 in 30, of defective items, then the proportion of defective items in samples of size 25, say, will form a distribution called the *sampling distribution of the proportion*. This distribution will be examined in a later chapter.

To sum up, then, if we have a single sample and have calculated its mean $\bar{x}$ in order to use it to estimate the population mean μ, this can only be done if we first have some knowledge of the sampling distribution of the mean. Similarly, to estimate the population variance σ^2, we must first know about the sampling distribution of the variance.

We shall start by establishing some facts about the sampling distribution of the mean.

Note: From now on we shall often refer to the population as the 'parent' population for the good reason that it gives 'birth' to sampling distributions.

12.3   The sampling distribution of the mean

We shall start with a simple numerical example.

A metal disc is marked with the numbers 3 and 4 on opposite sides. It is biased so that when it is tossed, the probability of getting a 3 is 2/5 and the probability of getting a 4 is 3/5. We shall take these two numbers 3 and 4 as our parent population, with distribution:

X	3	4
Probability	2/5	3/5

The mean $\mu = \dfrac{2}{5} \times 3 + \dfrac{3}{5} \times 4 = 3.6$

and the variance $\sigma^2 = \dfrac{2}{5} \times 9 + \dfrac{3}{5} \times 16 - (3.6)^2 = 0.24$.

If, now, we throw this disc *twice*, all the possible 'samples' of two scores from the population are (3, 3), (3, 4), (4, 3), (4, 4) and the probabilities that these will turn up are 4/25, 6/25, 6/25, 9/25 respectively.

This means that $\bar{X}$, the means of these samples will have the 'sampling distribution'

$\bar{X}$	3	3.5	3.5	4
Probability	4/25	6/25	6/25	9/25

Calculating the mean of this sampling distribution gives

$$E(\bar{X}) = \frac{4}{25} \times 3 + \frac{6}{25} \times 3.5 + \frac{6}{25} \times 3.5 + \frac{9}{25} \times 4 = 3.6.$$

In other words the mean of the sampling distribution is the *same* as the mean of the parent population, a result which is not entirely unexpected.

The variance of the sampling distribution gives

$$\text{Var}(\bar{X}) = \frac{4}{25} \times 9 + \frac{6}{25} \times 12.25 + \frac{6}{25} \times 12.25 + \frac{9}{25} \times 16 - (3.6)^2 = 0.12,$$

which is exactly *half* the variance of the parent population.

On the evidence of this example, it looks as though, for samples of size *two*, if the population mean is μ and its variance is σ^2, then the mean of the sampling distribution of the mean is also μ, but its variance is $\sigma^2/2$. We shall now confirm that this is so.

If the score on the first throw is X_1 and the score on the second throw is X_2, these are obviously independent of each other, and are therefore *independent random variables* and the distribution of each is the population distribution with mean μ and variance σ^2, say.

Now for any two random variables

$$E(X_1 + X_2) = E(X_1) + E(X_2) = 2\mu$$

so that $E(\bar{X}) = E[\tfrac{1}{2}(X_1 + X_2)] = \tfrac{1}{2}E(X_1 + X_2) = \mu.$

Similarly for two *independent* random variables

$$\mathrm{Var}(X_1 + X_2) = \mathrm{Var}(X_1) + \mathrm{Var}(X_2)$$

so that $\mathrm{Var}(\bar{X}) = \mathrm{Var}\,\tfrac{1}{2}(X_1 + X_2) = \tfrac{1}{4}\mathrm{Var}(X_1 + X_2) = \tfrac{1}{4} \times 2\sigma^2$

$$= \frac{\sigma^2}{2}.$$

Before dealing with the general case, one important comment must be made. In the example we have just given, the results of the two throws *had* to be independent because of the physical method of choosing the sample. This is not true of sampling in general. Suppose, for example, the population consists of ten pieces of card marked with numbers as follows:

If X is the score when one is picked out at random, the distribution of X is

X	21	34	17	18	12	11	42
Probability	2/10	1/10	3/10	1/10	1/10	1/10	1/10

Now if we pick a sample of three, say, it makes a difference whether or not the first one chosen is replaced, and similarly for the second one. Suppose the first one picked, X_1, is 17 (with probability 3/10) and is not replaced, then the population from which the second, X_2, is picked has been altered and the probability that X_2 is 34, say, is 1/9 and not 1/10 as it would have been if X_1 is replaced. In these circumstances X_2 is not independent of X_1.

In what follows we assume sampling *with replacement* so that the observations forming a sample are independent. Such sampling is called *simple random sampling*. (If the parent population is very large, sampling without replacement may be regarded as almost the same as sampling with replacement.)

In simple random sampling, if $X_1, X_2, \ldots, X_n$ are n observations making up the sample, taken from a population distribution whose mean is μ and variance σ^2, then $X_1, X_2, \ldots, X_n$ are independent random variables, *each of which has the same distribution as the parent population*. This important idea will be used to establish the general result that follows. Notice that the method is merely a natural extension of the method used above for a sample of two.

Let $\bar{X}$ be the mean of such a sample of size n, so that

$$\bar{X} = \frac{1}{n}(X_1 + X_2 + \ldots + X_n).$$

The mean of the sampling distribution of $\bar{X}$ is

$$E(\bar{X}) = E\left[\frac{1}{n}(X_1 + X_2 + \ldots + X_n)\right]$$

$$= \frac{1}{n}[E(X_1) + E(X_2) + \ldots + E(X_n)] \quad \text{(by the usual rules of expectation algebra).}$$

But, as we noted above, $E(X_1) = E(X_2) = \ldots = E(X_n) = \mu$.

Hence
$$E(\bar{X}) = \frac{1}{n}(\mu + \mu + \ldots + \mu) = \frac{1}{n}(n\mu)$$

so that
$$E(\bar{X}) = \mu.$$

The procedure for variance is similar.

$$\text{Var}(\bar{X}) = \text{Var}\left[\frac{1}{n}(X_1 + X_2 + \ldots + X_n)\right]$$

$$= \frac{1}{n^2} \times [\text{Var}(X_1) + \text{Var}(X_2) + \ldots + \text{Var}(X_n)]$$

$$\text{(since } X_1, X_2, \ldots, X_n \text{ are \textit{independent})}$$

But
$$\text{Var}(X_1) = \text{Var}(X_2) = \ldots = \text{Var}(X_n) = \sigma^2.$$

Hence
$$\text{Var}(\bar{X}) = \frac{1}{n^2}(\sigma^2 + \sigma^2 + \ldots + \sigma^2) = \frac{n\sigma^2}{n^2}$$

so that
$$\text{Var}(\bar{X}) = \frac{\sigma^2}{n}.$$

We have established therefore, the very important results that

the sampling distribution of the mean has mean μ and variance σ^2/n.

Since $\sigma^2/n < \sigma^2$, the distribution of sample means, while centering on μ, has a smaller variance than the parent distribution so that it is less spread out. This is what we would expect, because a sample mean is an average so that the larger or smaller individual observations will tend to be 'ironed out' in the calculation of the averages.

Standard error

A special name is given to the *standard deviation* of the sampling distribution of a statistic. It is called its *standard error*. So, in the case of the mean, we say that

$$\text{the standard error of the mean} = \frac{\sigma}{\sqrt{n}}.$$

EXERCISE 12a

1.   The scores in an aptitude test taken by a large number of students are distributed around a mean of 100 with variance 120. If random samples of 36 students were picked what would be the mean, the variance and the standard deviation (standard error) of $\bar{X}$, the mean scores of the student samples?

2.   Samples of size n are taken from a distribution whose variance is 110.25. The standard error of the sample means is 2.1. What is the value of n?

3.   A large number of pieces of card are placed in a container. A quarter of them are marked with the number 4 and the remainder are marked with the number 2.
   If X is the number showing when a single card is drawn, calculate μ and σ^2, the mean and variance of X.
   Samples of two cards are now drawn (with replacement), the possible samples being (2, 4), (4, 2), (2, 2), (4, 4). Draw up a sampling distribution of the mean, $\bar{X}$, of the samples (as on page 229) and use it to calculate the mean and variance of $\bar{X}$. Check that your answers agree with the results on page 231.

4.   If, as in question 3, a container contains cards marked 1 to 6 in equal proportions, calculate μ and σ^2 for X, the number showing when a single card is drawn at random.
   Use a possibility space diagram (see page 66) to draw up the sampling distribution of the mean $\bar{X}$ of the numbers showing on samples of two cards drawn at random (with replacement) and use it to calculate the mean and variance of $\bar{X}$.

5.   Repeat question 3 for samples of three cards.

12.4   Estimation and unbiased estimates

We are now in a position to make an estimate of the parameter μ of a parent population, supposing that all we have to go on is a single sample. Such an estimate is called a 'point estimate'.

Suppose, for example, that a random sample of 200 households in a certain area are found to watch television for an average of 15.3 hours per week and this value of $\bar{X}$ is used as an estimate of the population mean, μ. It is likely that we would instinctively feel that this is a reasonable estimate. On the other hand if the mean of a sample of 10 were used, we would be less happy about it.

These instinctive ideas must now be expressed in mathematical form, which means that we must establish some criteria for accepting a sample statistic as a 'good' estimate.

As we have seen, the sample means $\bar{X}$ cluster round the population mean μ, since the distribution of sample means has mean μ. Putting it another way, the average values of $\bar{X}$ for all possible samples gives the true value of μ:

$$E(\bar{X}) = \mu.$$

When *any* statistic has this property – that the mean of its sampling distribution is the same as the true population parameter – we say that the estimate of the parameter using the sample statistic is *unbiased*.

Consider another statistic, a proportion. Although we have not yet discussed the sampling distribution of a proportion it turns out that if, for example, the proportion in the population owning colour TV sets is π, the distribution of P, the proportion in samples of a given size who own colour TV is such that

$$E(P) = \pi.$$

Here, too, the estimate based on a sample will be unbiased.

In general, then, if a statistic 'T' is calculated from a single sample and the corresponding population parameter is 'θ', T is called an *unbiased* estimate of θ if

$$E(T) = \theta.$$

At this point, you may well be wondering whether there is any common statistic that is *biased*. We will therefore anticipate the next section by saying that if a sample *variance* is taken as an estimate of the population variance (σ^2), the estimate will not be unbiased, since it turns out that

$$E(\text{sample variance}) \text{ is } not \text{ equal to } \sigma^2.$$

Notation

The following notation is often used for unbiased estimates:

$$\text{estimated mean } \hat{\mu} \qquad \text{estimated proportion } \hat{\pi}$$
$$\text{estimated variance } \hat{\sigma}^2$$

so that for an unbiased estimate of the mean

$$E(\bar{X}) = \mu \quad \text{and we write} \quad \hat{\mu} = \bar{X}.$$

For any statistic 'T' with corresponding population parameter θ, T is an unbiased estimate of θ if

$$E(T) = \theta \quad \text{and we write} \quad \hat{\theta} = T.$$

Variance

The question must now be asked, is an unbiased estimate necessarily a 'good' one? The answer is 'No'.

At the beginning of this section we talked about basing an estimate of the average time spent watching TV on samples of different sizes. Now whether the sample is large or small, the estimate will be unbiased. The proof on page 231 shows this. But the distribution of the means of samples of 200 has a much smaller *variance* ($\sigma^2/200$) than that of the distribution for samples of 10 ($\sigma^2/10$). This means that the former distribution is much less spread out than the latter, and the mean of a small sample taken at random might well be far from the true population mean. In that sense the small sample estimate is not a useful one. This matter will be discussed more fully later in this chapter (page 240). Meanwhile we must turn our attention to the sampling distribution of the variance.

12.5   The sampling distribution of the variance

If a machine is producing metal nuts whose inside diameter, X is the important dimension, its performance will vary slightly and this variability will be measured by the variance σ^2. In order to check on this variance we shall suppose that a sample of 20 nuts is taken from the output and the variance of the inside diameters of this sample calculated. How can this be used to get an unbiased estimate of σ^2.

As with the mean, the answer to this question involves examining the sampling distribution of the variance and finding its expectation. We shall do this shortly, but the mathematics is rather tricky and some of you may prefer, on a first reading, to omit it.

It turns out that an unbiased estimate of σ^2 is *not* the sample variance, since E(sample variance) $\neq \sigma^2$.

In fact
$$E\left(\frac{n}{n-1} \times \text{sample variance}\right) = \sigma^2$$

and we therefore take $[n/(n-1) \times \text{sample variance}]$ as our unbiased estimate:

$$\hat{\sigma}^2 = \frac{n}{n-1} \times \text{sample variance.}$$

If, then, the 20 nuts mentioned above give a sample variance of 0.01^2 mm^2, an unbiased estimate for σ^2 is

$$\frac{20}{19} \times 0.01^2 = 0.0103^2 \text{ mm}^2.$$

In other words, the sample variance is an underestimate of the population variance and we allow for this fact by multiplying it by the factor $n/(n-1)$.

It follows that if the sample consists of the observations $X_1, X_2, \ldots, X_n$,

$$\text{the sample variance} = \frac{\Sigma (X_i - \bar{X})^2}{n}$$

and
$$\hat{\sigma}^2 = \frac{\Sigma (X_i - \bar{X})^2}{n-1} \quad (\text{summing for } i = 1 \text{ to } n).$$

[It is for this reason that the sample variance is sometimes *defined* as $\Sigma (X_i - \bar{X})^2/(n-1)$.]

The symbol for this unbiased estimate is s^2, so that, summing up, we have

$$E(s^2) = \sigma^2, \text{ where } s^2 = \frac{\Sigma (X_i - \bar{X})^2}{n-1} = \frac{n}{n-1} \times \text{sample variance.}$$

Note: We have been using the word 'estimate' up to now, but an estimated function like $\Sigma (X_i - \bar{X})^2/(n-1)$ is usually called an *estimator,* and the word estimate should strictly be used for the value of the estimator in a particular case. If, for example, the value of the sample variance is 0.4 and $n = 10$, the unbiased estimator is $\Sigma n/(n-1) \times \text{sample variance}$, and the corresponding unbiased estimate in this case is $(10/9) \times 0.4 \approx 0.44$.

There is another convention that estimators are written with capital letters and estimates (particular values) with lower case (small letters). However, it is customary to denote this special and important estimator by s^2 and not to distinguish it from its value in a particular case.

EXAMPLE 1   A sample of four matchboxes taken from the output of a filling machine contained 52, 56, 53 and 51 matches. Find unbiased estimates for the mean and variance of the output.

Here $n = 4, \bar{x} = \dfrac{212}{4} \Rightarrow \hat{\mu} = 53.$

As $\bar{x}$ is an integer and the sample is a small one, it is convenient to use

$$s^2 = \frac{\Sigma (x - 53)^2}{4 - 1} = \frac{14}{3}.$$

EXAMPLE 2   Estimate the population variance from the sample with recorded values

$$10.4 \quad 10.3 \quad 11.5 \quad 9.8 \quad 10.4.$$

Working with the coded distribution

$$0.4 \quad 0.3 \quad 1.5 \quad -0.2 \quad 0.4$$

the mean $\bar{x} = 0.48$.
    Using the formula

$$\Sigma \frac{x^2}{n} - (\bar{x})^2,$$

the sample variance $= [0.4^2 + 0.3^2 + 1.5^2 + (-0.2)^2 + 0.4^2]/5 - (0.48)^2$

$$= 0.3096$$

Hence $\hat{\sigma}^2 = s^2 = \dfrac{5}{4} \times 0.3096 = 0.387.$

EXAMPLE 3   (In this example we shall check the results we have stated in a simple numerical case.)
    A population consist of the numbers 1, 2, 3 printed on equal numbers of cards in a container. The distribution of X, the number obtained when a card is drawn at random is thus:

X	1	2	3
Probability	1/3	1/3	1/3

Find the variance of each of the possible samples of *two* drawn (with replacement) from the container and find the mean of the resulting distribution of sample variances.

The possible samples are

(X_1, X_2):   (1,1)   (1,2)   (1,3)   (2,1)   (2,2)   (2,3)   (3,1)   (3,2)   (3,3)

with means:

$\bar{X}$ 1   1.5   2   1.5   2   2.5   2   2.5   3 and variances $= \frac{1}{2}[(X_1 - \bar{X})^2 + (X_2 - \bar{X})^2]$:

$$0 \qquad 0.25 \quad 1 \qquad 0.25 \quad 0 \qquad 0.25 \quad 1 \qquad 0.25 \quad 0.$$

Hence $E(\text{sample variance}) = 3/9 = 1/3$.

Now $\mu = E(X) = 2$ and $\sigma^2 = \text{Var}(X) = \frac{1}{3}[(1-2)^2 + (2-2)^2 + (3-2)^2] = 2/3$.

In this case, $E(\text{sample variance}) \neq \sigma^2$, but, with $n = 2$,

$$E\left(\frac{n}{n-1} \times \text{sample variance}\right) = E(2 \times \text{sample variance}) = 2 \times \frac{1}{3} = \sigma^2.$$

EXAMPLE 4★   If $X_1, X_2, \ldots, X_n$ is a random sample from the distribution of the random variable X whose density function is

$$f(x) = \begin{cases} \dfrac{x}{\theta^2} e^{-x/\theta}, & x \geqslant 0 \quad (\theta > 0) \\ 0 & , \quad \text{otherwise}. \end{cases}$$

show that

$$T = \sum_{i=1}^{n} (X_i/2n)$$

is an unbiased estimator of the parameter θ of the parent distribution.

$$E(T) = E[\Sigma \ (X_i/2n)] = \frac{1}{2n} \Sigma \ [E(X_i)]$$

$$= \frac{1}{2n} \times n \times E(X), \quad \text{since } E(X_i) = E(X), \quad i = 1, 2, \ldots, n.$$

Now $E(X) = \displaystyle\int_0^\infty \frac{x^2}{\theta} e^{-x/\theta} dx,$

which, after integration by parts

$$= 2 \int_0^\infty e^{-x/\theta} \ dx = \left[-2\theta e^{-x/\theta} \right]_0^\infty$$

$$= 2\theta.$$

Hence $E(T) = \dfrac{1}{2} \times 2\theta = \theta.$

In other words T is an unbiased estimator of θ.

A *proof* that $s^2 = [n/(n-1)] \times \text{sample variance}$ appears after the following exercise.

EXERCISE 12b

1. Find unbiased estimates of the population mean and variance using the following sample observations:

$$28 \quad 32 \quad 38 \quad 57 \quad 46 \quad 31 \quad 65 \quad 44 \quad 40 \quad 49.$$

2.   A random sample taken from a population with mean μ and variance σ^2 yields the following values:

$$2.88 \quad 2.64 \quad 2.98 \quad 3.16 \quad 2.86 \quad 3.00.$$

Find unbiased estimates of μ and σ^2.

3.   Calculate unbiased estimates of the mean and variance of a population using the following recorded observations from a sample:

X	0	1	2	3	4	5
f	3	7	25	16	2	1

4.   If the variance of a sample of size n is used as an approximation for an unbiased estimate of the population variance, what will the percentage error be?

5.   A random sample of nine observations is drawn from a population having mean μ and variance σ^2. The sample mean is 6 and the unbiased estimate of σ^2 based on this sample is 9. Given that an additional random observation from the population distribution has the value 5, find unbiased estimates of μ and σ^2 based on all ten observations.   (JMB)

6.   A random sample of independent observations taken from a distribution having variance σ^2 yielded the values $x_1, x_2, \ldots, x_n$. Explain what is meant by the statement that s^2 is an unbiased estimate of σ^2, where

$$(n-1)s^2 = \sum_{i=1}^{n} (x_i - \bar{x})^2,$$

$\bar{x}$ being the sample mean.
   Show that, when $n = 2$,

$$s^2 = \tfrac{1}{2}(x_1 - x_2)^2.$$

Two pairs of numbers are chosen at random with replacement from the set 1, 3, 5. By considering all possible choices of pairs of numbers, verify that in this case the statistic s^2 defined above is an unbiased estimate of the distribution variance.   (JMB)

★Proof that $s^2 = [n/(n-1)] \times$ sample variance

Suppose a sample of observations $X_1, X_2, \ldots, X_n$ is taken from a population whose mean is μ and variance is σ^2.
   The sample variance $= (1/n)\Sigma\,(X - \bar{X})^2$, and the variances of all possible samples of size n will form a distribution, the sampling distribution of the variance.
   We shall find the mean of this distribution, E(sample variance).
   As the algebra is complicated we shall first state some of the basic facts that will be needed.

1. $E(X + Y + \ldots) = E(X) + E(Y) + \ldots$ or
   $$E[\Sigma\,(X)] = \Sigma[E(X)], \quad \text{that is, } E \text{ and } \Sigma \text{ may be interchanged.}$$

2. $\mathrm{Var}(X) = E(X^2) - [E(X)]^2 = E(X^2) - \mu^2$
   $\Rightarrow E(X^2) = \mathrm{Var}(X) + \mu^2 = \sigma^2 + \mu^2.$

3. $\mathrm{Var}(\bar{X}) = E(\bar{X}^2) - [E(\bar{X})]^2 = E(\bar{X}^2) - \mu^2 \quad$ (since $E(\bar{X}) = \mu$)
   $\Rightarrow E(\bar{X}^2) = \mathrm{Var}(\bar{X}) + \mu^2$
   $\qquad\qquad = \sigma^2/n + \mu^2 \quad$ (since $\mathrm{Var}(\bar{X}) = \sigma^2/n$).

Proof

$$\text{Sample variance} = \frac{1}{n} \Sigma (X - \bar{X})^2 \Rightarrow E(\text{sample variance}) = \frac{1}{n} E[\Sigma (X - \bar{X})^2]$$

$$E(\text{sample variance}) = \frac{1}{n} E[\Sigma X^2 - 2\bar{X} \Sigma X + n\bar{X}^2]$$

$$= \frac{1}{n} E[\Sigma X^2 - 2n\bar{X}^2 + n\bar{X}^2] \quad (\text{since } \Sigma X = n\bar{X})$$

$$= \frac{1}{n} E[\Sigma X^2 - n\bar{X}^2]$$

$$= \frac{1}{n} \{\Sigma [E(X^2)] - nE(\bar{X}^2)]\} \quad (\text{using 1})$$

$$= \frac{1}{n} \left\{n(\sigma^2 + \mu^2) - n\left(\frac{\sigma^2}{n} + \mu^2\right)\right\} \quad (\text{using 2 and 3})$$

$$= \frac{1}{n} \{(n-1) \sigma^2\} = \frac{n-1}{n} \sigma^2.$$

Thus we have proved that the distribution of sample variances has mean

$$\frac{n-1}{n} \sigma^2.$$

In other words the sample variances do not cluster round the true population variance, σ^2, but round $(n-1)\sigma^2/n$, and this therefore gives a *biased* estimate.

If, however, we multiply both sides of the formula

$$E(\text{sample variance}) = \frac{n-1}{n} \sigma^2 \quad \text{by} \quad \frac{n}{n-1},$$

remembering that $kE(X) = E(kX)$, we get

$$E\left(\frac{n}{n-1} \times \text{sample variance}\right) = \sigma^2.$$

So the values of the statistic $[n/(n-1)] \times$ sample variance *do* centre on the true population variance, and we take this statistic as an unbiased estimate of σ^2.

(After this rather heavy piece of algebra, you will be relieved to know that we do not need to investigate the *variance* of the sampling distribution of the variance, as it is not needed in elementary work.)

12.6   Point estimation using pooled samples

So far we have obtained our estimates from a single sample. Suppose now that we have two samples: one, size n_1, from which

$$\bar{X}_1 = \frac{\Sigma X_1}{n_1} \quad \text{and} \quad s_1^2 = \frac{\Sigma (X_1 - \bar{X}_1)^2}{n_1 - 1}$$

are calculated, and the second, from which

$$\bar{X}_2 = \frac{\Sigma X_2}{n_2} \quad \text{and} \quad s_2^2 = \frac{\Sigma (X_2 - \bar{X}_2)^2}{n_2 - 1}$$

are calculated.

We want to use the information obtained by pooling these samples to estimate the population mean μ and the population variation σ^2.

For the mean, since the total number of observations in the two samples together is $n_1 + n_2$, and the sum of the observations is $\Sigma X_1 + \Sigma X_2 = n_1 \bar{X}_1 + n_2 \bar{X}_2$, the mean of the pooled sample will be

$$\frac{n_1 \bar{X}_1 + n_2 \bar{X}_2}{n_1 + n_2}$$

and this is an unbiased estimator of the population mean, as it is the mean of a single pooled sample.

For the variance, the approach is different.

Pooling the squared differences from the means of the two samples gives

$$\Sigma (X_1 - \bar{X}_1)^2 + \Sigma (X_2 - \bar{X}_2)^2.$$

Now $\quad E[\Sigma (X_1 - \bar{X}_1)^2 + \Sigma (X_2 - \bar{X}_2)^2]$

$$= E[(n_1 - 1)s_1^2 + (n_2 - 1)s_2^2]$$

$$= (n_1 - 1)\sigma^2 + (n_2 - 1)\sigma^2 \quad [E(s_1^2) = E(s_2^2) = \sigma^2]$$

$$= (n_1 + n_2 - 2)\sigma^2.$$

Hence $\quad E\left(\dfrac{\Sigma (X_1 - \bar{X}_1)^2 + \Sigma (X_2 - \bar{X}_2)^2}{n_1 + n_2 - 2} \right) = \sigma^2.$

In other words

$$\frac{\Sigma (X_1 - \bar{X}_1)^2 + \Sigma (X_2 - \bar{X}_2)^2}{n_1 + n_2 - 2}$$

is an unbiased estimator of σ^2, with the alternative form

$$\frac{(n_1 - 1)s_1^2 + (n_2 - 1)s_2^2}{n_1 + n_2 - 2}.$$

This important result, which will often be used in sampling theory, may be generalised for more than two samples as follows.

For samples of sizes $n_1, n_2, \ldots, n_k$ where

$$s_1^2 = \frac{\Sigma (X_1 - \bar{X}_1)^2}{n_1 - 1}$$

and $\quad s_2^2, \ldots, s_k^2$ are similarly defined, an unbiased estimator for σ^2 is

$$s^2 = \frac{(n_1 - 1)s_1^2 + (n_2 - 1)s_2^2 + \ldots + (n_k - 1)s_k^2}{n_1 + n_2 + \ldots + n_k - k}.$$

12.7   Consistency and efficiency of estimators

Clearly estimates arrived at from a single sample are, by themselves, of limited value, and we shall investigate later a method which gives some idea of the *range* of values for parameters that might be considered reasonable.

There is, however, a way of assessing whether a particular estimator is 'consistent'.

For a given sample size, if the variance of the sampling distribution of the statistic T is large, the value of the statistic could well differ from its expected value by a considerable amount.

If we take T as an unbiased estimator for θ, then T is said to be a *consistent* estimator if the variance of $T \to 0$ as $n \to \infty$.

In the case of the sampling distribution of the *mean*, since $E(\bar{X}) = \mu$ and $\text{Var}(\bar{X}) = \sigma^2/n$, which $\to 0$ as $n \to \infty$, $\bar{X}$ is therefore both an unbiased and consistent estimator.

(An estimator which is both unbiased and consistent is sometimes described as 'good'.)

A sample may, however, be used to obtain a *set* of unbiased estimates, or two samples may be available.

Suppose, for example, a sample $X_1, X_2, \ldots, X_n$ is available. We could take as an unbiased estimator for μ, a single observation, X_1, say, since $E(X_1) = \mu$.

Alternatively we could take $(X_1 + X_2 + X_3)/3$, since

$$E(X_1 + X_2 + X_3)/3 = \frac{1}{3}[E(X_1) + E(X_2) + E(X_3)] = \frac{1}{3} \cdot 3\mu = \mu.$$

Now $\text{Var}(X_1) = \sigma^2$ and $\text{Var}(X_1 + X_2 + X_3)/3 = \frac{1}{9} \cdot 3\sigma^2 = \frac{\sigma^2}{3}.$

The second alternative is said to be a more *efficient* estimator than the first, since its variance is smaller.

It should be stressed that, given a sample of observations, there is often more than one way of using them to obtain an unbiased estimator of a parameter (see, for example, exercise 12c, questions 2 and 3). Given several estimators, the *best*, or the most efficient, is the one with the smallest variance.

EXERCISE 12c

1.   A sample of six values $X_1, X_2, \ldots, X_6$ is taken from a population whose mean is μ and variance is σ^2.

Determine which of the following estimators of σ^2 is unbiased:
(a) $(X_1 + X_3 + X_5)/4$ (b) $(2X_1 + 2X_4 + 3X_6)/7$.

2.   Random observations X_1, X_2, X_3 are taken from the distribution of a random variable X which has mean μ (to be estimated) and variance σ^2. Two possible unbiased estimators are

$$\hat{\mu} = (X_1 + X_2 + X_3)/3 \quad \text{and} \quad \hat{\mu} = (X_1 + 2X_2)/3.$$

By comparing their variances, determine which is the best or most efficient.

3.   If $X_1, X_2, \ldots, X_n$ is a sample taken from some parent population, prove that

$$k_1 X_1 + k_2 X_2 + \ldots + k_n X_n$$

is an unbiased estimator for the mean if the k_i are *any* numbers that satisfy

$$\sum_{i=1}^{n} k_i = 1.$$

(It can be shown that the best estimator, that is, the one with the least variance, is given by putting $k_i = 1/n$ for all i, so that the estimator is $\bar{X}$.)

12.8   The nature of the sampling distribution of the mean

Consider again the example on page 232. A random sample of 200 households yielded a mean weekly time of 15.3 hours spent watching TV. Although we took this as an unbiased estimate of the population mean, how confident can we be about doing this? Questions like this, to be useful, must be framed more precisely in mathematical terms, so we might ask specifically: within what *range* of values can we be 95 % sure that the true population mean lies? (The figure 95 % is taken as expressing a high degree of confidence. We might ask the same question using, say, 90 % or 99 %. For the moment we shall not discuss this point further, but it will turn out to be an important one.)

The method of arriving at an answer to this sort of question will be described in the next chapter, but it is clear that we must first know more about the sampling distribution of the mean than that its mean is μ and that its variance is σ^2/n. In fact we need to know what *kind* of distribution it is.

The same applies to problems involving the *variance* but we shall not be examining the nature of the sampling distribution of the variance (or rather a closely related distribution) until much later (see chapter 24).

There are two cases to be considered. We shall state the results first and discuss their justification afterwards.

1.   *The distribution of $\bar{X}$ when the parent population is normally distributed.*

If a random sample $X_1, X_2, \ldots, X_n$ of size n is drawn from a normal distribution $N(\mu, \sigma^2)$, the distribution of $\bar{X}$ is *also* normal. Since we know that $E(\bar{X}) = \mu$ and $\text{Var}(\bar{X}) = \sigma^2/n$

$$\text{the distribution of } \bar{X} \text{ is } N(\mu, \sigma^2/n).$$

This result is not really unexpected; the equivalent result for a non-normal population *is*, though, as we shall now see.

2.   *The distribution of $\bar{X}$ when the parent population is not normally distributed.*

If the sample size is *large*, the distribution of X is *approximately* normal, no *matter what the distribution of the parent population is*, so that

$$\text{the distribution of } \bar{X} \approx N(\mu, \sigma^2/n).$$

This very remarkable result is a special case of the *central limit theorem* (see page 214). It may be expressed more precisely as follows:

If samples of size n are taken from *any* population with mean μ and variance σ^2, the distribution of $\bar{X}$, the sample mean, tends to a normal distribution $N(\mu, \sigma^2/n)$ as $n \to \infty$.

(There are, in fact, some exceptional populations for which this is not true, but they will not concern us in this book.)

Justification of the distribution of $\bar{X}$

1.   *When the parent population is normal*

If you look back at page 204, you will recall that if any two independent random variables are normally distributed, then so is $X + Y$, with the extension that if, $X_1, X_2, \ldots, X_n$ are each normally distributed so is their sum.

In other words, if X_1 is $N(\mu_1, \sigma_1^2)$, X_2 is $N(\mu_2, \sigma_2^2)$ and so on, the distribution of $X_1 + X_2 + \ldots + X_n$ is normal with

$$\text{mean} \quad \mu_1 + \mu_2 + \ldots + \mu_n \quad \text{and variance} \quad \sigma_1^2 + \sigma_2^2 + \ldots + \sigma_n^2.$$

Now if $X_1, X_2, \ldots, X_n$ are observations forming a sample from $N(\mu, \sigma^2)$ each has the *same* distribution so that the distribution of $X_1 + X_2 + \ldots + X_n$ is $N(n\mu, n\sigma^2)$.

It follows that the distribution of

$$\bar{X} = \frac{1}{n}(X_1 + X_2 + \ldots + X_n) \quad \text{is} \quad N(\mu, \sigma^2/n)$$

[using $E(kX) = kE(X)$ and $\text{Var}(kX) = k^2\text{Var}(X)$].

2.   *When the parent population is not normal*

In the discussion on page 214 of the central limit theorem, it was stated (without proof at that stage) that if $X_1, X_2, \ldots, X_n$ are independent random variables with the same non-normal distribution with mean μ and variance σ^2, then the distribution of $X_1 + X_2 + \ldots + X_n$ is approximately normal for large values of n, $N(n\mu, n\sigma^2)$.

Using the same argument as in the previous section it follows that

$$\text{the distribution of } \bar{X} \approx N(\mu, \sigma^2/n).$$

12.9   Applications of the central limit theorem

In applications, the size of n needed for the normal distribution to be a good approximation (when the parent population is *not* normal) depends on the degree of skewness of the parent population.

As a rough guide, a sample size of about 30 will serve if the distribution is only slightly skew. For a very skew distribution, (the exponential distribution for example) a much larger sample size is needed – of the order of hundreds.

The figures below give some idea of the comparison between the population distribution and the sampling distribution of the mean in three cases. The curves are, of course, idealised models for the actual distributions.

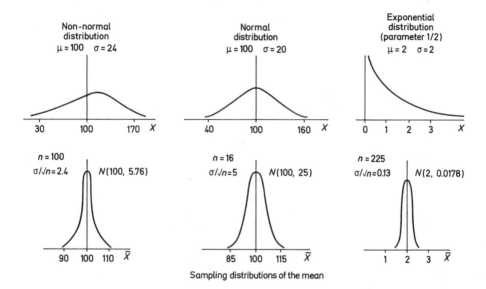

Sampling distributions of the mean

To sum up, then. If the population distribution is normal we use the fact that the distribution of $\bar{X}$ is $N(\mu, \sigma^2/n)$ for any size of sample. Standardising in the usual way, the distribution of

$$\frac{\bar{X} - \mu}{\sigma/\sqrt{n}} \text{ is } N(0, 1).$$

If the population distribution is not normal, the distribution of $\bar{X}$ is approximately $N(\mu, \sigma^2/n)$ for large samples and the distribution of

$$\frac{\bar{X} - \mu}{\sigma/\sqrt{n}} \text{ is approximately } N(0, 1).$$

The examples and exercises that follow are perhaps rather artificial, but they are designed to help you become familiar with the distribution of $\bar{X}$ in different contexts, before tackling the more practical applications in the following chapters.

EXAMPLE 5 The heights of a large group of men are distributed with mean 175 cm and standard deviation 6 cm. A sample of 100 men is taken. What is the probability that the mean height of the sample is less than 176?

The population variable X has mean 175 and variance 36, so that the sampling distribution of the mean $\bar{X}$ has mean 175 and variance $36/100 = 0.36$.
As the sample is large the distribution of $\bar{X} \approx N(175, 0.36)$.

Hence $P(\bar{X} < 176) = \Phi\left(\dfrac{176 - 175}{0.6}\right) = 0.953$ to 3 dec. places.

EXAMPLE 6   The maximum load for a lorry used for carrying workmen to building site is 2700 kg. The weights of the large number of men who may be using the lorry have a mean of 86 kg and variance 121 kg². What is the probability that a random sample of 30 men will exceed the maximum load?

We are here concerned with the *total* weight of a sample, and it may be tackled in two ways.

1.   If the total weight of a sample of 30 is greater than 2700, it is equivalent to saying that the mean $\bar{X}$ of the sample is greater than 2700/30 kg = 90 kg.

Even if we do not assume that the population distribution is normal, we can take the distribution of $\bar{X}$ as being approximately normal,

$$N(86, 11^2/30).$$

Hence $P(\bar{X} > 90) \approx 1 - \Phi\left(\dfrac{90-86}{11/\sqrt{30}}\right) = 1 - \Phi(1.99) = 0.023$ to 3 dec. places.

2.   Alternatively we can work with the sample *total* weight as random variable.

The distribution of X, the individual weights has mean 86 and variance 121.
If Y is the total weight of 30 men, then by the central limit theorem, the distribution of Y will be approximately normal with mean $30 \times 86 = 2580$ and variance $30 \times 121 = 3630$.

Hence $P(Y > 2700) \approx 1 - \Phi\left(\dfrac{2700-2580}{\sqrt{3630}}\right) = 1 - \Phi\left(\dfrac{120}{60.25}\right) = 0.023.$

EXAMPLE 7   The random variables X, Y are independent and normally distributed, X being $N(2,4)$ and Y being $N(3,25)$. A sample of 10 observations is taken from the distribution of X and a sample of 15 from the distribution of Y. Find $P(\bar{X} < \bar{Y})$.

Although the samples are small, since the parent populations are *normal*, the sampling distributions of the means will also be normal.
The distribution of $\bar{X}$ is $N(2, 4/10)$ and of $\bar{Y}$ is $N(3, 25/15)$. Hence the distribution of $\bar{Y} - \bar{X}$ is $N(3 - 2, 25/15 + 4/10)$ or $N(1, 31/15)$.

$$P(\bar{X} < \bar{Y}) = P(\bar{Y} - \bar{X} > 0) = 1 - \Phi\left(\frac{0-1}{\sqrt{(31/15)}}\right) = 0.76 \text{ to 2 dec. places.}$$

EXAMPLE 8   The life, X hours, of an electrical component has an exponential distribution with parameter $\lambda = 0.02$, that is, its density function is

$$f(x) = 0.02e^{-0.02x}, \quad x \geqslant 0.$$

A sample of 200 components is taken. What is the probability that the mean life of the sample exceeds 56 hours?

The mean and variance of an exponential distribution with parameter λ are $1/\lambda$ and $1/\lambda^2$ respectively (see page 185).

In this case, then, the population mean is 50 and the population variance is 2500. The distribution of $\bar{X}$ is therefore taken to be approximately normal with mean 50 and variance 2500/200, that is, $N(50, 12.5)$.

$$\text{Hence} \quad P(\bar{X} > 56) = 1 - \Phi\left(\frac{56 - 50}{\sqrt{12.5}}\right) = 0.045 \quad \text{to 3 dec. places.}$$

EXERCISE 12d

1. A machine fills cartons in such a way that the distribution of weights of the cartons is normal with mean 1.5 kg with a standard deviation of 0.2 kg.
   If a sample of 9 is taken from the output of the machine find the probability that the mean weight of the sample will be less than 1.4 kg. What will be the probability that the mean weight of a sample of 16 will be less than 1.4 kg?

2. A sample of size 5 is taken from a normal distribution $N(12, 4)$. What is the probability that the sample mean exceeds 13?

3. The mean time spent by forty-year old adult males watching television on Sunday nights is 3.1 hours with a standard deviation of 0.8 hours. The distribution is normal. What is the probability that in a random sample of 100 forty-year old adult males the mean time watching television on Sunday nights is between 3.0 and 3.2 hours?

4. A sample of 25 is taken from $N(28, 9)$. What is the probability that the sample mean is between 26.8 and 28.3?

5. From the binomial distribution $(40, \frac{1}{2})$ a random sample of 100 is taken. Estimate the probability that the sample mean is between 19.3 and 20.5.

6. A random sample of 20 is taken from a Poisson distribution with mean 30. Find the approximate probability that the sample mean will be greater than 33.

7. A rectangular distribution has density function

$$f(x) = \begin{cases} 1/8, & 0 \leqslant x \leqslant 8 \\ 0, & \text{otherwise.} \end{cases}$$

A random sample of size 40 is taken. Find the approximate probability that the sample mean will be less than 4.5.
   (For the mean and variance of a uniform distribution see page 183, exercise 10d, question 2.)

8. The distribution of X has mean 20 and variance 8, and the distribution of Y has mean 40 and variance 4. Both distributions are normal.
   If a sample of 10 is taken from the first and a sample of 15 from the second, find the expectation and variance of: (a) $\bar{X} + \bar{Y}$, (b) $\bar{Y} - \bar{X}$, where $\bar{X}$ and $\bar{Y}$ are sample means.

9. Independent samples of sizes 10 and 15 are taken from $N(20, 3)$. What is the probability that the sample means differ by more than 0.3?

10. If the distributions of X and Y are independent normal distributions $N(50, 16)$ and $N(40, 16)$ respectively, and random samples of 36 are taken from each, what are the distributions of: (a) $\bar{X} + \bar{Y}$, (b) $\bar{X} - \bar{Y}$, (c) $2\bar{X} + 3\bar{Y}$?

11. A large number of samples of size n are taken from $N(50, 20)$. Ninety-nine per cent of the sample means are less than 55. Estimate n.

12. Six independent observations $\{X_i\}$ are taken from a normal distribution with mean 2 and standard deviation 2 and twelve independent observations $\{Y_j\}$ are taken from a normal distribution with mean 3 and standard deviation 4.

Find $P(\bar{X} \leqslant 0.5\bar{Y}-2)$ and $P(\bar{X} \geqslant 2-0.5\bar{Y})$, where $\bar{X}$ and $\bar{Y}$ are the means of the observations from the two distributions respectively. (MEI)

13.★ The continuous random variable X has density function

$$f(x) = \begin{cases} \lambda x(1-x)^3, & 0 \leqslant x \leqslant 1, \\ 0 & , \quad \text{elsewhere.} \end{cases}$$

Find the mean and standard deviation of the distribution of X.

Calculate the probability that X is more than 0.935 standard deviations above the mean.

What is the probability that the mean $\bar{X}$ of 100 independent observations from the distribution will be without 10% of the mean? (MEI)

EXERCISE 12e (miscellaneous)

1. A random variable X has a probability density function given by

$$f(x) = \begin{cases} \dfrac{1}{\theta}e^{-x/\theta}, & x \geqslant 0, \text{where } \theta > 0. \\ 0 & , \quad \text{otherwise.} \end{cases}$$

A random sample $X_1, X_2, \ldots, X_n$ is taken from a population with the above distribution. The estimator T is defined by

$$T = k \sum_{i=1}^{n} X_i^2, \quad \text{where } k \text{ is a constant.}$$

Find the value of k such that T is an unbiased estimator of θ^2. (AEB 1981)

2. Let $\bar{X}$ denote the mean of a random sample of 80 observations of the random variable X, whose probability density function f is given by

$$f(x) = \tfrac{3}{4}x(2-x), \quad 0 < x < 2$$
$$f(x) = 0 \qquad , \quad \text{otherwise.}$$

Using an appropriate approximation to the sampling distribution of $\bar{X}$, calculate the probabilities that $\bar{X}$ will be: (a) less than 1, (b) less than 0.95. (JMB)

3. The random variable X has normal distribution $N(\mu, 1)$. If a random sample of size n is taken from this distribution, what is the smallest value of n that will ensure that

$$P(\mu-0.2 < \bar{X} < \mu+0.2) \geqslant 0.95.$$

4. One number is to be drawn at random from the set $\{1, 2, 3, 4\}$. Denoting the drawn number by X, calculate the mean μ and the variance σ^2 of X.

Three numbers are to be drawn at random *with replacement* from the above set. Let $\bar{X}$ denote the mean and M the median of the three numbers drawn. (The median is the middle number when the three numbers are ranked according to size.)

(a) Write down the values of the mean and the variance of $\bar{X}$.

(b) Show that $P(M = 4) = 5/32$ and calculate the values of $P(M = r)$, $r = 1, 2, 3$. Hence verify that M is an unbiased estimator of μ and calculate its variance. (JMB)

5.   A random sample of size n is taken from a population with zero mean and variance σ^2. The members of this sample are $x_1, x_2, \ldots, x_n$, and $\bar{x}$ is defined by

$$n\bar{x} = (x_1 + x_2 + \ldots + x_n) = \Sigma\, x_i.$$

Show from first principles that the variance of $\bar{x}$ is σ^2/n. (*You may assume that x has zero mean.*)

The losses in weight in kilograms of 110 persons after taking a slimming course are denoted by $x_1, x_2, \ldots, x_{110}$, and satisfy

$$\Sigma\, x_i = 40, \qquad \Sigma\, x_i^2 = 475.$$

Calculate the mean m and standard deviation s of these losses. Assuming that these readings are a random sample from a population with *zero* mean and your calculated standard deviation s, estimate the probability that the mean loss for a random sample of size 110 is greater than m.   (MEI)

6. X and Y are independent random variables, each having a Poisson distribution with parameter μ. Show that both $(X+Y)/2$ and $(X-Y)^2/2$ are unbiased estimators of μ.   (O)

7.   Given that X and Y are independent random variables, prove that

$$\mathrm{var}(aX + bY) = a^2\,\mathrm{var}(X) + b^2\,\mathrm{var}(Y),$$

where a and b are real constants.

The random variable $\bar{X}$ is defined by

$$\bar{X} = \frac{1}{8}\sum_{i=1}^{8} X_i,$$

where $X_1, X_2, \ldots, X_8$ are independent random variables each having a normal distribution with mean 1 and variance 6. Similarly, the random variable $\bar{Y}$ is defined by

$$\bar{Y} = \frac{1}{4}\sum_{j=1}^{4} Y_j,$$

where Y_1, Y_2, Y_3, Y_4 are independent random variables each having a normal distribution with mean 2 and variance 4. Find, to three decimal places, the probabilities that:
(a) $|\bar{Y}| \geqslant 1$,   (b) $2\bar{X} \geqslant 1 - \bar{Y}$,   (c) $2\bar{X} \leqslant \bar{Y} - 1$.   (C)

8.   A company packs sugar into bags. The masses of the packed bags are normally distributed with a standard deviation of 0.05 kg. The mean mass varies from day to day, but always lies between 0.94 kg and 1.02 kg. Each day a random sample of n bags is taken from the output, and the day's output is rejected if the sample mean is less than m kg, where n and m are to be chosen to satisfy two criteria. These criteria are:
(a) When the mean mass is actually 1.02 kg, there is a probability of at least 95 % that the sample mean mass is greater than m kg.
(b) When the mean mass is actually 0.94 kg, there is a probability of at most 5 % that the sample mean mass is greater than m kg.
Determine suitable values for n and m.   (MEI)

9.   If $X_1, X_2, \ldots, X_n$ is a random sample from a population which is normally distributed with mean μ and variance σ^2, what is the distribution of the sample mean $\bar{X}$? Write down an unbiased estimator of σ^2.

A steel disc has its radius measured on n independent occasions. The results $R_1, R_2, \ldots, R_n$ may be assumed to be a random sample from a normal distribution with mean r, the true radius, and variance σ^2, where r and σ^2 are both unknown. Derive an unbiased estimator for the circumference of the disc. Is this estimator consistent? Justify your reply. Show that the estimator $T = \pi\bar{R}^2$ is a biased estimator for the area of the disc. Modify T in order to produce an unbiased estimator for the area.   (AEB 1982)

10.   Three independent random samples, each of ten observations, were drawn from a distribution having unknown mean μ and unknown variance σ^2. The means of the three samples were calculated as 1.6, 2.5 and 1.9, respectively. Combine the given information on all three samples to obtain an unbiased estimate of μ and an unbiased estimate of σ^2.

Given further that the sum of the squares of the ten values in the first sample was 28.44, in the second sample was 71.07 and in the third sample was 42.24, find a better unbiased estimate of σ^2, explaining why you regard it as an improvement on your first estimate. (JMB)

11.★ The random variable X has probability density function given by

$$f(x) = \begin{cases} \theta x^{\theta-1}, & 0 < x < 1, \quad \text{where } \theta > 0 \\ 0 & , \quad \text{otherwise.} \end{cases}$$

Derive the mean and variance of X.

A random sample $X_1, X_2, \ldots, X_n$ is taken from a population with the above distribution. Write down the mean and variance of $\bar{X}$, the sample mean. If $\bar{X}$ is used to estimate θ in the above distribution and $E(\bar{X}) = \theta + b(\theta)$, then $b(\theta)$ is called the bias of X. Calculate this bias.

The mean square error of $\bar{X}$ in estimating θ is defined by

$$E(\bar{X} - \theta)^2 = E(\bar{X}^2) - 2\theta E(\bar{X}) + \theta^2.$$

Verify that for this example mean square error of $\bar{X} = \text{variance}(\bar{X}) + b^2(\theta)$.   (AEB 1980)

12.★ Define an unbiased estimator. Explain why in choosing between unbiased estimators of a parameter, the one with the least variance is preferable.

$\bar{x}_1$ and $\bar{x}_2$ are the means of random samples of sizes n_1 and n_2 from each of two populations, which have normal distributions with unknown means μ_1 and μ_2 and known variances σ_1^2 and σ_2^2 respectively. Show that $\bar{x}_1 - \bar{x}_2$ is an unbiased estimator of $\mu_1 - \mu_2$ and write down the variance of this estimator.

Given that $n_1 + n_2 = n$, where n is a fixed number, show that the variance of $\bar{x}_1 - \bar{x}_2$ is a minimum when $n_1 : n_2 = \sigma_1 : \sigma_2$. (You may assume that n_1 and n_2 are so large that they may be treated as continuous variables in determining the minimum.) Find these values of n_1 and n_2 when $\sigma_1^2 = 0.0036$, $\sigma_2^2 = 0.0064$, $n = 70$.   (JMB)

13 Hypothesis tests and confidence intervals (means): 1

13.1 Hypotheses and significance levels

We are now ready to answer some of the central questions that arise when we want to draw conclusions from statistical observations. If you look back at examples 1 and 2 outlined in chapter 1, you will see what is meant by this, and indeed we will deal specifically with the second example very shortly, justifying the answer that was given there.

Suppose that over a number of years a manufacturing process has been turning out batteries for pocket calculators, whose life is normally distributed with mean (μ) 600 hours and variance (σ^2) 10 000 hours^2 (standard deviation = 100 hours). In an attempt to improve the product, the process is adjusted and during the trials of the new batteries a sample of 50 is found to have a mean life $\bar{x} = 620$ hours. The manufacturer will obviously want to decide whether the improved mean life of this sample arose by chance or whether it provides reasonable evidence that the new process does in fact produce an improved product.

In order to decide which of these two conclusions is likely to be correct, we first state two *hypotheses*. The first is that the new method has made no difference, that the population mean is still 600, and that the improved sample mean could well have arisen by chance. This is usually called the *null hypothesis* and we write:

$$H_0 : \mu = 600.$$

The other one, called the *alternative hypothesis*, is that the new process has improved the population mean life, and is written:

$$H_1 : \mu > 600.$$

Now it might be thought that this restricts the alternative hypothesis unduly. Might not the new mean be either greater or less than 600? Here common sense suggests that those who devised the new process would not have claimed that it was an improvement if there were any chance that it produced a shorter-lived battery. We shall assume, therefore, that this is the correct alternative hypothesis for this particular case.

The procedure now is to set up a model *on the assumption that H_0 is true* and calculate the probability that a sample mean of 620 will turn up by chance. If this probability is *small*, we reject the null hypothesis in favour of the alternative one.

But what is small? Let us say, for the moment, that a probability of 0.05 (5%) or less is small in this context. This is a figure often used by statisticians, and although it is obviously an arbitrary one, there are ways of deciding whether 5% or, say, 1% or 10% is appropriate in any given circumstances (see chapter 16).

Having decided that we shall use a probability of 5% or less as meaning 'unlikely', we must make one further assumption — that the population variance remains unaltered under the new process, whatever may have happened to the population mean. This is usually a reasonable assumption in the kind of examples we shall be considering, and indeed without this assumption it would not be easy to proceed.

Suppose, then, that H_0 is true, so that the distribution of X, the individual lives of batteries, is still

$$N(600, 100^2)$$

under the new process, and the distribution of $\bar{X}$ is therefore

$$N(600, 100^2/50).$$

Standardising, the distribution of the standardised variable

$$Z = \frac{\bar{X} - 600}{100/\sqrt{50}} \quad \text{is } N(0, 1).$$

Now, from the normal tables we know that there is a 5% probability that a value of Z will be greater than 1.64 since

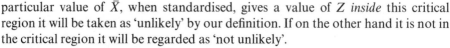

$$\Phi(1.64) = 0.95.$$

$Z > 1.64$ therefore defines what we will call a 'critical region'. If any particular value of $\bar{X}$, when standardised, gives a value of Z *inside* this critical region it will be taken as 'unlikely' by our definition. If on the other hand it is not in the critical region it will be regarded as 'not unlikely'.

In our example the value of Z for $\bar{x} = 620$ is

$$z = \frac{620 - 600}{100/\sqrt{50}} = 1.41.$$

This is *not* in the critical region and we therefore conclude that there is no convincing evidence against H_0 being true; H_0 is thus accepted.

The usual way of stating this conclusion is that the null hypothesis is accepted on the grounds that $\bar{X}$ is *not significantly different from 600 at the 5% level.*

This process of testing a hypothesis is called a *significance test*, and the standardised variable

$$Z = \frac{\bar{X} - \mu}{\sigma/\sqrt{n}}$$

is called the *test statistic*; 5% is the *significance level.*

Now that we have explained the method in some detail and defined the terms used, we shall show how to set out the reasoning more concisely, using example 2 in chapter 1, where the question (now slightly amplified) was as follows.

EXAMPLE 1   At a large American university it has always been the custom to select a random sample of 50 students from each new year's intake, weigh them and calculate the mean weight of the sample. In 1942 the mean weight was $\bar{x} = 70.4$ kg.

Now the distribution of the weights of the whole intake of first-year students over many previous years had been found to be normal with mean $\mu = 67.2$ kg and variance $\sigma^2 = 80$ kg^2. The authorities therefore asked themselves whether this increase in mean weight for the sample arose by chance or whether it meant that the intake for 1942 were significantly heavier than before, possibly reflecting an improved national diet since the Depression of the early 1930s.

The alternative hypotheses to be tested are

$$H_0 : \mu = 67.2 \quad \text{as against} \quad H_1 : \mu > 67.2.$$

The wording of the problem shows that the alternative hypothesis is that the population weight had *increased*.

We shall again take 5% as the significance level.

Under H_0 (this is a short way of saying 'assuming that H_0 is true') the value of the test statistic

$$Z = \frac{\bar{X} - 67.2}{8.94/\sqrt{50}} \quad (\sigma = \sqrt{80} \approx 8.94)$$

when $\bar{x} = 70.4$ is $\dfrac{70.4 - 67.2}{8.94/\sqrt{50}} = 2.93.$

This is *in* the 5% critical region, and we conclude that there is evidence that H_0 is not true.

H_0 is therefore rejected at the 5% level.

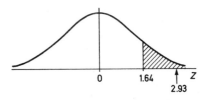

One-tail and two-tail tests

The last example, in which the alternative hypothesis was

$$H_1 : \mu > \mu_0 \text{ (a stated value)}$$

was an instance of what is called a *one-tail* test, since the critical region only appears in one 'tail' of the normal curve. This critical region is therefore defined by

$$Z > 1.64.$$

A *two-tail* test is one in which the appropriate alternative hypothesis is

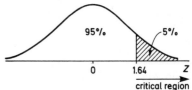

$$H_1 : \mu \neq \mu_0$$

which means that the population mean might be either greater or less than μ_0.

In that case, if the significance level is 5%, the critical region occupies two tails (each of $2\frac{1}{2}$%) of the normal curve.

From the normal tables, since

$$\Phi(1.96) = 0.975$$

$$= 0.95 + 0.025$$

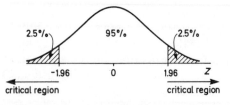

the critical region is defined by $Z < -1.96$ and $Z > 1.96$.

If, then, the standardised sample mean lies in either of these two tails, H_0 will be rejected at the 5 % level.

The choice of 5 % to determine the critical region is, as we have said, arbitrary, and the figure used in any given case will indicate how strong the evidence is for or against accepting the null hypothesis. All that we can say about this at the moment is that the critical region should be chosen sensibly, which means that it must be small enough, so that it is unlikely that the observed result will be in it unless H_0 is false. On the other hand, if it is too small, a null hypothesis which is in fact false will stand a good chance of being accepted.

If, at this point, you suspect that the choice of significance level, and therefore of critical region, is a matter of careful judgement, you will be right. The matter is discussed very fully in chapter 16. Meanwhile it will be found that the significance level to be used is usually specified; if not it will be safe to use 5 %.

In the example that follows, a two-tail test with a significance level of 5 % is illustrated.

EXAMPLE 2   At an army training camp an experiment is made to see whether the time taken to undertake a specified physical task will be affected if soldiers are given a special and unusual diet for a period of time before attempting the task.

Over a long period of time it has been found that the time taken for the standard task is normally distributed with mean 50 s and variance 100 s².

A sample of 20 soldiers is given the special diet for a week, after which the mean time taken by this sample is 46 s. Is this significant evidence that the special diet has made any difference to their performance? Use a significance level of 5 %.

The question does not ask whether the diet has resulted in an *improvement* in performance, but whether it has made *any difference*. We therefore take as the hypothesis to be tested

$$H_0 : \mu = 50 \quad \text{as against} \quad H_1 : \mu \neq 50.$$

At the 5 % level the critical region is, as we have shown,

$$Z < -1.96 \quad \text{and} \quad Z > 1.96.$$

Under H_0 the distribution of Z is

$$N(50, 100/20)$$

Now when $\bar{x} = 46$, the value of Z is

$$z = \frac{46 - 50}{10/\sqrt{20}} = -1.79.$$

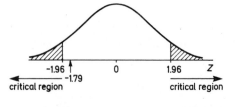

Since this is not in the critical region we say that there is no convincing evidence that the diet has made a difference and H_0 is accepted at the 5 % level.

Critical regions for different significance levels

In the exercises that follow various significance levels are laid down and it may help you to look at two examples.

1.   *A one-tail test with a significance level of 10 %*

If the alternative hypothesis is H_1: pop. mean $> \mu_0$, the critical region is $Z > 1.28$ since $\Phi(1.28) = 0.9$.

If the alternative hypothesis is H_1: pop. mean $< \mu_0$, the critical region is $Z < -1.28$.

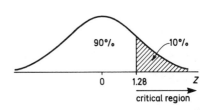

You may occasionally come across another way of expressing the hypotheses for a one-tail test (see Appendix on page 274 for a discussion of this slightly tricky point).

2.   *A two-tail test with a significance level of 1 %*

If the alternative hypothesis is H_1: pop. mean $\neq \mu_0$ the critical region consists of two tails of $\frac{1}{2}\%$ and is therefore given by

$$Z < -2.58 \quad \text{and} \quad Z > 2.58,$$

since $\Phi(2.58) = 0.995$

$$= 0.99 + 0.005.$$

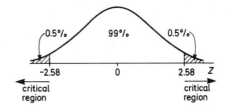

EXERCISE 13a

1.   The diameters of beer cans produced by a metal container manufacturing company are distributed normally with standard deviation 0.3 mm and mean thought to be 70 mm. A sample of 10 is inspected and it is found that the sample mean is 70.2 mm.
 Is this significantly different from 70 mm at the 5 % level?

2.   The times taken for 10 g specimens of washing powder to remove a standard stain from a fabric have, over a long period of time, been normally distributed with mean 163 s and standard deviation 12 s. A sample of 16 specimens are treated with powder to which a 'miracle ingredient' has been added and the mean time for the sample is 159 s. Is this significant evidence, at the 5 % level, that the ingredient produces an improvement?

3.   In an examination taken by a large number of pupils the mean mark was 64.5 with a standard deviation 8. A sample of 200 scripts arrived late from abroad and the mean mark for this sample was 65.5. Is this mean mark significantly:
(a) different from 64.5,
(b) better than 64.5.
at the 5 % level? Assume a normal distribution.

4.   The mean weight of twenty-five-year-old men in a certain country is 78 kg with a standard deviation of 9 kg. A random sample of 100 men of this age taken from one particular town has mean 75.5 kg. Using a 1 % level of significance, test whether the mean

weight of twenty-five-year-old men in this town is different from the overall mean for the country. (Assume a normal distribution.)

5.   In a psychological experiment a memory test was taken by a large number of students. The resulting scores were normally distributed with mean 100 and standard deviation 12.
    An analysis of a sample of 30 of these students gave the following results:

Score	101	102	103	104	105	106	107
Frequency	2	4	6	8	5	3	2

On this evidence determine whether the mean score of this group was significantly different from 100, using a 5% significance level.

6.   The marks of students taking a Law Society examination were normally distributed with mean 61 and standard deviation 20. It was noticed that for the scripts marked by one examiner, a random sample of 400, the mean mark was 58.5.
    Is there reason: (a) at the 5% significance level, (b) at the 1% level, to think he was marking too strictly?

7.   The masses of cereals in a packet are assumed to be normally distributed with mean 250 g and standard deviation 2.5 g. A routine inspection of a sample of 30 packets shows that the mean mass for this sample is 251 g.
    Using a 2% significance level, test whether this sample mean is significantly:
(a) different from the assumed mean,
(b) greater than the assumed mean.

8.   All ten-year-old children in a large city were given a verbal reasoning test. The scores were assumed to be normally distributed, and the mean score was 64 with standard deviation 8.
    One school, 54 of whom took the test, obtained a mean score of 68. Can one be sure that these children's performance is significantly better than average, or could the result have arisen by chance?   (C)

13.2   Testing a hypothesis about the difference between the means of two normal distributions with known variances

We now consider the problem of comparing the means of *two* populations, by taking a sample from each, not necessarily of the same size, and calculating the difference between the means of these samples.
    The method is set out in full in the next example.

EXAMPLE 3   Two machines are designed to cut lengths of tubing. One machine works to a standard deviation of 0.01 cm (σ_1) and the other to a standard deviation of 0.013 cm (σ_2).
    A sample of 10 from the first machine has a mean length 8.250 cm ($\bar{x}_1$) and a sample of 12 from the other machine has a mean length 8.244 cm ($\bar{x}_2$).
    Is this significant evidence that the two machines are producing tubes of different mean lengths?

The null hypothesis here is that there is *no* difference between the two population means, so that, calling these means μ_1 and μ_2:

$$H_0: \mu_1 - \mu_2 = 0 \quad \text{as against} \quad H_i: \mu_1 - \mu_2 \neq 0.$$

From the wording of the question we must use a two-tail test, and we shall take a significance level of 5%.

Now the distribution of X_1 and X_2, the lengths of the tubes from the two machines are respectively

$$N(\mu_1, \sigma_1^2) \quad \text{or} \quad N(\mu_1, 0.01^2)$$

and $\qquad\qquad N(\mu_2, \sigma_2^2) \quad \text{or} \quad N(\mu_2, 0.013^2)$.

It follows that the distributions of $\bar{X}_1$ and $\bar{X}_2$, the sample means, are respectively

$$N(\mu_1, \sigma_1^2/n_1) \quad \text{or} \quad N(\mu_1, 0.01^2/10)$$

and $\qquad\qquad N(\mu_2, \sigma_2^2/n_2) \quad \text{or} \quad N(\mu_2, 0.013^2/12)$,

where n_1 and n_2 are the sample sizes.

Hence the distribution of $\bar{X}_1 - \bar{X}_2$ is

$$N(\mu_1 - \mu_2, 0.01^2/10 + 0.013^2/12) = N(\mu_1 - \mu_2, 0.0049^2).$$

We are now ready to test the null hypothesis.

Under $H_0 : \mu_1 - \mu_2 = 0$, the distribution of $\bar{X}_1 - \bar{X}_2$ is

$$N(0, 0.0049^2)$$

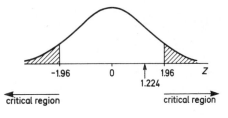

and we want to know whether the sample value of $\bar{X}_1 - \bar{X}_2$, when standardised, is in the critical region which, as we know, is given by $|z| > 1.96$.

Now $\bar{X}_1 - \bar{X}_2$ has the value $\bar{x}_1 - \bar{x}_2 = 8.250 - 8.244 = 0.006$, which, when standardised, gives

$$z = \frac{0.006}{0.0049} = 1.224.$$

This is not in the critical region, so H_0 is accepted and we conclude that there is no strong evidence to show that the machines are producing tubes of different mean lengths.

Summarising this method formally, remembering that it is used when the parent distributions are *normal with known variances*:

The population distributions of X_1 and X_2 are respectively

$$N(\mu_1, \sigma_1^2) \quad \text{and} \quad N(\mu_2, \sigma_2^2)$$

To test whether μ_1 and μ_2 are the same, samples of n_1 and n_2 are taken, giving means $\bar{X}_1$ and $\bar{X}_2$ respectively.

The hypotheses are

$$H_0 : \mu_1 - \mu_2 = 0 \quad \text{as against} \quad H_1 : \mu_1 - \mu_2 \neq 0 \quad \text{(for a two-tail test)}$$

$$\text{or} \quad H_1 : \mu_1 - \mu_2 > 0 \quad \text{(for a one-tail test)}.$$

Since the distribution of $\bar{X}_1 - \bar{X}_2$ is $N(\mu_1 - \mu_2, \sigma_1^2/n_1 + \sigma_2^2/n_2)$ under $H_0 : \mu_1 - \mu_2 = 0$, this distribution is $N(0, \sigma_1^2/n_1 + \sigma_2^2/n_2)$, and the test statistic used is

$$Z = \frac{(\bar{X}_1 - \bar{X}_2) - 0}{\sqrt{(\sigma_1^2/n_1 + \sigma_2^2/n_2)}}.$$

Note: It will often be the case that the population variances are the same, in which case the method is similar, and the test statistic takes the form

$$Z = \frac{(\bar{X}_1 - \bar{X}_2) - 0}{\sigma\sqrt{(1/n_1 + 1/n_2)}}$$

where σ^2 is the common variance.

EXERCISE 13b

1. In an examination taken by a large number of students the marks were normally distributed with standard deviation 20. Two schools each entered 100 candidates and the mean marks for the two schools were 57.5 and 52.5.
    Is the difference between these results significant at the 5 % level?

2. An investigation into the ages of mothers at the birth of their first children produces the following information (ages are decimals, not years and months):
    In 1978 the distribution of ages had standard deviation 2.8 years and the average age of a sample of 200 mothers was 22.36 years.
    In 1972 the distribution of ages had standard deviation 2.5 years and the average age of 150 mothers was 21.2 years.
    Do these figures suggest that there is a significant difference in the average age of mothers producing their first children in the years 1972 and 1978? (Normal distributions may be assumed.)   (SUJB)

3. The egg-laying capacities of two sets of hens chosen from a large number were tested with the following results:

Group	Number of hens	Mean number of eggs per bird
A	50	55
B	40	46

The standard deviation for the whole population of hens is known to be 20. Estimate, using a 5 % significance level, whether the difference between the means is significant.   (C)

4. Components are produced by two machines. The lengths of the components produced by each machine are normally distributed. In each case the mean is supposed to be 20 cm. However, the lengths of components produced by the first machine have variance 0.08 cm², while for the second machine this variance is 0.05 cm².
    Random samples of six components from the first machine and seven from the second were found to have the following lengths (in centimetres):

First machine		19.5		20.1		20.0		19.7		19.9		19.4	
Second machine	20.2		19.6		20.1		20.3		20.1		20.2		20.1

Examine the data for evidence, significant at the 1 % level, that the mean lengths of components produced by the two machines differ.   (MEI)

13.3   Testing hypotheses about means when the samples have to be large

If you have mastered the methods for testing hypotheses about a mean (or the difference between two means) that have just been described, where the parent population/s) were *normal* and the variance/s were *known*, you will find that when these conditions do not exist, the methods are similar, provided the problem involves *large samples*.

There are two types of situation to be considered:

1.   If the nature of the population distribution is *unknown*, but its *variance is known*.

It is here that the central limit theorem comes into play. As we saw in the last chapter, provided the sample size is *large enough* (and the meaning of this phrase was discussed on page 242) the distribution of the sample mean $\bar{X}$ is

$$\text{approximately} \quad N(\mu, \sigma^2/n)$$

and we may therefore still use the test statistic

$$Z = \frac{\bar{X} - \mu}{\sigma/\sqrt{n}}$$

for testing a hypothesis about a mean.

In fact the only comment that need be made is that some caution should be exercised in accepting or rejecting a null hypothesis if the value of the test statistic turns out to be close to a border of the critical region.

Similarly, in testing a hypothesis about the difference between two means, when the population distributions, although not normal, have known variances, the method outlined on page 254 may be applied, *provided the samples are large*.

2.   If the population distribution is *normal*, but its *variance is unknown*.

This is a common situation, and is dealt with by an approximate significance test, using an *estimated value of σ^2*, in place of σ^2. The argument is this.

An unbiased estimate of σ^2 is $s^2 = \dfrac{\Sigma (x - \bar{x})^2}{n-1}$. (See page 234.)

Now if n is large (greater than 30, say) this is not very different from σ^2, so that it seems reasonable to say that the test statistic

$$\frac{\bar{X} - \mu}{s/\sqrt{n}} \quad \text{may be used} \quad \left(\text{in place of } \frac{\bar{X} - \mu}{\sigma/\sqrt{n}} \right).$$

In fact it will be proved later that the distribution of this statistic *is* approximately $N(0, 1)$ so long as the population distribution is normal, or nearly normal, and the sample is large.

EXAMPLE 4   Certain electronic components have lives which are distributed normally, with mean thought to be 2610 hours. A sample of 100 has mean 2580 hours and standard deviation 120 hours. Is this significant evidence that the population mean is, in fact 2610 hours?

The sample variance $= 120^2$.

An unbiased estimate of σ^2 is

$$s^2 = (120)^2 \times \frac{100}{99} = (120.6)^2.$$

We shall take as hypotheses

$$H_0: \mu = 2610 \quad \text{as against} \quad H_1: \mu \neq 2610$$

and use a 5% level for the test.

Under H_0, the distribution of

$$\frac{\bar{X} - \mu}{s/\sqrt{n}} \approx N(0, 1),$$

since n is large.

With $\bar{x} = 2580$, the value of this statistic is

$$\frac{2580 - 2610}{120.6/\sqrt{100}} = -2.49.$$

This is in the critical region ($|Z| > 1.96$) for a two-tail 5% test, so we reject H_0 at this level.

The method of testing the *difference between two means*, using samples from each of two *normal* distributions with *unknown variances* is also similar to the method shown on page 254.

Instead of using the test statistic

$$\frac{(\bar{X}_1 - \bar{X}_2) - 0}{\sqrt{(\sigma_1^2/n_1 + \sigma_2^2/n_2)}}$$

we use

$$Z = \frac{(\bar{X}_1 - \bar{X}_2) - 0}{\sqrt{(s_1^2/n_1 + s_2^2/n_2)}}$$

where s_1^2 and s_2^2 are the two *estimated* population variances.

EXAMPLE 5   The weekly wages received by a random sample of 50 labourers employed by a construction company had a mean of £63.30 and a standard deviation £1.20 (1979 figures). It was thought by members of this workforce that a company employing workers on a similar job in the neighbourhood paid higher wages than theirs.

On investigation it turned out that the mean weekly wage of a random sample of 50 from the second company had a mean of £63.90 and a standard deviation of £2.20.

Using a 5% significance level, do a test to discover whether the men from the first company could be right in thinking the second company paid a higher mean weekly wage. (The distributions are assumed to be normal.)

If the population means are μ_1 and μ_2 the hypotheses are

$$H_0: \mu_1 - \mu_2 = 0 \quad \text{as against} \quad H_1: \mu_1 - \mu_2 < 0.$$

(From the wording of the question, this involves a one-tail test.)
  The unknown population variances are estimated to be

$$s_1^2 = (1.20)^2 \times \frac{50}{49} = (1.212)^2$$

and

$$s_2^2 = (2.20)^2 \times \frac{50}{49} = (2.222)^2.$$

For a 5% one-tail test, the critical
region is given by

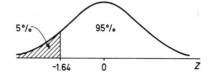

$$Z < -1.64.$$

With $\bar{x}_1 = 63.30$ and $\bar{x}_2 = 63.90$, the value of

$$Z = \frac{(\bar{X}_1 - \bar{X}_2) - 0}{\sqrt{(s_1^2/n_1 + s_2^2/n_2)}} \quad \text{is} \quad \frac{-0.60}{\sqrt{[(1.212)^2/50 + (2.222)^2/50]}}$$

$$= -1.69.$$

Since this is in the critical region, H_0 is rejected and it looks as though the complaining workmen have a case as far as mean wage levels are concerned.

EXERCISE 13c

1.   A mail-order firm sells parcels of assorted toys and advertises that the mean retail value of the contents of a parcel is £3.50. After receiving several complaints from customers, a government inspector purchased 100 of these parcels and found that the mean retail value of the toys per parcel was £3.38 with a standard deviation of £0.48.
  Use a statistical argument to discuss whether or not the inspector is justified in upholding the customers' complaints. (Take a 5% level of significance.)   (JMB)

2.   In one county in England, a random sample of 225 twelve-year-old boys and 250 twelve-year-old girls were given an arithmetic test. The average mark for the boys was 57 with a standard deviation of 12, while the average for the girls was 60 with standard deviation of 15.
  Assuming that the distributions are normal, does this provide evidence at the 2% level that twelve-year-old girls are superior to twelve-year-old boys at arithmetic?   (SUJB)

3.   (a) A certain type of battery for calculators is said to last for 2000 hours. A sample of 200 of these batteries was tested: the mean life was 1995 hours and the standard deviation of the lives was 25.5 hours. Use these data to test the hypothesis that the population mean life is 2000 hours against the alternative hypothesis that it is less than 2000 hours. State what level of significance you are using in the test.
  (b) Two types of battery were compared for the length of time they lasted. The data obtained are summarised in the table.

Battery type	Number tested	Sample mean	Sample s.d.
A	200	1995	25.5
B	150	2005	32.8

Test the hypothesis that the populations from which these samples were drawn have equal means against the alternative hypothesis of unequal means. State the level of significance you are using.   (C)

4.   A group (*A*) of 50 art students were given a perception test which involved recording the time taken by each student to identify shapes and the time taken to match colours. A control group (*C*) of 50 other students were given the same tests. For each group the mean and standard deviation was calculated with the following results:

Group	Time to identify shapes (minutes)	Time to match colours (minutes)
A	Mean 3.1 s.d. 1.1	Mean 7.2 s.d. 2.3
C	Mean 3.7 s.d. 1.5	Mean 7.9 s.d. 2.5

Using a 5% significance level determine whether:
(a) the difference in mean times for identifying shapes is significant,
(b) the difference in mean times for matching colours is significant.
   Is the difference in mean times for identifying shapes significant at the 2% level?   (C)

5.   The following table relates to the number of machine breakdowns per day in a large factory in a period of 70 days.

Number of breakdowns per day	0	1	2	3	4	5	6	7	8	9	10 and over
Number of days	1	5	10	13	16	11	7	4	2	1	0

Calculate the mean number of breakdowns and the variance from this sample.
   If these observations may be taken as a random sample from a large number of days, estimate the population variance and test the hypothesis that the population mean is: (a) 3.5, (b) 4.5.   (MEI)

13.4★ Testing the assumption that two samples may be regarded as coming from the same normal distribution

We include here, for convenience, a test that will be of great importance later.
   Suppose observations are made relating to toxic levels in two groups of babies, one group being 'at risk' as a result of being born with some special medical condition, and the other being a 'normal' group used as a control group for comparison.
   The observations are given in suitable coded units as follows:

Group	n	Σx	Σx^2
Control	30	120.9	521.51
Risk	140	773.1	4645.53

where n is the number of babies in each group, $\Sigma\,x$ is the sum of the observations and $\Sigma\,x^2$ is the sum of the squares of the observations.

What we want to test is whether there is any difference between the distributions, of which these are two samples. To be more specific, the test is:

Can these two sets of observations be regarded as coming from the *same normal distribution*, in other words from normal distributions with the *same mean and variance*?

At this stage we are not in a position to test whether the parent variances are the same, but if we make the assumption that there *is* a common variance (an assumption for which there is an appropriate test, as will be seen much later in this book) an estimate of this common variance may be calculated using the formula

$$\hat{\sigma}^2 = \frac{(n_1-1)s_1^2 + (n_2-1)s_2^2}{n_1+n_2-2} \quad \text{(see page 239)}.$$

It can be shown that if n_1 and n_2 are large it is permissible to use the test statistic

$$Z = \frac{(\bar{X}_1 - \bar{X}_2) - 0}{\hat{\sigma}\sqrt{(1/n_1 + 1/n_2)}}$$

since its distribution is approximately $N(0, 1)$.

We shall now use this test for our example.

The mean of the control group $\bar{x}_1 = 120.9/30 = 4.03$

The mean of the group at risk $\bar{x}_2 = 733.1/140 = 5.522$

$$s_1^2 = \Sigma\,[(x_1 - \bar{x}_1)^2]/(n_1 - 1)$$

$$\Rightarrow (n_1 - 1)s_1^2 = \Sigma\,(x_1 - \bar{x}_1)^2 = \Sigma\,(x_1^2) - n_1\,\bar{x}_1^2$$

$$= 521.51 - 30(4.03)^2 = 34.283.$$

Similarly $(n_2 - 1)s_2^2 = 4645.53 - 140(5.522)^2 = 376.582.$

Hence $\hat{\sigma}^2 = \dfrac{34.283 + 376.582}{30 + 140 - 2} = 2.45.$

The hypotheses to be tested are

$$H_0 : \mu_1 - \mu_2 = 0 \quad \text{as against} \quad H_1 : \mu_1 - \mu_2 \neq 0.$$

Under H_0 the distribution of $\bar{X}_1 - \bar{X}_2$ is approximately

$$N[0, \hat{\sigma}^2(1/n_1 + 1/n_2)] = N(0, 0.1428).$$

The value of

$$\frac{(\bar{X}_1 - \bar{X}_2) - 0}{\hat{\sigma}(1/n_1 + 1/n_2)}$$

for these samples is

$$\frac{1.492}{\sqrt{0.1428}} = 3.95$$

so that H_0 is rejected.

EXERCISE 13d

1. Two samples of steel from different sources were tested for carbon content and the results in appropriate units were:

Size of sample	Mean carbon content in sample	Sample s.d.
25	9.01	0.04
30	9.03	0.03

Assuming a common population variance test the hypothesis that the samples may be regarded as coming from the same normal distribution. (Use a 5 % significance level.)

2. The random variable X is the breaking strength in tonnes of a chain link. Two samples were tested to breaking point, the second sample being tested a week after the first one, in order to check on consistency in the manufacturing process. The results were as shown in the following table where Σx is the sum of breaking strengths, and Σx^2 is the sum of the squares of breaking strengths.

Sample	Sample size	Σx	Σx^2
First	50	75	150
Second	70	140	300

Assuming a common population variance, determine whether there is evidence that there is a difference between the mean breaking strengths of the populations from which the samples are taken. (Use a 1 % significance level.)

3. A large number of tomato plants are grown under controlled conditions. Half of the plants, chosen at random, are treated with a new fertiliser, and the other half are treated with a standard fertiliser. Random samples of 100 plants are selected from each half, and records are kept of the total crop mass of each plant.
    For those treated with the new fertiliser, the crop masses (in suitable units) are summarised by the figures

$$\Sigma x = 1030.0, \qquad \Sigma x^2 = 11\,045.59.$$

The corresponding figures for those plants treated with the standard fertiliser are

$$\Sigma y = 990.0, \qquad \Sigma y^2 = 10\,079.19.$$

Treating the sample as a large sample from a normal distribution, and assuming that the population variances of both distributions are equal, obtain a two-sample pooled estimate of the common population variance.
    Assuming that it is impossible for the new fertiliser to be less efficacious than the old fertiliser, test whether the results provide significant evidence (at the 3 % level) that the new fertiliser is associated with a greater mean crop mass, stating clearly your null and alternative hypotheses.   (C)

13.5   Confidence intervals

Let us look again at a typical question of the kind that a hypothesis test is designed to answer.
    The times taken for 10 g specimens of washing powder to remove a standardised stain from a fabric have, over a long period of time, been normally distributed with

mean $\mu = 163$ s and standard deviation 12 s. A sample of 16 specimens are treated with powder to which a so-called 'miracle ingredient' has been added, and the mean time for the sample is 159 s. Is this significant evidence that the ingredient produces an improvement?

What is being asked here is whether or not it is reasonable to conclude that the population mean μ for the treated fabric is, or is not, equal to some *stated* value (163 s).

We shall now consider a different sort of situation in which a population mean is *unknown* and where we want to come to some reasonable conclusion about it, based on the evidence of a single sample.

As we have seen, we may often take the sample mean as a good 'point estimate' of the population mean, since on average, it gives the correct result $[E(\bar{X}) = \mu]$. Nevertheless this point estimate is almost certainly not exactly equal to μ. The question we must ask now is: how confident can we be about this estimate? Or, to be more specific, within what *range of values* can we be reasonably sure that μ lies? This range of values, when we find it, will be called a *confidence interval*.

Suppose the breaking strain of a certain type of steel bar is normally distributed with variance 0.16 kN. The mean breaking strain is unknown, but a sample of 8 bars gives a mean of 13.5 kN. If we take 13.5 kN as a point estimate for μ, we want to use it to find a range of values (confidence interval) about which we can claim with reasonable confidence that it contains the true population mean.

As usual the words 'reasonable confidence' are too imprecise and we must first state a figure which represents the degree of confidence we are prepared to accept. The figure 95 % is the one often used by statisticians.

Now we know that if a single sample of size n is taken from a normal distribution $N(\mu, \sigma^2)$ the distribution of $\bar{X}$ is $N(\mu, \sigma^2/n)$, and that the distribution of

$$Z = \frac{\bar{X} - \mu}{\sigma/\sqrt{n}} \quad \text{is } N(0, 1).$$

This means that 95 % of the distribution of Z lies between $Z = -1.96$ and $Z = 1.96$.

In other words

$$P(-1.96 \leqslant Z \leqslant 1.96) = 0.95$$

$$\Rightarrow P\left(-1.96 \leqslant \frac{\bar{X} - \mu}{\sigma/\sqrt{n}} \leqslant 1.96\right) = 0.95. \quad (1)$$

We shall now turn the statement inside the bracket round into a different form.

The inequality $\dfrac{\bar{X} - \mu}{\sigma/\sqrt{n}} \leqslant 1.96$ can be re-arranged as

$$\mu \geqslant \bar{X} - 1.96\sigma/\sqrt{n}$$

and the inequality $\quad -1.96 \leqslant \dfrac{\bar{X} - \mu}{\sigma/\sqrt{n}}$

is similarly equivalent to

$$\mu \leqslant \bar{X} + 1.96\sigma/\sqrt{n}.$$

Combining these, we can rewrite the statement (1) as

$$P\left(\bar{X} - 1.96\frac{\sigma}{\sqrt{n}} \leqslant \mu \leqslant \bar{X} + 1.96\frac{\sigma}{\sqrt{n}}\right) = 0.95.$$

The interval $\bar{X} - 1.96\dfrac{\sigma}{\sqrt{n}}$ to $\bar{X} + 1.96\dfrac{\sigma}{\sqrt{n}}$

is called a *95 % confidence interval for* μ, or, strictly a *central* 95 % confidence interval as it is symmetrical about $\bar{X}$, and the end points are called *95 % confidence limits* for μ.

In our example $\sigma = 0.4$, $n = 8$ and $\bar{x} = 13.5$ so that the 95 % confidence limits are

$$\left(13.5 - 1.96 \times \frac{0.4}{\sqrt{8}}, \quad 13.5 + 1.96 \times \frac{0.4}{\sqrt{8}}\right)$$

or (13.22, 13.78).

We say, therefore, that there is a *95 % probability that this interval contains* μ.

The meaning of a confidence interval

It is important that you should realise what is meant by that last, carefully worded statement. First of all μ is not a variable; although unknown it is *fixed*. It is the confidence interval that varies, depending on the value of $\bar{x}$, the sample mean.

Suppose, for example, that the (unknown) value happens to be 13.6.

For different values of the sample means, the confidence intervals are shown in the following table:

$\bar{x}$	Confidence interval	
13.12	(12.84, 13.40)	
13.50	(13.22, 13.78)	(as in the example above)
13.65	(13.37, 13.93)	
13.91	(13.63, 14.19)	

The second and third intervals do contain μ (13.6), while the other two do not. This is illustrated in the diagram.

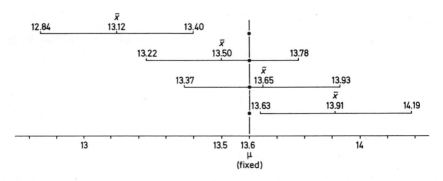

The statement that there is a 95 % probability that a confidence interval contains μ means that, in the long run, a confidence interval will contain μ 19 times out of 20. A gambler might be prepared to bet 19 to 1 against the population mean lying outside any particular confidence interval.

To sum up, then, when we said in our example that a 95 % confidence interval for μ is (13.23, 13.78), it means that we are 95 % *confident that μ lies in this interval.*

The figure 95 %, although often used, may be replaced by 99 %, say, if we want to be more confident.

99 % *confidence limits* are defined in a similar way with 1.96 replaced by 2.58, since

$$\Phi(2.58) = 0.995.$$

In our example, 99 % confidence limits for μ are

$$\bar{x} \pm 2.58 \times \frac{0.4}{\sqrt{8}} \quad \text{or} \quad (13.14, 13.86).$$

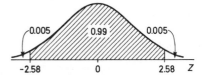

In general a K % confidence interval is

$$\left(\bar{X} - z \cdot \frac{\sigma}{\sqrt{n}} \; , \; \bar{X} + z \cdot \frac{\sigma}{\sqrt{n}} \right)$$

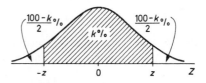

where the appropriate value of z is taken from the normal tables.

Confidence interval for the mean of any distribution with known variance (large sample)

The method we have illustrated by an example where the population was normally distributed may be used in cases where the distribution is *not normal*, provided the sample is *large*. The distribution of the test statistic

$$Z = \frac{\bar{X} - \mu}{\sigma / \sqrt{n}} \quad \text{is then approximately} \quad N(0, 1).$$

As a rough guide, n should be greater than about 30 for a not-too-skew distribution (see page 242).

Confidence interval for the mean of a normal distribution with unknown variance (large sample)

Here, too, there is no new method to be learned. The procedure is exactly the same except that σ^2 has to be replaced by an unbiased estimate, given by

$$s^2 = \frac{\Sigma (x - \bar{x})^2}{n-1} = \frac{n}{n-1} \times \text{sample variance.}$$

(It is proved in a later chapter that the distribution of $(\bar{X} - \mu)/(s/\sqrt{n})$ is also approximately normal $N(0, 1)$ for large values of n. In fact little is lost by using the sample variance instead of s^2. The additional error does not make much difference when we are already using an approximate distribution. If, however, the sample is small an alternative method is needed. This is described in the next chapter.)

EXAMPLE 6   A random sample of 100 capacitors, each of nominal capacitance $2\,\mu F$, was taken from a very large batch. The capacitance of each capacitor in the sample was measured. The results are summarised in the following table.

Capacitance (μF) (mid-interval value)	1.85	1.90	1.95	2.00	2.05	2.10	2.15
Number of capacitors	2	12	23	31	20	10	2

Assuming that these data refer to a random sample from a normal population, determine 95% confidence limits for the mean capacitance of capacitors in the distribution.

By the usual method the sample mean $\bar{x} = 1.9965$ and the sample variance $= 4.16 \times 10^{-3}$.

$$\text{Hence}\quad s^2 = \frac{100}{99} \times 4.16 \times 10^{-3} \Rightarrow s = 0.0648 \quad\text{and}\quad s/\sqrt{n} = \frac{0.0648}{10} = 0.00648.$$

Ninety-five per cent confidence limits are therefore $\bar{x} \pm 1.96 \times s/\sqrt{n} = 1.9965 \pm 1.96 \times 0.00648$, giving a confidence interval

$$(1.98, 2.01) \text{ to 2 dec. places.}$$

EXERCISE 13e

1.   A sample of size 16 is taken from a normal distribution with variance 49. The sample mean is 142. Find: (a) 95% confidence limits, (b) 90% confidence limits, for the mean of the distribution.

2.   A normal distribution has variance 0.24. A random sample has the following values:

$$25.4 \quad 26.6 \quad 25.8 \quad 26.0 \quad 26.2.$$

Find 98% confidence limits for the mean of the distribution.

3.   The standard deviation of the weights of all twenty-one-year-old males is 10 kg. A random sample of 100 men had mean weight 73.6 kg. Find a 99% confidence interval for the mean weight of all twenty-one-year-old males.

4.   A sample of 400 students from a large university had an average weight 74.5 kg and standard deviation 3.1 kg. Find 95% confidence limits for the average weight of all students in the college. The distribution of student's weights is assumed to be normal.

5.   The table below gives the distribution of age in years at marriage of 175 males:

Age (mid-interval value)	17.5	22.5	27.5	32.5	37.5	42.5	47.5	52.5	57.5	62.5
Frequency	28	68	43	18	9	4	2	1	0	2

Calculate the mean and standard deviation of these ages.
If these ages are taken to be a random sample from a normal distribution, find 95% confidence limits for the population mean.   (MEI)

6.   The distribution of measurements of thicknesses of a random sample of yarns produced in a textile mill is shown in the following table.

Yarn thickness (mid-interval value) (microns)	Frequency
72.5	6
77.5	18
82.5	32
87.5	57
92.5	102
97.5	51
102.5	25
107.5	9

Use these data to determine approximate 95 % confidence limits for the population mean.
(MEI)

Confidence interval for the difference between two means (large samples)

A large manufacturing organisation is divided into two factories, A and B. In A, a sample of 100 workers were found to take an average time of 30 minutes to complete a given task. The sample variance was 9 minutes^2. For the same task in factory B a sample of 120 produced a mean time of 27 minutes with variance 4 minutes^2.

An unbiased estimate of the difference between the two population means $\mu_1 - \mu_2$ is the difference between the two sample mean $\bar{x}_1 - \bar{x}_2 = 3$ minutes.

Suppose the management wanted to find a 95 % confidence interval for $\mu_1 - \mu_2$, using the data from the samples:

$$\text{Factory } A \quad n_1 = 100, \qquad \bar{x}_1 = 30, \quad \text{var}(x_1) = 9$$

$$\text{Factory } B \quad n_2 = 120, \qquad \bar{x}_2 = 27, \quad \text{var}(x_2) = 4.$$

Using the same argument as for a single mean, since the distribution of

$$\bar{X}_1 - \bar{X}_2 \quad \text{is} \quad N(\mu_1 - \mu_2, \sigma_1^2/n_1 + \sigma_2^2/n_2)$$

95 % confidence limits for $\mu_1 - \mu_2$ are

$$(\bar{x}_1 - \bar{x}_2) \pm 1.96 \sqrt{\left(\frac{\sigma_1^2}{n_1} + \frac{\sigma_2^2}{n_2}\right)}.$$

Although the population variances σ_1^2 and σ_2^2 are not known, *estimated* values s_1^2 and s_2^2 will give a reasonable approximation since n_1 and n_2 are large.

$$s_1^2 = 9 \times \frac{100}{99} = 9.09, \qquad s_2^2 = 4 \times \frac{120}{119} = 4.03.$$

Substituting these values we arrive at the confidence limits

$$(30 - 27) \pm 1.96 \sqrt{\left(\frac{9.09}{100} + \frac{4.03}{120}\right)}$$

and a confidence interval (2.30, 3.69).

EXERCISE 13f

1. Two samples of the same plant were cultivated under different conditions, one in a greenhouse and the other outdoors. The first sample, of 200 plants, grew to a mean height of 23 cm with standard deviation 4 cm; the other sample, of 150, grew to a mean height of 22 cm with standard deviation of 6 cm.
    Determine: (a) a 90% confidence interval for the difference between the mean population heights, $\mu_1 - \mu_2$, (b) a 98% confidence interval for $\mu_1 - \mu_2$.

2. One hundred and sixty students from one part of a large city gained an average mark of 59% in A-level mathematics, with a standard deviation of 12. One hundred students from another part of the city, who took the same examination gained an average of 53% with a standard deviation of 10. Find a 95% confidence interval for the difference between the mean marks for students taking the same examination in the two parts of the city.
    ★Find also a 95% confidence interval for this difference if it is assumed that the two populations have a common variance.

13.6 Summary of hypothesis tests and confidence intervals for means

You have now met a fair number of results and it may be helpful to summarise them briefly.

Hypothesis test for a single mean

The basic test statistic is $Z = \dfrac{\bar{X} - \mu}{\sigma/\sqrt{n}}$.

1. This may be used for *any* size of sample from a normal distribution with known variance, and for a large sample even when the distribution is not normal.

2. For a normal distribution with unknown variance, replace σ^2 by an estimated value using

$$s^2 = \frac{n}{n-1} \times \text{sample variance} = \frac{\Sigma (x - \bar{x})^2}{n-1}.$$

Confidence limits for a single mean

Ninety-five per cent confidence limits for μ are $\bar{X} \pm 1.96 \dfrac{\sigma}{\sqrt{n}}$.

The remarks in 1 and 2 apply here also.

Hypothesis test for the difference between two means

The basic test statistic is $Z = \dfrac{(\bar{X}_1 - \bar{X}_2) - 0}{\sqrt{(\sigma_1^2/n_1 + \sigma_2^2/n_2)}}$.

1. This may be used for any size of samples from normal distributions, and for large samples even when the distributions are not normal.

2.   For normal distributions with unknown variances, replace σ_1^2 and σ_2^2 by s_1^2 and s_2^2, where

$$s_1^2 = \frac{\Sigma (x_1 - \bar{x}_1)^2}{n_1 - 1}, \qquad s_2^2 = \frac{\Sigma (x_2 - \bar{x}_2)^2}{n_2 - 1}.$$

3.$\star$ In the special case when the populations are assumed to have a common variance the test statistic is

$$Z = \frac{(\bar{X}_1 - \bar{X}_2) - 0}{\hat{\sigma}\sqrt{(1/n_1 + 1/n_2)}}$$

where $\hat{\sigma}^2$ is the estimate arrived at by pooling the two samples and using the formula

$$\hat{\sigma}^2 = \frac{(n_1 - 1)s_1^2 + (n_2 - 1)s_2^2}{n_1 + n_2 - 2}.$$

Confidence limits for the difference between two means

Ninety-five per cent confidence limits for $\mu_1 - \mu_2$ are

$$(\bar{X}_1 - \bar{X}_2) \pm 1.96 \sqrt{\left(\frac{\sigma_1^2}{n_1} + \frac{\sigma_2^2}{n_2}\right)}.$$

The remarks in 1 and 2 apply here also.

Significance testing and confidence intervals

It is quite easy to confuse the techniques of significance testing and estimating a confidence interval as there is some similarity between them. Problems should therefore be read carefully before being attempted. In the next example, the two methods are used in the two parts of the question in order to make clear the distinction between them.

EXAMPLE 7   A reading test taken by nine-year-old children in this country is known to give scores which are distributed with a standard deviation of 8. Obtain 95 % confidence limits for the population mean if a randomly selected group of 400 children aged 9 years produced a mean score of 85.4.

  If previous wide experience had produced a mean score of 86, interpret the significance of the value obtained from the group of 400.

  The distribution is not stated to be normal, but even if it were fairly skew, the size of sample (400) is large enough for the distribution of sample means to be approximately normal. (In fact, it is likely that this distribution would be near-normal.)

$$\sigma = 8, \qquad n = 400, \qquad \bar{x} = 85.4.$$

Ninety-five per cent confidence limits for μ are $\bar{x} \pm 1.96 \times \sigma/\sqrt{n}$ or

$$85.4 \pm 1.96 \times 8/20 = 85.4 \pm 0.784$$

and the 95 % confidence interval is (84.62, 86.18) to 2 dec. places.

The second part is about significance, and the hypotheses to be tested are:

$$H_0: \mu = 86 \quad \text{as against} \quad H_1: \mu \neq 86.$$

In other words we are testing whether the sample can be regarded as coming from a population with mean 86. The alternatives are that there might have been a change in the population mean since the country-wide tests were done, or there could have been something special about the sample.

Under H_0, the distribution of $Z = \dfrac{\bar{X} - 86}{8/20} \approx N(0, 1)$.

Now when $\bar{x} = 85.4$, $\quad z = \dfrac{85.4 - 86}{0.4} = -1.5.$

Since this is less than 1.96, we conclude that H_0 can be accepted at the 5 % level.

★Confidence interval for a function of a parameter

Having obtained a confidence interval for an unknown parameter, it is sometimes possible to obtain from it a confidence interval for a particular *function* of this parameter.

For an example, where the parameter is a mean and the function is 'reliability', see the Appendix on page 275.

EXERCISE 13g (miscellaneous)

1.   A manufacturer is producing resistances normally distributed about a mean of 50.0 ohms with a standard deviation of 1.0 ohm. To check the consistency of his product, he periodically extracts 100 resistances and measures them. He finds that one such sample has a mean of 50.24 ohms. Has there been any significant change in his product?

2.   The diameters of 25 steel rods are measured and found to have a mean of 0.980 cm and a standard deviation of 0.015 cm. Assuming this is a sample from a normal distribution with the same variance, find 99 % confidence limits for the population mean.   (O & C)

3.   An examination is taken by a large number of candidates. The marks of a random sample of 50 candidates were classified in intervals of 10 marks as follows:

Mid-interval value	4.5	14.5	24.5	34.5	44.5	54.5	64.5	74.5	84.5	94.5
Frequency	1	0	3	10	9	9	7	5	3	3

Use this sample of marks to estimate the population mean and variance. Calculate 95 % confidence limits for the population mean.   (L)

4.   The following table shows the observed frequency distribution of the number of matches per box in a random sample of 100 boxes of a particular brand:

Number of matches	46	47	48	49	50	51	52
Number of boxes	4	15	24	30	16	8	3

Calculate unbiased estimates of the mean and variance of the number of matches per box of this brand. Determine approximate 90 % confidence limits for the mean number of matches per box.

The number of matches per box in a random sample of 50 boxes of another brand gave unbiased estimates 49.4 and 2.17, respectively, of the mean and variance of the number of matches per box of this brand. Using a 5% significance level test the hypothesis that the average number of matches per box are the same for the two brands. (JMB)

5.   Explain what is meant by (a) the null hypothesis, (b) the significance level of a test.
   A random sample x_i $(i = 1, 2, \ldots, 10)$ is chosen from a normal population:

$$\Sigma\, x = 20.$$

Test the null hypothesis that the mean $\mu = 0.5$ against the alternative hypothesis that $\mu > 0.5$, assuming that the population variance $\sigma^2 = 3$. Use a 1% significance level. (O)

6.   As part of a training course 50 soldiers were given a rifle-shooting test after spending the previous 24 hours on a low diet of food. A week later 40 of them were given the same test, but on this occasion they had eaten normally. The results were as follows:

	First test	Second test
Mean score	92.5	94.5
Standard deviation	5	4

Was the improved result on the second test significantly better: (a) at the 5% level, (b) at the 1% level? (C)

7.   A scientist wishing to make inferences concerning the mean of a population takes a random sample of 50 observations from the population. He obtains values $x_1, x_2, \ldots, x_{50}$ from which he calculates

$$\bar{x} = \left(\sum_{1}^{50} x_i \right) \bigg/ 50 = 3.4, \qquad \sum_{1}^{50} (x_i - \bar{x})^2 = 392.$$

Find a 95% confidence interval for the population mean.
   Later the scientist becomes interested in another population and he wishes to test whether the means of the two populations are equal. He takes a random sample of 40 observations from the second population and he obtains values $y_1, y_2, \ldots, y_{40}$ such that

$$\bar{y} = \left(\sum_{1}^{40} y_j \right) \bigg/ 40 = 4.5, \qquad \sum_{1}^{40} (y_j - \bar{y})^2 = 312.$$

Use a 5% significance level to test the hypothesis that the means of the two populations are equal. (JMB)

8.   Independent observations $X_1, X_2, \ldots, X_{25}$ are taken from the same normal distribution which has mean μ and variance 16. Independent observations $Y_1, Y_2, \ldots, Y_{64}$ are taken from another normal distribution with mean λ and variance 16. If $\bar{x} = 2.0$ and $\bar{y} = 3.4$, show that $\bar{x}$ lies outside the 95% confidence interval for λ whereas $\bar{y}$ lies inside the 95% confidence interval for μ.
   Test the null hypothesis that $\mu = \lambda$ at a 5% level of significance. (MEI)

9.   The mass of jam in a jar is x grams, the mass of the jar is y grams and the mass of the lid is z grams. x, y and z are independent and normally distributed with means 502, 90 and 10 and standard deviations 0.60, 0.24 and 0.07 respectively. Deduce the mean and standard deviation of the total mass of the jar full of jam, complete with lid.
   These full jars are packed in boxes of 25. Find the mean and standard deviation of the mass of the total contents of a box.
   The average mass of jam per jar in a box is found to be 502.2 grams. Test, at the 5% level, whether or not this is: (a) significantly different from the expected value, (b) significantly greater than the expected value. (JMB)

10.   A random sample of 60 car drivers living in a town were asked to state the age at which they learned to drive. A sample of 60 car drivers from a country district were asked the same question. From their answers the following figures were obtained.

Area	Mean age at which they learned to drive	s.d.
Town	20.4	2.8
Country	19.4	2.1

Determine whether these two means are significantly different at the 5% level.   (C)

11.   A large group of sunflowers is growing in the shady side of a garden. A random sample of 36 of these sunflowers is measured. The sample mean height is found to be 2.86 m, and the sample standard deviation is found to be 0.60 m. Treating the sample as a large sample and assuming the heights to be normally distributed, give a symmetric 99% confidence interval for the mean height of the sunflowers in the shady side of the garden.

A second group of sunflowers is growing in the sunny side of the garden. A random sample of 26 of these flowers is measured. The sample mean height is found to be 3.29 m and the sample standard deviation is found to be 0.90 m. Treating the samples as large samples from normal distributions having the same variance, test whether the results provide significant evidence (at the 5% level) that the sunny-side sunflowers grow taller, on average, than the shady-side sunflowers.   (C)

12.   Twenty-five independent observations are taken from a normal distribution with mean μ_1 and variance 9. Another twenty-five independent observations are taken from a separate normal distribution with mean μ_2 and variance 16. Construct 95% confidence intervals for μ_1 and for μ_2 when the mean from the first sample is 3.5 and the mean from the second sample is 5.0.

Show that:
(a) the null hypothesis $\mu_1 = 5$ is rejected by a significance test at the 5% level,
(b) the null hypothesis $\mu_1 = \mu_2$ is accepted by a significance test at the 5% level.   (MEI)

13.★ Two independent random samples, of n_1 and n_2 observations are drawn from normal distributions with common variance σ^2. If s_1^2 and s_2^2 are unbiased estimates of σ^2 based on the first and second sample, respectively, show that $(n_1 s_1^2 + n_2 s_2^2)/(n_1 + n_2)$ is an unbiased estimate of σ^2.

Two makes of car safety belts, A and B, have breaking strengths which are normally distributed with the same variance. A sample of 140 belts of make A and a sample of 220 belts of make B were tested, and the sample means, and sums of squares about the means, of the breaking strengths (in standard units) were $(2685, 1.9 \times 10^4)$ for make A and $(2680, 3.4 \times 10^4)$ for make B. Examine whether these results provide significant evidence that belts of make A are stronger, on average, than belts of make B.   (JMB)

14.   The power consumption of a certain brand of light bulb is nominally 100 watts, and may be assumed to follow a normal distribution with mean 100 watts and standard deviation 2 watts. Calculate the probabilities:
(a) that 3 bulbs selected at random each have a power consumption exceeding 101 watts,
(b) that 3 bulbs selected at random have a total power consumption exceeding 303 watts,
(c) that the mean consumption of 50 bulbs selected at random is less than 99.5 watts.

Following a change in the manufacturing process, a sample of 100 bulbs was tested. Denoting the power consumption in watts of a bulb by w, it was found that $\Sigma (w - 100) = 52$ and $\Sigma (w - 100)^2 = 93$. Calculate a 99% confidence interval for the new mean power consumption.   (C)

15.★ A random sample of 27 individuals from the population of young men aged 18 and of high intelligence have foot lengths (in centimetres to the nearest centimetre) as summarised below.

Foot length (cm)	24	25	26	27	28	29	30
Number with this foot length	1	2	3	9	6	5	1

Obtain the sample mean and show that the unbiased estimate of the population variance, based on this sample, is 2.00. Obtain a 96% confidence interval for the mean foot length of this type of person.

A random sample of 48 individuals from the population of young men aged 18 and of moderate intelligence have foot lengths summarised by $\bar{x} = 26.6$, $\Sigma (x_i - \bar{x})^2 = 123.20$.

A complex genetic theory suggests that persons of high intelligence have a greater foot length than do those of moderate intelligence. The two samples described above may be assumed to have been drawn at random from independent normal distributions having a common variance. Obtain an unbiased two-sample estimate of this common variance. Treating the samples as large samples, test this genetic theory, using a significance level of 1% and stating clearly the hypotheses under comparison. (C)

16.   A group of 10 tomato plants was grown with a certain fertiliser and the mean yield was found to be 1.200 kg. Records show that the same variety of plant grown under similar conditions but without fertiliser shows a mean yield of 1.160 kg with standard deviation 0.070 kg. Stating your null and alternative hypotheses clearly, test at the 5% level of significance whether it can be claimed that the fertiliser improves the yield.

A further sample of 6 plants treated with the fertiliser gave a mean yield of 1.168 kg. Pool the results of the two samples to see whether or not the extra sample alters the previous results. (Assume a normal distribution.) (SUJB)

17.   The sum of the lengths of 100 articles manufactured in one section of a factory is 305 cm and the sum of their squares is 1225 cm². The sum of the lengths and sum of squares of the lengths of 50 such articles manufactured in a different section are 180 cm and 746 cm². Derive an approximate 95% symmetrical confidence interval for the difference between the two distribution means.

Test, at the 5% level of significance, the hypotheses that:
(a) the mean length of articles manufactured in the first section of the factory is 3 cm;
(b) the difference between the means of the lengths of articles manufactured in the two
    sections is zero. (JMB)

18.★ An investigator is planning an experiment to compare the means of two normally distributed populations, each of which is known to have variance 4.5 square units. The investigator decides that he can afford to take a fixed total of $2N$ sample observations. Show that in order to minimise the width of the 95% confidence interval for the difference between the two population means, he should take N observations from each population. In this case, find the smallest value of N for the width of the confidence interval to be less than 1.5 units.

The investigator actually took a random sample of 100 observations from each of the two populations. Calculate, to three decimal places, the probability that the two observed sample means will differ by at most 0.3 units in each of the cases when:
(a) the two population means are equal,
(b) the mean of one population is one unit more than the mean of the other popula-
    tion. (JMB)

19.★ Random samples are to be taken from two independent normal distributions which have means μ_1, μ_2 and standard deviations σ_1, σ_2 respectively.

The samples are of size $2n$ and n respectively and the corresponding sample means are denoted by $\bar{X}_1, \bar{X}_2$. Write down expressions for $E(\bar{X}_1 - \bar{X}_2)$ and $\text{Var}(\bar{X}_1 - \bar{X}_2)$. (*contd.*)

A researcher knows that $\sigma_1 = 2$ and $\sigma_2 = \sqrt{3}$. Samples are taken as above with $n = 5$ in order to test the null hypothesis $\mu_1 = \mu_2$ against the alternative hypothesis $\mu_1 \neq \mu_2$.

The actual values of μ_1 and μ_2 (not known to the researcher) are 5 and 4 respectively. Find the probability that, when testing at the 5% level of significance, the null hypothesis will be accepted.

Find the least value of n for which the probability of accepting the null hypothesis would be less than 5%.   (C)

20.★ A random sample of size 100 is drawn from a normal population of unknown mean μ and known variance σ^2. Show that if m is the sample mean then there is a probability 0.95 that the interval $(m - 0.196\sigma, m + 0.196\sigma)$ contains μ; thus this interval is a 95% confidence interval for μ.

A random sample of size 80 is drawn from a normal population of unknown mean λ and known variance σ^2. Write down a 95% confidence interval for λ in terms of the sample mean m'.

The two samples are to be used to test the hypothesis that $\lambda = \mu$. The hypothesis will be rejected only if the 95% confidence intervals for λ and μ do not overlap. Show, that, if the hypothesis is correct, there is a probability of about 0.006 that this procedure leads to the wrong conclusion.   (MEI)

21.   The heights of 100 men in a random sample were measured to the nearest centimetre, giving a sample mean and standard deviation of 190 cm and 10 cm respectively. Determine 95% confidence limits for the population mean to 3 significant figures, explaining precisely what is meant by the term 'confidence limit'.

In another random sample the heights of 800 women were measured, giving a sample mean of 175 cm and sample standard deviation of 5 cm. Find 95% confidence limits for the difference between the two population means to 3 significant figures. (Assume both populations are normally and independently distributed.)   (MEI)

13.7   Appendix

One-tail tests

As we have already seen, a one-tail test is written in the form

$$H_0: \mu = 120 \text{ (say)}, \quad \text{as against} \quad H_1: \mu > 120$$

and the critical region for a 5% significance level is

$$Z > 1.64.$$

It could be argued that this procedure ignores the fact that it *might* happen that we are wrong in assuming that these are correct alternatives, and that μ could be less than 120. In that case H_0 and H_1 as we have stated them do not cover all possibilities, and the correct hypotheses should be

$$H_0: \mu \leqslant 120 \quad \text{as against} \quad H_1: \mu > 120.$$

If you come across hypotheses expressed in this form, you may wonder what value of μ to take for the test statistic

$$Z = \frac{\bar{X} - \mu}{\sigma / \sqrt{n}}.$$

The answer is that you should take $\mu = 120$ as usual, for the following reason.

Suppose

$$\frac{\bar{x}-120}{\sigma/\sqrt{n}}$$

is in the critical region for $\mu = 120$ and is therefore greater than 1.64, then '$\mu = 120$' is rejected as unlikely.

For $H_0: \mu < 120$, μ might be 110 (say), then, since

$$\frac{\bar{x}-110}{\sigma/\sqrt{n}} > \frac{\bar{x}-120}{\sigma/\sqrt{n}},$$

$$\frac{\bar{x}-110}{\sigma/\sqrt{n}}$$

will also be in the critical region (for $\mu = 120$) and H_0 will still be rejected.

In other words the two ways of stating the hypotheses are equivalent, but it is recommended that you should normally stick to the usual form.

Confidence interval for 'reliability'

If the life of an electronic component is X, its *reliability* R is defined as follows:

$$R(x) = P(X > x),$$

the probability that it will still be operating at time x.

Suppose the life X is distributed normally with known variance 5 hours^2 and unknown mean μ hours, and that we have a sample of 100 components whose mean life is 52 hours. From this information 95 % confidence limits for the mean are

$$52 \pm 1.96 \times \frac{\sqrt{5}}{10} \quad \text{or} \quad (51.56, 52.43).$$

In other words $P(51.56 \leqslant \mu \leqslant 52.42) = 0.95$.

We shall now state the problem.

The value of R is a function of x. The problem will be to find 95 % confidence limits for R, when x has a particular value, say $x = 50$, that is, to find 95 % confidence limits for the *probability that the component will still be operating after 50 hours*.

Now $$R(x) = P(X > x) = 1 - \Phi\left(\frac{x-\mu}{\sigma}\right)$$

so that $$R(50) = 1 - \Phi\left(\frac{50-\mu}{\sqrt{5}}\right).$$

$R(50)$ is thus a function of μ, and it will be seen that as μ increases, so does R.

In fact, as μ increases from 51.56 to 52.43,

$$R(50) \text{ increases from } 1 - \Phi\left(\frac{50-51.56}{\sqrt{5}}\right) \text{ to } 1 - \Phi\left(\frac{50-52.43}{\sqrt{5}}\right),$$

or from 0.76 to 0.86.

Hence $P[0.76 \leqslant R(50) \leqslant 0.86] = P(51.56 \leqslant \mu \leqslant 52.43) = 0.95$

that is, a 95 % confidence interval for $R(50)$ is $(0.76, 0.86)$.

14 Hypothesis tests and confidence intervals (means): 2

14.1 Small samples

The methods used for testing hypotheses about means and establishing confidence intervals for means described in the last chapter are applicable, as we saw, to two kinds of situation:

1. when the parent distribution has a *known variance*, in which case the sample could be *any size*,

2. when the parent distribution has a variance which is *not known*, only if the sample is a *large* one.

Consider the case of a car manufacturer who is thinking of placing an order for a certain kind of safety belt and has been told by the supplier of the belt that its mean breaking strength is 2700 standard units. In order to test this claim he subjects a sample of 10 to a test and finds that the sample mean is 2640, with variance 2.1×10^4. He does not know what the variance of the parent distribution is, and he is obliged to rely on a small sample for his test, since testing to 'destruction' with a large sample, say of 150, might be more expensive than is acceptable. What is he to do?

In the nature of things, too, small samples may be all that are available. A researcher may wish to investigate the effect on a child's educational development of two factors, heredity and environment, using identical twins who happen to have been separated at an early age and brought up in different environments. Clearly the number of such twins is unlikely to be large.

In this chapter, therefore, we shall extend the work on significance tests and confidence intervals to cover cases where only small samples are available and the population variance is unknown. You will find that the methods are much the same as for large samples but that a new test statistic is required.

For *large* samples we used the fact that the distribution of

$$Z = \frac{\bar{X} - \mu}{s/\sqrt{n}}$$

is approximately $N(0, 1)$ where s^2 is the unbiased estimate of σ^2, calculated by the formula

$$\hat{\sigma}^2 = s^2 = \frac{\Sigma (x - \bar{x})^2}{n - 1}$$

and we justified this procedure (without proof) by pointing out that, for a large value of n, s^2 would be a good estimate for σ^2.

Now if n is *small* the distribution of $(\bar{X} - \mu)/(s/\sqrt{n})$ is far from being approximately normally distributed and we must therefore give a full description of this distribution before applying it in particular cases.

14.2   The *t*-distribution

The statistic $(\bar{X} - \mu)/(s/\sqrt{n})$ is denoted by 't', and its distribution was first discovered by W. S. Gosset (1876–1937), who worked as a statistician for the firm of Guinness in Dublin. Employees of this firm were not allowed to publish under their own names so he wrote under the pseudonym of 'Student'. The distribution was therefore called 'Student's *t*-distribution'. Apart from making this important discovery, Gosset was the first statistician to apply the Poisson distribution in a biological problem. It concerned, unsurprisingly, yeast cells.

(Traditionally a small 't' has been used for this statistic and we shall therefore make no distinction between the statistic and any particular value obtained by substituting $\bar{x}$ for $\bar{X}$).

The most important difference between the *t*-distribution and the normal distribution is that the density function, $f(t)$, of the *t*-distribution is *different for each value of n*. We shall not quote the complicated density function here, but if you are interested, you will find it in the Appendix on page 293. The distribution assumes that the parent distribution is *normal*.

The other new feature that is always used to describe any particular *t*-distribution is *the number of degrees of freedom*. It is written as v and is one less than the sample size n, so that

$$v = n - 1.$$

(For the moment this is all you need know about this formidable-sounding expression, although in due course you should read the explanation on page 287.)

The distribution used when $n = 5$ is therefore called the *t*-distribution with 4 degrees of freedom, or t_4.

In the diagram below the shapes of the graph of the density function for two

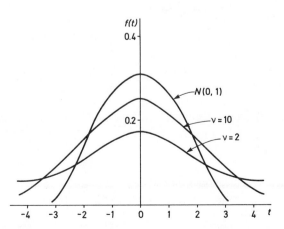

different values of v have been sketched, together with the normal distribution $N(0, 1)$ for comparison. It will be seen that they have a symmetrical bell-shaped form, as does the normal distribution, and if you wonder what happens if n, and therefore v, is large, we shall state here, and prove later, that *for large values of n the distribution of t approximates closely to the normal*. Meanwhile the first step towards using the t-distribution is to learn how to use the associated t-table.

Using the *t*-table

On page 519 a table called the t-table gives what are called 'percentage points' for different values of v. Part of this table is reproduced below.

P	20	10	5	2	1	0.2	0.1
$v = 1$	·	·	·	·	·	·	·
2	·	·	·	·	·	·	·
3	1.64	2.35	3.18	4.54	5.84	10.2	12.9
4	1.53	2.13	2.78	3.75	4.60	7.18	8.61

If $v = 4$ and $P = 10$, $t_4 = 2.13$.

This means that the *shaded* area is 10% of the area under the graph, that is,

$$P(|t| > 2.13) = 0.10$$

or

$$P(t > 2.13) = 0.05.$$

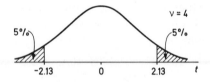

In other words the values of t are tabulated specifically for *two-tail tests*. Thus:

If $v = 4$, for a 5% level *two-tail test* ($P = 5$), $t_4 = 2.78$ and the critical region starts at $t = \pm 2.78$;

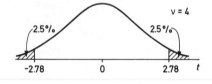

If $v = 4$, for a 1% level *two-tail test* ($P = 1$), $t_4 = 4.60$ and the critical region starts at $t = \pm 4.60$;

If $v = 3$, for a 5% level *one-tail test* ($P = 10$), $t_3 = 2.35$ and the critical region is $t > 2.35$;

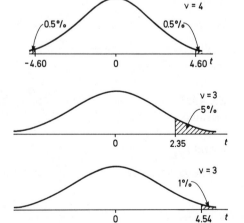

If $v = 3$, for a 1% level *one-tail test* ($P = 2$), $t_3 = 4.54$ and the critical region is $t > 4.54$.

14.3   Significance tests for means and differences between two means (paired and unpaired): small samples

Hypothesis test for the mean of a normal distribution (small sample)

We shall now show how the *t*-distribution enables us to deal with the example quoted at the beginning of this chapter.

EXAMPLE 1   Car safety belts supplied to a manufacturer are described by the supplier as having a mean breaking strength of 2700 (in standard units). A sample of 10 have mean breaking strength 2640 with variance 2.1×10^4. Assuming the population distribution to be normal, is this sample mean significantly different from the claimed mean of 2700? Is it significantly lower? (The manufacturer is more likely to want to know the answer to the second question, but we shall deal with both so as to illustrate the method.)

The data are $\bar{x} = 2640$, $n = 10$, sample variance $= 2.1 \times 10^4$, and we shall first test

$$H_0 : \mu = 2700 \quad \text{as against} \quad H_1 : \mu \neq 2700.$$

We first calculate the estimated population variance

$$s^2 = \text{sample variance} \times \frac{n}{n-1} = 2.1 \times 10^4 \times \frac{10}{9} = 2.33 \times 10^4$$

so that $s = 1.53 \times 10^2 = 153$.

Under H_0, the distribution of $(\bar{X} - \mu)/(s/\sqrt{n})$ is t_9 $(v = 10 - 1)$ and from the table, the critical 5% region for a two-tail test $(P = 5)$ starts at $t_9 = \pm 2.26$.

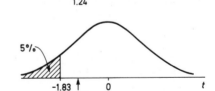

Our value of $t = \dfrac{2640 - 2700}{153/\sqrt{10}}$

$$= -1.24.$$

This is not in the critical region and we conclude that H_0 may be accepted.

In order to answer the second question we test the hypothesis

$$H_0 : \mu = 2700 \quad \text{as against} \quad H_1 : \mu < 2700.$$

The critical 5% region for a one-tail test $(P = 10)$ starts at $t_9 = \pm 1.83$, so that H_0 is still accepted.

EXAMPLE 2   A sample of four matchboxes from the output of a filling machine (assumed to be normally distributed) contains 53, 57, 54 and 52 matches. Is the sample mean significantly greater than the mean of 51, for which the machine is meant to be set? Use a 5% significance level.

The sample mean $\bar{x} = 54$ and the sample variance,

$$\frac{\Sigma (x - 54)^2}{4} = \frac{1 + 9 + 0 + 4}{4} = \frac{14}{4}.$$

Hence $\hat{\sigma}^2 = s^2 = \dfrac{14}{4} \times \dfrac{4}{3} = 4.67 \Rightarrow s = 2.16.$

Testing $H_0: \mu = 51$ as against $H_1: \mu > 51$, we know that under H_0 the distribution of $(\bar{X} - 51)/(2.16/\sqrt{4})$ is t_3 (3 degrees of freedom).

With $\bar{x} = 54$ the value of this statistic is 2.78.

Now for $v = 3$, the critical region for a one-tail test at 5% ($P = 10$) starts at $t_3 = 2.35$.

Since $2.78 > 2.35$, we reject H_0 and conclude that the setting of the machine may have to be adjusted.

EXERCISE 14a

1.   A sample of 10 observations taken from a normal distribution has mean 5.1 and variance 0.09. Test the hypothesis that the population mean is 4.8, using a 5% significance level.

2.   For the following sample statistics, taken from a normal distribution, test whether the sample mean is significantly different from the assumed population mean of 28.
(a) $n = 8$, $\bar{x} = 26$ standard deviation $= 6$.
(b) $n = 8$, $\bar{x} = 26$ standard deviation $= 2$.
(Use a 5% significance level).

3.   It is thought that the length of a certain kind of lizard is normally distributed with mean 17.3 cm. When a sample of 21 is examined the mean length turns out to be 16.5 and the sample variance is 5.6 cm^2. Use a 1% significance level to test whether there is evidence to suggest that the population mean is in fact 17.3 cm.

4.   The claim is made by a supplier of rope that it has a mean breaking strength of 3584 newtons. When a sample of 6 ropes are tested by a customer, the sample mean breaking strength turned out to be 3472 newtons with a standard deviation of 65 newtons. Does this provide evidence at: (a) the 5% level, (b) the 1% level, that the population mean is in fact less than 3584 newtons?

5.   A salesman claims that a new brand of table-tennis ball will withstand, on average, a force of 49 N without breaking. The secretary of a club (somewhat sceptical about the truth of salesmen's claims) puts his hand into a large carton of the balls and withdraws six of them. He measures the breaking force for each of these balls with the following results:

42.3 N, 35.6 N, 49.0 N, 51.2 N, 37.9 N, 40.1 N.

The secretary has decided to buy some of the new balls if the mean of his sample is consistent with the salesman's claim at the 5% level of significance. Stating any assumptions you make, determine whether the secretary buys the balls.   (MEI)

6.   A manufacturer of matches claims that there are, on average, 50 matches in each of the boxes he makes. A customer buys 8 of these boxes and counts the number, x, of matches in each box. He finds that $\Sigma\,x = 384$ and $\Sigma\,x^2 = 18460$.

Assuming that these boxes are a random sample from the manufacturer's production, carry out a t-test, at the $2\frac{1}{2}\%$ level, to determine whether the mean is significantly less than 50. State any assumptions that you need to make.   (C)

Hypothesis test for the difference between two means from normal distributions (small samples)

There are two types of problem to be considered here, the first of which is very similar to the one we demonstrated for large samples on page 261. In fact the procedure is identical, except that the t statistic is used in place of the Z statistic.

EXAMPLE 3   Strength tests on samples of two types of fibre gave the following results, in appropriate units:

Type 1	38	27	34	25		
Type 2	34	37	35	40	30	34

Determine, at the 5% level of significance, whether the difference in the means of the two samples is significant. Normal distributions are assumed.

For small samples it is necessary to assume that the parent population variances are the same and to estimate this common variance using the formula for pooled data:

$$\hat{\sigma}^2 = \frac{(n_1-1)s_1^2+(n_2-1)s_2^2}{n_1+n_2-2} = \frac{\Sigma\,(x_1-\bar{x}_1)^2+\Sigma\,(x_2-\bar{x}_2)^2}{n_1+n_2-2}.$$

The number of degrees of freedom for the t-test is then

$$n_1+n_2-2.$$

This is justified later (page 288).

Calling the observations from the parent populations X_1 (Type 1) and X_2 (Type 2), we are concerned with the difference between the sample means $\bar{X}_1$ and $\bar{X}_2$. With the usual notation

$n_1 = 4,$ $\bar{x}_1 = 124/4 = 31,$ $\Sigma\,(x_1-\bar{x}_1)^2 = 7^2+(-4)^2+3^2+(-6)^2 = 110$

$n_2 = 6,$ $\bar{x}_2 = 210/6 = 35,$ $\Sigma\,(x_2-\bar{x}_2)^2 = (-1)^2+2^2+0^2+5^2+(-5)^2$
$$+(-1)^2 = 56.$$

Assuming a common variance

$$\hat{\sigma}^2 = \frac{110+56}{4+6-2} = 20.75 \Rightarrow \hat{\sigma} = 4.555.$$

The hypotheses to be tested are $H_0:\mu_1-\mu_2 = 0$ as against $H_1:\mu_1-\mu_2 \neq 0$, where μ_1 and μ_2 are the means of the two populations. Under H_0, the distribution of

$$\frac{(\bar{X}_1-\bar{X}_2)-0}{\hat{\sigma}\sqrt{(1/n_1+1/n_2)}}\text{ is `}t\text{' with }6+4-2\text{ degrees of freedom.}$$

With $\bar{x}_1 = 31, \bar{x}_2 = 35$ $t = \dfrac{(-4)-0}{4.555\sqrt{(1/4+1/6)}} = -1.36.$

The critical region for $v = 8$ starts at $t_8 = \pm2.31$, so we conclude that the difference between the sample means is not significant.

Note: Whether or not it is reasonable to assume a common population variance can be decided by a preliminary test (see page 475).

Hypothesis test for the difference between two means from normal distributions when the samples (small) are paired

The word 'paired' in the heading needs explanation and implies that a different technique is involved from the one used in the last example. We will distinguish between the two by considering a particular case.

Suppose a metallurgist wants to analyse the percentage of impurity in a quantity of silver and, having at his disposal two different methods of analysis, A and B, wants to compare the two. One course open to him would be to take a sample of, say, eight quantities of silver and find the mean percentage of impurity in this sample by method A. He could then take *another* sample of eight quantities of silver and find the mean percentage of impurity in this second sample by method B. The results might be as follows:

Sample	Mean %	Sample variance	Size of sample
A	$\bar{x} = 3.66$	0.09	8
B	$\bar{y} = 3.28$	0.07	8

and the metallurgist could then test the significance of the difference between these two sample means using the method we have just described.

Now this procedure could be criticised on the grounds that the two methods have been tested on *different* samples and there might, unknown to the researcher, be some difference between the silver from which the two samples have been taken. It would surely be better if method A and method B were each applied to the *same* sample. If this were done the researcher would be analysing results of *paired* samples, which would provide a more useful comparison. Let us suppose that the figures obtained are:

Sample number	1	2	3	4	5	6	7	8
% of impurity using method A (X)	2.6	4.8	3.6	4.4	4.0	5.0	1.9	2.3
% of impurity using method B (Y)	2.1	3.3	3.4	4.4	4.1	4.5	2.1	1.5

The question to be answered is this: do the two methods give equivalent results on average? In other words is the difference between the two methods significant? We shall assume a normal distribution.

As we are only concerned with the difference between two methods applied to the *same* sample we first calculate the difference $D = X - Y$ for each of the eight members of the sample:

D	0.5	1.5	0.2	0	− 0.1	0.5	− 0.2	0.8

From now on we have no further use for the original data and will work with D as random variable.

The null hypothesis is that there is no difference between the means of X and Y. If we write $\mu_D = \mu_X - \mu_Y$

$$H_0 : \mu_D = 0 \quad \text{as against} \quad H_1 : \mu_D \neq 0.$$

We now proceed exactly as in example 1.
The sample mean $\bar{d} = 0.4$ and the sample variance

$$= \frac{0.1^2 + 1.1^2 + (-0.2)^2 + (-0.4)^2 + 0.5^2 + 0.1^2 + (-0.6)^2 + 0.4^2}{8}$$

$$= 0.275,$$

so that $\hat{\sigma}_D^2$ (the estimated variance of the distribution of D)

$$= s_D^2 = 0.275 \times 8/7 = 0.314 \Rightarrow s_D = 0.561.$$

Under H_0 the distribution of

$$t = \frac{\bar{D} - 0}{s_D/\sqrt{n}}$$

is t_7 ($v = 8 - 1$) and the value of t with $\bar{d} = 0.4$, $s_D = 0.561$ and $n = 8$ is

$$t = 2.02.$$

For a 5% two-tail test the critical region for t_7 starts at $t = \pm 2.36$, and we conclude that there is no significant difference between μ_X and μ_Y and there is therefore no convincing evidence that the two methods A and B give different results.

Note: In any particular case you should take care to recognise when a paired-sample test is called for. The way in which the data are laid out almost always gives the clue.

To summarise the method:

If paired samples

$$\begin{cases} X_1, X_2, \ldots, X_n \\ Y_1, Y_2, \ldots, Y_n \end{cases}$$

are given, work with the random variable $D = X_i - Y_i$ ($i = 1$ to n).
    Calculate $\bar{d}$ and s_D^2, the estimated population variance of D, using

$$s_D^2 = \frac{n}{n-1} \times \text{variance of the } D \text{ sample}$$

$$= \frac{\Sigma (d - \bar{d})^2}{n-1}.$$

The hypothesis to be tested is

$$H_0: \mu_D = 0 \quad \text{as against} \quad H_1: \mu_D \neq 0,$$

and the test statistic is $(\bar{D} - 0)/(s_D/\sqrt{n})$, whose distribution is t_{n-1}.

EXERCISE 14b

1.   Two factories A and B produce similar material. Each sends a sample (5 pieces from A and 7 pieces from B) to a public analyst. He measures the strength of the pieces with the following results:

Factory A	21.0	21.5	21.7	21.0	22.2		
Factory B	20.3	21.1	21.4	21.1	20.8	20.4	20.8

Is there any evidence that the materials from the two factories differ in strength? (Normal distributions may be assumed.)   (MEI)

2.   The only visible difference between wild and farmed trout is the greater average length of the latter. I have kept a record of the lengths of six wild trout recently caught as follows:

Length (cm)	26.8	26.0	25.7	25.6	24.9	24.3

I am offered four trout which are claimed to be farm-reared and their lengths (in centimetres) are:

$$27.1, \quad 27.5, \quad 26.4, \quad 26.1.$$

Apply a suitable t-test, stating clearly your null and alternative hypotheses, to determine whether I can reasonably accept the statement that the four are farm-reared.   (C)

3.   To test a new chicken additive, eight hens were given the normal food for three weeks and then were given the normal food together with the special additive for the next three weeks.
The number of eggs laid by each hen was as follows:

Hen	1	2	3	4	5	6	7	8
Fed with normal food, number of eggs	14	15	16	15	16	15	17	18
Fed with food plus additive, number of eggs	15	16	16	16	17	17	18	18

Perform a paired-sample t-test at the 5% level to investigate whether or not the additive results in a greater mean number of eggs.   (SUJB)

4.   In order to compare two methods for finding the percentage of iron in a compound, each of ten different compounds was analysed by both methods, and the results are given below.

Compound number	1	2	3	4	5	6	7	8	9	10
Method A %	13.3	17.6	4.1	17.3	10.1	3.7	5.1	7.9	8.7	11.6
Method B %	13.4	17.9	4.1	17.0	10.3	4.0	5.1	8.0	8.8	12.0

Use Student's t-distribution to test, at the 5% level of significance, the hypothesis that the two methods of analysis differ.
State the assumptions that are necessary to justify the use of this distribution.   (MEI)

5.   Eleven students sat a chemistry examination consisting of one theory and one practical paper. Their marks out of 100 are given in the table below.

Student	A	B	C	D	E	F	G	H	I	J	K
Theory mark	30	42	49	50	63	38	43	36	54	42	26
Practical mark	52	58	42	67	94	68	22	34	55	48	17

Use the t-test for paired values, at the 5% level of significance, to test the hypothesis of no difference in the marks on the two papers.   (AEB 1980)

6.   Two groups of rodents were used in an experiment to investigate the possible toxicity of a cosmetic preparation. One group was fed with the standard preparation and the other

group was given a special preparation with a reduced concentration of the suspected ingredient. Analyse the following weights and report your conclusions.

Rodent	Body weights (g × 100)						
Fed standard preparation	36	66	41	42	34	43	
Fed special preparation	26	57	33	40	40	44	34

(MEI)

14.4 Confidence intervals using the *t*-distribution

The procedure for confidence intervals when samples are small is the same as for large samples, except that the t statistic replaces the Z statistic.

EXAMPLE 4 The heights of five students, aged 20, taken from a large normal population of students of this age, are (in centimetres)

$$180 \quad 176 \quad 178 \quad 182 \quad 185$$

Find 95 % confidence limits for the mean of the population.

From these figures, the sample mean $\bar{x} = 180.2$.
Subtracting 180, the sample variance

$$= \{(0)^2 + (-4)^2 + (-2)^2 + (2)^2 + (3)^2\}/5 - (0.2)^2 = 6.56.$$

Hence $s^2 = 6.56 \times 5/4 = 8.2 \Rightarrow s = 2.86.$

For $n = 5$, $v = 4$ and at the 95 % level ($P = 5$), $t_4 = 2.78$.
The confidence limits are therefore $180.2 \pm 2.78 \times 2.86/\sqrt{5}$ and the confidence interval is (176.6, 183.8).
As would be expected, when n is small, the confidence interval is large.

EXAMPLE 5 Find a 95 % confidence interval for the difference between the means of two normal distributions with the same variance when the sample statistics are:

$$Sample \ 1 \quad \bar{x}_1 = 4.1, \quad n_1 = 8, \quad variance = 1.7$$

$$Sample \ 2 \quad \bar{x}_2 = 3.3, \quad n_2 = 6, \quad variance = 1.2.$$

The unbiased estimate for the common variance

$$= \frac{\Sigma (x_1 - \bar{x}_1)^2 + \Sigma (x_2 - \bar{x}_2)^2}{n_1 + n_2 - 2}$$

$$= \frac{8\text{var}(x_1) + 6\text{var}(x_2)}{8 + 6 - 2} = \frac{13.6 + 7.2}{12} = 1.733 \Rightarrow \hat{\sigma} = 1.316.$$

For $v = 8 + 6 - 2 = 12$, the 95 % ($P = 5$) value of t^{12} is 2.18.
Hence the 95 % confidence limits are

$$(4.1 - 3.3) \pm 2.18 \times 1.316 \sqrt{\left(\frac{1}{8} + \frac{1}{6}\right)}$$

and a 95 % confidence interval is $(-0.75, 2.35)$.

EXERCISE 14c

1. The protein intakes for a sample of six prisoners-of-war was observed to be (in grams/day):

86   82   66   66   94   104

Calculate a 95 % confidence interval for the average protein intake of prisoners-of-war.

2. The lengths of the femur in a sample of nine small mammals of a certain type are (in millimetres):

9.2   10.6   11.3   9.7   8.9   9.6   10.1   12.3   10.0

Calculate a 98 % confidence interval for the mean femur length for these mammals.

3. In order to investigate the lengths of components produced by a machine, ten components were selected at random from a large batch. Their lengths (in metres) were 1.125, 1.127, 1.125, 1.124, 1.131, 1.132, 1.125, 1.129, 1.123 and 1.129.

Use this sample to calculate symmetric 95 % confidence limits for the mean length of components in the batch, stating any assumptions that you need to make.   (C)

4. Find a 95 % confidence interval for the difference between the means of two normal distributions with the same variance, based on the following two samples:

Sample A	102	96	98	127	131	105	91
Sample B	72	84	87	99	79	65	

5. The fat content (measured on a suitable scale) may be determined by two different methods. Each of eight meat specimens was divided into two and the two halves were allocated to the different methods at random. The results were:

Method	Fat content of meat							
1	23	69	47	17	35	44	61	44
2	21	64	46	17	31	50	58	40

From this sample, derive symmetric 99 % confidence limits for the mean amount by which the fat content assessed by method 1 exceeds that assessed by method 2.   (C)

A final word on large and small sample tests

At the end of this chapter there are a number of mixed examples, including observations based on both large and small samples. It is quite likely that if you have worked through this chapter and the last one you may feel some uncertainty as to whether to use the Z statistic or the t statistic in any particular case. If so you should find the following simple rules helpful:

1. If σ^2 is *known* the Z statistic is used for any size of sample.

2.   If σ^2 is *unknown* the t statistic is used for any size of sample, but for large samples, t and Z are alternatives.

(The reason for this is that *when n is large the t-distribution* approximates closely to $N(0, 1)$, a fact that was mentioned in passing on page 278.)

As an example, if $n = 120$, the critical two-tail region at 5% $(P = 5)$ starts at $t_{119} = \pm 1.98$. For $N(0, 1)$ it starts at $z = \pm 1.96$.

A proof of the general result is given in the Appendix on page 293.

Note: For small samples the parent population must be normal.

14.5   The meaning of 'degrees of freedom'

Up to now we have used the expression 'degrees of freedom' as a kind of label attached to a particular t-distribution, but the idea is such an important one that you should understand how it arises, especially as we shall meet it in other contexts later in the book.

For an explanation of the phrase when it is used in connection with the t-distribution, we consider a simple example.

The t statistic is defined as $(\bar{X} - \mu)/(s/\sqrt{n})$ and in using it we have first to calculate

$$s^2 = \frac{\Sigma\,(x - \bar{x})^2}{n - 1}.$$

Suppose that $n = 4$ and the sample observations are

$$x_1 = 53 \qquad x_2 = 56 \qquad x_3 = 55 \qquad x_4 = 52$$

with mean $\bar{x} = 54$.

The *sum* of the deviations from $\bar{x}$

$$= (x_1 - \bar{x}) + (x_2 - \bar{x}) + (x_3 - \bar{x}) + (x_4 - \bar{x})$$

$$= \sum_{i=1}^{4} x_i - 4\bar{x} = 4\left[\frac{\Sigma\,x_i}{4} - \bar{x}\right] = 0, \quad \text{by definition of } \bar{x}.$$

In other words the deviations are not independent, since *their sum has to be zero*. This means that once we have calculated three of them:

$$x_1 - \bar{x} = -1, \qquad x_2 - \bar{x} = 2, \qquad x_3 - \bar{x} = 1,$$

the fourth *must* be -2.

In this sense only *three* of the four deviations used in calculating s^2 are independent.

In the general case, when the sample size is n,

$$\sum_{i=1}^{n} (x_i - \bar{x})$$

has to be zero since

$$\sum_{i=1}^{n} (x_i - \bar{x}) = \sum_{i=1}^{n} x_i - n\bar{x} = n\bar{x} - n\bar{x} = 0$$

and we say that only $n-1$ of the deviations are independent, or that there are $n-1$ *degrees of freedom*. In other words, the fact that the sum of the deviations has to be zero acts as a 'constraint' on the calculation of s^2.

Similarly, for the difference between two means, we use the statistic

$$t = \frac{(\bar{X}_1 - \bar{X}_2) - 0}{\hat{\sigma}\sqrt{(1/n_1 + 1/n_2)}}$$

where

$$\hat{\sigma}^2 = \frac{(n_1 - 1)s_1^2 + (n_2 - 1)s_2^2}{n_1 + n_2 - 2}.$$

Now in calculating s_1^2 there are n_1 observations of which only $n_1 - 1$ are independent, and in calculating s_2^2 there are n_2 observations of which only $n_2 - 1$ are independent. Hence, of the $n_1 + n_2$ deviations involved, only $(n_1 - 1) + (n_2 - 1)$ are independent. We say, therefore, that the number of degrees of freedom is $n_1 + n_2 - 2$.

It will be found that, in general, for any sample statistic, *the number of degrees of freedom = the sample size minus the number of constraints involved in the calculation of the statistic.*

EXERCISE 14d (miscellaneous)

(This exercise includes examples using both large and small samples.)

1. A manufacturer is producing resistances normally distributed about a mean value of 50.0 ohms with a standard deviation of 1 ohm. To check the constancy of his product, he periodically extracts 100 resistors and measures them. He finds that one such sample has a mean of 50.24 ohms. Has there been any significant change in his product?

2. The weekly amounts of liquid raw material used for a chemical process at a factory are known to be independent and normally distributed with a mean of 1500 litres and a standard deviation of 200 litres.

The amounts of raw material used per week over four successive weeks were 1576, 1624, 1586 and 1678 litres respectively. Use a 5% significance level to test whether these data provide enough evidence to conclude that the mean amount used per week has increased from 1500 litres. (JMB)

3. In a chemical plant the acid content of the effluent from the factory is measured frequently. From 400 measurements the acid content in grams per 100 litres of effluent is recorded in the following frequency distribution:

Acid content	12	13	14	15	16	Total
Frequency	5	52	235	74	34	400

(a) Assuming a normal distribution of the acid content, give 95% confidence limits for the mean acid content.
(b) Is the result consistent with a mean acid content of 14.13 grams per 100 litres obtained from tests over several years? (O & C)

4. Describe briefly what is meant by the 'sampling distribution of a statistic', illustrating your remarks by reference to the sampling distribution of the mean of a random sample of n observations from a normal distribution with mean μ and variance σ^2.

Weights of a certain type of biscuit are normally distributed with mean 0.5 oz and standard deviation 0.05 oz. Random samples of 16 biscuits are chosen to produce for sale packets of nominal weight 8 oz. Calculate the probability that the contents of such a packet weigh less than 7.5 oz.

Similar biscuits made in a different mould are also sold in packets of 16. Again the weights of individual biscuits are normally distributed with standard deviation 0.05 oz. A random sample of 30 packets had a sample average net weight of 7.9 oz. Examine whether this is significant evidence that packets of biscuits produced in this new mould have mean weight different from the others. (JMB)

5. (a) The weights of all the pupils in a school are taken and are distributed normally with standard deviation 6 kg. The separate boys' and girls' weights are also distributed normally, the boys with standard deviation 5 kg and the girls 8 kg.

The weights of a random sample of ten children are taken (in kilograms):

Boys	46	42	52	44	56	36
Girls	45	53	38	36		

Determine 95% confidence limits for the mean weight of: (i) the boys, (ii) the girls, (iii) the whole school.

(b) The mean number of letters delivered to the School Office and to the Principal's Office were recorded over varying periods of time:

Office	*Number of days recorded*	*Mean number of letters*	*Standard deviation*
School	63	48.1	6.24
Principal	52	44.6	8.56

Is there a significant difference between the means? (Use a 5% level of significance.) (SUJB)

6. A random sample of size 16 is taken from a normal population of unknown mean and variance. The members of the sample are $x_1, x_2, \ldots, x_{16}$ and

$$\Sigma x_i = 454, \qquad \Sigma x_i^2 = 13\,026.$$

Determine the sample mean and variance.

Examine whether there is evidence to reject, at the 5% level of significance, the hypothesis that the population mean is 30, using the following two methods:
(a) a method using the normal distribution;
(b) a method using Student's t-distribution.

State which of (a) or (b) is preferable in this case, giving your reasons.

Determine, by the most accurate method known to you, a 95% confidence interval for the population mean. (MEI)

7. The life in hours of a particular kind of battery has a normal distribution with variance 16. From a large batch of these batteries a sample is selected and tested. Their lives turned out to be:

$$56.4 \quad 52.1 \quad 49.5 \quad 56.4 \quad 56.0 \quad 48.1 \quad 54.5 \quad 47.8$$
$$58.0 \quad 48.4 \quad 53.9 \quad 46.7$$

Calculate 95% confidence limits for the mean life of the population.

8.   A study of the effect of weathering on the radioactivity of granite was made by selecting a random sample of weathered rocks and another random sample of freshly exposed rocks. Measurements were made of the amount of radioactivity emitted in periods of one minute and the following data were obtained. Examine the data for evidence of the effect of weathering and find a 95 % confidence interval for the difference in mean radioactivity.

Rocks	Counts per minute						
Fresh	225	188	165	211	178	171	199
Weathered	177	170	140	132	168	172	170

(MEI)

9.   Mass-produced washers have thicknesses which are normally distributed with mean 3 mm and standard deviation 0.2 mm.
(a) Find, correct to three decimal places, the probability that the mean thickness of a random sample of 4 washers will lie between 2.9 mm and 3.1 mm.
(b) During a check on the manufacturing process, a random sample of 25 washers is taken from production and the mean thickness $\bar{x}$ mm is calculated. Find the interval in which the value of $\bar{x}$ must lie in order that the hypothesis that the production mean thickness is 3 mm will not be rejected when the significance level is 5 %. (JMB)

10.   In order to compare the colour retention properties after washing of two methods of colouring yarn, eight pieces were cut into two and washed after the methods had been used. Using a standard measure of colour change, the following results were obtained:

Method 1	13.7	19.1	14.2	17.1	11.8	12.2	14.0	11.2
Method 2	13.9	21.4	14.3	17.2	13.0	13.5	16.1	11.7

Is there any evidence that the two methods produce different results. (Normal distributions to be assumed.) (MEI)

11.   The mass, in grams, of marmalade in a randomly chosen jar from a large batch filled by a machine is normally distributed. The mean μ for a batch can be adjusted but the standard deviation is always 4. Each batch produced is tested by determining the mean mass $\bar{x}$ of the marmalade in a random sample of n jars. Under the null-hypothesis $\mu = 454$, a batch is rejected if $\bar{x}$ is such that there is significant evidence, at the 1 % level, that the value of μ is less than 454. The sample size is chosen so that, in the case $\mu = 452$, the probability of rejecting the batch is at least 0.95. Find the smallest value of n to achieve this, and the corresponding critical value of $\bar{x}$ for rejection. (C)

12.   Show that, if $X_1, X_2, \ldots, X_n$ are n independent random variables with mean μ and variance σ^2, then $\Sigma \, X_i/n$ has mean μ and variance σ^2/n.
  The diameters of 25 steel rods are measured and found to have a mean of 0.980 cm and a standard deviation of 0.015 cm. Assuming this is a sample from a normal distribution with the same variance, find 99 % confidence limits for the population mean. (O & C)

13.   The manufacturers of a product A wish to compare the mass of the contents of a packet of their product with that of a packet of a competitor's product B which is nominally of equal mass. For 100 packets of each product the mass x grams of each packet is measured. The following data are collected.

	Product A	Product B
$\Sigma \, x$	45 551	45 404
$\Sigma \, (x - \bar{x})^2$	3 300	2 475

Find a symmetrical two-sided 95 % confidence interval for the amount by which the mean mass of a packet of *A* exceeds that of a packet of *B*. State whether the observed difference of the sample means is significant at the 5 % level.   (JMB)

14.   An experiment was performed to compare two methods of analysing the percentage impurity in a certain chemical. Tests were made on twelve different samples of the chemical, each sample being analysed by both methods. The results of the experiment are given in the following table:

Sample number	*Percentage impurity*		*Sample number*	*Percentage impurity*	
	Method A	*Method B*		*Method A*	*Method B*
1	2.12	2.26	7	2.56	2.65
2	2.45	2.45	8	2.41	2.36
3	2.43	2.46	9	2.41	2.50
4	2.51	2.44	10	2.46	2.52
5	2.42	2.55	11	2.43	2.48
6	2.44	2.56	12	2.38	2.42

Test, at both the 5 % and 1 % levels of significance, the hypothesis that on the average the two methods of analysis are equivalent.   (MEI)

15.   (a) The heights of a random sample of 1304 Scotsmen had mean 68.545 in with standard deviation 2.480 in. The heights of a random sample of 6194 Englishmen had mean 67.438 in with standard deviation 2.548 in.
   Test whether these samples suggest that there is a significant difference between the mean heights of Scotsmen and Englishmen.
   (b) To test the effect of using a special seed drill, ten plots of land were sown with an ordinary seed drill and ten with the special seed drill. The twenty plots of land were of equal size and were chosen in pairs, two neighbouring plots forming a pair. The decision as to which of the two neighbouring plots was to be treated with the special seed drill was made by tossing a coin. The yields of the plots were subsequently compared; the coded results are shown in the table below:

Special seed drill	8.0	8.4	8.0	6.4	8.6	7.7	7.7	5.6	5.6	6.2
Ordinary seed drill	5.6	7.4	7.3	6.4	7.5	6.1	6.6	6.0	5.5	5.5

Test whether there is a significant difference between the yields for the two types of drill.   (MEI)

16.   Small amounts of a certain potentially hazardous chemical are discharged in the effluent from a factory. The percentage by volume of this chemical in the effluent is critical, and the plant management endeavour to ensure that this does not on average exceed 1.7 (a figure somewhat below the maximum allowed by public health regulations).
   Twelve random determinations of this percentage by volume gave the following results:

$$1.64 \quad 1.70 \quad 1.53 \quad 1.85 \quad 1.78 \quad 1.64$$
$$1.87 \quad 1.60 \quad 1.66 \quad 1.73 \quad 1.72 \quad 1.77$$

Test at the 5 % level of significance whether it appears that the plant management are meeting their target, and provide a 99 % confidence bound for the average percentage by volume of the chemical in the effluent.   (MEI)

17.   To test the effectiveness of a new powdered milk for babies a manufacturer used ten pairs of identical twins, each twin having the same birth weight as his brother or sister. Their

diets were identical for a period of one month, except that one twin was fed with the old powder and the other with the new. The weights in kilograms after one month were as follows:

Twins	1	2	3	4	5	6	7	8	9	10
New food	4.32	5.62	5.24	4.12	5.74	4.36	5.34	4.76	5.26	4.97
Old food	4.30	5.43	5.26	4.18	5.41	4.22	5.39	4.61	5.14	4.86

Perform a paired-sample t-test to investigate whether or not the new food results in a greater mean weight than the old. What assumptions must be made in order for the t-test to be valid?

Obtain 95% confidence limits for the difference in mean weights of twins fed on the two foods. (SUJB)

18. A large consignment of similarly graded apples arrived at a company's warehouse for distribution to retail outlets. Two varieties were chosen and a random sample of each had their masses, in grams, measured. The results are tabulated below.

Variety 1 (Granny Delicious)	110.5	89.6	89.1	85.6	115.0	98.2
	113.1	92.0	104.3	100.7	97.5	106.1
Variety 2 (Golden Smiths)	125.6	118.3	118.0	110.8	116.5	108.7
	108.2	104.4	114.4	98.4	111.2	

Assuming that these independent samples came from underlying normal populations, use a 5% level of significance to test the hypothesis that the population means are the same. (AEB 1982)

19. (a) When a confidence interval is constructed for a parameter, one measure is how narrow it is. Now if $X_1, X_2, \ldots, X_n$ is a random sample from a normal distribution with mean μ and variance σ^2, both unknown, and a γ% confidence interval is constructed for μ, how does the width of the interval change (other things being equal) if: (i) n becomes larger, (ii) γ becomes larger, (iii) σ^2 becomes larger? (Justify your replies briefly.)

(b) During a particular week, 13 babies were born in a maternity unit. Part of the standard procedure is to measure the length of the baby. Given below is a list of the lengths in centimetres of the babies born in this particular week.

49 50 45 51 47 49 48 54 53 55 45 50 48

Assuming that this sample came from an underlying normal population, calculate a 95% confidence interval for the mean of the population. (AEB 1980)

20. The mean and variance of a set of 64 independent readings of a change in blood pressure are 8 units and 63 square units respectively. Assuming that the observations come from a normal distribution, give a 95% confidence interval for its mean, and hence test for difference from zero.

It is subsequently discovered that when the experiment was conducted the positive and negative changes were recorded separately, and that the above data refer to the positive changes. There were also 36 negative changes whose mean and variance were -10 units and 64 square units. Give a 95% confidence interval for the mean of the distribution taking all the observations into account, and again test for difference from zero. (O)

21. A man has two apple trees, one growing in his front garden and the other in his back garden. A random sample of 5 apples taken from the front garden tree has masses (in grams) summarised by $\Sigma x = 550$, $\Sigma x^2 = 61\,720$. An independent random sample of 6 samples taken from the back garden tree has masses summarised by $\Sigma y = 540$, $\Sigma y^2 = 50\,350$. For

each tree the distribution of the mass of an apple may be assumed to be normal, and the two distributions may be assumed to have the same variance but not necessarily the same mean. Obtain an unbiased estimate of each of the separate population means and also obtain the unbiased two-sample estimate of the common variance. Explain what is meant by 'an unbiased estimate'.

The man believes that the apples from the front garden are on average heavier than those from the tree in his back garden. Taking the null hypothesis to be that the two distributions have equal means, and taking their common variance to be $330\,\text{g}^2$, test, at the 1% significance level, whether there is evidence to support the man's belief. State clearly your alternative hypothesis and the result of your test.   (C)

22.★ Holes are drilled in steel plates by a machine. A random sample of 10 holes is selected and the diameters $x_1, x_2, \ldots, x_{10}$ are measured in millimetres. It is found that $\Sigma\, x_i = 94$, $\Sigma\, (x_i - \bar{x})^2 = 0.27$. Estimate the population variance and hence find 99% confidence limits for the population mean.

The machine is then adjusted to make smaller holes and the diameters $y_1, y_2, \ldots, y_{10}$ are measured in millimetres. It is found that $\Sigma\, y_i = 76$, $\Sigma\, (y_i - \bar{y})^2 = 0.51$. Assuming the population variance has remained constant, find an overall estimate of the population standard deviation, and use this to find revised confidence limits for the population mean before the adjustment was made.   (O & C)

14.6 Appendix

The density function of the distribution of 't'

The density function is

$$f(t) = K(1 + t^2/v)^{-\frac{1}{2}(v+1)}, \quad -\infty < t < \infty.$$

where $v = n - 1$.

The constant K depends on n and is chosen so that $\displaystyle\int_{-\infty}^{\infty} f(t)\, dt = 1.$

The 't'-distribution when n is large

When n is large the Z statistic $\dfrac{\bar{X} - \mu}{\sigma/\sqrt{n}}$

and the t statistic $\dfrac{\bar{X} - \mu}{s/\sqrt{n}}$

are alternatives since

$$\text{the } t\text{-distribution} \to N(0, 1) \quad \text{as } n \to \infty.$$

Proof

$$f(t) = K(1 + t^2/v)^{-\frac{1}{2}(v+1)}$$
$$= K[(1 + t^2/v)^v]^{-\frac{1}{2}}(1 + t^2/v)^{-\frac{1}{2}}.$$

Now $(1 + t^2/v)^v \rightarrow e^{t^2}$ as $v \rightarrow \infty$ $\left[\lim_{n \rightarrow \infty} (1 + x/n)^n = e^x \text{ see page 146} \right]$

and $(1 + t^2/v) \rightarrow 1$ as $v \rightarrow \infty$

so that $f(t) \rightarrow K \cdot e^{-\frac{1}{2}t^2}$ as $v \rightarrow \infty$.

But $\displaystyle\int_{-\infty}^{\infty} f(t) \, dt = 1 \Rightarrow K = 1/\sqrt{(2\pi)}$.   (See page 223.)

Hence $f(t) \rightarrow \dfrac{1}{\sqrt{(2\pi)}} e^{-\frac{1}{2}t^2}$ which is the density function of $N(0, 1)$.

15 Tests for proportions and medians: sign tests

The amount of space devoted, in the last two chapters, to testing hypotheses about means and establishing confidence intervals for means reflects their importance in interpreting data from samples. Moreover, the fact that the methods involved the use of continuous distributions (normal and 't') made it easy to demonstrate ideas like 'significance level' and 'one-tail test' diagrammatically.

In this chapter the work will be extended to other parameters such as proportion and median.

The procedure will, in some cases, entail using the normal distribution and will, therefore be familiar to you. We shall start, however, with a *discrete* distribution, the binomial, which, when the sample size is small, demands its own special treatment.

15.1 A hypothesis test of a proportion using the binomial distribution (small sample)

Suppose that the proportion of people who survive after suffering from a particular disease is thought to be $\frac{1}{4}$ – a figure arrived at after long experience. A medical research team experiments with a new kind of treatment, as a result of which 4 sufferers survive out of a sample of 10. Can this be regarded as a significant improvement?

Making a judgement in this case amounts to testing a proportion when the only sample available is a *small* one, and it will be remembered (see page 207) that we cannot use the normal approximation unless the sample is large.

Calling the proportion in the population π, the competing hypotheses are

$$H_0 : \pi = \tfrac{1}{4} \quad \text{as against} \quad H_1 : \pi > \tfrac{1}{4}$$

and, assuming H_0 to be true the model is $B(10, \tfrac{1}{4})$.

Under H_0, the probability that the number surviving, X, takes the values $0, 1, \ldots, 10$ will be given by

$$P(X = r) = \binom{10}{r} (\tfrac{1}{4})^r (\tfrac{3}{4})^{10-r}$$

and the resulting probability distribution is:

X	0	1	2	3	4	5	6	7	8	9	10
Probability	0.056	0.188	0.282	0.250	0.146	0.058	0.016	0.003	0.000	0.000	0.000

the figures being rounded to three decimal places.

For a one-tail test with a significance level of 5 % we need to find an *integer a* such that

$$P(X \geqslant a) < 0.05.$$

(As X is a discrete variable it is unlikely that this probability is *equal* to 0.05). Looking at the table we see that

$$P(X \geqslant 6) = 0.016 + 0.003 = 0.019$$

and that $$P(X \geqslant 5) = 0.058 + 0.016 + 0.003 = 0.077.$$

We must therefore take as the critical region $X \geqslant 6$, since $0.019 < 0.05$ and $0.077 > 0.05$.

As $X = 4$ in our example it is not in the critical region and H_0 is accepted.

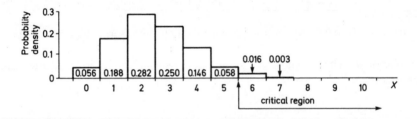

(If we wanted a more stringent test, using a critical region for a 1 % significance level, this region would be given by $X \geqslant 7$, since $P(X \geqslant 6) = 0.019 > 0.01$, but $P(X \geqslant 7) = 0.003 < 0.01$).

(Examples of this test are to be found in exercise 15a, questions 1 to 4 on page 301.)

15.2   Hypothesis test for a proportion (large samples)

For a large sample, the procedure is less cumbersome and more familiar as the following example will show.

A manufacturer states that 94 % of the electrical components delivered to retailers have a life of at least 500 hours. One of his client retailers tests a batch of 200 and finds that the proportion that have the life claimed by the manufacturer is 90 %. The problem is to decide whether, on this evidence, the manufacturer's claim is still likely to be true, or whether the product has deteriorated. In other words, is the sample proportion $p(0.90)$ significantly less than the claimed population proportion $\pi(0.94)$?

This can be dealt with in two ways.

The first method involves the sampling distribution of the proportion, which is easily derived from the binomial distribution.

The sampling distribution of the proportion

Suppose the proportion in a population having some particular attribute (such as being colour blind) is π, then if a random sample of size n is taken, X, the *number* in the sample with this attribute will be binomial with mean $n\pi$ and variance $n\pi(1-\pi)$.

If P is the *proportion* in the sample with the attribute,

$$P = X/n$$

then the distribution of P will also be binomial, since the probability that $X = r$ is the same as the probability that $P = r/n$. Division of the random variable by n makes no difference to the probabilities.

Moreover $E(P) = E(X/n) = (1/n)E(X) = (1/n)n\pi = \pi$

and $\text{Var}(P) = \text{Var}(X/n) = (1/n^2)\text{Var}(X) = (1/n^2)n\pi(1-\pi)$

$$= \pi(1-\pi)/n.$$

The sampling distribution of P is therefore binomial with mean π and variance $\pi(1-\pi)/n$, and if n is large we may use the normal approximation, with the test statistic

$$Z = \frac{P-\pi}{\sqrt{[\pi(1-\pi)/n]}}.$$

In our example the appropriate hypotheses are:

$$H_0 : \pi = 0.94 \quad \text{as against} \quad H_1 : \pi < 0.94$$

and the sample value of P is $p = 0.90$; $n = 200$. We shall use a 5% significance level.

Under H_0, the distribution of P is approximately normal with mean 0.94 and variance $0.94 \times 0.06/200$.

Strictly speaking, a continuity correction $0.5/n$ (in this case 0.0025) should be applied, but for large values of n, this will make little difference unless the standardised value of p turns out to be near the critical value. Without a continuity correction, the value of the test statistic

$$z = \frac{0.90 - 0.94}{\sqrt{0.94 \times 0.06/200}}$$

$$= -2.38.$$

This is well inside the critical region, which starts at $z = -1.64$ for a one-tail 5% test. H_0 is therefore rejected and we conclude that there is some reason to doubt the manufacturer's claim.

In this example, if p had been 0.91, the value of z without a continuity correction would have been -1.79, in which case it would have been advisable to repeat the calculation with the correction.

The alternative method of doing this problem is to use the distribution of X, the *number* satisfying the manufacturer's claim, when the sample number $= 200 \times 0.90 = 180$. Under H_0, the distribution of X is $B(200, 0.94)$ and the normal approximation to this distribution will give the result we have just found.

Testing a hypothesis about the difference between two proportions

Once again we shall take a particular example to explain the procedure.

Out of a random sample of 270 men, aged 40, 130 were found to need to wear spectacles at all times. The corresponding number in a random sample of 420 women was 180. Are these proportions significantly different?

We shall call the proportion in the samples P_1 and P_2 for men and women respectively. The proportions in the populations are π_1 and π_2.

As with the test for the difference between two means, the null hypothesis here will be

$$H_0: \pi_1 = \pi_2 \ (\text{or } \pi_1 - \pi_2 = 0) \quad \text{as against} \quad H_1: \pi_1 \neq \pi_2.$$

Now under H_0, since $\pi_1 - \pi_2 = 0$, the distribution of $P_1 - P_2$ will be approximately normal with mean 0 and variance

$$\frac{\pi(1-\pi)}{n_1} + \frac{\pi(1-\pi)}{n_2} \quad \text{or} \quad \pi(1-\pi)\left(\frac{1}{n_1} + \frac{1}{n_2}\right)$$

where π is the assumed *common population proportion*.

But as this is not known we have to make an unbiased estimate, using the data from the *two samples combined*. Since the total number wearing spectacles is $130 + 180$ and the number in the combined samples is $270 + 420$, the proportion in the combined samples is $310/690 = 0.449$.

Hence, since $E(P) = \pi$ (see page 297)

$$\hat{\pi} = 0.449.$$

Using this value, the distribution is approximately

$$N\{0, 0.449(1 - 0.449)\,[(1/270 + 1/420)]\} = N(0, 0.001505).$$

(Using $\hat{\pi}$ instead of π does of course produce another source of error, but we may assume that, as in the case of testing a mean, when we may have to use s^2 instead of σ^2, the additional possible error is not important for large samples.)

The value of $P_1 - P_2$ in this case is

$$130/270 - 180/420 = 0.0529$$

and the value of the test statistic is

$$z = \frac{0.0529 - 0}{\sqrt{0.001505}} = 1.36.$$

On this evidence, then, since $1.36 < 1.96$, there is no significant difference (at the 5% level) between the two (unknown) population proportions.

Note: A continuity correction was not used here, and as it turned out that z was well outside the critical region, the use of the correction would have made no difference to the conclusion.

15.3   Summary of hypothesis tests of proportions

When the sample is small, the binomial distribution has to be used.

When the sample is large, the normal approximation to the binomial is used, and the test statistic is

$$Z = \frac{P - \pi}{\sqrt{[\pi(1-\pi)/n]}} .$$

To test a hypothesis about the difference between two proportions the test statistic is

$$Z = \frac{(P_1 - P_2) - 0}{\sqrt{[\hat{\pi}(1-\hat{\pi})(1/n_1 + 1/n_2)]}}$$

where $\hat{\pi}$ is the estimated value of π arrived at by combining the observations from both samples.

Note: If, say, the sample observations were given in this form

$$\text{First sample} \quad n_1 = 100 \qquad p_1 = 0.4$$
$$\text{Second sample} \quad n_2 = 150 \qquad p_2 = 0.5$$

$$\hat{\pi} \text{ will be } \frac{n_1 p_1 + n_2 p_2}{n_1 + n_2} = 0.46.$$

(For an alternative method of testing the significance of the difference between two proportions using the χ^2 distribution, see pages 457–8.)

At this point, before going on to the next section, you should tackle some of questions 5 to 9 in exercise 15a.

15.4   Confidence interval for a proportion (large sample)

In chapter 1 a number of examples were given of the kind of problem that statistical mathematics is designed to deal with. We are now in a position to look at the first of these and justify what was stated there to be the conclusion.

A sample of 200 people were interviewed by an opinion-polling organisation. Eighty in the sample declared their intention of voting for the Radical Party in the next local council election. The rest said they would vote for the Ratepayers' Party. There were no other parties in the field. The question posed was: what conclusion could be drawn about the voting intentions of the populations of the locality as a whole?

By now you will know that the most that can be deduced from these sample results is a confidence interval; we shall therefore calculate 95 % confidence limits for the population proportion π, based on the sample proportion $p = 0.4$ in a sample of size $n = 200$. As usual, we assume the sample is a random one.

The procedure is similar to finding a confidence interval of a mean.

If P is the proportion in a single sample who intend to vote Radical, the distribution of P is approximately

$$N[\pi, \pi(1-\pi)/n].$$

Since we do not know the value of π, we have to be content with using the sample value, p, as an unbiased estimate. ($E(P) = \pi$, so that $\hat{\pi} = p$.) The confidence limits are thus taken to be

$$p \pm 1.96\sqrt{[\hat{\pi}(1-\hat{\pi})/n]} \quad \text{or} \quad p \pm 1.96\sqrt{[p(1-p)/n]}.$$

In this case $p = 80/200 = 0.4$, giving confidence limits for the proportion who intend to vote Radical,

$$0.4 \pm 1.96 \times \sqrt{(0.4 \times 0.6)/200}$$

and a confidence interval $(0.33, 0.48)$.

Note: You may have noticed that the formula for confidence limits, while bearing some similarity to the one for a mean, contains p in the variance. It has to be said that although the expression seems (and is) reasonable, it cannot in fact be derived in the way we obtained the equivalent formula for a mean on page 263. This is because the statement

$$P\left(-1.96 \leqslant \frac{P-\pi}{\sqrt{\pi(1-\pi)/n}} \leqslant 1.96\right) = 0.95$$

cannot be manipulated algebraically to give the confidence limits quoted above, since π appears in the formula for variance. Nevertheless, you may take it that the formula can be justified, although the proof is difficult.

Confidence interval for the difference between two proportions

We shall complete this section on proportions by dealing with a problem that involves finding a 95% confidence interval for the *difference* between two population means.

A random sample of 153 solicitors aged over 40 were asked if they were satisfied with the English practice of using lay magistrates to administer justice in the lower courts; 112 said they were satisfied. When the same question was put to a random sample of 182 solicitors aged under 30, 108 were satisfied.

The sample proportions are $p_1 = 112/153 = 0.732$ and $p_2 = 108/182 = 0.593$.

Suppose we want to find 95% confidence limits for $\pi_1 - \pi_2$ where π_1 and π_2 are the two population proportions for the two age groups.

If P_1 and P_2 are the sample proportions when the sample sizes are n_1 and n_2 the distribution of $P_1 - P_2$ is approximately

$$N\{\pi_1 - \pi_2, [\pi_1(1-\pi_1)]/n_1 + [\pi_2(1-\pi_2)]/n_2\}.$$

Using p_1 and p_2 as estimates of π_1 and π_2, 95% confidence limits for $\pi_1 - \pi_2$ will be

$$(p_1 - p_2) \pm 1.96\sqrt{[p_1(1-p_1)/n_1 + p_2(1-p_2)/n_2]}.$$

In our example, substituting $p_1 = 0.732$, $p_2 = 0.593$ the limits are

$$(0.732 - 0.593) \pm 1.96\sqrt{[(0.732 \times 0.268)/153 + (0.593 \times 0.407)/182]}$$

giving a confidence interval of $(0.04, 0.24)$ to 2 dec. places.

EXERCISE 15a

1.   A coin is tossed 12 times and a head comes up 9 times. Use a 5% significance level to test the null hypothesis that the coin is in fact unbiased.

2.   In a horticultural experiment the survival rate for a certain type of plant is found to be 4 in 10. When the conditions are altered only 2 out of a single sample of 12 survive. Is this significant evidence that the new conditions have produced a lower survival rate? Use a 5% level.

3.   At a medical inspection, two people out of a sample of 50 were found to be colour blind. Given that the proportion of the population who are colour blind is $1/100$, use a binomial distribution to test whether the sample result is significantly high.

4.   A man is a darts player and likes to play in his local pub after he has had his regular two pints of beer. His level of skill is measured by the fact that when he aims at the bull's-eye he only gets it once in three shots.
   A friend suggests that his performance might improve if he took three pints of beer instead of two. On the first evening that he tries this experiment he gets a bull's-eye 8 times out of 15 throws. Is this evidence of a significant improvement?

5.   In 480 throws of a die a six is obtained 60 times. Is this significant evidence that the die is biased?

6.   The proportion of brown-eyed people in a community is thought to be $1/10$. In a random sample of 400, 47 were found to be brown-eyed. Is this proportion significantly different from $1/10$?

7.   In one district of a political constituency, out of 500 people canvassed 320 were found to be firm supporters of the Labour party. In another district 400 out of 700 canvassed were Labour supporters. Are these proportions significantly different?

8.   Of a random sample of 50 shoppers in a certain city store 13 lived more than 10 miles from the city centre. Of a random sample of 50 shoppers from another store in the same city 9 lived more than 10 miles from the city centre. Stating your null and alternative hypotheses and using a significance level of 5%:
(a) test that the true proportion in both stores could be 0.15;
(b) show that the two samples do not offer evidence of a difference in proportions between
     the two stores.   (SUJB)

9.   In a new course last year, a university department received applications from 450 home students and 50 overseas students. This year, after an increase in the fees charged to overseas students, the corresponding numbers were 550 and 40. Test whether there is a significant difference in the proportion of overseas students applying for admission between the two years.
   Recalculate your answer if the figures of 450 and 50 are average figures over a long period.   (SMP)

10.   Treatment with a drug resulted in 24 cures out of 72 who had a certain ailment. Calculate an approximate 95% confidence interval for the proportion of cures in a population suffering from the same ailment who might be treated with the drug.

11.   Twenty out of 100 people questioned revealed that they had dreamed about a member of the Royal family during the past year. Within what limits could it be claimed, with 95% confidence, that the proportion of the population who had had a similar dream might lie?

12.   A certain country in a fairy tale is populated by elves and fairies. A random sample of

100 fairies were each asked the question 'Do you believe in people?': 72 fairies replied 'Yes' and the remainder replied 'No'. Give an approximate 98% confidence interval for the proportion of fairies that say they believe in people.

A random sample of 62 elves were each asked the same question: 54 elves replied 'Yes' and the remainder replied 'No'. A question of interest is whether the proportion of fairies that say they believe in people differs from the proportion of elves that say they believe in people. Assuming that these proportions are equal, obtain an unbiased estimate of the common proportion. Using this estimate, test the question of interest at the 10% significance level, stating clearly your null and alternative hypotheses.   (C)

13. In answer to a question submitted to a random sample of 1100 people: 'Are you in favour of prison sentences for drunken drivers who cause serious injury or death?' Four hundred and seventy-three said 'Yes' and the rest said 'No'. A little later an independent sample of 1100 people were asked the slightly amended question: 'Are you in favour of *mandatory* prison sentences for drunken drivers who cause serious injury or death?'. This time 352 said 'Yes' and the rest said 'No'.

Find an approximate 95% confidence interval for the difference between the proportions of those who said 'Yes' on the two occasions.

14. In order to throw light on the question whether two 16th century texts whose authorship is unknown might have been written by the same person a stylistic analysis is undertaken. The method is to take a passage of narrative from each text and find, for each passage: (a) the rate of occurrence of the word 'the', (b) the rate of occurrence of the word 'that', (c) the proportion of occurrences of 'the' which are followed by a noun. The results are as follows:

Test for	*Text A*	*Text B*
Number of words	2341	2462
Occurrences of 'the'	130	323
Occurrences of 'that'	52	31
'The' followed by noun	37	39

Calculate for each text:
 (i) the proportion of occurrences of 'the',
 (ii) the proportion of occurrences of 'that',
(iii) the proportion of occurrences of 'the' which are followed by a noun.

For each of these pairs of proportions find whether the difference is significant.

15.5   Non-parametric tests

In all the hypothesis tests we have described so far it has been necessary to assume that either the population distribution is known or, if this is not the case, that the size of sample is large enough for the central limit theorem to apply. For the small sample *t*-test, in particular, the population distribution has to be normal.

Consider now some examples where these conditions do not apply. The distribution of the ages at which people die of a specific disease is heavily skewed and therefore cannot be normal. The same is true of the ages at which people get married. If then, using data from a small sample, we want to test a hypothesis about the population mean, the *t*-test cannot be used. In these cirumstances there are various tests available and in none of them is any assumption made about the distribution of the population. They are therefore called *distribution-free* or *non-*

parametric tests, since they do not test parameters of known distributions. We shall look at three of these, all of which use the *median* rather than the mean.

15.6   The sign test for a median

As an example of a distribution which is not known to be normal we shall take the times (to the nearest half minute) that Post Office customers have to wait before being served, after a new queueing system has been introduced. The observations were made at a time on a Friday when the office is always especially busy.

A sample of 12 customers produced the following times (to the nearest half-minute:

$$6\tfrac{1}{2} \quad 7 \quad 8 \quad 5 \quad 7\tfrac{1}{2} \quad 9 \quad 8\tfrac{1}{2} \quad 4 \quad 7 \quad 9\tfrac{1}{2} \quad 8 \quad 4\tfrac{1}{2}$$

The Post Office authorities claim, on the basis of experience of many previous Friday peak times under the new system, that the median time is 6 minutes. Some customers are inclined to challenge this figure, so that the competing hypotheses to be tested are

$$H_0: \text{median} = 6 \quad \text{as against} \quad H_i: \text{median} > 6.$$

The first stage is to replace the given figures by $+$ or $-$, depending on whether the time is greater or less than the median:

$$+ \quad + \quad + \quad - \quad + \quad + \quad + \quad - \quad + \quad + \quad + \quad -$$

Now for a continuous distribution, the median has the convenient property that, by definition, if a member of the population is picked at random it is equally likely to be above or below the median. Under H_0, therefore, the distribution of X, *the number of $+$ signs* should be

$$B(12, \tfrac{1}{2})$$

and we may proceed, as on page 296, to define a critical region for a test with a 5 % significance level. To do this it is only necessary to calculate the binomial probabilities at the tail end.

$$P(X = 9) = \binom{12}{9}(\tfrac{1}{2})^{12} = 0.054, \qquad P(X = 10) = \binom{12}{10}(\tfrac{1}{2})^{12} = 0.016$$

$$P(X = 11) = \binom{12}{11}(\tfrac{1}{2})^{12} = 0.003, \qquad P(X = 12) = (\tfrac{1}{2})^{12} = 0.000$$

to 3 dec. places.

Hence $P(X \geqslant 9) = 0.073$ and $P(X \geqslant 10) = 0.019,$

and since $P(X \geqslant 10) < 0.05$ we take the 5 % critical region to be $X \geqslant 10$.

For the sample, $X = 9$, which is not in the critical region, so that H_0 is accepted at the 5 % level.

Note:
1.   Care should be taken over the critical region. If, in the above example the

alternative hypothesis had been

$$H_1 : \text{median} \neq 6$$

a two-tail symmetrical critical region would have been needed.

Now $P(X \geqslant 10) = P(X \leqslant 2) = 0.019$

so that $P(X \geqslant 10 \text{ or } X \leqslant 2) = 0.038$, which is still less than 0.05.

The critical region at 5 % would therefore have to be $X \geqslant 10$ or $X \leqslant 2$ since

$$P(X \geqslant 9 \text{ or } X \leqslant 3) = 0.146,$$

which is more than 0.05.

This 'trial and error' method of calculating the critical region may be avoided by using a table giving cumulative probabilities for $B(n, \frac{1}{2})$.

2.   An alternative method for large samples is to use the normal approximation to the binomial distribution. Even with values of n not much greater than 10, the approximation will be a good one since $p = \frac{1}{2}$. We will use this method in the next example.

3.   If a sample value turns out to be *equal* to the hypothetical median (and this could happen even for a continuous distribution when measurements are not exact), this value must be discarded and the sample size therefore reduced. This will also be demonstrated in the next example.

EXAMPLE 1   The times that secretaries stay as employees in a large corporation is thought to have a median value of 46 months. A sample of 14 secretaries give the following results:

   40   64   48   38   72   80   46   52   60   39   55   38   53   56.

Is this significant evidence that the median time is no longer 46 months?

We take as hypotheses H_0: median $= 46$ and H_1: median $\neq 46$. The signs are:

   $-$ $+$ $+$ $-$ $+$ $+$ 0 $+$ $+$ $-$ $+$ $-$ $+$ $+$

and we discard the zero value.

$$X, \text{ the number of } - \text{ signs, is 4 out of 13,}$$

and under H_0 the distribution of X is $B(13, \frac{1}{2})$.

Using the approximation of a normal distribution with

$$\text{mean} = 13 \times \tfrac{1}{2} = 6.5 \quad \text{and} \quad \text{variance} = 13 \times \tfrac{1}{2} \times \tfrac{1}{2} = 3.25$$

the test statistic $z = (4.5 - 6.5)/\sqrt{3.25} = -1.11$, using a continuity correction.

This is greater than the critical 5 % value of -1.96 for a two-tail test and H_0 is accepted.

(The corrected value of X is taken to be 4.5 as we need to know whether or not $X = 4$ is inside the critical region.)

In this example we could, of course, have equally well considered the distribution of $+$ signs.

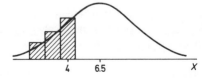

15.7   The sign test for two paired samples

As the sign test for a single sample corresponds to the test for a mean, it is to be expected that there will be an equivalent sign test for paired samples corresponding to the paired-sample t-test.

We shall suppose that a biological experiment was carried out in which 9 rats were made to complete some task and the times taken were measured, first under normal room temperature then with the temperature raised $10°$. The times (in seconds) were:

Normal temperature X	61	55	41	63	49	53	45	45	48
Temperature raised Y	59	54	43	60	47	54	47	44	44

Is this significant evidence that the difference in temperature made any difference to performance times?

Just as in using the t-test a common variance is assumed, so in what follows we make the assumption that the two underlying populations may differ only in one respect, the average as represented by the median.

As with the paired-sample t-test we work with the differences $D = X - Y$:

$$D: \quad 2 \quad 1 \quad -2 \quad 3 \quad 2 \quad -1 \quad -2 \quad 1 \quad 4$$

and the null hypothesis is that the median of the population of differences is 0.

$$H_0: \text{median of differences} = 0, \qquad H_1: \text{median of differences} \neq 0.$$

Our sample gives the following $+$ and $-$ values:

$$+ \quad + \quad - \quad + \quad + \quad - \quad - \quad + \quad +$$

and the number of $+$ signs, S, is 6 out of 9.

Under H_0 the distribution of S is $B(9, \frac{1}{2})$, and as before we calculate a critical region for a 5% significance level two-tail test. Looking at the two tail ends we find that

$$P(S = 9) = P(S = 0) = (\tfrac{1}{2})^9 = 0.002$$

$$P(S = 8) = P(S = 1) = 9(\tfrac{1}{2})^9 = 0.018$$

$$P(S = 7) = P(S = 2) = \binom{9}{2}(\tfrac{1}{2})^9 = 0.070 \quad \text{to 3 dec. places.}$$

Since $P(S \geqslant 7 \text{ or } S \leqslant 2)$ is obviously greater than 0.05, and

$$P(S \geqslant 8 \text{ or } S \leqslant 1) = 2(0.018 + 0.002) = 0.04 < 0.05$$

we take the latter as critical region.

$S = 6$ is not in this region so that H_0 is accepted at the 5% level.

EXERCISE 15b

1.   If the distribution of X is $B(8, \frac{1}{2})$, what range of values of X should be rejected at the 5% level for: (a) a one-tail test, (b) a two-tail test?

2.   If the distribution of X is $B(11, \frac{1}{2})$, what range of values of X should be rejected for: (a) a one-tail test at the 5% level, (b) a one-tail test at the 1% level?

3.   The times spent watching television during a week by 12 pupils at school were

$$15 \quad 19 \quad 20 \quad 12 \quad 18 \quad 18 \quad 16 \quad 9 \quad 30 \quad 16 \quad 25 \quad 31$$

Use the sign test at the 5 % level to test whether the median time for pupils at this age is 16 hours, as against the alternative hypothesis that it is greater than 16.

4.   Entrants to a college are given a standardised test in which the median mark over the years has turned out to be 55. One year a group of seven students from the new entry gain the following marks:

$$71 \quad 15 \quad 43 \quad 48 \quad 65 \quad 13 \quad 42$$

Test whether the median mark for the population from which this group comes is different from 55. Use a 5 % significance level.

5.   The median number of toys assembled in a day by women employed in a factory is 65. A new employee produces the following numbers in 10 successive days.

$$50 \quad 62 \quad 49 \quad 58 \quad 67 \quad 70 \quad 53 \quad 54 \quad 71 \quad 69$$

Does this provide evidence that the median value for the new employee is less than 65. Use a 5 % significance level.

6.   A shopkeeper finds that the number of sweets in 15 packets of a particular kind of sweet is:

$$28 \quad 27 \quad 29 \quad 22 \quad 25 \quad 21 \quad 29 \quad 32$$
$$34 \quad 26 \quad 31 \quad 28 \quad 30 \quad 26 \quad 28$$

Test the null hypothesis that the median number claimed by the supplier as 30 against the alternative hypothesis that it is not 30, using the normal approximation to the binomial distribution and a 1 % significance level.

7.   In a certain country, the ages of men convicted for the first time of violent crime have a median of 24. A sample from the north of the country gave the following figures:

$$22 \quad 28 \quad 19 \quad 16 \quad 18 \quad 20 \quad 20 \quad 35 \quad 21 \quad 23$$
$$32 \quad 22 \quad 17 \quad 19 \quad 30 \quad 28 \quad 23 \quad 19 \quad 17 \quad 30$$
$$29 \quad 16 \quad 17 \quad 18 \quad 20 \quad 28 \quad 23 \quad 22 \quad 32 \quad 21$$

Use the normal approximation to the binomial distribution and a 5 % significance level to test whether or not the median age for the north is less than 24.

8.   In a test of petrol consumption using two different brands 12 cars of the same model travel on a given course with 20 litres of brand A until they run out of petrol. The process is then repeated with brand B. The results are shown in the table.

Using a binomial distribution and a 5 % significance level test the hypothesis that brand B does not give a better median consumption than brand A.

Car	Kilometres run	
	Using brand A	Using brand B
1	181	184
2	165	165
3	193	195
4	160	164
5	180	176
6	194	201
7	203	206
8	201	205
9	184	189
10	158	165
11	168	165
12	179	179

9.   The percentage of impurity in 14 samples of chemical solution is measured by two different methods. The results are:

Sample	% recorded	
	Method A	Method B
1	46	32
2	32	24
3	25	13
4	11	5
5	39	26
6	14	20
7	31	19
8	18	7
9	15	3
10	34	38
11	9	1
12	44	29
13	41	34
14	27	13

Use a sign test and the normal approximation to the binomial distribution (with a 1% significance level) to find whether the two methods give significantly different results.

15.8   The Wilcoxon signed-rank test (T-test)

It may well have occurred to you that the sign test is a bit crude, ignoring, as it does the *magnitudes* of the differences from the median. In our next example we show how these magnitudes may be taken into account, using a method devised by Frank Wilcoxon in 1945.

EXAMPLE 2   The number of hours worked per week, including overtime, by 17 employees in a munitions factory during 1943 were as follows:

$$29 \quad 52 \quad 38 \quad 72 \quad 60 \quad 55 \quad 45 \quad 64 \quad 80$$
$$41 \quad 66 \quad 74 \quad 36 \quad 61 \quad 48 \quad 53 \quad 59$$

Test the hypothesis that the median is 46 against the alternative hypothesis that it is greater than 46.

The hypotheses are

$$H_0: \text{median} = 46, \qquad H_1: \text{median} > 46.$$

The actual differences between these values and the supposed median are:

$$-17 \quad +6 \quad -8 \quad +26 \quad +14 \quad +9 \quad -1 \quad +18 \quad +34$$
$$-5 \quad +20 \quad +28 \quad -10 \quad +15 \quad +2 \quad +7 \quad +13$$

If we now assign ranks, in ascending order, to these differences, ignoring the signs:

Difference	-1	$+2$	-5	$+6$	$+7$	-8	$+9$	-10	$+13$
Rank	1	2	3	4	5	6	7	8	9

Difference	$+14$	$+15$	-17	$+18$	$+20$	$+26$	$+28$	$+34$
Rank	10	11	12	13	14	15	16	17

it will be seen that the larger differences occur mainly on the $+$ side of the median.

Note: In assigning ranks, if any are tied each is given the rank which is the average of the ranks they would have had if these had been different. For example if the 5th and 6th differences are equal they are given the rank 5.5; if the 9th, 10th and 11th are equal they are ranked 10th.

The next stage is to find T^-, the sum of the ranks with negative sign, and T^+, the sum of the ranks with positive sign.

$$T^- = 1+3+6+8+12 = 30.$$

$$T^+ = 2+4+5+7+9+10+11+13+14+15+16+17 = 123.$$

Now if the null hypothesis is true, we would expect the hypothetical population of differences to be equally distributed on both sides of the median, so that the ranks and therefore the sum of the ranks on each side would be the same. In a random sample this balance might well be upset, but the probability of this occurring by chance should be small. If, then, we call the *smaller* of these sums T, the smaller the value of T, the stronger will be the evidence for rejecting H_0.

In this example, then, we need to know what is the probability of getting a value of T as small as 30 if the null hypothesis is true. Theoretically, it would be possible to work out the number of arrangements of ranks giving $T = 0$, $T = 1$, $T = 2$, etc. but there are two less cumbersome alternatives.

1.   *If the sample size $n > 15$*, the distribution of T is approximately normal with

$$\text{mean} = [n(n+1)]/4 \quad \text{and} \quad \text{variance} = [n(n+1)(2n+1)]/24$$

and the approximation holds good even when the allotted ranks are not all integers.

The proof of this result is difficult and depends on a generalisation of the central limit theorem. The derivation of the mean and variance is, however, reasonably easy and you will find a proof in the Appendix on page 316.

In our example, with $n = 17$, the mean $= 76.5$ and the variance $= 446.25$, so that the value of the test statistic is

$$z = (31.5 - 76.5)/\sqrt{446.25} = -2.13,$$

using a continuity correction. This is less than the critical one-tail value of -1.64 at the 5% level, and we therefore reject H_0.

2.   *For samples of any size up to 30* an alternative method is to use the table on page 517, which gives critical values of T for different values of n. We illustrate this, and the use of tied ranks, by looking again at the example on page 303, where the sample observations were waiting times at a Post Office:

$$6\tfrac{1}{2} \quad 7 \quad 8 \quad 5 \quad 7\tfrac{1}{2} \quad 9 \quad 8\tfrac{1}{2} \quad 4 \quad 7 \quad 9\tfrac{1}{2} \quad 8 \quad 4\tfrac{1}{2}$$

and the hypotheses were H_0: median $= 6$ against H_1: median > 6.

The differences and their associated ranks are:

Difference	$+\frac{1}{2}$	$+1$	$+2$	-1	$+1\frac{1}{2}$	$+3$	$+2\frac{1}{2}$	-2	$+1$	$+3\frac{1}{2}$	$+2$	$-1\frac{1}{2}$
Rank	$+1$	$+3$	$+8$	-3	$+5.5$	$+11$	$+10$	-8	$+3$	$+12$	$+8$	-5.5

giving $T^- = 16.5$, $T^+ = 61.5$ so that $T = 16.5$.

From the table, for $n = 12$ the 5% critical value for a one-tail test is 17, which means that values of $T \leqslant 17$ are too small for H_0 to be accepted. In this case, then, when rank as well as sign is taken into account H_0 is rejected.

The fact that in using the sign test H_0 was accepted is explained by the fact that in the more sensitive signed-rank test the value of T was near the critical 5% value of 17. The small-sample sign test, on the other hand, involved using the binomial distribution which, in this example, produced a critical region $(X \geqslant 10)$ corresponding to a probability of 0.034, which is considerably less than 0.05.

15.9   The Wilcoxon signed-rank test for paired samples

The procedure for using this test for paired samples is, of course the same as for a single sample. The single observations are replaced by the differences between each pair of observations and the null hypothesis is

$$H_0: \text{the median of the differences} = 0$$

as on page 304.

EXERCISE 15c

1.   Eighteen test areas were given a standard thickness of coating with a new paint and the drying times were recorded in minutes, with the following results:

114	105	118	106	100	119	126	130	103
97	112	102	93	121	101	102	95	96

Use the signed-rank test and a normal distribution to test the hypothesis that the median drying time is 110 minutes against the alternative hypothesis that it is less than 110 minutes.

2.   The times taken by 20 athletes to complete an annual marathon race were as follows (to the nearest minute):

126	144	116	138	129	113	101	141	125	133
121	149	113	117	97	173	159	118	140	135

The median time for all competitors over past years had been 137 minutes. Use the signed-rank test and a normal distribution to test whether the median time for competitors is any different from 137 minutes.

3.   Apply a signed rank test to the data in exercise 15b, question 3, using the table of critical values on page 517.

4.   Apply a signed-rank test to the data in exercise 15b, question 8, using the table of critical values on page 517.

5. The metal content of 5 trucks of ore is measured by two different methods. The percentages were:

Method	Truck				
	1	2	3	4	5
1	16	30	19	2	25
2	21	32	18	6	23

Use a signed-rank test and the table of critical values on page 517 to test the hypothesis that there is no difference between the two methods against the alternative hypothesis that method 2 gives a greater metal content than method 1. (Use a 5% significance level.)

15.10   The Wilcoxon rank-sum test for unpaired samples

In this test, as in the last one, there are two methods, one for small samples and one for large ones.

EXAMPLE 3   The time, in months, taken to complete the construction of 4 submarines by a Swedish shipyard, and the times taken by a Norwegian shipyard to complete 5 submarines are recorded below. The submarines were all identical. Is this evidence that the Swedish shipyard is quicker than the Norwegian yard?

Swedish yard	31	33	37	39	
Norwegian yard	32	40	42	44	46

The first step is to rank these figures, taken together, in ascending order:

Combined figures	31	32	33	37	39	40	42	44	46
Rank	1	2	3	4	5	6	7	8	9

and to underline the ranks of the figures in the *smaller sample*.

The statistic used in this test is the sum of the ranks for the smaller sample:

$$W = 1+3+4+5 = 13.$$

The null hypothesis, as in the paired-sign test, is that there is no difference between the underlying populations of the two distributions and that each is 'balanced' round the same median. This means that the ranks are comparable, and therefore so are the sums of the ranks. Clearly the smaller the value of W in a sample, the greater the evidence for a lack of balance and therefore for rejecting the null hypothesis. In this example, from the wording, the test will be one-tailed, and we shall take a 5% significance level.

What, then, is the probability that in this case W is 13 or less and what is the critical region for W for accepting or rejecting H_0?

For a problem like this there is no particular difficulty in finding the relevant part of the probability distribution of W, its 'tail', as we can easily list the various arrangements of four ranks that give $W = 10, 11, 12$ and 13:

					W
	1	2	3	4	10
	1	2	3	5	11
	1	2	3	6	12
Rank	1	2	3	7	13
	1	2	4	5	12
	1	2	4	6	13
	1	3	4	5	13

Since the number of arrangements of the 4 ranks in 9 'slots' is

$$\binom{9}{4} = 126,$$

the tail of the probability distribution of W is:

W	10	11	12	13
Probability	1/126	1/126	2/126	3/126

so that

$$P(W \leqslant 11) = 2/126 = 0.016$$

$$P(W \leqslant 12) = 4/126 = 0.032 < 0.05$$

$$P(W \leqslant 13) = 7/126 = 0.056 > 0.05.$$

For a 5% one-tail test the critical region is, therefore,

$$P(W \leqslant 12)$$

and H_0 is accepted.

The method described in the last example would clearly be very cumbersome for larger values of W. However, if the sample sizes are n_1 (the smaller) and n_2 (the larger), it can be proved that if $n_2 > 7$, a good approximation to the distribution of W is the normal distribution with

$$\text{mean} = \tfrac{1}{2}n_1(n_1+n_2+1) \quad \text{and} \quad \text{variance} = \tfrac{1}{12}n_1 n_2(n_1+n_2+1).$$

EXAMPLE 4   A driving instructor noted the number of hours tuition he gave 4 men and 8 women before they passed the driving test. None of them had previous experience of driving. Do the following figures provide evidence that there is any difference between men and women in this respect?

Men	12	15	16	21				
Women	13	18	19	23	25	26	27	28

We shall take the usual null hypothesis that there is no difference and use a two-tail test.

The combined figures with their ranks are:

12	13	15	16	18	19	21	23	25	26	27	28
1	2	3	4	5	6	7	8	9	10	11	12

with the men's ranks underlined.

The value of the statistic $W = 1+3+4+7 = 15$.

With $n_1 = 4$ and $n_2 = 8$, we use the normal approximation with

$$\text{mean} = \tfrac{1}{2} \times 4 \times 13 = 26 \quad \text{and} \quad \text{variance} = \tfrac{1}{12} \times 4 \times 8 \times 13 = 34.67.$$

The standardised normal statistic

$$z = (15.5 - 26)/\sqrt{34.67} = -1.78 > -1.96,$$

so that H_0 is accepted at the 5% level for a two-tail test.

EXERCISE 15d

1.   For samples of sizes 3 and 4, derive the probabilities that the rank-sum statistic $W = 6, 7,$ 8 and 9, using the listing method on pages 310–11 and state the critical region for a one-tail test with a significance level of: (a) 5%, (b) 10%.

2.   The percentage of monosodium glutamate flavouring in two brands of packeted tomato soup powder were found to be as follows:

Brand A	3.1	1.4	4.5	2.8	5.3
Brand B	1.1	1.9	0.9		

Use the rank-sum test with a 5% significance level to examine the claim that the percentage is higher in brand A.

3.   The number of red blood cells (measured in millions per cubic millimetre) for 8 men and 10 women were found to be:

Men	4.21	4.90	5.51	5.31	5.29	4.84	5.34	4.73	
Women	5.03	3.96	4.71	5.34	4.90	5.42	4.02	4.38	4.13	5.31

Use the rank-sum test with a 5% significance level to test the hypothesis that there is no difference between men and women as far as red blood cells are concerned.

4.   The scores in a test taken by two groups of pupils, 10 in each, from the same age group in two large neighbouring schools were:

School A	84	78	95	80	79	76	89	98	73	87
School B	67	92	78	81	88	75	90	78	85	79

Use the rank-sum test with a 1% significance level to test whether there is any difference between the performances of the two schools.

EXERCISE 15e (miscellaneous)

1.   An investigation is being carried out into the possible authorship of an ancient Greek text. Its style suggests that it might have been written by Proclus (5th century AD). One test consists of counting the number of occurrences of the Greek word for 'not' and to compare the proportion of these occurrences with the proportion to be found in works known to be by Proclus.

  In the anonymous text there are 3140 words, and 'not' occurs 52 times. Does this give a proportion significantly different from the proportion 0.013 in Proclus' works?

2.   In the Blackheath district of London, out of 203 married men who were questioned, 41 claimed that they helped with washing-up after meals at least twice weekly. When the same

question was put to 186 married men in the neighbouring district of Deptford, 30 claimed to do so.
(a) Are the two proportions significantly different?
(b) Test the hypothesis that the proportion in Deptford is significantly lower.
(Use a 5% significance level in both cases.)

3. 'Murphy's law' states that when a piece of buttered toast is accidentally dropped on the floor there is a 0.9 probability that it will land with the buttered side downwards.
    A scientist persuaded his young son, who was being tiresome, to conduct an experiment under controlled conditions, as a result of which 142 out of 200 pieces of buttered cardboard landed buttered side down. Does this result cast significant doubt on the truth of Murphy's law?

4. Out of a sample of 524 people, 192 said that they always watched televised darts matches if possible.
    Find: (a) 95%, (b) 98% confidence limits for the proportion in the population of viewers who do the same.

5. The owner of a large apple orchard states that 10% of the apples on the trees in his orchard have been attacked by birds. A random sample of 2500 apples is picked and 274 apples are found to have been attacked by birds. Test, at the 8% significance level, whether there is significant evidence that the owner has understated the proportion of the apples on the trees in his orchard that have been attacked. State your hypotheses clearly.
    A second random sample of 3500 apples is picked at the same time as the first sample, and 338 apples are found to have been attacked by birds. Pooling the information provided by the samples, obtain an approximate symmetric 96% confidence interval for the proportion of apples in the orchard that have been attacked by birds. (C)

6. An insurance company has sold a certain specialised form of insurance to a large number of customers and wishes to investigate the proportion p of them who are women. Records for a random sample of 100 customers are examined. State the distribution of the number of women in samples of this size and give the mean and variance of the normal distribution that may be used to approximate this. Given that there are in fact 32 women in the sample, use a 1% significance test to determine whether it is reasonable to assume that this type of insurance appeals equally to men and to women. Provide also an approximate two-sided 95% confidence interval for p. (MEI)

7.★ The Notax Party have a candidate in a by-election and $100p$% of the whole electorate support this candidate. Suppose that, before the election, a random sample of size n is taken of the electorate, and $100R$% of the sample say that they support the Notax Party candidate. Write down the values of the mean and variance of R in terms of p and n.
    Owen Money, the leader of the Notax Party, has decided to mount a massive publicity campaign provided that at least 35% of the electorate support his candidate. In a sample survey of 200 voters, 80 said that they would support the Notax Party candidate. Test, at the 5% significance level, the hypothesis that $p = 0.35$, the alternative hypothesis being that $p > 0.35$.
    Given that in fact $p = 0.4$, estimate the probability that analysis of the results of a second sample survey of 200 voters, with the same null and alternative hypotheses as above, will lead to acceptance of the null hypothesis (even though it is false). (MEI)

8.★ A random sample of n_1 individuals is drawn from a very large population in which a proportion θ of the individuals have a particular attribute A. If p_1 denotes the proportion of individuals in the sample having the attribute A, show that the variance of p_1 cannot exceed $1/4n_1$.
    Another random sample of n_2 individuals is independently drawn from the same population, and a proportion p_2 of the individuals in this sample have the attribute A. Show that, in the two samples combined, the proportion of individuals (p) having the attribute A has mean θ.

Assuming that the variance of $p_1 - p_2$ is equal to the sum of the variances of p_1 and p_2, show that if p is used as an estimate of θ, the corresponding estimate of the variance of $p_1 - p_2$ is given by

$$v = p(1-p)(1/n_1 + 1/n_2).$$

In the remaining part of this question you may assume that for p_1 and p_2 as defined above the distribution of $p_1 - p_2$ may be approximated by a normal distribution having variance v.

In a survey it was found that there were 33 smokers in a random sample of 100 doctors, and 55 smokers in a random sample of 120 lawyers. Discuss statistically the extent to which the results of this survey confirm or refute the claim that the proportion of doctors who smoke is less than the proportion of lawyers who smoke.   (JMB)

9.   The times that 18 visitors to a Social Security Office had to wait before receiving attention were, in minutes:

$$
\begin{array}{ccccccccc}
20 & 23 & 24 & 23 & 21 & 22 & 29 & 27 & 24 \\
25 & 22 & 26 & 23 & 20 & 23 & 25 & 22 & 25
\end{array}
$$

The office manager claimed that, on average, the median time is 23 minutes. Use the sign test, with a 2% significance level to test this claim.

10.   On the 2nd July 1980, the incoming mail in each of 12 selected towns was randomly divided into two similar lots prior to sorting. In each town one lot was then sorted by the traditional hand sorting method, the other by a new Electronic Post Code Sensor Device (EPCSD). The times taken in hours, to complete these jobs are recorded below.

Town	A	B	C	D	E	F	G	H	I	J	K	L
Hand sort time	4.3	4.1	5.6	4.0	5.9	4.9	4.3	5.4	5.6	5.2	6.1	4.7
EPCSD sort time	3.7	5.3	4.5	3.1	4.8	5.0	3.5	4.9	4.6	4.1	5.7	3.5

Use the sign test and a 5% significance level to test the null hypothesis of no difference in the times against the alternative hypothesis that the EPCSD method is quicker. (AEB 1982)

11.   State the assumptions on which the sign test is based.

For samples of 12 items from two populations, obtain critical values for the number of + (or −) signs obtained for:
(a) a two-tail test at the 5% significance level,
(b) a one-tail test at the 5% significance level.

Twelve pairs of identical twins are allocated to two training programmes, A and B, (one twin of each pair to each programme). At the end of the training, the twins are observed and, for each pair, the better performer is determined. The results are as follows:

Pair	1	2	3	4	5	6	7	8	9	10	11	12
Better performer	A	A	B	A	A	A	B	A	A	B	A	A

Do these observations provide evidence that programme A is a better training programme than programme B?   (C)

12.   Twenty-five students in a physical training course were selected to be tested for their ability in performing a number of standard exercises soon after they had started on the course. One month later they were re-tested, using the same exercises. The scores were:

Student	1	2	3	4	5	6	7	8	9	10	11	12	13
1st score	56	43	59	62	38	49	53	37	71	53	47	39	37
2nd score	67	58	58	75	47	51	52	49	75	59	56	41	42

Student	14	15	16	17	18	19	20	21	22	23	24	25
1st score	68	27	68	75	42	53	61	56	58	35	46	37
2nd score	65	31	72	84	45	54	65	61	57	39	49	39

Use the sign test to determine whether these results suggest that students on the course will on average perform better after a month.

13.   Use the Wilcoxon signed-rank test on the data given in question 10.

14.   The times taken by 10 cars to travel from Hammersmith Broadway to Piccadilly Circus starting at 4.30 pm on a Friday were recorded. At 4.30 pm on the following Monday the times taken by the same cars to travel the same route were recorded. The results were:

Car	1	2	3	4	5	6	7	8	9	10
Friday times (minutes)	38	40	41	49	44	42	36	43	48	39
Monday times (minutes)	46	39	44	51	40	46	42	48	42	46

Use the Wilcoxon signed-rank test, with a 5% significance level, to determine whether there is any difference between Monday and Friday as far as times are concerned.

15.   State the necessary assumptions for Wilcoxon's T-test and state what difference between samples it measures.

The growth of plants with and without an artificial fertiliser is compared by using 30 matched pairs of seedlings and feeding one of each pair. The measured heights (centimetres) four weeks later are given in the table:

Pair	1	2	3	4	5	6	7	8	9	10	11	12	13	14	15
With	15	13	19	20	12	16	18	19	20	12	21	15	13	16	18
Without	10	14	17	17	11	12	15	21	16	15	19	9	11	15	14

Pair	16	17	18	19	20	21	22	23	24	25	26	27	28	29	30
With	9	11	10	17	18	15	9	14	21	12	19	15	17	11	19
Without	16	12	13	16	10	11	13	14	15	12	11	13	22	13	14

Do these observations suggest that the artificial fertiliser improves growth rates?

[Mean $T = n(n+1)/4$; variance $T = n(n+1)(2n+1)/24$.]   (C)

16.   Each woman in a group of 5 undertook the same course of exercises for a month, the course being designed to help them to reduce weight. During the same period each woman in another group of 4 followed the same course of dieting, with the same object. The resulting weight-losses for the two groups were:

Exercises	3.2	6.4	2.5	5.7	2.0
Dieting	3.0	1.6	3.9	4.2	

Use the Wilcoxon rank-sum test with a 5% significance level to test whether there was any difference between the two methods.

15.11 Appendix

The mean and variance of the distribution of T

Let T be the sum of the $-$ ranks for a sample of size n, where the ranks are integers $1, 2, 3, \ldots$.

If we write 0 where the sign of a rank is $+$

and r where the rank r is $-$,

the possible arrangements of the signed ranks for $T = 0$, $T = 1$, $T = 2$, $T = 3$, etc. are:

Ranks

	1	2	3	4	5	6	
$T = 0$	0	0	0	0	0	0	... (all the rest 0)
$T = 1$	1	0	0	0	0	0	...
$T = 2$	0	2	0	0	0	0	...
$T = 3$	0	0	3	0	0	0	...
	1	2	0	0	0	0	...
$T = 4$	0	0	0	4	0	0	...
	1	0	3	0	0	0	...

and so on.

T is therefore the sum of n independent variables

$$X_1, X_2, \ldots, X_n$$

where X_1 can either be 0 or 1, X_2 can either be 0 or 2 etc., and

$$P(X_r = 0) = \tfrac{1}{2}, \qquad P(X_r = r) = \tfrac{1}{2}.$$

Now $E(X_r) = \tfrac{1}{2} \cdot 0 + \tfrac{1}{2} \cdot r = \tfrac{1}{2}r$

$$\Rightarrow E(T) = \sum_{r=1}^{n} E(X_r) = \frac{1}{2} \sum_{r=1}^{n} r = \frac{1}{2}\left[\frac{n(n+1)}{2}\right] = \frac{n(n+1)}{4}.$$

Similarly

$$\text{Var}(X_r) = E(X_r^2) - (\tfrac{1}{2}r)^2 = (\tfrac{1}{2} \cdot 0^2 + \tfrac{1}{2} \cdot r^2) - (\tfrac{1}{2}r)^2 = \tfrac{1}{4}r^2$$

$$\Rightarrow \text{Var}(T) = \sum_{r=1}^{n} \text{Var}(X_r) = \frac{1}{4} \sum_{r=1}^{n} r^2 = \frac{1}{4}\left[\frac{n(n+1)(2n+1)}{6}\right] = \frac{n(n+1)(2n+1)}{24}.$$

Note: Unless a high proportion of ranks are tied, these formulae for $E(T)$ and $\text{Var}(T)$ may still be used as approximations when the ranks are not all integers.

16 Possible errors in significance testing

16.1 Type I and Type II errors

In all tests of significance the conclusions are stated, of course, in terms of probability, which means that there is a possibility that we are wrong in accepting or rejecting a null hypothesis. In this chapter we shall consider the kinds of error that can be made in significance testing and possible ways of limiting those errors. We shall also consider ways of assessing how good a test is for its particular purpose.

As a first example we shall take a *one-tail test* of a mean.

Suppose an electronic component produced at a factory has a mean life of 1000 hours and standard deviation 200 hours. After some modification to the process, a sample of 100 components gives a mean life of 1060. Is this significantly better than 1000 hours?

The alternative hypotheses are

$$H_0: \mu = 1000 \quad \text{as against} \quad H_1: \mu > 1000.$$

Adopting a 5% significance level and making a slight change in the usual procedure, the critical region for accepting or rejecting H_0 starts at

$$\bar{x} = 1000 + 1.64 \times 200/\sqrt{100} = 1033$$

$$\left(\text{i.e. } z = \frac{\bar{x} - 1000}{200/\sqrt{100}} = 1.64 \right).$$

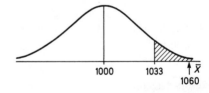

If, as in this case, the sample mean $\bar{x} = 1060$, H_0 is rejected at the 5% level.

Now clearly there is a possibility that we shall reject H_0 when it is true. The choice of sample may have been an 'unlucky' one. The probability of this happening is 5%, the significance level chosen for the test.

This kind of error is called a *Type I error* (*rejecting H_0 when it is true.*)

The probability of making a Type I error is called α, so that

$$\alpha = \text{the significance level of the test.}$$

In the present example $\alpha = 0.05$.

Now suppose the sample mean turned out to be 1020. H_0 would then be accepted on this test, even though it is possible that H_1 is in fact true. To be specific, suppose that μ has risen to 1040.

As the figure shows, the probability of this happening (accepting H_0 when it is false and μ is 1040) is

$$\Phi\left(\frac{1033-1040}{200/\sqrt{100}}\right) = \Phi(-0.35) = 0.363.$$

This kind of error is called a *Type II error* (*accepting H_0 when it is false*). The probability of committing a Type II error is called β.

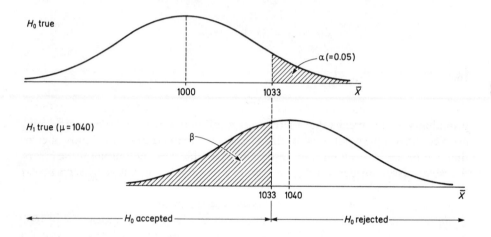

To sum up:

Type I error: rejecting H_0 when it is true

$$P(\text{Type I error}) = \alpha \quad \text{(the significance level of the test)}.$$

Type II error: accepting H_0 when it is false (and H_1 *is true*)

$$P(\text{Type II error}) = \beta.$$

(In our example, if $\mu = 1040$ and the sample mean is 1020, a Type II error would be made.)

Factors affecting β

α, the significance level, is obviously arbitrary, and β *depends on* α (as well as on the true value of μ).

Suppose that, in our example, α is set at 0.01 (1 %) instead of at 0.05 (5 %). The critical value of $\bar{X}$ for accepting or rejecting H_0 will now be

$$1000+2.33 \times 200/\sqrt{100} = 1047.$$

If, then, H_1 is true and $\mu = 1047$,

$$\beta = \Phi\left(\frac{1047-1040}{200/\sqrt{100}}\right) = 0.637.$$

As this example shows, *reduction of α increases β* (other things being equal; see figure (a) below).

And, of course, *increasing α reduces β*.

If, on the other hand, α is kept at 0.05 and the *sample size* is increased to 200, then the critical value of $\bar{X}$ for accepting or rejecting H_0 is

$$1000 + 1.64 \times \frac{200}{\sqrt{200}} = 1023.$$

If H_1 is true and $\mu = 1040$,

$$\beta = \Phi\left(\frac{1023 - 1040}{200/\sqrt{200}}\right) = \Phi(-1.63) = 0.05.$$

In other words, *for a fixed value of α, an increase in the sample size n, decreases β* (see figure (b) below).

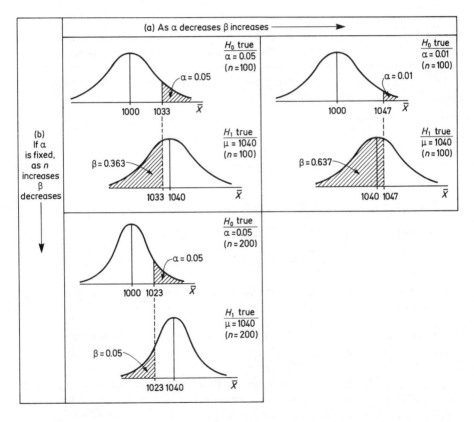

Since β depends on α in the way we have shown, the reduction of α to avoid the risk of a Type I error increases β and therefore the risk of a Type II error. (In our example the reduction of α from 0.05 to 0.01 resulted in increasing β from 0.363 to 0.637, which is an unacceptably large value.)

In practice, the decision as to what to do in a specific case depends on which of the errors is the more important. To help in the decision a new idea – the operating characteristic – is introduced.

16.2   Operating characteristic function of a test

The practical problem of setting up a test so as to take account of β obviously involves a more explicit framing of the alternative hypothesis H_1, allowing for various possible values of μ.

In our example, where the null hypothesis is $H_0: \mu = 1000$, the alternative hypothesis might be

$$H_1: \mu = \text{any value from 1001 to 1070.}$$

For selected values of μ in this range the corresponding values of β are given in the table below, with $\alpha = 0.05$ so that the critical value of $\bar{X}$ for accepting or rejecting H_0 is 1033.

μ	1001	1010	1020	1030	1040	1050	1060	1070
$z = \dfrac{1033 - \mu}{200/\sqrt{100}}$	1.6	1.15	0.65	0.15	-0.35	-0.85	-1.35	-1.85
$\beta = \Phi(z)$	0.945	0.875	0.742	0.560	0.363	0.198	0.089	0.032

The curve obtained by plotting β for various values of μ is shown below. It is called the *operating characteristic* curve.

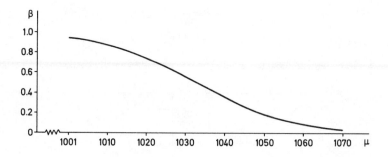

[The operating characteristic function is defined as P(accepting H_0 when it is false, *for a given μ*).]

The curve illustrates the fact that as μ increases the probability of making a Type II error, that is, incorrectly accepting H_0, (or failing to detect that H_1 is true) decreases.

Now suppose, for example, it is important that a difference between the assumed mean (1000) and an actual mean of, say, 1025 should be detected by the test.

From the operating characteristic curve, the value of $\beta = 0.65$ and the probability of failing to detect the difference is clearly too high. In order to reduce it there are two possibilities:

1.   *α could be increased*, to say, 0.10. (As we saw above, as α increases, β decreases.)

This would only be acceptable if the consequent increase in the risk of a Type I error (rejecting H_0 when it is true) mattered less than reducing the risk of a Type II error.

2.   *The sample size, n, could be increased*, which, as we saw results in a decrease in β.

If, for example n is increased from 100 to 400, the critical value $\bar{X}$ for accepting or rejecting H_0 is, for $\alpha = 0.05$,

$$1000 + 1.64 \times 200 / \sqrt{400} = 1016.$$

The value of β for $\mu = 1025$ is now

$$\Phi\left(\frac{1016 - 1025}{200 / \sqrt{400}}\right) = \Phi(-0.86) = 0.195.$$

With a sample size of 400, the probability of detecting a mean of 1025 (or more) is therefore increased.

Alternatively we could lay down the value of β, and say that we want, while keeping α at 0.05, to choose the sample size n so that the probability of failing to detect a mean of 1025 or more is no greater than 0.1 (that is $\beta = 0.1$ when $\mu = 1025$), say.

The critical value of $\bar{X}$ for accepting or rejecting H_0 is given by

$$\Phi\left(\frac{\bar{x} - 1000}{200 / \sqrt{n}}\right) = 0.95$$

and $\beta = \Phi\left(\dfrac{\bar{x} - 1025}{200 / \sqrt{n}}\right) = 0.10.$

Hence $\dfrac{\bar{x} - 1000}{200 / \sqrt{n}} = 1.64$

and $\dfrac{\bar{x} - 1025}{200 / \sqrt{n}} = -1.28.$

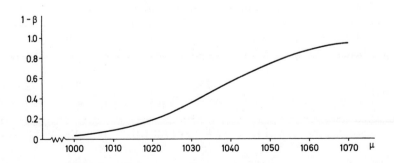

The solution of these equations is $\bar{x} = 1014$ and $n = 366$, so that a sample size of 366 will satisfy the conditions we have set.

16.3   The power of a test

The power function of a test is defined as $1 - P(\text{Type II error}) = 1 - \beta$.
It is therefore the probability of *correctly rejecting H_0* for a given μ. ·

In our example the power curve is:

The power curve, like the operating characteristic, depends on the sample size. If the power curve rises steeply, a small change in μ results in a sharp change in $1 - \beta$. In other words, the power is a measure of how well the test performs.

16.4   Operating characteristic for a two-tail test of a mean

If in our example the hypotheses to be tested were

$$H_0: \mu = 1000 \quad \text{as against} \quad H_1: \mu \neq 1000$$

where $\sigma = 200$, the sample size is 100, and the significance level (α) is set at 0.05 as before, the critical values of $\bar{X}$ for accepting or rejecting H_0 would be

$$1000 \pm 1.96 \times \frac{200}{\sqrt{100}}.$$

In other words H_0 would be rejected if $\bar{x} < 961$ or $\bar{x} > 1039$.

If the mean μ were in fact 1010, say, the probability of making a Type II error (accepting H_0 when it is false) is (see figure)

$$\beta = \Phi\left(\frac{1039 - 1010}{200/\sqrt{100}}\right) - \Phi\left(\frac{961 - 1010}{200/\sqrt{100}}\right) = 0.926 - 0.007 = 0.919.$$

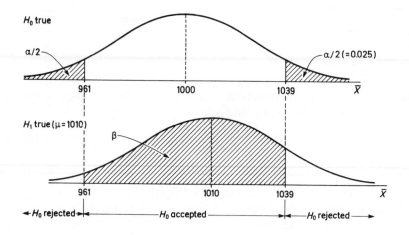

The values of β for different values of μ are set out in the table below:

μ	1070	1060	1050	1040	1030	1020	1010	1001
β	0.061	0.147	0.291	0.480	0.674	0.827	0.919	0.948
μ	999	990	980	970	960	950	940	930
β	0.948	0.919	0.827	0.674	0.480	0.291	0.147	0.061

(The limiting value 0.95 for $\mu = 1000$ is omitted as it is the probability of correctly accepting H_0.)

The operating characteristic for this test is thus:

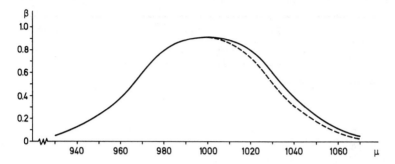

Comparison with the dotted curve, which is the operating characteristic for the one-tail test using $H_1: \mu > 1000$, shows that the one-tail test is better than the two-tail test when $\mu > 1000$, since it gives a smaller probability of a Type II error for a given value of μ.

16.5   Operating characteristic for the test of a proportion

EXAMPLE 1   Ninety per cent of the output of a factory making electronic components satisfy a stringent reliability test. After the installation of a new manufacturing process it is hoped that the proportion has increased. A sample of 50 produced by the new process is tested. Find the value of α if the critical test for rejecting the hypothesis that there has been no improvement is that the number of satisfactory components in the sample exceeds 48. If the true proportion of satisfactory components is now π, calculate the values of β for different values of π and draw the operating characteristic for this test.

The hypotheses to be tested are

$$H_0: \pi = 0.90 \quad \text{as against} \quad H_1: \pi > 0.90.$$

The significance level α is determined by the criterion that H_0 is to be rejected if the number of satisfactory components in the sample X exceeds 48.
Now if H_0 is true, using the binomial distribution,

$$P(X > 48) = 50(0.9)^{49}(0.1) + (0.9)^{50} = (0.9)^{49}(5 + 0.9)$$

$$= 0.034,$$

so that α (the probability that H_0 is rejected, though it is true) $= 0.034$.
However, if H_1 is true and the true proportion is π ($\neq 0.9$)

then $\beta = P(H_0 \text{ is accepted though it is false})$

$$= P(X \leqslant 48)$$

$$= 1 - P(X > 48)$$

$$= 1 - [50\pi^{49}(1 - \pi) + \pi^{50}]$$

$$= 1 - \pi^{49}(50 - 49\pi).$$

This is the operating characteristic function and its value for different values of π is given in the table below.

π	0.92	0.93	0.94	0.96	0.98	0.99	0.995
β	0.917	0.873	0.810	0.600	0.264	0.090	0.026

The operating characteristic curve is:

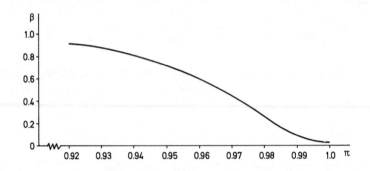

EXERCISE 16a (miscellaneous)

1.   Independent observations $x_1, x_2, \ldots, x_n$ are taken from the normally distributed population $N(\mu, \sigma^2)$ where the value of σ^2 is known to be 4. The null hypothesis value of μ is zero and the only other possible value of μ is 1.

   α denotes the probability that the null hypothesis will be rejected when in fact it is true; β denotes the probability that the null hypothesis will be accepted when in fact it is false.
(a) Suppose $n = 25$ and that it is decided to reject the null hypothesis if the observed sample mean $\bar{x}$ exceeds 0.4. Determine α and β.
(b) Suppose $n = 4$ and that it is required that $\alpha = 0.05$. Determine the constant k where observed values of $\bar{X}$ exceeding k will lead to rejection of the null hypothesis. Also find the value of β.
(c) Suppose it is required that $\alpha = \beta = 0.01$. Determine the necessary sample size n and the constant k where observed values of $\bar{X}$ exceeding k will lead to rejection of the null hypothesis. (MEI)

2.   Two different computer systems, A and B, are being compared in respect of the time it takes to run typical jobs. A convenient unit of measurement has been devised, combining central processor time, input/output time, etc. For both systems it has been found that the variations in run time from job to job are well modelled by normal distributions, whose variances may be taken as 4.2 for system A and 3.6 for system B; the means of these distributions μ_A and μ_B are unknown.

   A random sample of 10 jobs run on system A was found to give an average run time of 8.6 units, and a random sample of 12 jobs run on system B gave an average of 7.4 units.
(a) Test at the 5% level of significance the null hypothesis $H_0: \mu_A = \mu_B$ against the alternative hypothesis $H_1: \mu_A > \mu_B$.
(b) Now suppose that in fact system A is 2 units slower than system B on average, that is $\mu_A - \mu_B = 2$. Find the probability that the test procedure in (a) wrongly leads to the acceptance of H_0.
(c) Discuss very briefly whether you think the probability in (b) is sufficiently small and how it could be made smaller. (MEI)

3. A coin is tossed 10 times. The null hypothesis is that the coin is unbiased, that is, that π (the probability of a head) is $\frac{1}{2}$. This hypothesis is rejected if there are 8, 9 or 10 heads.

Find the value of α (the probability of a Type I error), and of β (the probability of a Type II error) if π is in fact 2/3.

4. Explain what is meant by the terms: Type I error, Type II error.

When a production process is working correctly, only 1 % of the items produced are faulty. From a large batch, a random sample of 200 items is selected and tested. Using a Poisson approximation, determine, correct to three significant figures, the probability that more than 4 items in the sample are faulty.

Occasionally, the production ceases to work correctly, the proportion of faulty items then being 3 %. Devise a test procedure to determine whether the process is working correctly, using a random sample of 200 items and an approximate 5 % significance level. State clearly the hypotheses being tested. State also the conditions specified by your procedure under which the null hypothesis is accepted. Determine, correct to three significant figures the probability associated with the Type II error for your procedure. (C)

5. A bag is assumed to contain 10 red balls and 90 black balls. Four balls are drawn at random, *without replacement*, and the null hypothesis that the assumption is correct is accepted if all four are black.

If the alternative hypothesis is that there are, in fact 30 red balls and 70 black balls, find the probability of making Type I and Type II errors.

6. A coin has probability p of falling heads when tossed. Write down, in terms of p, expressions, which need not be simplified, for: (a) the probability that if the coin is tossed four times all four tosses will fall alike (that is all heads or all tails), (b) the probability that if the coin is tossed seven times six or more of the tosses will fall alike.

The following two procedures have been suggested for testing whether the coin is unbiased.

Procedure 1: Toss the coin four times and conclude that it is biased if all four tosses fall alike.

Procedure 2: Toss the coin several times and conclude that it is biased if six or more of the tosses fall alike.

Show that the two procedures are equally likely to lead to the conclusion that the coin is biased when in fact it is unbiased.

Determine which of the two procedures is the less likely to lead to the conclusion that the coin is unbiased when in fact the probability of a head in any toss is equal to 2/3. (JMB)

7. Explain the use of the Poisson distribution as an approximation to the binomial distribution. Under what conditions is the approximation a good one?

A single sampling plan for sampling inspection is specified with sample size 50 and acceptance number 2. Calculate the probability of the producer rejecting satisfactory batches and the probability of the consumer accepting unsatisfactory batches if the acceptable quality level is 1 % defective and the rejectable quality level is 8 % defective. (MEI)

8. Dating of archaeological specimens is a difficult task. It is known that specimens emit a certain type of radioactive particle; the number of particles emitted in n minutes having a Poisson distribution with parameter $n\lambda$, where the value of λ depends upon the age of the specimen.

Two hypotheses concerning the age of one particular specimen are put forward:

H_A: specimen is 7000 years old (in which case $\lambda = 1.0$)

H_B: specimen is 15 000 years old (in which case $\lambda = 4.0$).

It is decided to count the number, X, of radioactive particles emitted in n minutes and

accept H_A (and reject H_B) if $X \leqslant 1$,

and accept H_B (and reject H_A) if $X \geqslant 2$.

If $n = 1$, what is: (a) the probability of rejecting H_A when H_A is in fact true, (b) the probability of rejecting H_B when H_B is in fact true?

If the probability of rejecting H_B when H_B is in fact true is to be less than 0.001, show that the minimum number of complete minutes for which counting should be recorded is three. What is the corresponding probability of rejecting H_A when H_A is in fact true?   (AEB 1980)

9.   A manufacturer of dice makes fair dice and also slightly biased dice which are to be used for demonstration purposes in the teaching of probability and statistics. The probability distributions of the scores for the two types of die are shown in the table.

Score	1	2	3	4	5	6
Probability for fair dice	1/6	1/6	1/6	1/6	1/6	1/6
Probability for biased dice	1/10	1/10	1/5	1/5	1/5	1/5

Calculate the expectation and variance of the score for each type.

Unfortunately some dice have been made without distinguishing marks to show whether they are fair or biased. The manufacturer decides to test such dice as follows: each die is thrown 100 times and the mean score $\bar{x}$ is calculated; if $\bar{x} > 3.7$, the die is classified as biased but, if $\bar{x} \leqslant 3.7$, it is classified as fair. Find the probability that a fair die is wrongly classified as a result of this procedure.

To improve the test procedure the manufacturer increases the number of throws from 100 to N, where N is chosen to make as close as possible to 0.001 the probability of wrongly classifying a biased die as fair. Find the value of N.   (C)

10.   A machine has been producing metal die-castings with a mean tensile strength of 200 N and standard deviation 20 N. After an adjustment to the machine, a sample of 100 is tested. If a 0.01 level of significance is set, what is the critical value of the mean tensile strength of a sample for accepting or rejecting the hypothesis that there has been no improvement? Find the probability that this hypothesis will be accepted even though the mean tensile strength has in fact risen to 210 N.

For values of the new mean tensile strength ranging from 195 N to 220 N, draw an operating characteristic curve for this test.

11.★ A random variable X has normal distribution with standard deviation 2. In order to test $H_0 : \mu = 30$ as against $H_1 : \mu < 30$, a sample of size n is taken. H_0 is rejected if the sample mean $\bar{X}$ is less than some critical value.
(a) If the significance level is 0.05, find this critical value.
(b) Find the value of n if the significance level is 0.015 and the probability of accepting H_0 when μ is, in fact, 27, is 0.01.
(c) Taking a significance level of 0.05, and sample size 16, with alternative hypotheses $H_0 : \mu = 30$ and $H_1 : \mu \neq 30$, draw the operating characteristic.

12.   A sampling scheme for testing the output of a machine works in the following way. A random sample of 8 items is taken from an hour's output and tested. The output is accepted if there are no defective items. If there are 3 or more defective items the output is rejected. If there are 1 or 2 defective items a second sample of 8 items is taken and if there are fewer than 3 defective items in the *combined* sample of 16 the output is accepted. Otherwise it is rejected.

Show that, if the probability that a randomly chosen item is defective is p, the probability that an hour's output will be accepted is

$$(1-p)^8[1+8p(1-p)^7+92p^2(1-p)^6].$$

Hence sketch the operating characteristic curve for this sampling scheme for $p = 0.1, 0.2, 0.3, 0.4, 0.5$.

13.   A company buys replacement bulbs for slide projectors from a sub-contractor in large batches. Some of these replacement bulbs are faulty; the company has been using a sampling

inspection scheme, under which a random sample of 8 bulbs is taken from each batch, and each of these bulbs tested, the batch being accepted if and only if no defectives are found.

Assuming that each bulb in a batch has, independently, probability p of being faulty, state the distribution of the number of defectives in a sample. Determine the probability of accepting a batch for $p = 0.05, 0.1, 0.15, 0.2, 0.3, 0.4$ and hence draw a rough plot of the operating characteristic for this inspection scheme for values of p between 0.05 and 0.4.

A new sampling inspection scheme is proposed in which a random sample of 15 bulbs from each batch will be tested, the batch being accepted if and only if *at most* 1 defective is found. With the same assumption as before, determine the probability of accepting a batch for each of the above values of p, and draw on the same diagram as before a rough plot of the operating characteristic for this scheme.

Suppose it is considered that the proportion of defectives is unlikely to exceed 10 %; which inspection scheme is to be preferred, and why? Would you still prefer the same scheme if the proportion of defectives was expected to be much higher, say about one-third or more? Justify your answer.   (MEI)

17* Maximum likelihood estimation

17.1 Maximum likelihood estimation for single parameter distributions, discrete and continuous

When 'point estimation' of parameters was defined in chapter 12, the discussion was restricted to the problem of finding estimates which are unbiased, consistent and so on. You may have found this approach rather limiting and might have wondered whether there was a more general method of making estimates of parameters from sample observations that would seem intuitively reasonable. In fact there are several, one of which, the method of maximum likelihood, is the subject of this chapter. Developed by R. A. Fisher in 1930, it has the merit, not only of giving 'reasonable' estimates, but of doing so by an approach which seems entirely natural.

Suppose a random variable X has a Poisson distribution with unknown parameter μ, where X is the number of accidents reported at a police station on any given day, and on four successive days the number of accidents reported are 2, 4, 0 and 3.

In order to estimate μ, the question we ask is this. What value of μ is *more likely than any other* to produce this set of sample observations?

Now if μ is fixed $P(X = 2) = e^{-\mu}\mu^2/2!$. If, on the other hand, μ is not known and the value $X = 2$ turns up we shall call the values of $e^{-\mu}\mu^2/2!$, for different values of μ, *likelihoods*.

Calling the sample values X_1, X_2, X_3 and X_4, it is natural, therefore, to define the likelihood that

$$X_1 = 2, \ X_2 = 4, \ X_3 = 0 \text{ and } X_4 = 3$$

for any particular value of μ as the *joint probability*

$$P(X_1 = 2) \cdot P(X_2 = 4) \cdot P(X_3 = 0) \cdot P(X_4 = 3) = \frac{e^{-\mu}\mu^2}{2!} \times \frac{e^{-\mu}\mu^4}{4!} \times e^{-\mu} \times \frac{e^{-\mu}\mu^3}{3!}.$$

This will be written as $L(2, 4, 0, 3 \,|\, \mu)$, or $L(\mu)$ for short, and we define the maximum likelihood estimate of μ as the value which maximises $L(\mu)$.

Simplifying the expression for the likelihood,

$$L(\mu) = \frac{e^{-4\mu}\mu^9}{2!4!3!}$$

$$\Rightarrow \ln[L(\mu)] = -4\mu + 9\ln\mu - \ln(2!4!3!).$$

Since $\ln L$ increases as L increases, their maxima will coincide, and we therefore differentiate $\ln L$ twice:

$$d(\ln L)/d\mu = -4 + 9/\mu \quad \text{and} \quad d^2(\ln L)/d\mu^2 = -9/\mu^2.$$

Hence the maximum likelihood estimate is given by $\mu = 9/4$, so that $\hat{\mu} = 2.25$.

In this simple example, the value of μ is, in fact, equal to the sample mean. This will not always be the case. The important point to notice is that in using this method, logarithmic differentiation is usually advisable.

Definition of a maximum likelihood estimate

We now define 'likelihood' formally for distributions with a single parameter.

If sample values $x_1, x_2, \ldots, x_n$ are taken from the distribution of a random variable X with parameter θ, the likelihood that $X = x_1$, $X = x_2, \ldots$, $X = x_n$ for a particular value of θ is written as

$$L(x_1, x_2, \ldots, x_n \mid \theta) \quad \text{or} \quad L(\theta) \text{ for short.}$$

For a *discrete* variable,

$$L(\theta) = \text{the joint probability} \quad P(X_1 = x_1) \cdot P(X_2 = x_2) \ldots P(X_n = x_n).$$

The maximum likelihood estimate of θ, $\hat{\theta}$, is the value which maximises $L(\theta)$. As usual, the expression 'maximum likelihood estimator' is used for a general formula, and 'maximum likelihood estimate' for the value in a particular case. Generally the maximum likelihood estimate is found by calculus, but it is always understood to mean the value which gives the greatest likelihood (see example 5).

EXAMPLE 1   The probability of getting a 'head' when a biased coin is tossed is p. X is the number of tosses until a head turns up. In six experiments a head shows 3 times on the first throw, once on the third throw and twice on the fourth throw. Find the maximum likelihood estimate of p.

The frequency distribution of the random variable X is

X	1	2	3	4
f	3	0	1	2

and the distribution of X is geometric. Hence

$$P(X = 1) = p, \qquad P(X = 3) = (1-p)^2 p, \qquad P(X = 4) = (1-p)^3 p.$$

The probability that $X = 1$ three times is therefore p^3, and the probability that $X = 4$ twice is $[(1-p)^3 p]^2$, so that

$$L(p) = p^3 \times [(1-p)^2 p] \times [(1-p)^3 p]^2 = p^6 (1-p)^8.$$
$$\Rightarrow \ln L(p) = 6 \ln p + 8 \ln(1-p).$$

$$d(\ln L)/dp = \frac{6}{p} - \frac{8}{1-p} \quad \text{and} \quad d^2(\ln L)/dp^2 = -\frac{6}{p^2} - \frac{8}{(1-p)^2}.$$

Equating $d(\ln L)/dp$ to zero, gives

$$\frac{6}{p} = \frac{8}{1-p} \Rightarrow p = 3/7,$$

and, since $d^2(\ln L)/dp^2$ is negative, this gives a maximum.
    Hence $\hat{p} = 3/7$.

EXAMPLE 2   The distribution of X, the number of daily landings on a private airfield of aircraft coming from the north, is Poisson with parameter λ. For aircraft coming from the south, Y, the number of daily landings, is Poisson with parameter 1.6λ. On one particular day there are a landings from the north and b landings from the south. Find the maximum likelihood estimator of λ.

    For landings from the north, $P(X = a) = e^{-\lambda}\lambda^a/a!$ and for landings from the south, $P(Y = b) = e^{-1.6\lambda}(1.6\lambda)^b/b!$

    Hence $L(\lambda) = e^{-\lambda}\lambda^a/a! \times e^{-1.6\lambda}(1.6\lambda)^b/b!$

$$\Rightarrow \ln L = -\lambda + a\ln\lambda - 1.6\lambda + b\ln(1.6\lambda) - \ln a! - \ln b!$$

$$= -2.6\lambda + (a+b)\ln\lambda + \text{a constant.}$$

$$d(\ln L)/d\lambda = -2.6 + (a+b)/\lambda \quad \text{and} \quad d^2(\ln L)/d\lambda^2 = -(a+b)/\lambda^2.$$

$$\Rightarrow \hat{\lambda} = 5(a+b)/13.$$

    This approach seems, intuitively, to be one which should give, in some sense, a good estimate. The question that obviously arises is this: How does it compare with point estimation, as described in chapter 12, a method which was developed much earlier by Karl Pearson? The answer is that in some cases, the two methods give the same result, as in examples 1 and 2. In cases where they do not, the maximum likelihood estimator is usually superior. It can be shown that it is both *consistent* and more *efficient* in the senses defined on page 240, than any other estimator if the sample is large. [A third method, 'least squares' will be described in chapter 19.]

EXERCISE 17a

1.   The random variable X has a Poisson distribution. A random sample of n values of X consists of

$$x_1, x_2, \ldots, x_n.$$

Prove that the maximum likelihood estimator of the parameter μ is

$$\bar{x} = \sum_{i=1}^{n} x_i/n.$$

2.   The proportions of 'ordinary' orange-juice and 'low-calorie' orange-juice produced by a manufacturer are $\frac{1}{2} + \frac{1}{2}p$ and $\frac{1}{2} - \frac{1}{2}p$ respectively.
    In a sample of $a+b$, a were ordinary and b were low-calorie. Find the maximum likelihood estimator of p. $(a > b)$.

3.   The distribution of the discrete random variable X is binomial, $B(5, \pi)$, where π is unknown. Write down an expression for the likelihood that $X = 3$, $L(3|\pi)$, and deduce the maximum likelihood estimate of π.

4.   A binomial distribution $B(n, \pi)$ has an unknown parameter, π, the probability of 'success' in a single trial. Find an expression for $L(r \mid \pi)$, the likelihood of r successes in n trials. For fixed values of r and n, find the value of π which maximises $L(r \mid \pi)$.

5.   In the output from a firm making inexpensive pocket calculators, it is believed that a proportion π have some slight defect in the casing. Samples of 5 are taken on four different occasions and the numbers with this defect were found to be 3, 0, 1 and 4. Find the maximum likelihood estimate of π.

6.   The distribution of the random variable X is geometric with parameter p, so that

$$P(X = r) = (1-p)^{r-1} \cdot p.$$

In a series of experiments the following frequency distribution was obtained:

X	1	2	3	> 3
f	a	b	c	0

In other words X was equal to 1 on a occasions, and so on.
   Use this information to find the maximum likelihood estimator of p.

7.   The distribution of the discrete random variable X has probability generating function

$$G(t) = \left(\frac{1-p}{1-pt}\right)^4.$$

A random sample of values of five values of X were:

$$0 \quad 1 \quad 2 \quad 0 \quad 0.$$

Find the maximum likelihood estimate of p.

8.   The discrete random variable X can take any integral value with the following probabilities:

$$P(X = 0) = k, \qquad P(X = r) = k(1-\lambda)^r/r! \quad (r = 1, 2, \ldots).$$

Prove that $k = e^{\lambda - 1}$.
   If a sample of size $a+b+c$ gives $X = 0$ a times, $X = 1$ b times and $X = 2$ c times, find the maximum likelihood estimator of λ.

9.   In a breeding experiment the four different types of offspring are expected to occur in the ratios of

$$p : 2 - p : 1 - p : 1 + p,$$

where p is unknown except that $0 \leqslant p \leqslant 0.5$. The observed numbers of offspring are a, b, c, d respectively.
(a) Show that the overall probability of obtaining these results is given by e^L where

$$L = a \ln p + b \ln(2-p) + c \ln(1-p) + d \ln(1+p) + K$$

and K does not depend on p.
(b) Show that $d^2 L/dp^2$ is negative for all values of p.
(c) Find an equation for p which is satisfied when L has a maximum value.
(d) If $a = 1, b = 23, c = 5, d = 11$ verify that the maximum likelihood estimate of p is 0.1, to one place of decimals.
(e) Improve this estimate by one application of Newton's method.   (O & C)

Maximum likelihood estimation for a continuous distribution

Suppose sample values $x_1, x_2, \ldots, x_n$ are taken from the distribution of the continuous random variable X whose density function is $f(x)$, with parameter θ. Although $P(X = x_r)$ is not defined for a continuous distribution, $f(x)$ is the probability per unit value of X so that it is natural to define the likelihood

$$L(x_1, x_2, \ldots, x_n \mid \theta)$$

as the *joint probability density*

$$L(\theta) = f(x_1) \cdot f(x_2) \ldots f(x_n).$$

EXAMPLE 3   The time to failure of a light bulb is given by the exponential distribution

$$f(t) = \lambda e^{-\lambda t}, \quad t \geqslant 0.$$

A sample of five gives times to failure of t_1, t_2, t_3, t_4, t_5. Estimate λ by the method of maximum likelihood.

$$L(\lambda) = \lambda e^{-\lambda t_1} \cdot \lambda e^{-\lambda t_2} \cdot \lambda e^{-\lambda t_3} \cdot \lambda e^{-\lambda t_4} \cdot \lambda e^{-\lambda t_5}$$

$$= \lambda^5 \cdot e^{-\lambda(t_1 + t_2 + t_3 + t_4 + t_5)}.$$

$$\ln(L) = 5 \ln \lambda - \lambda \sum_{i=1}^{5} t_i \Rightarrow d(\ln L)/d\lambda = 5/\lambda - \sum_{i=1}^{5} t_i$$

and $d^2(\ln L)/d\lambda^2 = -5/\lambda^2$.

Hence $\hat{\lambda} = 5 \left/ \left(\sum_{i=1}^{5} t_i \right) \right. = 1/\bar{t}$.

EXAMPLE 4   Single observations x_1 and x_2 are taken from the distributions of each of two independent random variables X_1 and X_2 which are respectively $N(0, \sigma^2)$ and $N(0, 2\sigma^2)$. Find the maximum likelihood estimator of σ^2.

The density functions of the two distributions are

$$\frac{1}{\sigma \sqrt{(2\pi)}} e^{-\frac{1}{2}x^2/\sigma^2} \quad , \quad \frac{1}{\sqrt{(2)}\sigma \sqrt{(2\pi)}} e^{-\frac{1}{2}x^2/2\sigma^2} \quad \text{(see page 224)}$$

so that $L(x_1, x_2 \mid \sigma) = \dfrac{1}{\sqrt{(2)}\sigma^2 (2\pi)} \cdot e^{-\frac{1}{2}x_1^2/\sigma^2} \times e^{-\frac{1}{2}x_2^2/2\sigma^2}$.

$$\ln(L) = -\ln(2\pi\sqrt{2}) - 2 \ln \sigma - \tfrac{1}{2}(x_1^2/\sigma^2 + x_2^2/2\sigma^2)$$

$$= -2 \ln \sigma - \frac{1}{4\sigma^2}(2x_1^2 + x_2^2) + \text{a constant.}$$

$$d(\ln L)/d\sigma = -\frac{2}{\sigma} + \frac{1}{2\sigma^3}(2x_1^2 + x_2^2).$$

When this is equated to zero, $\sigma^2 = \tfrac{1}{4}(2x_1^2 + x_2^2)$, and the sign of $d^2(\ln L)/d\sigma^2$ is easily shown to be negative.

Hence $\hat{\sigma}^2 = \tfrac{1}{4}(2x_1^2 + x_2^2)$.

EXAMPLE 5★ The random variable X has density function

$$f(x) = \begin{cases} \lambda e^{-\lambda(x-\alpha)}, & x > \alpha \geqslant 0 \\ 0 & , \quad \text{otherwise.} \end{cases}$$

λ is known, but α is not. Find a maximum likelihood estimator for α if a sample of four values of X are x_1, x_2, x_3, x_4.

$$L(\alpha) = \lambda^5 e^{-\lambda \Sigma (x_i - \alpha)} \quad \text{and} \quad \ln L = 5 \ln \lambda - \lambda \Sigma (x_i - \alpha).$$

Now this is linear in α and therefore does not give a maximum by differentiation. We must look for the value of α which gives the greatest value of L by sketching L as a function of α.

To do this we first note that, since $0 \leqslant \alpha < x$, α must be less than or equal to the smallest value of x in the sample, which we will write as $\min(x_1, x_2, x_3, x_4)$.

At $\alpha = 0$, $L = \lambda^5 e^{-\lambda \Sigma x_i} = L_0$, say.

At $\alpha = \min(x_1, x_2, x_3, x_4)$, $L > L_0$.

The sketch can now be completed, and it will be seen that the maximum likelihood estimator of α must be

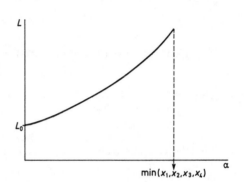

$$\hat{\alpha} = \min(x_1, x_2, x_3, x_4).$$

17.2   Two-parameter distributions

If the distribution has two parameters, we use the fact that a necessary condition for a maximum value of the likelihood is that the partial differential coefficients of L with respect to each parameter must be zero (see page 370). As the next example shows, this leads to two simultaneous equations for the two estimators.

EXAMPLE 6   Find maximum likelihood estimators for the mean and variance of the random variable X, whose distribution is $N(\mu, \sigma^2)$, given sample values $X = x_1$, x_2 and x_3.

The density function of X is $(1/\sigma\sqrt{(2\pi)})e^{-\frac{1}{2}(x-\mu)^2/\sigma^2}$, (see page 224) so that the likelihood

$$L(x_1, x_2, x_3 \mid \mu, \sigma) = \left(\frac{1}{\sigma\sqrt{(2\pi)}}\right)^3 \times \exp\left[-\frac{1}{2\sigma^2} \sum_{i=1}^{3} (x_i - \mu)^2\right]$$

$$\Rightarrow \ln L(\mu, \sigma) = 3\ln[1/\sqrt{(2\pi)}] - 3\ln \sigma - \frac{1}{2\sigma^2} \sum_{i=1}^{3} (x_i - \mu)^2.$$

Hence $\partial(\ln L)/\partial \mu = \dfrac{1}{2\sigma^2}\left[2\sum_{i=1}^{3}(x_i - \mu)\right] = \dfrac{1}{\sigma^2}\sum_{i=1}^{3}(x_i - \mu)$ \hfill (1)

and $\partial(\ln L)/\partial \sigma = -\dfrac{3}{\sigma} + \dfrac{1}{\sigma^3} \times \sum_{i=1}^{3}(x_i - \mu)^2.$ \hfill (2)

Equating (1) to zero gives

$$\sum_{i=1}^{3} (x_i - \mu) = 0 \Rightarrow \sum_{i=1}^{3} x_i = 3\mu \Rightarrow \mu = \sum_{i=1}^{3} x_i/3 = \bar{x}.$$

Substituting $\mu = \bar{x}$ in (2) and equating to zero gives

$$\sigma^2 = \frac{1}{3} \sum_{i=1}^{3} (x_i - \bar{x})^2 = \mathrm{var}(x).$$

Thus $\hat{\mu} = \bar{x}$ and $\hat{\sigma}^2 = \mathrm{var}(x)$.

In this case $\hat{\mu}$ is unbiased, since $E(\bar{X}) = \mu$, but $\hat{\sigma}^2$ is not unbiased, since $E[\mathrm{Var}(X)] = (n-1)\sigma^2/n$ and not σ^2.

EXERCISE 17b

1.   A random variable X has probability density function

$$f(x) = (\lambda + 1)x^{\lambda}, \quad 0 < x < 1.$$

Show that the maximum likelihood estimator of λ, based on sample values $x_1, x_2, \ldots, x_n$ is

$$-1 - n \bigg/ \left(\sum_{i=1}^{n} \ln x_i \right).$$

Find the maximum likelihood estimate if the sample values are

$$0.40 \quad 0.25 \quad 0.80 \quad 0.61 \quad 0.45 \quad 0.53.$$

2.   The random variable X has normal distribution $N(\mu, 1)$. A random sample of observations from this distribution have values x_1, x_2, x_3, x_4 and x_5.

Find the maximum likelihood estimator of the parameter μ.

3. T, the time to failure in hours of an electronic device, has the probability density function

$$f(t) = \begin{cases} \lambda e^{-\lambda(t - t_0)}, & t > t_0 > 0 \\ 0 & , \quad \text{otherwise,} \end{cases}$$

where t_0 is known and λ is unknown.

A number n of these devices are tested and give times to failure of $t_1, t_2, \ldots, t_n$.

Show that the maximum likelihood estimator of λ is $1/(\bar{t} - t_0)$.

EXERCISE 17c (*miscellaneous*)

1.   In respect of a certain inherited characteristic the people of a certain country fall into three types called A, B and C which are expected to occur in the ratio of

$$p^2 : 2p(1-p) : (1-p)^2$$

where p is unknown except that $0 \leqslant p \leqslant 1$. A sample of n people chosen at random contains a people from group A. A special test applied to the remainder identified b people of group B and c people of group C, but the other $(n-a-b-c)$ people were only known to be either group B or group C.

(a) If L is the overall probability of obtaining these results show that

$$\ln L = (2a+b)\ln p + (n-a+c)\ln(1-p) + (n-a-b-c)\ln(1+p) + K$$

where K is independent of p.

(b) Show that when $\ln L$ has a maximum value, p satisfies a quadratic equation, and find it.

(c) Hence find the maximum likelihood estimate of p when $a = 2$, $b = 4$, $c = 5$ and $n = 50$.   (O & C)

2. The random variable X has density function

$$f(x) = \begin{cases} k\lambda x e^{-\lambda x^2}, & 0 \leqslant x < \infty \quad (\lambda > 0) \\ 0 & , \quad \text{otherwise.} \end{cases}$$

Three sample values of X are $x_1 = 3$, $x_2 = 4$, $x_3 = 5$. Find the maximum likelihood estimate of λ.

3. State the conditions for the function

$$f(t) = a_0 + a_1 t + a_2 t^2 + \dots$$

to be a probability generating function.

The random variable R takes all non-negative integer values and its probability generating function is

$$1 + k - \sqrt{(1 - pt)}$$

where p and k are constants independent of t and $0 < p < 1$. Show that $p + k^2 = 1$ and find in terms of p the probability that: (a) $R = 0$, (b) $R = 1$, (c) $R = 2$.

Find also the mean value of R.

In 100 independent trials the values obtained for R were:

R	0	1	2	> 2
Frequency	70	25	5	0

Use the principle of maximum likelihood to estimate the value of p.   (O & C)

4. The random variable X has probability density function

$$f(x) = \begin{cases} \dfrac{3}{4\alpha^3}(\alpha^2 - x^2), & -\alpha \leqslant x \leqslant \alpha \\ 0 & , \quad \text{elsewhere.} \end{cases}$$

A sample of two observations of X are x_1 and x_2. Show that if L is the likelihood of these observations

$$\ln L = K - 6 \ln \alpha + \ln(\alpha^2 - x_1^2) + \ln(\alpha^2 - x_2^2).$$

Find an equation for $\hat{\alpha}$, the maximum likelihood estimate of α, and verify that if $x_1 = 1$ and $x_2 = -1$, $\hat{\alpha} = \sqrt{3}$ satisfies this equation.

5. On three different occasions the numbers of telephone calls a, b, c passing through an exchange in a 5-minute period are recorded. Assuming these are independent observations from three Poisson distributions whose means are μ, $\lambda\mu$, $\lambda^2\mu$ respectively, show that the maximum likelihood estimates for μ and λ satisfy

$$\mu + \lambda\mu + \lambda^2\mu = a + b + c$$
$$\lambda\mu + 2\lambda^2\mu = b + 2c.$$

Find a quadratic equation for λ and hence find both λ and μ when $a = 21$, $b = 30$, $c = 57$.   (O & C)

6. Whenever a large group of people is exposed to an infectious disease, the number of cases is a random variable having a Poisson distribution with unknown mean μ. However, if there are no cases, the presence of the disease passes unnoticed. Show that the probability that an observed outbreak of the disease has r cases is

$$\mu^r / (e^\mu - 1)r!, \quad \text{where } r = 1, 2, \dots.$$

The number of cases observed in 5 outbreaks were 3, 2, 6, 3 and 1. Show that the likelihood of these observations is L where

$$\ln L = 15 \ln \mu - 5 \ln(e^\mu - 1) + K$$

with K independent of μ.

Find an equation for the maximum likelihood estimate of μ in the form

$$e^{-\mu} = A + B\mu,$$

where A and B are constants. By plotting a graph of $e^{-\mu}$ and drawing a suitable straight line on this graph, find an approximate value for the maximum likelihood estimate of μ.

Improve this approximation by a single application of Newton's method.   (O & C)

7.★ The random variable X is uniformly distributed from $x = 0$ to $x = \beta$, so that its density function is

$$f(x) = \begin{cases} 1/\beta, & 0 \leqslant x \leqslant \beta \\ 0, & \text{otherwise.} \end{cases}$$

A sample from the distribution has values $x_1, x_2, \ldots, x_n$. Find the maximum likelihood estimator of β (*Hint*: See example 5).

18 Correlation

18.1 Correlation

Up to now we have been concerned with analysing data associated with a single variable, but many statistical investigations have to do with the relationship between several variables. In this chapter and the next we shall be examining ways of handling data associated with *two* variables.

EXAMPLE 1 The following table shows the annual crime rate Y (measured in numbers per thousand of the population) for 13 towns whose size is X (measured in units of 100 000). 'Crime' is taken to mean serious or indictable crime.

X	0.7	1.1	2.0	2.5	3.3	4.1	4.4	4.5	4.8	5.3	5.8	6.5	7.0
Y	4.6	5.8	4.3	6.4	8.5	7.3	8.5	7.5	8.3	9.0	9.8	10.9	10.5

When these are plotted on a graph (see below), giving what is called a 'scatter diagram', it is apparent that there is a tendency for large values of Y to go with large values of X, and for small values of Y to go with small values of X, and that the

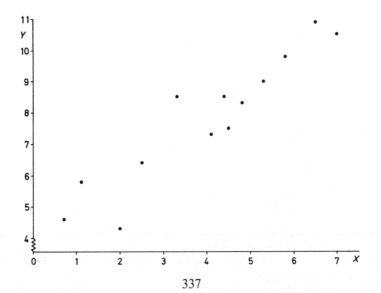

general trend is of a line with a positive gradient. In this sort of example we can say that there is a *linear* relationship between X and Y, though not, of course, an exact one as the points do not lie on a straight line. The problem is to find a way of defining, by some kind of numerical measure, the *degree* to which the values of X and Y are linearly related. In order to make any headway with this problem we must first look at some other possible kinds of scatter diagram.

EXAMPLE 2   In the next table we show the results of a ten-week experiment undertaken by a police authority on a stretch of motorway. The number of weekly accidents Y is recorded for different number of patrol cars X.

X	2	3	4	5	6	7	8	9	10	11
Y	51	36	46	27	26	22	33	16	19	18

The scatter diagram below shows that in this case large values of Y tend to go with small values of X and vice versa, but that the general trend is again linear, with a negative gradient this time.

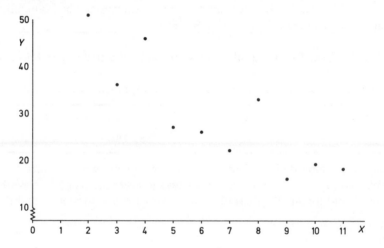

EXAMPLE 3   Suppose now that in a fourteen-week period a record is kept of the number of weekly accidents Y, together with the average age of the cars involved, X, measured in years:

X	0.6	1.0	1.7	2.3	3.0	3.0	4.3	5.1	6.0	6.4	7.1	7.3	8.0	8.2
Y	50	20	34	70	50	31	60	42	68	31	53	23	48	33

In this case the scatter diagram (at the top of page 339) shows no particular trend.

We want, therefore, as a measure of the degree of linear relationship between X and Y (which will be called *correlation*), some quantity (or coefficient) which will be relatively large in examples such as the two scatter diagrams above and small in examples like the third scatter diagram. It will also be convenient if its value is positive in the first and negative in the second so as to distinguish between the two.

This quantity, when we have defined it, will be called the *correlation coefficient*.

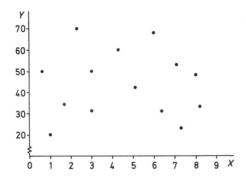

Correlation coefficient

As a first stage we divide each of these scatter diagrams into four quadrants, by lines going through the point representing the mean $(\bar{x}, \bar{y})$ of the point pairs. These are respectively, $(4.0, 7.8)$, $(6.5, 29.4)$ and $(4.57, 43.8)$.

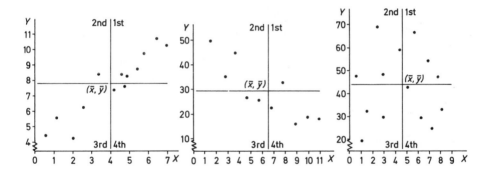

This gives us a simple way of distinguishing between the three cases.

In the first figure most of the points lie in the 1st and 3rd quadrants; in the second figure most of them lie in the 2nd and 4th quadrants; in the third figure the points are more or less evenly scattered between the quadrants.

To make the distinction more precise we now consider the sign of *the deviations from the mean* of each point in the diagrams. These deviations are $(x - \bar{x})$ and $(y - \bar{y})$.

Clearly, in the 1st quadrant both $(x - \bar{x})$ and $(y - \bar{y})$ will be positive; in the 3rd quadrant both will be negative; in the 2nd quadrant $(x - \bar{x})$ will be negative and $(y - \bar{y})$ positive; and in the 4th quadrant $(x - \bar{x})$ will be positive and $(y - \bar{y})$ negative.

This means that the product

$$(x - \bar{x})(y - \bar{y})$$

will be + in quadrants 1 and 3 and − in quadrants 2 and 4.

Looking again at our three scatter diagrams, it can be seen that in examples like the first one, the number of + values of $(x - \bar{x})(y - \bar{y})$ will exceed the number of − values, so that the *sum*

$$\Sigma (x-\bar{x})(y-\bar{y})$$

for all the points will be comparatively large and positive.

In examples like the second, the sum will be comparatively large and negative.

In examples like the third scatter diagram, the $+$ and $-$ values of $(x-\bar{x})(y-\bar{y})$ will tend to cancel out and the sum will be comparatively small.

At first sight, then, it looks as though we might take

$$\Sigma (x-\bar{x})(y-\bar{y})$$

as our measure of correlation. However, the magnitude of this quantity will depend on the units used so that if, for example, the values of X in the first example had been given in units of 10 000 instead of 100 000 each value of X would have been ten times larger. This would result in multiplying $\Sigma (\bar{x}-\bar{x})(y-\bar{y})$ by a factor of ten.

It is for this reason that the correlation is defined as:

$$r = \frac{\Sigma (x-\bar{x})(y-\bar{y})}{\sqrt{[\Sigma (x-\bar{x})^2 \cdot \Sigma (y-\bar{y})^2]}}.$$

This is independent of any change of origin or scale, since the deviations from the means, $(x-\bar{x})$ and $(y-\bar{y})$ will not be affected by a change of origin and a change of scale for X or Y or both will affect the numerator and denominator in the *same* way. For example, dividing each value of X by h and each value of Y by k will result in $\Sigma (x-\bar{x})(y-\bar{y})$ being divided by hk. Similarly $\sqrt{[\Sigma (x-\bar{x})^2 \Sigma (y-y)^2]}$ will be divided by $\sqrt{(h^2k^2)} = hk$, so that the value of r will be unchanged.

This formula for the correlation coefficient of a sample is called the *product-moment correlation coefficient* (p.m.c.c. for short). Using it in this form can be awkward if $\bar{x}$ and $\bar{y}$ are not exact. In practice, therefore, it is often more convenient to use one of several alternative forms. The first of these will introduce a new concept, that of *covariance*.

18.2   Covariance

The variance of x, is defined, as we know, as:

$$\text{var}(x) = \frac{\Sigma (x-\bar{x})^2}{n} \quad \text{with the alternative form} \quad \frac{\Sigma x^2}{n} - \bar{x}^2. \qquad (1)$$

We now define the *covariance* of x and y as:

$$\text{cov}(x,y) = \frac{\Sigma (x-\bar{x})(y-\bar{y})}{n} \quad \text{with the alternative form} \quad \frac{\Sigma xy}{n} - \bar{x} \cdot \bar{y}. \qquad (2)$$

Proof of the alternative form for cov(x, y)

$$\Sigma (x-\bar{x})(y-\bar{y}) = \Sigma xy - \bar{x} \Sigma y - \bar{y} \Sigma x + n\bar{x} \cdot \bar{y}$$

$$\Rightarrow \text{cov}(x,y) = \frac{\Sigma xy}{n} - \bar{x} \cdot \frac{\Sigma y}{n} - \bar{y} \frac{\Sigma x}{n} + \bar{x} \cdot \bar{y} = \frac{\Sigma xy}{n} - \bar{x} \cdot \bar{y} - \bar{y} \cdot \bar{x} + \bar{y} \cdot \bar{x}$$

$$= \frac{\Sigma xy}{n} - \bar{x} \cdot \bar{y}.$$

Notation for variance and covariance

As both variance and covariance play a large part in correlation, it is convenient to have a short way of writing them:

$$\text{var}(x) = \frac{\Sigma\,(x-\bar{x})^2}{n} = \frac{S_{xx}}{n}, \quad \text{where } S_{xx} = \Sigma\,(x-\bar{x})^2 \qquad (3)$$

$$\text{var}(y) = \frac{\Sigma\,(y-\bar{y})^2}{n} = \frac{S_{yy}}{n}, \quad \text{where } S_{yy} = \Sigma\,(y-\bar{y})^2$$

$$\text{cov}(x, y) = \frac{\Sigma\,(x-\bar{x})(y-\bar{y})}{n} = \frac{S_{xy}}{n}, \quad \text{where } S_{xy} = \Sigma\,(x-\bar{x})(y-\bar{y}). \qquad (4)$$

This enables us to write the formula for the product-moment correlation coefficient as

$$r = \frac{S_{xy}}{\sqrt{(S_{xx}S_{yy})}} \quad \left\{ = \frac{\Sigma\,(x-\bar{x})(y-\bar{y})}{\sqrt{[\Sigma\,(x-\bar{x})^2 \cdot \Sigma\,(y-\bar{y})^2]}} \right\}.$$

However, writing r in a short form like this merely helps us remember it. In actually calculating r, we generally use an alternative formula, obtained by substituting the alternative forms for $\text{var}(x)$ and $\text{cov}(x, y)$.

From (1) and (3) $S_{xx} = \Sigma\,x^2 - n\bar{x}^2$.

Similarly $S_{yy} = \Sigma\,y^2 - n\bar{y}^2$.

From (2) and (4) $S_{xy} = \Sigma\,xy - n\bar{x} \cdot \bar{y}$.

Hence:

$$r = \frac{\Sigma\,xy - n\bar{x}\cdot\bar{y}}{\sqrt{[(\Sigma\,x^2 - n\bar{x}^2)(\Sigma\,y^2 - n\bar{y}^2)]}} \quad \text{or} \quad \frac{\Sigma\,xy - [(\Sigma\,x)(\Sigma\,y)]/n}{\sqrt{[(\Sigma\,x^2 - (\Sigma\,x)^2/n)(\Sigma\,y^2 - (\Sigma\,y)^2/n)]}}.$$

Note: To avoid confusion between these different forms, you should stick to just one of them. In the examples which follow we shall use the last one.

EXAMPLE 4   Calculate the product-moment correlation coefficient for the following pairs of values of X and Y:

X	1	2	4	5	7
Y	2	5	4	7	8

The working is set out in a table:

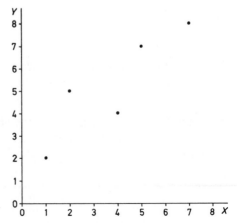

x	y	x^2	y^2	xy
1	2	1	4	2
2	5	4	25	10
4	4	16	16	16
5	7	25	49	35
7	8	49	64	56
19	26	95	158	119

$$r = \frac{\Sigma \, xy - [(\Sigma \, x)(\Sigma \, y)]/n}{\sqrt{[(\Sigma \, x^2 - (\Sigma \, x)^2/n)(\Sigma \, y^2 - (\Sigma \, y)^2/n)]}} = \frac{119 - (19)(26)/5}{\sqrt{[(95 - 19^2/5)(158 - 26^2/5)]}}$$

$= 0.89$ to 2 dec. places.

EXAMPLE 5   Use the data from example 2, illustrated in the scatter diagram, to calculate the p.m.c.c.

X	2	3	4	5	6	7	8	9	10	11
Y	51	36	46	27	26	22	33	16	19	18

Although there is no particular difficulty, with a calculator, in using the figures as they stand, the large numbers involved may be avoided by using coded values u, v, where $U = X - 6$ and $V = Y - 30$. *As we saw on page 340, change of origin and/or scale makes no difference to the value of* r.

x	y	u	v	u^2	v^2	uv
2	51	-4	21	16	441	-84
3	36	-3	6	9	36	-18
4	46	-2	16	4	256	-32
5	27	-1	-3	1	9	3
6	26	0	-4	0	16	0
7	22	1	-8	1	64	-8
8	33	2	3	4	9	6
9	16	3	-14	9	196	-42
10	19	4	-11	16	121	-44
11	18	5	-12	25	144	-60
		5	-6	85	1292	-279

Using these values and $n = 10$ we get

$$r = \frac{\Sigma \, uv - [(\Sigma \, u)(\Sigma \, v)]/n}{\sqrt{[(\Sigma \, u^2 - (\Sigma \, u)^2/n)(\Sigma \, v^2 - (\Sigma v)^2/n)]}}$$

$$= \frac{-279 - 5(-6)/10}{\sqrt{[(85 - 5^2/10)(1292 - (-6)^2/10]}}$$

$= -0.85.$

18.3   Possible values of r

Two questions arise immediately. What range of values can r take, and how is any particular value to be interpreted? The answer to the first of these questions is:

1.   *If the sample points lie exactly on a straight line,* $r = \pm 1$.
    $r = 1$ if the slope of the line is positive. This is called perfect positive correlation.
    $r = -1$ if the slope is· negative. This is called perfect negative (or inverse) correlation.

2.   *For any sample,* $-1 \leqslant r \leqslant 1$.
    Proofs of these statements, which are what we would expect, are given in the Appendix on pages 354–5.

In example 4, r is quite close to 1 and in example 5, r is quite close to −1, so we say, provisionally that there is a reasonably close linear relationship in both cases, describing the first as a fairly high positive correlation and the second as a fairly high negative correlation. If you complete question 7 in the following exercises, which refers to figure 3, you will find that the numerical value of r is small, indicating low correlation. At this stage we cannot say more, since we really need to know whether any particular value of r is close enough, numerically, to 1 to be considered significant. This matter will be touched on at the end of the chapter, and dealt with in greater detail in chapter 20.

18.4   Interpreting *r*

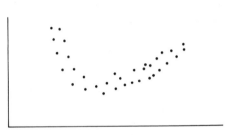

It should be emphasised that r only measures the degree of *linear* relationship between X and Y. A scatter diagram like the one here will give a very low value of r even though there is obviously a marked curvilinear trend.

The other important point to remember is that r is only a *mathematical* measure and that a high value does not imply that there is any *causal* link between the two variables. The possible link between proneness to lung cancer and smoking habits is often quoted in this connection, since the correlation between them is high. It could be that those who, for some other reason, are prone to lung cancer tend also, for the same reason, to need to smoke. The common factor might, for example, be stress. Similarly in our example showing a high correlation between the crime rate and size of town there might be some other factor, such as the relative size of the police force, that produced this apparent link. In other words, the fact that two phenomena go more or less in step with each other will only be a sign that there is a causative link if further investigation, such as seeing whether alteration of one phenomenon results in a similar change in the other, provides further evidence.

EXERCISE 18a

1.   Draw scatter diagrams for each of the following, and calculate the product-moment correlation coefficient.

(a)

X	1	3	4	5	7
Y	2	6	7	6	8

(b)

X	1	2	3	4	4	5	6	7	7
Y	2	5	3	5	2	7	4	3	5

(c)

X	10	12	10	9	11	14
Y	12	11	15	18	14	8

(d)

X	0.3	0.8	0.8	1.1	1.3	1.6	1.8
Y	23	23	17	20	13	17	15

2. Find the product-moment correlation coefficient for the following:

X	−3	−2	−1	0	1	2	3
Y	5	0	−3	−4	−3	0	5

What is the relationship between X and Y?

3.   The average price for each of eight successive months of a certain commodity in dollars in New York (y) and in pounds sterling in London (x) is given in the following table:

y	83	85	88	80	98	95	97	94
x	28	29	29	27	31	32	32	32

Calculate the product-moment correlation coefficient.   (O & C)

4.   The scores of ten soldiers who were given an objective test for general muscular coordination are given in the table below, together with their respective scores in a shooting test:

Muscular coordination	87	85	83	77	72	69	60	54	52	50
Shooting	75	94	80	92	61	73	78	70	62	74

Calculate the product-moment correlation coefficient.   (C)

5.   A sample of broad beans was examined, and for each bean its length and mass were measured and recorded as follows:

Mass (g)	0.7	1.2	0.9	1.4	1.2	1.1	1.0	0.9	1.0	0.8
Length (cm)	1.7	2.2	2.0	2.3	2.4	2.2	2.0	1.9	2.1	1.6

Calculate the product-moment correlation coefficient.   (O)

6.   Calculate the p.m.c.c. for the data in example 1, page 337.

7.   Calculate the p.m.c.c. for the data in example 3, page 338.

18.5   Calculating *r* from a grouped frequency table

If the data are given in grouped frequency form, the formulae for r given above apply, with $\Sigma\,x$, $\Sigma\,xy$, etc. replaced by $\Sigma\,fx$, $\Sigma\,fxy$ etc., and n replaced by $\Sigma\,f$.

EXAMPLE 6 The diagram opposite shows the relationship between X and Y in grouped frequency form; for example, when the mid-interval value of X is 25 there are 4 values of Y with mid-interval value of 45.

Calculate the product-moment correlation coefficient of this sample.

					X	
	x⧵y	5	15	25	35	45
	65				3	1
	55			2	2	
Y	45		2	4	2	
	35		3			
	25	1				

The figures are coded by letting $U = (X-25)/10$ and $V = (Y-45)/10$. The figures appear in the diagram below, together with the method of working.

In each 'cell' the top left-hand circled figure is the value of uv.

The bottom right-hand circled figure is fuv, obtained by multiplying uv by the frequency, f, in the centre of the cell.

The rest of the working is self-explanatory.

value of $uv = 1 \times 2$ — value of $fuv = 3 \times 2$

u⧵v	-2	-1	0	1	2	f	fv	fv²	fuv
2				② 3 ⑥	④ 1 ④	4	8	16	10(=6+4)
1			⓪ 2 ⓪	① 2 ②		4	4	4	2(= 0+2)
0		⓪ 2 ⓪	⓪ 4 ⓪	⓪ 2 ⓪		8	0	0	0
-1		① 3 ③				3	-3	3	3
-2	④ 1 ④					1	-2	4	4
f	1	5	6	7	1	Σf = 20	Σ fv =7	Σ fv²= 27	Σ fuv = 19
fu	-2	-5	0	7	2	Σ fu = 2			
fu²	4	5	0	7	4	Σ fu²= 20			
fuv	4	3	0	8	4	Σ fuv = 19			

Substituting in the formula

$$\frac{\Sigma\, fuv - [(\Sigma\, fu)(\Sigma\, fv)]/\Sigma\, f}{\sqrt{[(\Sigma\, fu^2 - (\Sigma\, fu)^2/\Sigma\, f)(\Sigma\, fv^2 - (\Sigma\, fv)^2/\Sigma\, f)]}}$$

$$r = \frac{19 - 2 \times 7/20}{\sqrt{(20 - 2^2/20)(27 - 7^2/20)}} = 0.83 \quad \text{to 2 dec. places.}$$

EXERCISE 18b

1. Calculate the product-moment correlation coefficient of X and Y from the following frequency distribution:

Y \ X	1	2	3	4
1	1			
2	2	4	3	
3		2	5	2
4				1

2. Calculate the product-moment correlation coefficient of X and Y from the given frequency table

		X						
		−3	−2	−1	0	1	2	3
	2	3	5	2	3			
	1	8	6	3	3	9		
Y	0		2	8	2	9	6	
	−1			7	1	9	9	4
	−2				8	4	6	5

(X denotes values of x which have been coded with origin 100 in intervals of 5 units. Y denotes values of y which have been coded with origin 50 in intervals of 10.)

Calculate also the correlation coefficient of x and y.

3. One hundred pupils take examination papers in Pure Mathematics and Applied Mathematics. The table below shows the results, where the values of X and Y are the mid-interval marks for Pure and Applied Mathematics respectively. For example, $X = 54.5$ stands for the interval 50–59. Calculate the product moments correlation coefficient.

		X					
		34.5	44.5	54.5	64.5	74.5	84.5
	34.5	3	5	4	—	—	—
	44.5	3	6	6	2	—	—
Y	54.5	—	5	9	5	2	—
	64.5	—	—	5	10	8	1
	74.5	—	—	—	5	6	5
	84.5	—	—	—	2	4	4

18.6 Rank correlation coefficient

Suppose two judges in a music competition are assessing the relative merits of nine violinists and, without alloting marks to each contestant, place them in the following order:

	A	B	C	D	E	F	G	H	J
Judge 1 (X)	1	2	3	4	5	6	7	8	9
Judge 2 (Y)	3	1	7	2	8	5	4	6	9

In order to measure the extent to which the judges are in agreement, it is natural to calculate the correlation coefficient between these two sets of *ranks*. Now when there are no tied places, the product-moment correlation coefficient takes a conveniently simple form. We will quote the result and then prove it.

If $d = x - y$ for each pair of ranks, and $n =$ the number of pairs, then

$$r_s = 1 - \frac{6 \, \Sigma \, d^2}{n(n^2 - 1)}.$$

To distinguish this formula from the p.m.c.c., it is called the *Spearman rank correlation coefficient* (after its author) and is written r_s. It can be seen immediately that the smaller the differences between the ranks the larger the value of r_s will be.

Proof

We shall use the form $r = S_{xy}/\sqrt{(S_{xx} S_{yy})}$, where $S_{xy} = \Sigma \, xy - n\bar{x}\bar{y}$, $S_{xx} = \Sigma \, x^2 - n\bar{x}^2$, $S_{yy} = \Sigma \, y^2 - n\bar{y}^2$.

Suppose there are n pairs of ranks, then each x_i and each y_i is one of the integers $1, 2, \ldots, n$. Hence

$$\Sigma \, x_i = n(n+1)/2 \Rightarrow \bar{x} = (n+1)/2.$$

Similarly $\bar{y} = (n+1)/2$.

$$\Sigma \, x_i^2 = n(n+1)(2n+1)/6 \Rightarrow S_{xx} = [n(n+1)(2n+1)/6] - [n(n+1)^2/4]$$

$$= n(n+1)(n-1)/12.$$

Hence

$$S_{xx} = S_{yy} = n(n^2 - 1)/12.$$

Let

$$d_i = x_i - y_i$$

Then

$$\Sigma \, d_i^2 = \Sigma \, x_i^2 + \Sigma \, y_i^2 - 2 \, \Sigma \, x_i y_i$$

$$= \frac{1}{3} n(n+1)(2n+1) - 2 \, \Sigma \, x_i y_i$$

so that

$$\Sigma \, x_i y_i = \frac{1}{6} n(n+1)(2n+1) - \frac{1}{2} \Sigma \, d_i^2$$

$$\Rightarrow S_{xy} = \frac{1}{6} n(n+1)(2n+1) - \frac{1}{2} \Sigma \, d_i^2 - \frac{1}{4} n(n+1)^2$$

$$= \frac{n(n^2 - 1)}{12} - \frac{1}{2} \Sigma \, d^2.$$

Hence

$$r = \frac{[n(n^2 - 1)/12] - \frac{1}{2} \Sigma\, d^2}{n(n^2 - 1)/12} = 1 - \frac{6 \Sigma\, d_i^2}{n(n^2 - 1)}.$$

Applying this result to our example:

x	y	$d = x - y$	d^2
1	3	-2	4
2	1	1	1
3	7	-4	16
4	2	2	4
5	8	-3	9
6	5	1	1
7	4	3	9
8	6	2	4
9	9	0	0
		$\Sigma\, d = 0$	$\Sigma\, d^2 = 48$

Hence $r_s = 1 - \dfrac{6 \times 48}{9(81 - 1)}$

$= 0.6.$

(It is advisable to check that $\Sigma\, d = 0$.)

You may have noticed that the derivation of this simplified formula depended on the ranks being integers. If there *are* tied ranks, this method may still be used provided there are not more than one or two of these. The procedure is to allocate average ranks to each of the tied ones in the following way:

Rank	1	2	3	3	5	6	6	6	9
Allotted rank	1	2	3.5	3.5	5	7	7	7	9

since the ranks available for the two 3 = places are 3 and 4, and the ranks available for the 6 = places are 6, 7 and 8.

(Should there be more than one or two tied ranks, the only alternative is to use the formula for r, the p.m.c.c., with these ranks as actual 'scores'.)

In this example the only data available were ranks. It may happen that when the actual values of X and Y are given, the values may be less important than the order in which they occur, in which case Spearman's rank order coefficient may be calculated after the values have been ranked.

EXAMPLE 7   Eight students each took two papers in an examination, one in physics and one in practical physics. Their percentages in the papers were:

	A	B	C	D	E	F	G	H
Physics (X)	67	73	50	75	42	44	75	40
Practical physics (Y)	82	70	64	78	52	64	80	64

Calculate Spearman's rank correlation coefficient.

Ranking these results gives:

x	4	3	5	1 = (1.5)	7	6	1 = (1.5)	8
y	1	4	5 = (6)	3	8	5 = (6)	2	5 = (6)

Rank for X x	Rank for Y y	d	d^2
4	1	3	9
3	4	−1	1
5	6	−1	1
1.5	3	−1.5	2.25
7	8	−1	1
6	6	0	0
1.5	2	−0.5	0.25
8	6	2	4
		$\Sigma d = 0$	$\Sigma d^2 = 18.5$

Hence:

$$r_s = 1 - \frac{6 \times 18.5}{8 \times 63}$$

$$= 0.78.$$

It should be noted that r_s can only be taken as an approximation to r if there is a high degree of correlation (see question 3 in exercise 18c).

EXERCISE 18c

1. In an international competition for pianists two judges A and B, placed the eight competitors in order of merit for (a) technical ability, (b) general musicianship, that is, sensitivity, interpretation of the music etc. The orders were as follows:

	Judge								
Technical ability	A	1	2	3	4	5	6	7	8
	B	2	3	1	7	4	5	6	8
General musicianship	A	1	2	3	4	5	6	7	8
	B	5	6	4	1	7	2	8	3

Calculate a coefficient of rank correlation for each of these sets of figures and comment on your results.   (C)

2. Nine men who regularly completed a daily newspaper's crossword puzzle agreed to keep a record of their average time for completing the puzzle over a period of a month. The table shows these average times T, to the nearest half-minute, set alongside the number of years N that each man had been a regular solver.

N (years)	1	2	4	5	6	9	10	12	14
T (mins)	8.5	10	4.5	12	7	9	8	6	4

Calculate a coefficient of rank correlation.   (C)

3. The examination marks for a group of ten students in a pure mathematics and a statistics paper are as shown.

Pure mathematics	89	73	57	53	51	49	47	44	42	38
Statistics	51	53	49	50	48	21	46	19	40	43

Find the product-moment correlation coefficient for the two sets of marks.

Place the marks in order of class position and calculate Spearman's rank correlation coefficient for the two papers.

The following is a quotation from a statistics text book. 'Rank correlation can be used to give a quick approximation to the product-moment correlation coefficient.' Comment on this in the light of your results.   (JMB)

4. Calculate Spearman's rank correlation coefficient for the data in example 5, on page 342.

5. In an investigation into the quality of twelve packages of bacon from a supermarket, the percentage of additives, X, was compared with the quality, given as a rank order, Y, assessed by a professional food taster. The results were:

X	6.8	7.6	8.4	10.9	10.9	11.8	13.3	16.5	17.0	18.2	18.2	18.8
Y	9	11	7	5	1	2	3	4	6	12	8	10

Calculate the rank correlation coefficient between X and Y.

18.7★ Testing the significance of r

In example (5) on page 342, where X was the number of police patrol cars on a motorway and Y the number of weekly accidents in ten successive weeks, we obtained a value of $r = -0.85$. On the face of it, this seems to be quite close to -1, but is it in fact significant? Suppose it had turned out to be -0.76, would this value still be close enough to imply correlation? To answer questions like this we need some way of *testing* whether or not a given value of r is significant.

Now the figures in example (5) were taken from a single *sample*. If the figures were recorded over a number of ten-weekly periods, with varying numbers of patrol cars the result would be a number of different values of r. These could be thought of as giving estimates of a 'true' correlation coefficient, the value for the underlying *population*. In the usual way we will refer to this by the Greek letter ρ, so that

$$\rho = \text{the population correlation coefficient.}$$

The broad question is: what does a given sample value of r tell us about ρ?

Commonsense will suggest that if $r \approx 0$ then ρ is probably zero, and that if r is close to 1 numerically then ρ is likely to be near 1; but we need something more precise than this, in other words a significance test.

Testing whether $\rho = 0$

For this test the assumption has to be made that X and Y are jointly normally distributed, which means that for any particular value of X, Y is normally distributed and vice versa. (More is said on this point in chapter 20).

The test statistic used is $r\sqrt{[(n-2)/(1-r^2)]}$, where n is the size of the sample and r is the sample p.m.c.c.

If $\rho = 0$, *the distribution of this statistic is 't' with $n-2$ degrees of freedom.* (The derivation of this distribution is discussed in chapter 20).

Assuming this result, we shall apply it to our example, where $r = -0.85$ and $n = 10$, taking as hypotheses

$$H_0 : \rho = 0 \quad \text{as against} \quad H_1 : \rho < 0.$$

The test is a one-tailed test as it is unlikely that there is a *positive* correlation between the number of police cars and the number of accidents on the motorway.

The value of $r\sqrt{[(n-2)/(1-r^2)]}$ is $-0.85\sqrt{\{(10-2)/[1-(-0.85)^2]\}} = -4.56$.

For a 5% one-tail test the critical $P = 10$ value of t when $v = 8$ is -1.86. ($P = 10$ because the t tables are calculated for two-tail tests.)

Since -4.56 is inside the critical area we conclude that H_0 should be rejected and that there *is* correlation.

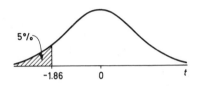

EXAMPLE 8   The correlation coefficient calculated from a sample of six observations is 0.62. Is this significantly different from 0?

Taking as hypotheses

$$H_0 : \rho = 0 \quad \text{as against} \quad H_1 : \rho \neq 0$$

the value of the test statistic is $0.62 \times \sqrt{\{(6-2)/[1-(0.62)^2]\}} = 1.58$.

The critical 5% ($P = 5$) value of t for $v = 4$ is 2.78, so that H_0 is accepted at the 5% level and we conclude that the value of r is not significantly different from zero.

Note: If we want to test the significance of Spearman's rank correlation coefficient, the same test statistic may be used when n is greater than 10, since the distribution of

$$r_s\sqrt{[(n-2)/(1-r_s^2)]}$$

is then approximately 't' with $n-2$ degrees of freedom.

Testing the significance of r when $\rho \neq 0$

When $\rho \neq 0$ another, more complicated, test statistic has to be used. It will be described later in chapter 20.

EXERCISE 18d★

1.   For a sample of 20 pairs of values of (X, Y) the product-moment correlation coefficient is calculated to be 0.83. Is this significantly different from zero? Use a 5% significance level.

2.   If $r = 0.58$ and $n = 8$, use a 5% significance level to test the null hypothesis $H_0 : \rho = 0$ against: (a) the alternative hypothesis $H_1 : \rho \neq 0$, (b) the alternative hypothesis $H_1 : \rho > 0$.

3.   Use the data of example 1 on page 337 to test the significance of the sample p.m.c.c., using a 5% level (a one-tail test). [See exercise 18(a), question 6.]

4.   Use the data of example 3 on page 338 to test the significance of the sample p.m.c.c., using a 5% level. [See exercise 18(a), question 7.]

5.   Test whether the rank correlation coefficient calculated in example 6 on page 345 is significantly different from 0 using: (a) a 5 % level, (b) a 1 % level.

EXERCISE 18e (miscellaneous)

1.   (a) The marks of eight candidates in English and mathematics are:

Candidate	1	2	3	4	5	6	7	8
English (x)	50	58	35	86	76	43	40	60
Mathematics (y)	65	72	54	82	32	74	40	53

Rank the results and hence find a rank correlation coefficient between the two sets of marks.

(b) Using the data in part (a), obtain the product-moment correlation coefficient.   (SUJB)

2.   Calculate the coefficient of correlation between the weight of the heart and the weight of the liver in mice, using the following data. What do you conclude from this analysis?

	Weight of organs (in units of 0.01 g)							
Heart	20	16	20	21	26	24	18	18
Liver	230	126	203	241	159	230	140	242

(MEI)

3.   Sketch scatter diagrams for which:
(a) the product-moment correlation coefficient is -1,
(b) Spearman's correlation coefficient is $+1$, but the product-moment correlation is *less than 1*.
    Five independent observations of the random variables X and Y were:

X	0	1	4	3	2
Y	11	8	5	4	7

Find:
(c) the sample product-moment correlation coefficient,
(d) Spearman's correlation coefficient.   (O & C)

4.   Calculate the product-moment correlation coefficient for X and Y from the following frequency distribution table:

Y \ X	1	2	3	4	5
1				1	3
2			2	4	2
3		5	1	3	
4		3			
5	4	2			

5. A sample of n pairs of values (X, Y) is observed and a correlation coefficient r is calculated.
(a) What does r indicate?
(b) What limitations are there on the values of r?
   Sketch scatter diagrams to illustrate:
(c) high positive correlation;
(d) high negative correlation;
(e) low correlation when X and Y are unrelated;
(f) low correlation when X and Y are related.
   Three historians A, B and C were interviewed for two posts of examiner and were given 12 essays to put into merit order according to criteria laid down. Their results were as follows:

Essay	1	2	3	4	5	6	7	8	9	10	11	12
A's order	5	8	4	11	10	9	7	3	2	12	1	6
B's order	7	4	1	9	8	12	11	6	3	10	2	5
C's order	4	11	2	8	10	12	6	1	5	9	3	7

By calculating suitable correlation coefficients, decide which pair show greatest agreement. (SUJB)

6. The Government-financed Industrial Research Unit selected twelve factories at random from all those engaged in heavy engineering in this country. They were similar in all respects except size of workforce. Each factory was asked to indicate the percentage of employees absent from work for at least one whole day during a particular week. The results are tabulated below.

Factory	A	B	C	D	E	F	G	H	I	J	K	L
Size of workforce (x) (thousands)	1.1	1.9	3.0	4.2	5.1	5.8	7.0	8.3	9.3	10.0	10.9	12.1
Percentage of absentees (y)	5	7	6	6	8	9	11	12	14	20	43	60

(a) Calculate the product-moment correlation coefficient for these data.
(b) Calculate Spearman's rank correlation coefficient. (AEB 1982)

7. Each of the variables x and y takes the values $1, 2, \ldots, n$ but not necessarily in the same order as each other. Prove that the covariance of x and y is

$$\operatorname{cov}(x, y) = \frac{n^2 - 1}{12} - \frac{1}{2n} \sum_{i=1}^{n} (x_i - y_i)^2.$$

Hence show that Spearman's rank correlation between x and y may be written as

$$1 - \frac{6}{n(n^2 - 1)} \sum_{i=1}^{n} (x_i - y_i)^2.$$

Seven army recruits $(A, B, \ldots, G)$ were given two separate aptitude tests. Their orders of merit in each test were

Order of merit	1st	2nd	3rd	4th	5th	6th	7th
1st test	G	F	A	D	B	C	E
2nd test	D	F	E	B	G	C	A

Find Spearman's coefficient of rank correlation between the two orders and comment briefly on the correlation obtained. (O & C)

8.   The following table gives the distribution of 3000 persons according to age and highest audible pitch in 1000 c.p.s.

	Number of persons					
Age (years)	5.5–	20.5–	35.5–	50.5–	65.5–	
Highest audible pitch (1000 cps)						Total
5–	5	9	11	19	10	54
11–	81	153	153	75	11	473
17–	1249	860	141	7		2257
23–	161	46	1			208
29–	6	2				8
Totals	1502	1070	306	101	21	3000

Calculate the coefficient of correlation between highest audible pitch and age. (MEI)

9.★ The pairs of values (x_1, y_1), $(x_2, y_2), \ldots, (x_n, y_n)$ have a product-moment correlation coefficient r and the sample standard deviations for the n x-values and for the n y-values are S_x and S_y respectively. Show that the product-moment correlation coefficient R between the quantities $ax + by + c$ and ky, where a, b, c, k are positive constants, is given by

$$R = \frac{arS_x + bS_y}{\sqrt{(a^2 S_x^2 + 2abrS_x S_y + b^2 S_y^2)}}.$$

By considering R^2 or otherwise, show that when $0 \leqslant r < 1$ then $r < R < 1$.
State the value of R for each of the three cases:
(a) $r = 1$,   (b) $r = -1, aS_x < bS_y$,   (c) $r = -1, aS_x > bS_y$. (JMB)

18.8   Appendix

Properties of r
1.   If the sample points lie on a straight line, $r = \pm 1$.

Proof

Suppose all the points lie on $y = a + bx$.
    Then $\Sigma y_i = na + b \Sigma x_i \Rightarrow \bar{y} = a + b\bar{x}$, that is $(\bar{x}, \bar{y})$ lies on the line, which can therefore be written

$$y - \bar{y} = b(x - \bar{x})$$

$$\Rightarrow y_i - \bar{y} = b(x_i - \bar{x}), \quad i = 1, 2, \ldots, n$$

$$\Rightarrow x_i y_i - x_i \bar{y} = bx_i^2 - bx_i \bar{x} \tag{1}$$

and

$$y_i^2 - y_i \bar{y} = bx_i y_i - by_i \bar{x}. \tag{2}$$

From (1), $\quad\quad \Sigma\ x_i\,y_i - n\bar{x}\bar{y} = b\ \Sigma x_i^2 - bn\bar{x}^2 \Rightarrow S_{xy} = bS_{xx}$.

From (2), $\quad\quad \Sigma\ y_i^2 - n\bar{y}^2 = b\ \Sigma\ x_i\,y_i - bn\bar{y}\bar{x} \Rightarrow S_{yy} = bS_{xy}$

Eliminating b, $S_{xy}^2/S_{xx}\,S_{yy} = 1 \Rightarrow S_{xy}/\sqrt{(S_{xx}\,S_{yy})} = \pm 1$.

If $b > 0$, S_{xy} is positive and $r = 1$ (perfect positive correlation).

If $b < 0$, S_{xy} is negative and $r = -1$ (perfect negative or inverse correlation).

2. *For any sample $r^2 \leqslant 1$.*

Proof

The expression $\Sigma\ [(x_i - \bar{x}) + \lambda(y_i - \bar{y})]^2 \geqslant 0$ for all values of λ.

$$\Rightarrow \lambda^2\ \Sigma\ (y_i - \bar{y})^2 + 2\lambda\ \Sigma\ (x_i - \bar{x})(y_i - \bar{y}) + \Sigma\ (x_i - \bar{x})^2 \geqslant 0 \text{ for all } \lambda.$$

For this to be true for this quadratic function of λ

$$\Sigma\ (x_i - \bar{x})(y_i - \bar{y}) - \Sigma\ (x_i - \bar{x})^2\ \Sigma\ (y_i - \bar{y})^2 \leqslant 0$$
$$\Rightarrow r^2 \leqslant 1.$$

19 Linear regression

In the last chapter we started with a scatter diagram of points (x, y), where X measured crime rate and Y the size of city, and found a method of calculating a numerical measure of the way in which X and Y were related. This measure was called the correlation coefficient r.

This coefficient r is defined in such a way that it measures the degree to which X and Y are *linearly* related, so that when r is nearly 1, X and Y are nearly linearly related, and when $r = \pm 1$, the points (x, y) lie on a straight line.

We now consider a different problem, in which it is known that the relationship between X and Y *is* linear, so that the data from an experiment or survey, when plotted on a graph do lie approximately on a straight line. The problem is to fit the best possible line to the observations, not roughly by eye on the graph, but by some reasonable method of calculation.

Such a line is called a *line of closest fit*.

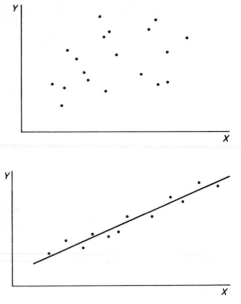

19.1 Line of closest fit

Let us look at an experiment in which one variable, X, is controlled, and the other, Y, is dependent on X.

Suppose four similar plots of land with uniform fertility are treated with predetermined amounts of fertiliser and the crop yield from each plot is recorded. Since the amount of fertiliser is controlled, it will be taken as the variable X; crop yield is dependent on X and will be called Y. (In this initial example we shall work with only four pairs of values for simplicity.)

$X = $ *amount of fertiliser* (g/m^2)	30	60	90	120
$Y = $ *crop yield* (tonnes/hectare)	4	5	9	10

Assuming that there is good reason to expect that Y and X are linearly related, the problem is to find the equation of the line

$$y = a + bx$$

that fits these points best.

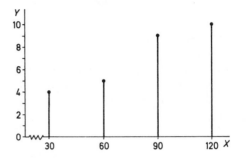

We shall call the points

$$(x_1, y_1), (x_2, y_2), (x_3, y_3), (x_4, y_4)$$

and the diagram shows one of these points (x_1, y_1), together with the 'true' line we are looking for.

Q_1 is where the point P_1 (x_1, y_1) would be if there were no experimental error, and we shall suppose that $P_1 Q_1 = d_1$, the vertical deviation of P_1 from the expected line.

Since Q_1 is on $y = a + bx$, its y-coordinate will be $a + bx_1$ so that

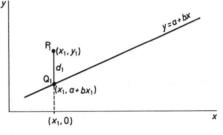

$$d_1 = y_1 - (a + bx_1).$$

The two diagrams opposite show the way the deviations for all four points might look in relation to the expected line. Our problem is to arrange for these deviations, taken together in some way, to be as small as possible.

Now we are only concerned with the absolute value of the deviations (regardless of sign) so that it seems reasonable to choose a and b so that

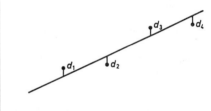

$$d_1^2 + d_2^2 + d_3^2 + d_4^2$$

is minimised.

(The vertical deviations are chosen, because, for any particular value of X, which is controlled, it is the difference between the observed and expected values of Y that matters.)

Calling this sum of squared deviations R, we are faced with finding values of a and b such that

$$R = [y_1 - (a + bx_1)]^2 + [(y_2 - (a + bx_2)]^2 + [(y_3 - (a + bx_3)]^2 + [(y_4 - (a + bx_4)]^2$$

is a minimum.

Now if there were n points, the expression for this 'sum of squares' could be written

$$R = \sum_{i=1}^{n} [y_i - (a + bx_i)]^2.$$

which is a function of *two* variables, a and b, the values of x and y being, of course numerical.

How is the minimum value of R to be found? It can be done by using calculus, but this involves partial differentiation, which you may not have met in your pure mathematics course. This method is given in the Appendix on page 370. Here we shall adopt a purely algebraic method, which looks formidable but is not in fact difficult.

Finding a and b to minimise R

Consider one of these vertical deviations

$$y_1 - (a + bx_1).$$

Adding $\bar{y} - \bar{y} \, (= 0)$ and $b\bar{x} - b\bar{x} \, (= 0)$, where $\bar{x}$ is the mean of the x's and $\bar{y}$ is the mean of the y's, and rearranging, it can be written as

$$(y_1 - \bar{y}) - b(x_1 - \bar{x}) + (\bar{y} - a - b\bar{x}).$$

If we write $y_1 - \bar{y} = Y_1$, $x_1 - \bar{x} = X_1$, and $\bar{y} - (a + b\bar{x}) = m$, then

$$y_1 - (a + bx_1) = Y_1 - bX_1 + m.$$

Hence $R = \displaystyle\sum_{i=1}^{n} (Y_i - bX_i + m)^2$

$$= \sum (Y_i^2 + b^2 X_i^2 + m^2 - 2bX_i \, Y_i - 2bmX_i + 2mY_i)$$

$$= \sum Y^2 + b^2 \sum X^2 + nm^2 - 2b \sum XY - 2bm \sum X + 2m \sum Y$$

(dropping the suffix i).

Now $\sum X = \sum Y = 0$, by definition of $\bar{x}$ and $\bar{y}$, so that R simplifies to

$$R = \sum Y^2 + b^2 \sum X^2 + nm^2 - 2b \sum XY.$$

Moreover

$$\sum X^2 = \sum (x - \bar{x})^2 = S_{xx}, \qquad \sum Y^2 = \sum (y - \bar{y})^2 = S_{yy}$$

and

$$\sum XY = \sum (x - \bar{x})(y - \bar{y}) = S_{xy},$$

as defined in the previous chapter.

Hence $R = S_{yy} + b^2 S_{xx} + nm^2 - 2bS_{xy}$

$$= S_{yy} + nm^2 + S_{xx}[b^2 - 2b(S_{xy}/S_{xx})].$$

Completing the square of the expression inside the bracket,

$$R = S_{yy} + nm^2 + S_{xx}\{[b - (S_{xy}/S_{xx})]^2 - (S_{xy}/S_{xx})^2\}$$
$$= S_{yy} + nm^2 + S_{xx}[b - (S_{xy}/S_{xx})]^2 - S_{xy}^2/S_{xx}$$
$$= S_{yy} + n[\bar{y} - (a + b\bar{x})]^2 + S_{xx}[b - (S_{xy}/S_{xx})]^2 - S_{xy}^2/S_{xx}.$$

Now when a and b vary, R will be least when each of the two brackets is zero, in other words when

$$\bar{y} - (a + b\bar{x}) = 0 \quad \text{and} \quad b - S_{xy}/S_{xx} = 0.$$

From these two equations we get the values of a and b we need:

$$b = S_{xy}/S_{xx} \quad \text{and} \quad a = \bar{y} - b\bar{x},$$

where, as in the last chapter,

$$S_{xy} = \Sigma (x - \bar{x})(y - \bar{y}) = \Sigma xy - n\bar{x}\bar{y}$$
$$S_{xx} = \Sigma (x - \bar{x})^2 = \Sigma x^2 - n\bar{x}^2.$$

Applying these results to the example we started with on page 356:

x	y	x^2	xy
30	4	900	120
60	5	3600	300
90	9	8100	810
120	10	14400	1200
300	28	27000	2430

$n = 4$

$\bar{x} = 300/4 = 75; \qquad \bar{y} = 28/4 = 7.$

$$S_{xx} = \Sigma x^2 - n\bar{x}^2 = 27000 - (4 \times 75^2) = 4500$$
$$S_{xy} = \Sigma xy - n\bar{x}\bar{y} = 2430 - (4 \times 75 \times 7) = 330.$$

$$\Rightarrow b = \frac{S_{xy}}{S_{xx}} = \frac{330}{4500} = 0.0733$$

and

$$a = \bar{y} - b\bar{x} = 7 - (0.0733 \times 75) = 1.50.$$

Hence the line of closest fit is

$$y = 1.50 + 0.073x$$

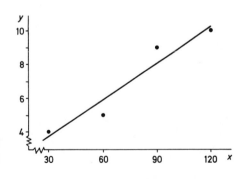

The graph shows the original four points together with the line of closest fit.

Note: Since $\bar{y} = a + b\bar{x}$, the line of closest fit $y = a + bx$ goes through the mean point $(\bar{x}, \bar{y})$, which is a result that would be expected. Also since it goes through $(\bar{x}, \bar{y})$ and has gradient b its equation may be written as

$$y - \bar{y} = b(x - \bar{x}).$$

In practice, it is generally best to use the form $y = a + bx$, calculating b by one of the alternative forms:

$$b = \frac{\Sigma\ xy - n\bar{x}\cdot\bar{y}}{\Sigma\ x^2 - n\bar{x}^2} = \frac{\Sigma\ xy - (\Sigma\ x)(\Sigma\ y)/n}{\Sigma\ x^2 - (\Sigma\ x)^2/n} = \frac{S_{xy}}{S_{xx}},$$

and calculating a by the relation

$$\bar{y} = a + b\bar{x} \Rightarrow a = \bar{y} - b\bar{x}.$$

The fact that there is more than one formula for b in common use can be confusing, and you will probably find it easier, as with the correlation coefficient, to get used to one only and stick to it. For the rest of this chapter we shall use the second form above.

The line of closest fit, when X is the controlled variable and Y is the dependent variable is called *the line of regression of y on x*, and the method we have just outlined is called *the method of least squares*. It seems, on the face of it, to be reasonable, and it will be found later (in the next chapter) that an entirely different approach leads to exactly the same result.

EXAMPLE 1   Find the regression line of y on x for the following eight point pairs:

$$(1,0)\quad(2,1)\quad(3,2)\quad(4,2)\quad(5,4)\quad(6,4)\quad(7,5)\quad(8,5)$$

The values of $\Sigma\ x, \Sigma\ y, \Sigma\ x^2$ and $\Sigma\ xy$ are shown in the following table:

x	y	x^2	xy
1	0	1	0
2	1	4	2
3	2	9	6
4	2	16	8
5	4	25	20
6	4	36	24
7	5	49	35
8	5	64	40
36	23	204	135

Using the formula

$$b = \frac{\Sigma\ xy - (\Sigma\ x)(\Sigma\ y)/n}{\Sigma\ x^2 - (\Sigma\ x)^2/n}$$

with $n = 8$, we get

$$b = \frac{135 - (36)(23)/8}{204 - (36)^2/8} = 0.75.$$

$$\bar{x} = 36/8 = 4.5$$

$$\bar{y} = 23/8 = 2.875$$

so that $a = \bar{y} - b\bar{x}$

$$= 2.875 - (0.75 \times 4.5)$$

$$= -0.5.$$

The line of closest fit or line of regression of y on x is therefore

$$y = -0.5 + 0.75x.$$

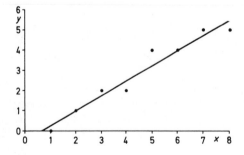

EXERCISE 19a

Find the regression line of y on x for each of the following sets of point pairs. Plot the points on a graph, together with the line that you have calculated.

1. $(1,2)\quad(2,5)\quad(3,4)\quad(4,7)\quad(5,8)\quad(6,11)$.

2. (2, 12) (2, 10) (4, 8) (5, 6) (6, 5) (6, 4) (8, 2).

3. (0.5, 10) (1, 15) (2, 15) (2.5, 25) (3, 30) (3.5, 30) (4, 35) (5, 40).

4. (−3, 0.15) (−2, 0.25) (−1, 0.28) (0, 0.36) (1, 0.45) (2, 0.46) (3, 0.54).

19.2   Use of coding in finding regression lines

The effect of coding is as follows:

1.   Subtraction of an arbitrary number from x or y or both makes no difference to S_{xy} and S_{xx}, since each of the deviations $(x - \bar{x})$ and $(y - \bar{y})$ is unaffected. [The same is true, of course, for addition.]

2.   If the scale is altered, so that, for instance; $U = (X - c)/h$ and $V = (Y - d)/k$, then

$$S_{uv} = S_{xy}/hk, \qquad S_{uu} = S_{xx}/h^2 \Rightarrow \frac{S_{xy}}{S_{xx}} = \frac{hk \cdot S_{uv}}{h^2 \cdot S_{uu}}.$$

The following example should make the method clear.

EXAMPLE 2   A wire is heated to various predetermined temperatures and its resistance measured at each temperature. The results are shown in the following table.

Temperature X (Kelvin)	180	200	220	240	260	280
Resistance Y (ohms × 10⁻³)	2.2	2.4	3.0	3.2	3.5	4.1

Calculate the regression line of y on x, and use it to estimate the resistance for a temperature of 290 Kelvin.

Using the coding $u = (X - 220)/20$, $v = Y - 2.0$, the table is:

u	v	u^2	uv
−2	0.2	4	−0.4
−1	0.4	1	−0.4
0	1.0	0	0
1	1.2	1	1.2
2	1.5	4	3.0
3	2.1	9	6.3
3	6.4	19	9.7

We shall use

$$b = \frac{\Sigma\, xy - (\Sigma\, x)(\Sigma\, y)/n}{\Sigma\, x^2 - (\Sigma\, x)^2/n}$$

and the corresponding formula for u and v,

$$\frac{S_{uv}}{S_{uu}} = \frac{\Sigma\, uv - (\Sigma\, u)(\Sigma\, v)/n}{\Sigma\, u^2 - (\Sigma\, u)^2/n}.$$

With $n = 6$, $\dfrac{S_{uv}}{S_{uu}} = \dfrac{9.7 - (3)(6.4)/6}{19 - (3)^2/6} = 0.3714.$

But the scale for X was reduced by a factor of 20, so that

$$b = \frac{S_{xy}}{S_{xx}} = \frac{20 \times 1 \times S_{uv}}{20^2 \times S_{uu}} = \frac{0.3714}{20} = 0.019.$$

Using $a = \bar{y} - b\bar{x}$, where $\bar{x} = 20\bar{u} + 220 = \left(20 \times \dfrac{3}{6}\right) + 220 = 230$,   and

$$\bar{y} = \bar{v} + 2.0 = \dfrac{6.4}{6} + 2.0 = 3.07,$$

$$a = 3.07 - (0.019 \times 230) = -1.30.$$

The regression line is therefore

$$y = -1.30 + 0.019x.$$

Putting $x = 290$ in this equation gives an estimated value for y of 4.2×10^{-3} ohms.

EXERCISE 19b

1. In the table, y is the mass (in grams) of potassium bromide which will dissolve in 100 grams of water at a temperature of $x\ °C$.

x	10	20	30	40	50
y	61	64	70	73	78

Find the equation of the regression line of y on x.   (SUJB)

2.   The table shows corresponding experimental values of two related variables, x and y.

x	15	20	25	30	35	40
y	17	19	24	34	33	41

Draw a scatter diagram and calculate the line of best fit. Add this line to your diagram.
Estimate the value of y when $x = 45$.

3.   Calculate the equation of the regression line of y on x for the following sample values:

x	25	30	35	40	45	50
y	58	50	45	38	28	22

4.   At a certain location, the percentage of sand in soil at different depths is given by the following table:

x (depth in mm)	25	35	45	55	65
y (% of sand)	92	86	84	79	70

Find the equation of the line of regression of y on x for these data. Use your equation to estimate the value of y at the surface and comment on your result.   (C)

5.   In the following table, W grams is the mass of a certain chemical substance which dissolved in water at $T\ °C$:

T	10	20	30	40	50	60	70	80	90
W	45	46	50	56	59	63	64	67	74

Calculate the equation of the line of regression of W on T. Use this equation to obtain a tentative value of W for $T = 56$.   (MEI)

6.   The following data show, in convenient units, the yield (y) of a chemical reaction at various different temperatures (x):

Temperature (x)	110	120	130	140	150	160	170
Yield (y)	2.1	2.3	3.1	3.4	2.9	3.5	3.3

Plot the data, and, assuming that a linear regression model is appropriate, estimate the regression line of yield on temperature.

Plot your estimated line on your graph, and indicate clearly on your graph the distances, the sum of whose squares is minimised by the linear regression procedure.   (MEI)

7.   Find the regression line of y on x for the following point pairs:

$$(130, 31.5) \quad (125, 31.4) \quad (120, 31.35) \quad (115, 31.26) \quad (110, 31.15).$$

8.   Excavations at two sites, one in Africa and the other in Europe, have revealed the skeletons of prehistoric men. The lengths, x and y (in suitable units), of two named bones from each of six of these skeletons are given below.

Source	x	y
Europe	181	78
Europe	175	64
Europe	178	71
Africa	163	70
Africa	166	73
Africa	160	67

Plot the numerical data on a scatter diagram and calculate the regression line of y on x.

An incomplete seventh skeleton is unearthed at the European site. The value of x for this skeleton is 180. Making suitable assumptions, which should be stated, give an estimate of the value of y for this skeleton based on the figures in the table above.   (C)

19.3   Correlation and regression

In our discussion of regression lines, one variable, X, was taken to be independent, with Y dependent on it. Suppose, however, that we have a sample of values of (X, Y) and there is thought to be some degree of correlation between them but there is no particular reason to regard Y as being dependent on X, rather than X being dependent on Y. In that case it will be possible to calculate *two* regression lines, one, regarding X as the independent variable,

$$y - \bar{y} = \frac{S_{xy}}{S_{xx}}(x - \bar{x}) \quad \text{the regression line of } y \text{ on } x,$$

the other, regarding Y as the independent variable,

$$x - \bar{x} = \frac{S_{xy}}{S_{yy}}(y - \bar{y}) \quad \text{the regression line of } x \text{ on } y.$$

The first would be used for predicting or interpolating y for a given value of x, and the second for predicting or interpolating x for a given value of y (see example 3 below).

These lines are, in general, different. If their gradients are m_1 and m_2 respectively,

$$m_1 = \frac{S_{xy}}{S_{xx}} \quad \text{and} \quad m_2 = \frac{S_{yy}}{S_{xy}}$$

so that $m_1/m_2 = \dfrac{(S_{xy})^2}{S_{xx}S_{yy}} = r^2$.

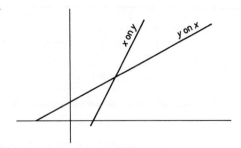

Thus, if $r^2 = 1$, the two lines coincide, and the larger the numerical value of r, the nearer the lines approach coincidence and the nearer the points are to having a linear trend.

The correlation coefficient r was, of course, defined as a measure of *linear* relationship. It does not follow, however, that if r is small there is no relationship of a non-linear kind. Consider, for example, the following points:

$$(-2,4) \quad (-1,1) \quad (0,0) \quad (1,1) \quad (2,4).$$

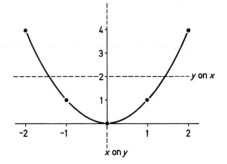

Here $r = 0$ and the two regression lines are $y = 2$ and $x = 0$, but the points are related by the *quadratic* relationship $y = x^2$.

In general, regression is a more useful concept than correlation. While the calculation of r tells us something, knowledge of the regression line tells us more. Apart from its use in prediction or interpolation it leads, as will be seen later, to a method of assessing the *significance* of r.

EXAMPLE 3   The table gives the scores in mathematics and English when a sample of eight students were tested in these subjects.

Mathematics (x)	70	48	26	46	40	62	50	58
English (y)	55	29	25	38	50	38	34	51

Calculate the lines of regression of y on x and x on y. Hence estimate the probable English score for a student whose mathematics score in this test was 80, and the probable mathematics score for a student who scored 20 in English.

Using the coding $U = X - 40$ and $V = Y - 30$ leads to the table:

u	v	u^2	v^2	uv
30	25	900	625	750
8	-1	64	1	-8
-14	-5	196	25	70
6	8	36	64	48
0	20	0	400	0
22	8	484	64	176
10	4	100	16	40
18	21	324	441	378
80	80	2104	1636	1454

$$\bar{u} = 80/8 = 10 \Rightarrow \bar{x} = \bar{u} + 40 = 50 \quad \text{and} \quad \bar{v} = 80/8 = 10 \Rightarrow \bar{y} = \bar{v} + 30 = 40.$$

$$\frac{S_{uv}}{S_{uu}} = \frac{\Sigma\,(uv) - (\Sigma\,u)\,(\Sigma\,v)/n}{\Sigma\,u^2 - (\Sigma\,u)^2/n} = \frac{1454 - (80)\,(80)/8}{2104 - (80)^2/8} = 0.50.$$

$$\Rightarrow \frac{S_{xy}}{S_{xx}} = 0.50, \quad \text{since the scales are unaltered.}$$

Hence the line of regression of y on x,

$$y - \bar{y} = \frac{S_{xy}}{S_{xx}}(x - \bar{x}) \quad \text{is} \quad y - 40 = 0.50(x - 50)$$

$$\Rightarrow y = 15 + 0.50x.$$

Similarly,

$$\frac{S_{uv}}{S_{vv}} = \frac{\Sigma\,(uv) - (\Sigma\,u)\,(\Sigma\,v)/n}{\Sigma\,v^2 - (\Sigma\,v)^2/n} = \frac{1454 - (80)\,(80)/8}{1636 - (80)^2/8} = 0.78$$

$$\Rightarrow \frac{S_{xy}}{S_{yy}} = 0.78,$$

and the line of regression of x on y,

$$x - \bar{x} = \frac{S_{xy}}{S_{yy}}(y - \bar{y}) \quad \text{is} \quad x - 50 = 0.78(y - 40)$$

$$\Rightarrow x = 18.8 + 0.78y.$$

Using the first equation, when $x = 80$ the estimated value of y is 55.
Using the second equation, when $y = 20$ the estimated value of x is $34.4 \approx 34$.

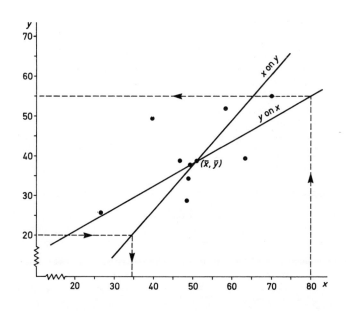

In general, extrapolation outside the range of the sample should be used with caution as the relationship between X and Y may well not be linear beyond a certain point.

EXERCISE 19c

1.   Determine, by calculation, the equation of the regression line of x on y based on the following information about 8 children:

Child	1	2	3	4	5	6	7	8
Arithmetic mark (x)	45	33	27	23	18	14	8	0
English mark (y)	31	33	18	20	19	9	13	1

(SUJB)

2.   For the following values of x and y, calculate:
(a) the regression line of y on x,
(b) the regression line of x on y.

x	55	57	52	58	61	59	57	56	60	58	54	53
y	48	48	46	51	50	48	47	45	48	49	45	46

Estimate the value of y when $x = 68$ and the value of x when $y = 40$.

3.   In two tests, one an objective 'intelligence' test and the other a test in computation given to eight pupils, the scores were as follows:

Objective test (x)	115	116	113	116	115	117	112	116
Computation test (y)	71	67	77	72	72	77	70	78

Use the regression lines of y on x and x on y to estimate the probable score in the computation test for a pupil who scored 100 in the objective test and the probable score in the objective test for a pupil whose score in the computation test was 85.

EXERCISE 19d (miscellaneous)

This set of exercises includes examples on both regression and correlation.

1.   In each of seven successive weeks, the number, N, of road accidents in a town and the number, P, of police cars on patrol are recorded. From the data in the table, obtain the equation of the line of regression of N on P.
     Calculate the product-moment correlation coefficient between N and P, and comment on the value you find.

P	2	5	3	3	6	4	5
N	40	31	42	40	33	36	30

(L)

2.   The results of an experiment to determine how the percentage sand content of soil y varies with depth in centimetres below ground level x are given in the following table.

x	0	6	12	18	24	30	36	42	48
y	80.6	63.0	64.3	62.5	57.5	59.2	40.8	46.9	37.6

Calculate:

(a) the covariance of x and y;

(b) the product-moment correlation coefficient of x and y;

(c) the equation of the line of regression of y on x.

Explain briefly why the product-moment correlation coefficient is preferable to the covariance as a measure of the association between x and y. (MEI)

3. Drums of hair shampoo are kept in storage for some time before being rebottled for retail sale. During storage, evaporation of part of the water content takes place and an examination of several drums gave the following results:

Drum number	1	2	3	4	5	6
Storage time (weeks)	2	5	7	9	12	13
Evaporation loss (ml)	38	57	65	73	84	91

Calculate the product-moment correlation coefficient, and hence comment on the appropriateness of using a linear regression model for predicting evaporation loss.

Find the line of regression of evaporation loss on storage time, and estimate the evaporation loss for a drum kept in storage for eleven weeks.

Why would you not expect to get good estimates from the line of regression for evaporation when the storage time is very long? In such a case, suggest, with reasons, whether the line of regression would underestimate or overestimate the evaporation loss. (MEI)

4. The following table gives the number of road vehicles x (unit 10^6) in use in September of each year and the number of road casualties y (unit 10^5) in each year:

Year	1952	1953	1954	1955	1956	1957	1958	1959
x	4.9	5.3	5.8	6.4	6.9	7.5	7.9	8.6
y	2.1	2.3	2.4	2.7	2.7	2.7	3.0	3.3

Find: (a) the line of regression of y on x,

(b) the correlation coefficient between x and y.

(O & C)

5. The following data relate the average maximum water level Y in metres at Cairo to the epochs X. Fit a line of regression of Y on X and forecast the mean level in the year 2000 AD. Compare this with a forecast based on the last four readings only and discuss the difference.

X mean date (AD)	750	950	1050	1150	1250	1350	1450	1600	1700	1800	1900
Y average max. level (m)	0.87	0.88	0.89	0.91	0.91	0.93	0.99	1.11	1.12	1.14	1.17

(The actual data have been modified slightly.) (MEI)

6. The 1973 and 1980 catalogue prices in pence of five British postage stamps are as follows:

1973 *price*	50	45	65	25	15
1980 *price*	500	350	600	500	120

(a) Plot these results on a scatter diagram.

(b) Write down the coordinates of one point through which the regression line of the 1980 price on the 1973 price must pass.

(c) Denoting the 1980 price by y and the 1973 price by x, calculate the regression line of y on x in the form $y = a+bx$. Draw this line on your graph. Determine the value of y when x has the value 20.   (C)

7.   In a regression calculation for five pairs of observations, one pair of values was lost when the data were filed. For the regression line of y on x the equation was calculated as

$$y = 2x - 0.1.$$

The recorded pairs of values are:

x	0.1	0.2	0.4	0.3
y	0.1	0.3	0.7	0.4

Find the missing pair of values, using the following date for the four pairs above:

$$\Sigma\, x = 1 \quad \Sigma\, x^2 = 0.3 \quad \Sigma\, xy = 0.47 \quad \Sigma\, y = 1.5. \tag{O \& C}$$

8.   You are investigating the relationship, which is approximately linear, between time and temperature in the course of a chemical reaction, and wish to estimate the time when a temperature of 50 °C should be reached. In order to avoid constant monitoring you decide to record the temperature at constant time intervals of 5 minutes and to fit a straight line to these observations. A preliminary experiment has suggested that the time should be between 80 and 90 minutes. Assuming you wish to make at most about 6 readings, which of the following procedures do you think better, and why?
(a) Start measuring at 60 minutes and stop when 50° is reached.
(b) Start measuring at 70 minutes and continue to 100 minutes.
   The following data arise from such an experiment.

Time (min)	70	75	80	85	90	95	100
Temperature (°C)	43.2	45.7	47.8	50.4	52.3	54.9	56.7

State whether it would be more appropriate to fit a regression line of time on temperature or of temperature on time, and justify your choice. What value does the line of your choice, fitted by the least squares procedure, give for the time at which 50 °C is reached?   (O)

9.   The following table gives data relating to tests on ten specimens of brass.

Specimen number	1	2	3	4	5	6	7	8	9	10
Hardness H (rockwell units)	57	49	59	45	54	51	49	57	46	55
Tensile strength (newtons)	76	69	83	64	74	73	66	79	65	80

Given that $\Sigma\, (H-45) = 72,$ $\Sigma\, (H-45)^2 = 734,$

$\Sigma\, (T-64) = 89,$ $\Sigma\, (T-64)^2 = 1197,$

and $\Sigma\, (H-45)(T-64) = 923,$

calculate the equation of the line of regression of H on T.
   A further specimen of brass has a tensile strength 75 N. What do you predict about its hardness?   (MEI)

10.   The following table gives the wheat yield (x) in millions of tonnes against the area planted (y) in millions of hectares for ten areas of the USSR in 1980:

Area	1	2	3	4	5	6	7	8	9	10
x	7.0	9.8	11.6	17.5	7.6	8.2	12.4	17.5	9.5	19.5
y	1.2	2.1	3.4	6.1	1.3	1.7	3.4	6.2	2.1	7.1

(a) Calculate the equation of the line of regression of y on x.
(b) Calculate the product-moment correlation coefficient.

11. The following table gives the ages and weekly wages of a sample of office workers in a large insurance office in 1983:

Age (x)	27	30	37	38	32	36	32
Wage (y) $(£)$	118	136	156	150	140	155	157
Age (x)	32	38	42	36	44	33	38
Wage (y) $(£)$	114	144	159	149	170	131	160

(a) Calculate the equation of the line of regression of y on x, and draw this line on a scatter diagram, showing the original data.
(b) Calculate the product-moment correlation coefficient.

12. In a certain state in the USA, radioactive wastes from a plutonium plant leaked into a river flowing into the ocean. For each of 9 counties bounded by the river or the ocean (or both), an index x of exposure to radioactivity was calculated. The table shows the values of x for each of these counties together with the corresponding values of y, the mortality from cancer per 100 000 person–years, for the period from 1959–1964 inclusive.

County	Index of exposure: x	Cancer mortality: y
1	8.34	210.3
2	6.41	177.9
3	3.41	129.9
4	3.83	162.3
5	2.57	130.1
6	11.64	207.5
7	1.25	113.5
8	2.49	147.1
9	1.62	137.5

Calculate the mean of x and y.
Plot the data on a scatter diagram.
By performing an appropriate calculation, which should be justified, estimate the cancer mortality per 100 000 person–years that would have been observed in a county with an index of exposure of 7.00.   (MEI)

13. To test the effect of a new drug, twelve patients were examined before the drug was administered and given an initial score (I) depending on the severity of various symptoms. After taking the drug they were examined again and given a final score (F). A decrease in score represented an improvement. The scores for the twelve patients are given in the table.

Patient	1	2	3	4	5	6	7	8	9	10	11	12
Score (I)	60	28	8	14	42	34	32	31	41	25	20	50
Score (F)	49	12	3	4	28	27	20	20	34	15	16	40

Calculate the equation of the line of regression of F on I.

On the average what improvement would you expect for a patient whose initial score was 30?   (MEI)

14.★ Corresponding values of the variables X (known exactly) and Y (subject to random variation) are shown in the table:

x	−3	−2	−1	0	0	1	1	2	3	3	4	4	4
y	10	11	11	12	13	12	14	13	15	18	16	19	23

Plot y against x on a graph and explain why the conditions for valid use of the regression line for y on x seem unlikely to be fulfilled. Assuming that an equation of the form $y = A(B^x)$ is a valid deduction from the data, estimate the values of A and B to three significant figures.   (C)

19.4   Appendix

In order to use calculus to find the values of a and b which give the minimum value of R, we need to give a short account of the way maxima and minima of functions of *two* variables are found.

Short account of finding maxima and minima of a function of two variables

For a function of a single variable $f(x)$ a necessary condition for a maximum or minimum at a point is that

$$\frac{df}{dx} = 0 \quad \text{at that point.}$$

If a function of two variables x and y is to have a maximum or minimum at a point (x_0, y_0) necessary conditions are that

$$\frac{\partial f}{\partial x} = 0 \quad and \quad \frac{\partial f}{\partial y} = 0 \quad \text{at } (x_0, y_0),$$

where $\partial f/\partial x$ is the result of differentiating the function with respect to x, *regarding y as a constant*, and $\partial f/\partial y$ is the result of differentiating the function with respect to y, *regarding x as a constant*.

Let us look at some simple examples.

1.   If $z = x^3 + 2xy + y^3$
$$\frac{\partial z}{\partial x} = 3x^2 + 2y + 0$$

$$\frac{\partial z}{\partial y} = 0 + 2x + 3y^2$$

2. If $z = (2x+y)^3$ $\qquad \dfrac{\partial z}{\partial x} = 3(2x+y)^2 \times 2$

$\qquad\qquad\qquad\qquad\quad \dfrac{\partial z}{\partial y} = 3(2x+y)^2 \times 1$

3. If $z = (5-a-3b)^2$ $\qquad \dfrac{\partial z}{\partial a} = 2(5-a-3b) \times (-1)$

$\qquad\qquad\qquad\qquad\qquad \dfrac{\partial z}{\partial b} = 2(5-a-3b) \times (-3).$

We now apply this idea to minimising R.

Minimising R

To minimise $R = \Sigma \, [y_i - (a+bx_i)]^2$, we use the necessary conditions $\partial R/\partial a = 0$, $\partial R/\partial b = 0$.

$$\frac{\partial R}{\partial a} = 2 \sum_{i=1}^{n} [y_i - (a+bx_i)] (-1).$$

Equating this to zero and dividing by -2 gives

$$\Sigma \, [y_i - (a+bx_i)] = 0$$
$$\Rightarrow \Sigma \, y_i - na - b \, \Sigma \, x_i = 0$$
$$\Rightarrow \Sigma \, y_i = na + b \, \Sigma \, x_i \qquad\qquad (1)$$

$$\frac{\partial R}{\partial b} = 2 \, \Sigma \, [y_i - (a+bx_i)] (-x_i)$$
$$= -2 \, \Sigma \, (x_i \, y_i - ax_i - bx_i^2).$$

Equating this to zero and dividing by -2 gives

$$\Sigma \, x_i \, y_i - a \, \Sigma \, x_i - b \, \Sigma \, x_i^2 = 0$$
$$\Rightarrow \Sigma \, x_i \, y_i = a \, \Sigma \, x_i + b \, \Sigma \, x_i^2. \qquad\qquad (2)$$

Equations (1) and (2) are called the *normal equations*.

Dividing (1) by n gives

$$\bar{y} = a + b\bar{x}.$$

To find b, we eliminate a from (1) and (2).

$(1) \times \Sigma \, x \Rightarrow \Sigma \, x \, \Sigma \, y = na \, \Sigma \, x + b(\Sigma \, x)^2$

$(2) \times n \Rightarrow n \, \Sigma \, (xy) = na \, \Sigma \, x + bn \, \Sigma \, x^2.$

Subtracting the first equation from the second,

$$n \sum xy - \sum x \sum y = b[n \sum x^2 - (\sum x)^2]$$

$$\Rightarrow b = \frac{n \sum xy - \sum x \sum y}{n \sum x^2 - (\sum x)^2} = \frac{\sum xy - n\bar{x}\bar{y}}{\sum x^2 - n\bar{x}^2} \quad \text{(dividing by } n\text{)}$$

Hence $b = \dfrac{S_{xy}}{S_{xx}}.$

20* Models for regression lines and correlation coefficients

20.1 The model for regression lines

Our treatment of regression in the last chapter only involved the calculation of lines of closest fit for random *samples* of points. We must now turn our attention to the *populations* from which the pairs of variables, X and Y, may be taken to come. In other words, we need a population model, so that we may, as with parameters of a single variable, test hypotheses and find confidence intervals. In particular, this will lead to the very important matter of testing the significance of a sample correlation coefficient that was touched on briefly on page 350.

We start by looking at an experiment in which wire is heated to various predetermined temperatures and the resistance is measured at each temperature. Temperature, X, is the controlled variable and resistance Y is the dependent variable. The results of a single experiment of this kind are:

X (°Kelvin)	x_1	x_2	x_3	x_4	x_5	x_6
	180	200	220	240	260	280
Y (ohms $\times 10^{-3}$)	1.9	2.5	2.9	3.4	3.6	4.1

If the experiment is repeated five times, using wire of the same kind for each run, we obtain a range of five values of Y for each value of X, resulting from measurement errors and other random factors. Each of these ranges should be regarded as coming from a distribution whose nature we shall consider shortly.

	x_1	x_2	x_3	x_4	x_5	x_6
X	180	200	220	240	260	280
Y	2.5	2.9	3.2	3.5	3.8	4.3
	2.3	2.8	3.0	3.4	3.6	4.1
	2.2	2.5	2.9	3.2	3.5	4.0
	1.9	2.4	2.8	3.1	3.2	3.8
	1.8	2.2	2.7	3.0	3.1	3.6

What we have, then, are six values of X:

$$x_1, x_2, x_3, x_4, x_5, x_6$$

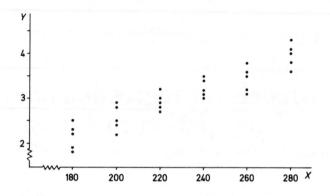

and associated with each of these is a random variable Y. We will call these random variables

$$Y_1, Y_2, Y_3, Y_4, Y_5, Y_6.$$

Now each of these random variables will be distributed about its mean:

$$E(Y_1) \quad E(Y_2) \quad E(Y_3) \quad E(Y_4) \quad E(Y_5) \quad E(Y_6)$$

and as the results seem to show a linear trend, we shall make the assumption that these means lie on a straight line. This line is to be thought of as the population regression line, as distinct from the sample regression lines of the previous chapter.

Suppose its equation is

$$E(Y) = \alpha + \beta x \qquad (1)$$

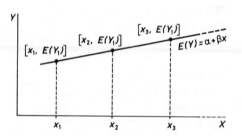

We now make three further assumptions:

1.   that the random variables Y_i ($i = 1$ to 6) are independent of each other;

2.   that each distribution has the *same variance* σ^2 so that the variability of each Y_i about its mean is the same;

3.   that the distribution of each Y_i is *normal*.

The diagram illustrating the situation has to have three axes: one for X, one for Y and one for $\phi(y)$, the normal density function for each Y.

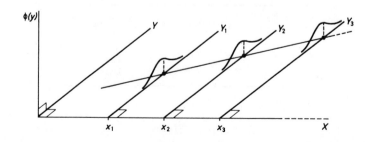

If we look again at one of the distributions, that of Y_i, then clearly, as the diagram shows, for any particular value of Y_i

$$y_i = E(Y_i) + \varepsilon_i \qquad (2)$$

where ε_i is the random error.

Putting (1) and (2) together we therefore have

$$y_i = \alpha + \beta x_i + \varepsilon_i$$

where, in accordance with our assumptions (2) and (3) the distribution of each ε_i is $N(0, \sigma^2)$ and the distribution of each Y_i is $N(\alpha + \beta x_i, \sigma^2)$. In fact these two distributions are identical apart from having different means.

This relation, then, is the model for the whole population distribution and reflects the fact that any particular value of Y may be regarded as having two components, one dependent on X in a linear way, and the other being random.

The next stage is to find a way of estimating α and β from sample observations.

20.2   The maximum likelihood method of estimating α and β

We have already seen in chapter 19 that in order to calculate the line of closest fit for a *sample* of points,

$$y = a + bx,$$

the values of a and b may be found using the method of least squares.

We shall now show, without using this arbitrary method, that the *maximum likelihood* estimates of α and β give the *same* results. In other words, given a sample (or, as in our present example, five samples)

$$\hat{\alpha} = a \quad \text{and} \quad \hat{\beta} = b$$

where $\quad b = \dfrac{S_{xy}}{S_{xx}} \quad$ and $\quad a = \bar{y} - b\bar{x}.$

Proof

Suppose a sample of n pairs of values (x_i, y_i) is taken and that

$$y_i = \alpha + \beta x_i + \varepsilon_i$$

according to our model. We thus have a sample of values of ε_i, and the density function for each ε_i is

$$\frac{1}{\sigma\sqrt{(2\pi)}} e^{-\frac{1}{2}\varepsilon_i^2/\sigma^2}.$$

Hence the joint probability density, or likelihood, is

$$L = (1/\sigma\sqrt{2\pi})^n \, e^{-\frac{1}{2}\,\Sigma\,\varepsilon_i^2/\sigma^2}$$

$$\Rightarrow \ln L = -\tfrac{1}{2}n \ln 2\pi - n \ln \sigma - \frac{1}{2\sigma^2} \Sigma \, (\varepsilon_i)^2$$

$$= -\tfrac{1}{2}n \ln 2\pi - n \ln \sigma - \frac{1}{2\sigma^2} \Sigma \, (y_i - \alpha - \beta x_i)^2.$$

Now this will be greatest when $\Sigma \, (y_i - \alpha - \beta x_i)^2$ is least, that is, when $\alpha = a$ and $\beta = b$, the *values found using the least squares method*.

(It can be seen that the assumption of a normal distribution for the ε_i is necessary for this proof.)

We have now justified the least squares method as a means of estimating α and β for our model, and it will be natural to call

$$y = \hat{\alpha} + \hat{\beta} x$$

the *estimated regression line* of y on x, α and β being the 'true' population values, corresponding to the sample values a and b.

20.3   Sampling distributions of *a* and *b*

The values of a and b will vary according to the sample; they will therefore form sampling distributions. These distributions will be discussed later, but we shall state here that the least squares estimates of α and β are, in fact, unbiased, since

$$E(a) = E(\hat{\alpha}) = \alpha \quad \text{and} \quad E(b) = E(\hat{\beta}) = \beta.$$

EXAMPLE 1   We now use the set of results given on page 373 to estimate the population regression line of y on x. These observations amount to a total sample of 30 pairs of points, five for each value of X. Coding: $U = X/100$ and $V = Y$.

u	v	u^2	uv
1.8	2.5, 2.3, 2.2, 1.9, 1.8 (total 10.7)	3.24	$10.7 \times 1.8 = 19.26$
2.0	2.9, 2.8, 2.5, 2.4, 2.2 (total 12.8)	4.00	$12.8 \times 2.0 = 25.6$
2.2	3.2, 3.0, 2.9, 2.8, 2.7 (total 14.6)	4.84	$14.6 \times 2.2 = 32.12$
2.4	3.5, 3.4, 3.2, 3.1, 3.0 (total 16.2)	5.76	$16.2 \times 2.4 = 38.88$
2.6	3.8, 3.6, 3.5, 3.2, 3.1 (total 17.2)	6.76	$17.2 \times 2.6 = 44.72$
2.8	4.3, 4.1, 4.0, 3.8, 3.6 (total 19.8)	7.84	$19.8 \times 2.8 = 55.44$
$\Sigma u = 13.8 \times 5$ $= 69$	$\Sigma v = 91.3$	$\Sigma u^2 = 32.44 \times 5$ $= 162.2$	$\Sigma uv = 216.02$

$$\frac{S_{uv}}{S_{uu}} = \frac{\Sigma\, uv - (\Sigma\, u)\,(\Sigma v)/n}{\Sigma\, u^2 - (\Sigma\, u)^2/n}, \quad \text{where } n = 30,$$

$$= \frac{216.02 - (69)\,(91.3)/30}{162.2 - (69)^2/30} = 1.723.$$

The scale for X was reduced by 100, which means that

$$b = \frac{S_{xy}}{S_{xx}} = \frac{S_{uv}}{S_{uu}} \times \frac{100}{100^2} = 0.01723.$$

$\bar{u} = 69/30 = 2.3 \Rightarrow \bar{x} = 2.3 \times 100 = 230; \bar{v} = 91.3/30 = 3.043 \Rightarrow \bar{y} = 3.043.$

Hence $a = \bar{y} - b\bar{x} = 3.043 - 0.01723 \times 230 = -0.920,$

and the estimated regression line of y on x is

$$y = -0.920 + 0.0172x.$$

20.4   Covariance and correlation for random variables

Having defined covariance and correlation for a sample, we now define them for random variables, that is, for a population of pairs (X, Y). They are the theoretical counterpart to the values defined for a sample, which is assumed to come from a theoretical probability distribution.

As usual with a probability distribution, the definitions are expressed in terms of expectation, so that, for example, $\Sigma\, x_i y_i / n$ will be replaced by $E(XY)$, $\Sigma\, (x_i - \bar{x})^2 / n$ by $\text{Var}(X)$.

Definitions

The covariance of two random variables X, Y

$$\text{Cov}(X, Y) = E(X - \mu_X)(Y - \mu_Y) \quad \text{writing } \mu_X \text{ for } E(X)$$

$$= E(XY) - \mu_X\, \mu_Y.$$

These are equivalent forms since

$$E(X - \mu_X)(Y - \mu_Y) = E(XY) - \mu_X\, E(Y) - \mu_Y\, E(X) + \mu_X\, \mu_Y$$

$$= E(XY) - \mu_X\, \mu_Y - \mu_Y\, \mu_X + \mu_X\, \mu_Y$$

$$= E(XY) - \mu_X\, \mu_Y.$$

The *correlation coefficient* between two random variables X, Y

$$\rho = \frac{\text{Cov}(X, Y)}{\sqrt{[\text{Var}(X) \cdot \text{Var}(Y)]}} \quad \text{where } \text{Var}(X) = E(X^2) - \mu_X^2.$$

As the size of a sample increases the value of the sample correlation coefficient, r, would be expected to tend towards ρ.

If $\rho = 0$, X and Y are said to be *uncorrelated*.

EXAMPLE 2   Calculate ρ for the joint probability distribution

X \ Y	0	1	2
0	0	0	0.2
1	0	0.2	0.2
2	0.4	0	0

(the figures in the cells now represent probabilities).

We shall use the same layout and method as for grouped frequency distributions (page 345).

Y \ X	0	1	2	p	pY	pY^2	pXY
0	0 0 0	0 0 0	0 0.2 0	0.2	0	0	0
1	0 0 0	1 0.2 0.2	2 0.2 0.4	0.4	0.4	0.4	0.6 ($=0.2+0.4$)
2	0 0.4 0	2 0 0	4 0 0	0.4	0.8	1.6	0
p	0.4	0.2	0.4	1.0	1.2	2.0	0.6
pX	0	0.2	0.8	1.0			
pX^2	0	0.2	1.6	1.8			
pXY	0	0.2	0.4	0.6			

$$\mu_X = E(X) = \Sigma\, pX = 1.0, \quad \mu_Y = E(Y) = \Sigma\, pY = 1.2.$$

$$\text{Var}(X) = \Sigma\, pX^2 - \mu_X^2 = 1.8 - 1 = 0.8.$$

$$\text{Var}(Y) = \Sigma\, pY^2 - \mu_Y^2 = 2.0 - 1.44 = 0.56.$$

$$\text{Cov}(X, Y) = \Sigma\, pXY - \mu_X\,\mu_Y = 0.6 - 1.0 \times 1.2 = -0.6.$$

$$\rho = \frac{-0.6}{\sqrt{(0.8 \times 0.56)}} = 0.896.$$

On pages 354–5 we proved that a sample correlation coefficient, r, must be numerically less than 1. The same must be true of course for ρ. A formal (and ingenious) proof appears in the Appendix on pages 397–8.

20.5   Dependent and independent variables

Two important results follow:

1.   If X and Y are independent, then they are uncorrelated, that is $\rho = 0$

Proof

When X and Y are independent, $E(XY) = E(X) \cdot E(Y)$ (see page 103). Hence $\text{Cov}(X, Y) = 0$ and $\rho = 0$.

The converse is *not* true, in general. It is possible for ρ to be zero without X and Y being independent.

Suppose, for example that X can take values 0 or 1, each with probability $\frac{1}{2}$ (tossing a coin, say, where a head scores 1 and a tail 0), and that Y has the same distribution.

Then $$E(X) = E(Y) = \tfrac{1}{2}.$$

If we now form two new variables $R = X + Y$ and $S = X - Y$,

$$\text{Cov}(R, S) = E(X^2 - Y^2) - E(X + Y)E(X - Y) = 0,$$

since $E(X - Y) = 0$ and $E(X^2 - Y^2) = 0$.

In other words R and S are uncorrelated. They are not, however, independent.

One necessary condition for the independence of two events A and B is that $P(A \mid B) = P(A)$ (see page 74). If we take A as the event $R = 1$ and B the event $S = 0$,

$$P(A) = P(X + Y = 1) = P(X = 1 \cap Y = 0) + P(X = 0 \cap Y = 1) = \tfrac{1}{2}.$$

But $P(A \mid B) = P(X + Y = 1 \mid X - Y = 0)$ which is zero, since if $X + Y = 1$ and $X - Y = 0$, $X = Y = \frac{1}{2}$, which can never occur. Hence R and S are dependent. (See also question 8 in exercise 20a.)

2.   If X and Y are not independent,

$$\text{Var}(X + Y) = \text{Var}(X) + \text{Var}(Y) \pm 2\text{Cov}(X, Y)$$
$$= \text{Var}(X) + \text{Var}(Y) \pm 2\rho\sqrt{[\text{Var}(X) \cdot \text{Var}(Y)]}.$$

Proof

$$V(X + Y) = E[X + Y - (\mu_X + \mu_Y)]^2 = E[(X - \mu_X) + (Y - \mu_Y)]^2$$
$$= E(X - \mu_X)^2 + E(Y - \mu_Y)^2 + 2E(X - \mu_X)(Y - \mu_Y)$$
$$= \text{Var}(X) + \text{Var}(Y) + 2\text{Cov}(X, Y)$$
$$= \text{Var}(X) + \text{Var}(Y) + 2\rho\sqrt{[\text{Var}(X) \cdot \text{Var}(Y)]}.$$

The proof for $\text{Var}(X - Y)$ is similar.

The equivalent result for sample values is:

$$\text{var}(x \pm y) = \text{var}(x) + \text{var}(y) \pm 2r\sqrt{[\text{var}(x) \cdot \text{var}(y)]}.$$

EXAMPLE 3   Prove that $\text{Cov}(X, X + Y) = \text{Var}(X) + \text{Cov}(X, Y)$

$$\text{Cov}(X, X + Y) = E(X^2 + XY) - E(X)E(X + Y)$$

$$= E(X^2)+E(XY)-E(X)[E(X)+E(Y)]$$
$$= E(X^2)-[E(X)]^2+E(XY)-E(X)E(Y)$$
$$= \mathrm{Var}(X)+\mathrm{Cov}(X, Y).$$

Note: It is often useful, since $\mathrm{Cov}(X, X) = \mathrm{Var}(X)$, to quote the result in this example in the form

$$\mathrm{Cov}(X, X+Y) = \mathrm{Cov}(X, X)+\mathrm{Cov}(X, Y) = \mathrm{Var}(X)+\mathrm{Cov}(X, Y).$$

It can easily be shown, similarly that

$$\mathrm{Cov}(X, Y+Z) = \mathrm{Cov}(X, Y)+\mathrm{Cov}(X,Z)$$
$$\mathrm{Cov}(X+Y, Z+T) = \mathrm{Cov}(X,Z)+\mathrm{Cov}(X, T)+\mathrm{Cov}(Y, Z)+\mathrm{Cov}(Y, T).$$

EXAMPLE 4 X, Y and Z are uncorrelated random variables with variance 4, 10 and 6 respectively. If $P = X+Y$ and $Q = Y+Z$, calculate the correlation coefficient between P and Q.

Quoting the above result,

$$\mathrm{Cov}(X+Y, Y+Z) = \mathrm{Cov}(X, Y)+\mathrm{Cov}(Y, Y)+\mathrm{Cov}(X, Z)+\mathrm{Cov}(Y, Z).$$

Hence $\mathrm{Cov}(P,Q)= \mathrm{Var}(Y)= 10$, since $\mathrm{Cov}(X, Y)= \mathrm{Cov}(X,Z)= \mathrm{Cov}(Y,Z)=0$.

$$\mathrm{Var}(P) = \mathrm{Var}(X+Y) = \mathrm{Var}(X)+\mathrm{Var}(Y) = 14,$$

since X and Y are uncorrelated.
Similarly $\mathrm{Var}(Q) = \mathrm{Var}(Y)+\mathrm{Var}(Z) = 16$

$$\Rightarrow \rho = \frac{10}{\sqrt{(14 \times 16)}} = 0.67.$$

EXERCISE 20a

1.   Find the correlation coefficient of the random variables X and Y whose joint probability distribution is given in the following table:

Y \ X	1	2	3
1	3/16	1/8	1/4
2	0	1/8	1/16
3	1/16	1/8	1/16

2.   The running times in minutes of a cross-city bus service, routed through the city centre, were observed over a long period. The following data were obtained:

Stage of journey	Mean running time	Standard deviation of running time
Terminus X to city centre	12	1
City centre to terminus Y	20	2

Given that the correlation coefficient of the running times on the two stages of the journey is $\frac{1}{2}$, find the mean and variance of the total time taken to complete a journey. If the scheduled running time is 33 minutes, estimate, using a normal distribution, the proportion of buses which take longer than the scheduled time. (MEI)

3.   The variances of the random variables X and Y are 4 and 9 respectively and their correlation coefficient is $1/3$.

If $U = X + Y$ and $V = X - Y$, find the correlation coefficient between U and V.

4.   The distribution of men's heights has a mean of 172 cm and standard deviation 6 cm. Assuming that the heights of brothers have a correlation coefficient 0.6, find the mean and variance of the difference in height between pairs of brothers.

5.   Prove that $\text{Cov}(X + Y, X + Z) = \text{Var}(X) + \text{Cov}(Y, Z) + \text{Cov}(X, Y) + \text{Cov}(X, Z)$, assuming only the definition of covariance.

6.   Independent random variables X_1 and X_2 each have the same distribution, with mean μ_X and variance σ_x^2. Y_1, Y_2 and Y_3 are mutually independent and have the same distribution with mean μ_Y and variance σ_Y^2. The correlation coefficient between each X_i and Y_j is ρ.

If $U = (X_1 + X_2)/2$ and $V = (Y_1 + Y_2 + Y_3)/3$, prove that the correlation coefficient of U and V is $\rho/\sqrt{6}$.

7.   The continuous random variable X has density function

$$f(x) = \begin{cases} kx, & 0 \leqslant x \leqslant 1 \\ 0, & \text{elsewhere.} \end{cases}$$

Calculate the correlation coefficient between X and X^2.

8. (X, Y) has the joint probability distribution shown in the table.

X \ Y	−1	0	1
−1	0.1	0.15	0.1
0	0.15	0	0.15
1	0.1	0.15	0.1

(a) Prove that X and Y are uncorrelated.
(b) Show that $P(X = -1 \text{ and } Y = 1) \neq P(X = -1) \cdot P(Y = 1)$, and hence that X and Y are not independent.

9.   Define and distinguish between independent and uncorrelated random variables.

X takes values 1,2,3,4,5,6 each with probability $1/6$, as in throwing a die; Y has the same distribution and is independent of X. For $T = X^2$, $U = X - Y$, $V = X + Y$ determine the coefficient of correlation between X and T, and between U and V.

Is either pair of variables independent? (MEI)

10.   Two people each spin two coins. The number of heads obtained are X and Y ($= 0,1,2$). Make a joint probability distribution table as in question 8.

Are X and Y correlated? Are they independent?

20.6　Confidence intervals for α and β

At the beginning of this chapter we showed that the maximum likelihood estimate of a regression line of y on x

$$y = \hat{\alpha} + \hat{\beta}x$$

is given by
$$\hat{\alpha} = a, \qquad \hat{\beta} = b$$

where $b = S_{xy}/S_{xx}$ and $a = \bar{y} - b\bar{x}$.

In example 1, for instance, a sample of 30 point pairs gave the values

$$a = -0.913 \quad \text{and} \quad b = 0.01723.$$

Bearing in mind that the values of a obtained from a number of samples of the same size will form a *sampling distribution*, as will the values of b, the problem we now confront is to find a way of establishing confidence intervals for α and β, the 'true' population parameters.

Our starting point is the model that we took for the distribution of point pairs (x_i, y_i):

$$y_i = \alpha + \beta x_i + \varepsilon_i$$

where the ε_i are mutually independent but each has the same normal distribution $N(0, \sigma^2)$.

For reasons that become clear later, the sampling distribution of b is the more important of the two and we will deal with this distribution first.

The sampling distribution of b

The result is this:

The distribution of b is $N(\beta, \sigma^2/S_{xx})$, where, as usual, $S_{xx} = \Sigma\,(x - \bar{x})^2$; and since $E(b) = \beta$, b is an unbiased estimate of β.

The proof of this result is complicated, but not difficult, and will be found in the Appendix on pages 398–9.

If σ^2 is *known* (which is not usually the case), we can easily find confidence limits for β, as the following example shows.

EXAMPLE 5　With the data of example 1 on page 376, find 95% confidence limits for β, given that the variance of each $\varepsilon_i = 0.04$.

With the coding $U = X/100$, $S_{uu} = 162.2 - 69^2/30 = 3.5$

$$\Rightarrow S_{xx} = 3.5 \times 100^2 = 35\,000.$$

Hence $\dfrac{\sigma^2}{S_{xx}} = \dfrac{0.04}{35\,000} = 1.143 \times 10^{-6}$

and the distribution of b is $N(\beta, 1.143 \times 10^{-6})$.

Ninety-five per cent confidence limits for β are therefore

$$0.0172 \pm 1.96 \times \sqrt{(1.143 \times 10^{-6})} = 0.0172 \pm 2.095 \times 10^{-3}$$

and a 95% confidence interval is $(0.0151, 0.0193)$.

★The sampling distribution of *a*

Here, too, we shall state the result, giving a proof in the Appendix on page 399.

The distribution of a is $N(\alpha, \sigma^2 \Sigma x^2/nS_{xx})$.

EXAMPLE 6   With the data of example 1 on page 376, find 95% confidence limits for α, given that the variance of each $\varepsilon_i = 0.04$.

As before, $S_{xx} = 3.5 \times 100^2$, and from the table

$$\Sigma u^2 = 162.2 \Rightarrow \Sigma x^2 = 162.2 \times 100^2;$$

so that with $n = 30$

$$\frac{\sigma^2 \Sigma x^2}{nS_{xx}} = \frac{0.04 \times 162.2}{30 \times 3.5} = 0.0618$$

and the distribution of a is $N(\alpha, 0.0618)$.

Ninety-five per cent confidence limits for α are $-0.193 \pm 1.96 \times \sqrt{0.0618}$ and a 95% confidence interval is $(-0.680, 0.294)$.

Confidence limits for *β* when σ^2 is unknown

When, as is usually the case, σ^2 is not known, we need to be able to make an unbiased estimate and then to know what the distribution of the corresponding test statistic is. Both of these involve complicated proofs beyond the scope of this book so we will have to be content with stating the results.

If we call the *minimum* value of the sum of squares

$$\Sigma (y_i - a - bx_i)^2,$$

M, then an unbiased estimate of σ^2 is

$$\hat{\sigma}^2 = \frac{M}{n-2}.$$

Now this minimum value, M, is the value obtained by putting $b = S_{xy}/S_{xx}$ and $a = \bar{y} - b\bar{x}$, and we will shortly prove that

$$M = S_{yy} - (S_{xy})^2/S_{xx}.$$

It follows that

$$\hat{\sigma}^2 = \frac{1}{n-2}[S_{yy} - (S_{xy})^2/S_{xx}].$$

With this estimate of σ^2, the distribution of

$$\frac{b - \beta}{\hat{\sigma}/\sqrt{S_{xx}}}$$

is 't' with $n-2$ degrees of freedom, and it follows that 95% confidence limits for β will be

$$b \pm t \cdot \hat{\sigma}/\sqrt{S_{xx}}$$

where t is the 5% ($P = 5$) value of t for $v = n-2$.

Proof that $M = S_{yy} - (S_{xy})^2/S_{xx}$

$$y_i - a - bx_i = y_i - \bar{y} + \bar{y} - a - bx_i = (y_i - \bar{y}) - b(x_i - \bar{x}), \quad [\bar{y} - a = b\bar{x}].$$

Hence $M = \Sigma [(y_i - \bar{y}) - b(x_i - \bar{x})]^2$

$$= \Sigma (y_i - \bar{y})^2 + b^2 \Sigma (x_i - \bar{x})^2 - 2b \Sigma (x_i - \bar{x})(y_i - \bar{y})$$

$$= S_{yy} + b^2 S_{xx} - 2b S_{xy}.$$

But $b = S_{xy}/S_{xx}$, so that M simplifies to $S_{yy} - (S_{xy})^2/S_{xx}$.

EXAMPLE 7   Calculate b and a 95% confidence interval for β from the following sample of point pairs:

$$(0.8, 0.5) \quad (0.9, 0.6) \quad (1.0, 0.8) \quad (1.1, 1.2) \quad (1.2, 1.3)$$

assuming a model $y_i = \alpha + \beta x_i + \varepsilon_i$, where the variance of each ε_i is σ^2.

x	y	x^2	y^2	xy
0.8	0.5	0.64	0.25	0.4
0.9	0.6	0.81	0.36	0.54
1.0	0.8	1.0	0.64	0.8
1.1	1.2	1.21	1.44	1.32
1.2	1.3	1.44	1.69	1.56
5.0	4.4	5.10	4.38	4.62

From the table,

$$S_{xx} = \Sigma x^2 - (\Sigma x)^2/n = 5.10 - (5)^2/5 = 0.10$$

$$S_{yy} = \Sigma y^2 - (\Sigma y)^2/n = 4.38 - (4.4)^2/5 = 0.508$$

$$S_{xy} = \Sigma xy - (\Sigma x)(\Sigma y)/n = 4.62 - (5)(4.4)/5 = 0.22.$$

Hence $b = S_{xy}/S_{xx} = 2.2$.

$$M = S_{yy} - (S_{xy})^2/S_{xx} = 0.508 - (0.22)^2/0.10 = 0.024$$

$$\Rightarrow \hat{\sigma}^2 = \frac{M}{n-2} = \frac{0.024}{3} = 0.008 \Rightarrow \hat{\sigma} = 0.0894.$$

It follows that $\hat{\sigma}/\sqrt{S_{xx}} = 0.0894/\sqrt{0.10} = 0.2827$.

For $n = 5$, $v = 3$, and the 5% ($P = 5$) value of $t_3 = 3.18$.

Hence 95% confidence limits for β are $2.2 \pm 3.18 \times 0.2827$ and the 95% confidence interval is $(1.30, 3.10)$.

★Confidence limits for α when σ^2 is unknown

Using $\hat{\sigma}^2 = \dfrac{M}{n-2}$, confidence limits for α are, similarly,

$$a \pm t \cdot \hat{\sigma} \sqrt{\left(\frac{\Sigma x^2}{n \cdot S_{xx}}\right)}$$

with the appropriate value of t.

EXERCISE 20b

1. For the data given below, assuming a model

$$y_i = \alpha + \beta x_i + \varepsilon_i$$

calculate b, the least squares estimate of β.
    If the variance of the ε_i is known to be 0.015, find 95 % confidence limits for β.

x	0	1	2	3	4
y	22.0	22.5	23.1	23.8	24.5

2. Use the data in question 1 to find 95 % confidence limits for β if the variance of the ε_i is not known.

3.★ Use the data in question 1 to calculate a, the least squares estimate of α and obtain 95 % confidence limits for α, if the variance of the ε_i is not known.

20.7   Testing the significance of a sample correlation coefficient

In chapter 18 on correlation we looked briefly at a method of testing the significance of a sample product-moment correlation coefficient. We are now in a position to justify what was stated there, without proof, and to deal with this important matter more fully.

There are, in fact, two kinds of test to be considered, the first of which answers the question:

Is r significantly different from zero?

which amounts to asking, simply, is r significant?

The clue to this test is the close link between r and b, whose distribution we have just been considering.

By definition $\quad r = \dfrac{S_{xy}}{\sqrt{(S_{xx} \cdot S_{yy})}} \quad$ and $\quad b = \dfrac{S_{xy}}{S_{xx}}$,

from which it follows that

$$b = r \sqrt{\left(\frac{S_{yy}}{S_{xx}}\right)}.$$

From this relationship, if either b or r is zero, the other will also be zero. The same will apply to the population parameters β and ρ. If, then, X and Y are independent, not only will ρ be zero (see page 379), but so will β be.

Now testing whether r is significantly different from zero is equivalent to testing the null hypothesis

$$H_0 : \rho = 0$$

which also implies that $\beta = 0$. Any such test will therefore involve b and β, and we shall be able to use our knowledge of the distribution of b to establish the following result:

Under the null hypothesis $H_0 : \rho = 0$, for a sample of size n with sample correlation coefficient r, the distribution of the statistic

$$r \sqrt{\left(\frac{n-2}{1-r^2} \right)}$$

is 't' with $n-2$ degrees of freedom.

It will be seen that this is a particularly convenient distribution as the statistic only involves the sample size, n, and the correlation coefficient, r.

Proof

The minimum value of the sum of squares, M, is given by

$$M = S_{yy} - b^2 S_{xx} \quad \text{(page 384)}.$$

Since
$$b^2 = r^2(S_{yy}/S_{xx}), \qquad S_{yy} = b^2 S_{xx}/r^2$$

$$\Rightarrow M = b^2 S_{xx}(1/r^2 - 1) = b^2 S_{xx}(1-r^2)/r^2$$

$$\Rightarrow \frac{r^2}{1-r^2} = \frac{b^2 S_{xx}}{M}$$

$$\Rightarrow \frac{r}{\sqrt{(1-r^2)}} = b \sqrt{\left(\frac{S_{xx}}{M} \right)}.$$

Now the unbiased estimate of σ^2 used in the distribution of b is

$$\hat{\sigma}^2 = \frac{M}{n-2} \Rightarrow M = (n-2)\hat{\sigma}^2.$$

Hence
$$\frac{r}{\sqrt{(1-r^2)}} = \frac{b}{\hat{\sigma}} \sqrt{\left(\frac{S_{xx}}{n-2} \right)}$$

$$\Rightarrow r \sqrt{\left(\frac{n-2}{1-r^2} \right)} = \frac{b}{\hat{\sigma}} \sqrt{S_{xx}} = \frac{b}{\hat{\sigma}/\sqrt{S_{xx}}}.$$

But the distribution of

$$\frac{b-\beta}{\hat{\sigma}/\sqrt{S_{xx}}}$$

is 't' with $n-2$ degrees freedom. It follows that under $H_0 : \rho = 0$ (and therefore $\beta = 0$), the distribution of

$$r \sqrt{\left(\frac{n-2}{1-r^2} \right)}$$

is 't' with $n-2$ degrees of freedom.

EXAMPLE 8 From a sample of 10 point pairs the correlation coefficient is calculated to be 0.68. Is this significantly different from 0 at the 5% level?

The hypotheses to be tested are

$$H_0: \rho = 0 \quad \text{as against} \quad H_i: \rho \neq 0.$$

For $r = 0.68$, $n = 10$ the statistic

$$r\sqrt{\left(\frac{n-2}{1-r^2}\right)} = 0.68\sqrt{\left(\frac{8}{1-(0.68)^2}\right)} = 2.62.$$

Under H_0, for $v = 8$, the critical value at 5% $(P = 5)$ for a two-tail test is $t_8 = 2.31$. Since $2.62 > 2.31$, our value of the statistic is in the critical region. H_0 is therefore rejected at this level and we say that r *is* significantly different from zero. There is, therefore, correlation.

Note:
1. For a distribution for which ρ could not be negative, a one-tail test should be used.

2. Spearman's rank correlation coefficient, r_s, may also be tested in this way so long as n is greater than 10.

An alternative test for the significance of r; confidence limits for ρ

An alternative test is available which enables us to test whether an observed sample value of r is consistent with the null hypothesis $H_0: \rho = \rho_0$ (some *stated* value, not necessarily zero). It also provides a means of finding confidence limits for ρ.
  (The statistic $r\sqrt{[(n-2)/(1-r^2)]}$ cannot be used under the null hypothesis unless $\rho_0 = 0$.)
  The sampling distribution of r itself is extremely complicated. However a transformed statistic, due to Fisher, has a normal distribution and provides what we want:

  Under $H_0: \rho = \rho_0$, if $Z = \frac{1}{2}\ln[(1+r)/(1-r)]$ and $\zeta = \frac{1}{2}\ln[(1+\rho)/(1-\rho)]$ then the distribution of Z is approximately $N[\zeta, 1/(n-3)]$.

Note:
1. The approximation is not very close for $n <$ about 50, although it is often used for small samples, for want of an alternative.

2. The derivation of this remarkable result is very complicated, and will not be pursued here.

3. An alternative form for Z is arc tanh r.

We shall now show how the distribution of this statistic may be used, either for significance testing or for finding confidence limits for ρ.

EXAMPLE 9 Is it reasonable to assume that the correlation coefficient of a population is 0.6 when a sample of size 120 has correlation coefficient 0.72?

The hypotheses to be tested are

$$H_0 : \rho = 0.6 \quad \text{as against} \quad H_1 : \rho \neq 0.6$$

$$\zeta = \frac{1}{2}\ln\left(\frac{1.6}{0.4}\right) = 0.6931$$

and the value of Z when $r = 0.72$ is

$$z = \frac{1}{2}\ln\left(\frac{1.72}{0.28}\right) = 0.9076.$$

Now under H_0, the distribution of Z is $N(0.6931, 1/117)$ or $N(0.6931, 0.00855)$.

The standardised value of Z is $(0.9076 - 0.6931)/\sqrt{(0.00855)} = 2.32$. As this is greater than 1.96, we conclude that z is significantly different from 0.6931 at the 5% level, so that r is significantly different from 0.6. H_0 is therefore rejected.

EXAMPLE 10   If two samples of 100 students take tests in mathematics and physics, and the correlation coefficients between the marks in the samples are 0.75 and 0.81 respectively, test the hypothesis that their population correlation coefficients are the same.

Let ρ_1 and ρ_2 be the theoretical correlation coefficients. The hypotheses to be tested are

$$H_0 : \rho_1 - \rho_2 = 0 \quad \text{as against} \quad H_1 : \rho_1 - \rho_2 \neq 0.$$

Let $\quad \zeta_1 = \frac{1}{2}\ln\left(\frac{1+\rho_1}{1-\rho_1}\right) \quad \text{and} \quad \zeta_2 = \frac{1}{2}\ln\left(\frac{1+\rho_2}{1-\rho_2}\right).$

The distribution of Z_1 is $N(\zeta_1, 1/97)$ and of Z_2 is $N(\zeta_2, 1/97)$. Under H_0 the distribution of $Z_1 - Z_2$ is $N(0, 2/97) = N(0, 0.0206)$.

$$\text{Now} \quad r_1 = 0.75 \Rightarrow z_1 = \frac{1}{2}\ln\left(\frac{1.75}{0.25}\right) = 0.973$$

$$r_2 = 0.81 \Rightarrow z_2 = \frac{1}{2}\ln\left(\frac{1.81}{0.19}\right) = 1.127.$$

so that the standardised value of

$$z_1 - z_2 = \frac{(0.973 - 1.127) - 0}{\sqrt{0.0206}}$$

$$= -1.073.$$

Since $1.073 < 1.96$, we conclude that the difference between z_1 and z_2 is not significant at the 5% level and H_0 is accepted.

EXAMPLE 11   A sample of 100 students in the first year of a large science faculty were given tests in mathematics and physics. The correlation coefficient between their marks in these subjects was 0.75. Find 95% confidence limits for ρ, the population correlation coefficient.

$$r = 0.75 \Rightarrow z = \frac{1}{2}\ln\left(\frac{1.75}{0.25}\right) = 0.973.$$

Ninety-five per cent confidence limits for ζ are

$$z \pm 1.96 \sqrt{\left(\frac{1}{n-3}\right)} = 0.973 \pm 1.96/\sqrt{97} = 0.973 \pm 0.199.$$

Hence a 95% confidence interval for

$$\frac{1}{2}\ln\left(\frac{1+\rho}{1-\rho}\right) \quad \text{is} \quad (0.744, 1.172),$$

and for

$$\left(\frac{1+\rho}{1-\rho}\right) \quad \text{is} \quad (4.702, 10.423),$$

and for ρ is

$$\left(\frac{4.702-1}{4.702+1}, \frac{10.423-1}{10.423+1}\right) \quad \text{or} \quad (0.65, 0.82).$$

Note: The distribution of $Z = \frac{1}{2}\ln[(1+r)/(1-r)]$ about mean $\zeta = \frac{1}{2}\ln[(1+\rho)/(1-\rho)]$ may be used to test $H_0: \rho = 0$. In that case $\zeta = 0$.

EXAMPLE 12   Use the Z-test for the data in example 8, namely test whether a sample correlation coefficient of 0.68, calculated from a sample of size 10, is significantly different from 0 at the 5% level.

The hypotheses to be tested are

$$H_0: \rho = 0 \quad \text{as against} \quad H_1: \rho \neq 0.$$

Under H_0 the distribution of $Z = \frac{1}{2}\ln[(1+r)/(1-r)]$ is $N(0, 1/7)$.

When $r = 0.68$

$$z = \frac{1}{2}\ln\left(\frac{1.68}{0.32}\right) = 0.829$$

and the standardised value of z is

$$\frac{0.829-0}{\sqrt{(1/7)}} = 2.19.$$

As $2.19 > 1.96$, H_0 is rejected.

It can be seen that this gives the same result as the *t*-test used in example 8, in spite of the small sample size.

EXERCISE 20c

1.   A correlation coefficient is calculated from a sample of 20 point pairs from a normal bivariate distribution. The value found is 0.45. Is this significant at the 5% level?

2.   Can a correlation coefficient of -0.23 be regarded as significant when $n = 100$?

3.   What is the smallest value of r for a sample of 25 point pairs that is significant at the 5% level?

4.   Is it reasonable to suppose that a population correlation coefficient is 0.3 when the correlation coefficient calculated from a sample of 90 is 0.42?

5.   Find 95% confidence limits for ρ if $r = 0.8$ when $n = 25$.

6.   From a sample of 12 pairs of observations the value of r is calculated as 0.38. Use the Z-test to find whether this is significantly different from: (a) 0, (b) 0.7.

7.   Test the hypothesis that two samples of 20 and 30 point-pairs with correlation coefficients 0.57 and 0.43 respectively are drawn from the same population.

8.   Find 99% confidence limits for ρ if a sample of size 40 gives a correlation coefficient of 0.6.

20.8★   Non-linear regression and multiple regression

We conclude this chapter with a short account of regression models which are not linear.

Suppose it seems likely that, for example, a *quadratic* regression curve is needed so that $y = a + bx + cx^2$ and the model for Y is

$$y_i = \alpha + \beta x_i + \gamma x_i^2 + \varepsilon_i$$

where the distribution of each ε_i is $N(0, \sigma^2)$. The estimated values of α, β, γ are then given by $\hat{\alpha} = a$, $\hat{\beta} = b$, $\hat{\gamma} = c$. It can be shown, by the same method as we used for linear regression, that the least square estimates coincide with the maximum likelihood estimates.

As in the linear case, the first step is to write down an expression for the sum of the squared deviations as on page 358.

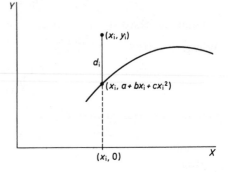

$$R = \Sigma \, d_i^2 = \Sigma \, [y_i - (a + bx_i + cx_i^2)]^2.$$

In order to find a, b and c to minimise R, it is necessary to use the calculus method, with partial differentiation, described on pages 370–2.

Equating $\partial R/\partial a$, $\partial R/\partial b$ and $\partial R/\partial c$ to zero gives the normal equations:

$$\Sigma \, y_i = na + b \, \Sigma \, x_i + c \, \Sigma \, x_i^2$$

$$\Sigma \, x_i y_i = a \, \Sigma \, x_i + b \, \Sigma \, x_i^2 + c \, \Sigma \, x_i^3$$

$$\Sigma \, x_i^2 y_i = a \, \Sigma \, x_i^2 + b \, \Sigma \, x_i^3 + c \, \Sigma \, x_i^4.$$

Substitution of the values of x_i, y_i from the observed sample leads to three simultaneous equations from which a, b and c may be calculated, giving $\hat{\alpha}$, $\hat{\beta}$ and $\hat{\gamma}$.

Another type of model is illustrated in the following example. It will be noticed that in it, nothing is said about the nature of the distribution of the ε_i. We needed to assume a *normal* distribution for the ε_i to demonstrate that the least squares method and maximum likelihood method give the same estimates in the above examples. This is not always the case, but the least squares method may be used so long as the ε_i have the same variance.

EXAMPLE 13   Use the method of least squares to estimate the constants α, β, γ in the model

$$y_i = \alpha + \beta \sin x_i + \gamma \cos x_i + \varepsilon_i \quad (i = 1, \ldots, 7)$$

using the following data. The errors $\{\varepsilon_i\}$ are independent with zero expectation and the same variance.

x_i:	0	$\pi/3$	$2\pi/3$	π	$4\pi/3$	$5\pi/3$	2π
y_i:	2.6	4.2	2.1	-1.4	-3.0	-1.0	2.7

(Plotting the points suggests that a sinusoidal model is appropriate in this case.) With the usual notation,

$$R = \Sigma \, [y_i - (a + b \sin x_i + c \cos x_i)]^2.$$

Equating $\partial R/\partial a$, $\partial R/\partial b$ and $\partial R/\partial c$ to zero gives the normal equations:

$$\Sigma \, y_i = na + b \, \Sigma \sin x_i + c \, \Sigma \cos x_i$$

$$\Sigma \, y_i \sin x_i = a \, \Sigma \sin x_i + b \, \Sigma \sin^2 x_i + c \, \Sigma \sin x_i \cos x_i$$

$$\Sigma \, y_i \cos x_i = a \, \Sigma \cos x_i + b \, \Sigma \sin x_i \cos x_i + c \, \Sigma \cos^2 x_i.$$

From the observed values

$$\Sigma \, y_i = 6.2, \ \Sigma \sin x_i = 0, \ \Sigma \cos x_i = 1, \ \Sigma \sin^2 x_i = 3.0,$$

$$\Sigma \cos^2 x_i = 4.0, \ \Sigma \sin x_i \cos x_i = 0, \ \Sigma \, y_i \sin x_i = 8.92, \ \Sigma \, y_i \cos x_i = 9.2.$$

Hence

$$6.2 = 7a + c$$

$$8.92 = 3b$$

$$9.3 = a + 3c.$$

The solution of these equations is $a = 0.47$, $b = 2.97$, $c = 2.91$, and the estimated regression curve of y on x is therefore

$$y = 0.47 + 2.97 \sin x + 2.91 \cos x.$$

Method of least squares for multiple regression

The method of least squares can also be used where there is more than one independent variable X and Z, and Y depends linearly on each. (X might measure rainfall, Z temperature and Y crop yield, for example.)

An appropriate model would then be

$$y_i = \alpha + \beta x_i + \gamma z_i + \varepsilon_i.$$

If the values of $E(Y)$ lie on the plane $y = \alpha + \beta x + \gamma z$, the estimated values of α, β, γ are found by minimising

$$R = \Sigma \, d_i^2 = \Sigma \, (y_i - a - b x_i - z_i)^2,$$

leading to the normal equations:

$$\Sigma \, y_i = na + b \, \Sigma \, x_i + c \, \Sigma \, z_i$$

$$\Sigma \, x_i y_i = a \, \Sigma \, x_i + b \, \Sigma \, x_i^2 + c \, \Sigma \, x_i z_i$$

$$\Sigma \, z_i y_i = a \, \Sigma \, z_i + b \, \Sigma \, z_i x_i + c \, \Sigma \, z_i^2.$$

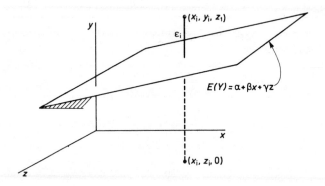

EXERCISE 20d★

1. Apply the method of least squares to fit the model

$$y_i = \alpha x_i + \beta x_i^2 + \varepsilon_i$$

for which $E(\varepsilon_i) = 0$ and $\mathrm{Var}(\varepsilon_i) = \sigma^2$ to the following data:

x_i	−3	−2	−1	0	1	2	3
y_i	20	8	2	0	5	16	30

Plot the original observations and the estimated regression line on a graph.   (MEI)

2. Use the given data to estimate the parameters α, β and γ in the model

$$y_i = \alpha + \beta x_i + \gamma \log_{10} x_i + \varepsilon_i,$$

where $\{\varepsilon_i\}$ are independent random variables of constant variance and zero mean.

x	2	4	6	8	10
y	27	20	15	10	5

(MEI)

3. Apply the method of least squares to fit the model

$$y_i = \alpha + \beta x_i + \gamma \sin(\tfrac{1}{2}\pi x_i) + \varepsilon_i$$

to the following data:

x_i	0	1	2	3	4	5	6	7
y_i	9.8	13.3	16.0	18.6	22.1	25.2	28.0	30.3

(MEI)

4. Some experiments were made in order to determine the effects of two types of catalyst at various temperatures on the yield of a chemical process.
The following results were obtained:

Temperature (°C − 50)	7	6	8	1	4	7	2
Catalyst	A	B	B	A	B	A	B
Yield (tonnes/hr)	14.8	12.9	9.9	12.0	10.1	14.7	12.0

Use the method of least squares to estimate the parameters α, β and γ in the model

$$y_i = \alpha + \beta x_i + \gamma z_i + \varepsilon_i \quad (i = 1, 2, \ldots, 7),$$

where y denotes the yield, x the temperature, ε an independently distributed error. The variable z_i has the value $+1$ when the catalyst is A and -1 when the catalyst is B.
Interpret the value found for γ. (MEI)

EXERCISE 20e (miscellaneous)

1. An unbiased cubical die is thrown twice. Let X_1 be the score on the first throw, X_2 the score on the second throw. Random variables Z and W are defined by

$$Z = X_1 + X_2, \qquad W = 3X_1 - 2X_2.$$

Find: (a) $E(W)$, (b) $Var(W)$, (c) $Cov(X_1, X_2)$, (d) $Cov(X_1, Z)$. (C)

2. A discrete random variable X takes values $-1, 0, 1$, and a discrete random variable Y takes values $0, 1, 2$. The joint probability distribution of X and Y is given in the following table, where $p \neq 0$.

(a) State, giving a reason, whether X and Y are independent.
(b) Show that $Cov(X, Y) = 0$.
(c) Given that $Var(Y) = 5/6$, find the values of p and q.

		Y		
		0	1	2
	−1	p	0	p
X	0	q	p	q
	1	p	0	p

(C)

3. The table below gives the inflation rate (x) and unemployment rate (y) for ten different countries during December 1979.

Country	A	B	C	D	E	F	G	H	I	J
Inflation rate x (%)	13.9	21.4	9.6	1.5	31.7	23.1	18.4	34.4	27.6	5.6
Unemployment rate y (%)	2.9	11.3	5.4	6.1	9.0	8.8	5.9	15.6	9.8	3.7

Calculate the product-moment correlation coefficient.

Assuming that the above data comprise a random sample from a population with a bivariate normal distribution, test at the 5% level of significance:

(a) the hypothesis $H_0: \rho = 0$, where ρ is the true population correlation coefficient (hypothesis tendered by the Workers Democratic Party),

(b) the hypothesis $H_0': \rho = 0.9$, where ρ is the true population correlation coefficient (hypothesis tendered by the Association of Financial Controllers.)   (AEB 1980)

4.   Temperature and humidity are measured at a meteorological station at midday on a hundred consecutive days. Suitably scaled measures of temperature (X) and humidity (Y) give the following data:

$$\Sigma x = 4 \quad \Sigma y = 12 \quad \Sigma x^2 = 302 \quad \Sigma xy = 70 \quad \Sigma y^2 = 715.$$

Calculate the estimate of the product-moment coefficient of X and Y from these data.

A certain theory states that temperature and humidity are unrelated. Does your result support this theory?

An alternative theory suggests that the correlation coefficient is 0.4. Does your result support this alternative theory?   (C)

5.   Calculate the coefficient of correlation between the packed cell volume and the red blood cell count using the following data from ten animals in order to see whether the quicker determination of packed cell volume can be used as a proxy for the slower r.b.c. count.

Animal	1	2	3	4	5	6	7	8	9	10
Packed cell volume (mm)	22	21	28	24	21	17	29	20	19	25
Red blood cell count (10^6)	3.3	3.2	4.7	3.7	3.5	2.9	4.8	3.1	3.3	4.3

Use the result that the distribution of $z = \frac{1}{2}\ln[(1+r)/(1-r)]$ is approximately normal with mean $\frac{1}{2}\ln[(1+\rho)/(1-\rho)]$ and variance $(n-3)^{-1}$ to obtain a 95% confidence interval for the population correlation coefficient, ρ, from your value of r.   (MEI)

6.   In a horticultural experiment the correlation coefficient between the mean number of plants produced by a certain size of plant and the amount of fertiliser used was 0.5. The sample size was 28.

In a later experiment, with atmospheric conditions slightly altered, a sample of 35 gave a correlation coefficient of 0.3.

Using a 5% significance level, test the hypothesis that the population correlation coefficients for these two experiments are the same.

7.   The marks gained by 10 students in a preliminary test and in the examination held a year later were:

Test (x)	34	24	25	26	35	39	26	30	34	37
Exam (y)	82	70	64	56	75	88	68	67	87	73

(a) Find the mean marks and also the standard deviations S_x and S_y.

(b) Find the correlation coefficient r. Does this suggest that the preliminary test is a useful guide to the students' performance in the examination?

(c) Find the equation of the regression line of y on x in the form $y = a + bx$.

(d) Calculate the standard error of b from the formula

$$\frac{S_y}{S_x}\sqrt{\left(\frac{1-r^2}{n-2}\right)}$$

and verify that a *t*-test (with $n-2$ degrees of freedom) of the difference between *b* and zero gives the same result as the test based on *r*. (O & C)

8.★ Use the method of least squares to estimate the constants *a*, *b*, *c* in

$$y = a + b\cos x + c\sin x,$$

from the following data, giving your answers to two decimal places.

x	0	$\pi/6$	$\pi/3$	$\pi/2$	$2\pi/3$	$5\pi/6$	π
y	2.0	1.2	0.2	-1.1	-1.2	0.1	0.6

(C)

9.★ Observations were made on the thickness of paint film obtained by two painters, Jim and Charlie, on each of two paints, Truwit and Newit.

	Thickness of film	
	Truwit	*Newit*
Jim	0.851	0.681
	0.863	0.690
	0.854	0.686
Charlie	0.775	0.636
	0.745	0.622
	0.769	0.649

The expected thickness of the paint film is

$$y = \mu + \alpha x + \beta z,$$

where *x* is $+1$ for Jim and -1 for Charlie; *z* is $+1$ for Truwit and -1 for Newit. Estimate the constants α, β, μ by using the method of least squares. (MEI)

10.★ $X_1, X_2, \ldots, X_n$ are *n* independent random variables with zero means and equal variances σ^2. The random variables *Y* and *Z* are defined by

$$Y = \sum_{i=1}^{n} a_i X_i, \qquad Z = \sum_{i=1}^{n} b_i X_i,$$

where $a_1, a_2, \ldots, a_n, b_1, b_2, \ldots, b_n$ are constants.

Show that the variance *V* of the random variable $Y + \lambda Z$ has its minimum value when the parameter λ has the value

$$-\sum_{i=1}^{n} a_i b_i \bigg/ \sum_{i=1}^{n} b_i^2.$$

Find this minimum value of *V* in terms of λ, the a_i and the b_i, and deduce that

$$\left(\sum_{i=1}^{n} a_i b_i \right)^2 \leqslant \left(\sum_{i=1}^{n} a_i^2 \right) \left(\sum_{i=1}^{n} b_i^2 \right).$$

(*Hint:* See Appendix, page 398.) (O & C)

11.★ The random variables X_i are independent with

$$E(X_i) = \mu_i \quad \text{and} \quad \text{Var}(X_i) = \sigma^2 \quad \text{for } i = 1, 2, 3, 4.$$

Linear combinations Y_1, Y_2 of these variables are defined by

$$Y_1 = \sum_{i=1}^{4} a_i X_i, \qquad Y_2 = \sum_{i=1}^{4} b_i X_i,$$

with constants a_i, b_i for $i = 1, 2, 3, 4$. Show that the condition for Y_1 and Y_2 to be uncorrelated is

$$\sum_{i=1}^{4} a_i b_i = 0.$$

Hence show that $\bar{X}$ and $X_i - \bar{X}$ are uncorrelated for $i = 1, 2, 3, 4$, where

$$\bar{X} = \frac{1}{4} \sum_{i=1}^{4} X_i, \quad \text{and} \quad \mu_i = \mu, \quad i = 1, 2, 3, 4.$$

12.★ (a) A regression line of y on x is to be found for the n points (x_i, y_i) $(i = 1, 2, \ldots, n)$ by assuming that each y_i is an independent observation from a normal distribution with mean $\alpha + \beta x_i$ and variance σ^2. Show that the maximum likelihood estimates $\hat{\alpha}$ and $\hat{\beta}$ of α and β are given by

$$\hat{\alpha} n + \hat{\beta} \sum x_i = \sum y_i$$
$$\hat{\alpha} \sum x_i + \hat{\beta} \sum x_i^2 = \sum x_i y_i.$$

(b) In addition there are another n points (x_i, z_i) where each z_i is an independent observation from a normal distribution with mean $\alpha + \beta x_i + \gamma$ and variance σ^2. By considering

$$n \frac{\partial L}{\partial \beta} - \sum x_i \frac{\partial L}{\partial \alpha},$$

where L is the log-likelihood for these data, show that the maximum likelihood estimate of β is the arithmetic mean of those obtained by considering the y's and z's separately. Find also the maximum likelihood estimates for α and γ in terms of $\hat{\beta}$, x_i, y_i, z_i and n.   (O & C)

13.★ The pairs of numbers (x_i, y_i), $i = 1, 2, \ldots, n$ are related by the equations $y_i = \alpha + \beta x_i + \varepsilon_i$ where α and β are unknown constants and the x_i are fixed known numbers. The ε_i are independent random variables, each with mean zero and variance σ^2. Obtain the least squares estimates a and b of α and β respectively in terms of the x_i and y_i.
    Show that

$$b = \beta + \frac{\sum (x_i - \bar{x})\varepsilon_i}{\sum (x_i - \bar{x})^2}$$

where $\bar{x} = \sum x_i/n$, and deduce the population mean value and variance of b in terms of β, σ^2 and the x_i.

(*Hint*: See Appendix on pages 398–9.)   (JMB)

14.★ The random variable X_i has expectation μ_i and variance σ_i^2 for $i = 1, 2$ and $\text{Cov}(X_1, X_2) = \rho$. Find the expectation and variance of $Z = aX_1 + bX_2$ for constants a and b.
    Under what conditions are $(X_1 + X_2)$ and $(X_1 - X_2)$ random variables for which the correlation coefficient is zero? What further conditions are needed for independence?   (MEI)

15.★ The regression line of y on x calculated from the n pairs of numbers (x_i, y_i), where $i = 1, 2, \ldots, n$, is $y = a + bx$. The pairs of numbers (t_i, u_i) are defined by the equations

$$t_i = \alpha + \beta x_i, \qquad u_i = \gamma + \delta y_i,$$

where $i = 1, 2, \ldots, n$ and β and δ are both non-zero.

Prove that the regression line of u on t has the same equation as that obtained by the elimination of x and y between

$$y = a + bx \qquad t = \alpha + \beta x \qquad u = \gamma + \delta y.$$

(O & C)

20.9   Appendix

Proof that $|\rho| \leqslant 1$

The method is to find an expression for λ, in terms of the variances and covariance of X and Y, which minimises the variance of $(Y - \lambda X)$.

By the definition of variance,

$$
\begin{aligned}
V(Y - \lambda X) &= E[(Y - \lambda X) - (\bar{Y} - \lambda \bar{X})]^2 \\
&= E[(Y - \bar{Y}) - \lambda(X - \bar{X})]^2 \\
&= E[(Y - \bar{Y})^2 - 2\lambda(X - \bar{X})(Y - \bar{Y}) + \lambda^2(X - \bar{X})^2] \\
&= E(Y - \bar{Y})^2 + \lambda^2 E(X - \bar{X})^2 - 2\lambda E(Y - \bar{Y})(X - \bar{X}) \\
&= \mathrm{Var}(Y) + \lambda^2 \mathrm{Var}(X) - 2\lambda \, \mathrm{Cov}(X, Y).
\end{aligned}
$$

Differentiating w.r.t. λ to find the value of λ which minimises this expression.

$$\mathrm{d}[\mathrm{Var}(Y - \lambda X)]/\mathrm{d}\lambda = 2\lambda \, \mathrm{Var}(X) - 2\,\mathrm{Cov}(X, Y),$$

which is zero when $\lambda = \mathrm{Cov}(X, Y)/\mathrm{Var}(X)$. This value of λ *minimises* $\mathrm{Var}(Y - \lambda X)$, since the second derivative $= 2\,\mathrm{Var}(X)$, which must be positive.

Now the value of $\mathrm{Var}(Y - \lambda X)$ when $\lambda = \mathrm{Cov}(X, Y)/\mathrm{Var}(X)$ is

$$\mathrm{Var}(Y) + \frac{[\mathrm{Cov}(X, Y)]^2}{\mathrm{Var}(X)} - 2 \times \frac{[\mathrm{Cov}(X, Y)]^2}{\mathrm{Var}(X)} = \mathrm{Var}(Y) - [\mathrm{Cov}(X, Y)]^2/\mathrm{Var}(X).$$

This must be greater than or equal to 0, since variance cannot be negative.

Hence $\mathrm{Var}(X) \cdot \mathrm{Var}(Y) \geqslant [\mathrm{Cov}(X, Y)]^2$

$$\Rightarrow \left| \frac{\mathrm{Cov}(X, Y)}{\sqrt{[\mathrm{Var}(X) \cdot \mathrm{Var}(Y)]}} \right| \leqslant 1 \Rightarrow |\rho| \leqslant 1.$$

The sampling distribution of *b*

$$b = \frac{\Sigma\,(x_i - \bar{x})(y_i - \bar{y})}{\Sigma\,(x_i - \bar{x})^2} = \frac{\Sigma\,(x_i - \bar{x})(y - \bar{y}_i)}{S_{xx}}$$

Now $\Sigma\,(x_i - \bar{x})(y_i - \bar{y}) = \Sigma\,(x_i - \bar{x})y_i - \bar{y}\,\Sigma\,(x_i - \bar{x})$

$$= \Sigma\,(x_i - \bar{x})y_i \quad [\text{since } \Sigma\,(x_i - \bar{x}) = 0]$$

$$= \Sigma\,(x_i - \bar{x})(\alpha + \beta x_i + \varepsilon_i)$$

$$= \alpha\,\Sigma\,(x_i - \bar{x}) + \beta\,\Sigma\,x_i(x_i - \bar{x}) + \Sigma\,\varepsilon_i(x_i - \bar{x})$$

$$= 0 + \beta[\Sigma\,x_i(x_i - \bar{x}) - \bar{x}\,\Sigma\,(x_i - \bar{x})] + \Sigma\,\varepsilon_i(x_i - \bar{x})$$

[since $\Sigma\,(x_i - \bar{x}) = 0$. *Note:* β has been multiplied by an extra zero term.]

$$= \beta\,\Sigma\,(x_i - \bar{x})^2 + \Sigma\,\varepsilon_i(x_i - \bar{x}) = \beta S_{xx} + \Sigma\,\varepsilon_i(x_i - \bar{x}).$$

Hence $b = \beta + \Sigma\,\varepsilon_i(x_i - \bar{x})/S_{xx}$

$$\Rightarrow b - \beta = \Sigma\,\varepsilon_i(x_i - \bar{x})/S_{xx}.$$

Now the values of x_i are *fixed*, so that if we write

$$k_i = \frac{x_i - \bar{x}}{S_{xx}}$$

k_i will also be fixed and

$$b - \beta = \Sigma\,k_i\varepsilon_i.$$

Now the distribution of each ε_i is $N(0, \sigma^2)$, and it follows that the distribution of $b - \beta$ is also normal, as it is the sum of a number of independent normal variables (see page 204).

$$E(b - \beta) = \Sigma\,k_i\,E(\varepsilon_i) = 0 \quad [\text{since } E(\varepsilon_i) = 0 \text{ for each } i]$$

$$\Rightarrow E(b) = \beta.$$

In other words $\hat{\beta} = b$, and b is an unbiased estimate of β.

$$\text{Var}(b - \beta) = \text{Var}(b) = \text{Var}(\Sigma\,k_i\varepsilon_i) = \Sigma\,k_i^2\,\text{Var}(\varepsilon_i) = \sigma^2\,\Sigma\,k_i^2$$

$$[\text{since Var}(\varepsilon_i) = \sigma^2].$$

Now $$k_i^2 = \frac{(x_i - \bar{x})^2}{(S_{xx})^2} \Rightarrow \Sigma\,k_i^2 = \frac{\Sigma\,(x_i - \bar{x})^2}{(S_{xx})^2}$$

$$= S_{xx}/(S_{xx})^2 = 1/S_{xx}.$$

Hence $\text{Var}(b) = \sigma^2/S_{xx}$.

Thus the distribution of $b - \beta$ is $N(0, \sigma^2/S_{xx})$ and the distribution of b is $N(\beta, \sigma^2/S_{xx})$.

The sampling distribution of *a*

$$y_i = \alpha + \beta x_i + \varepsilon_i \Rightarrow \bar{y} = \alpha + \beta\bar{x} + \Sigma\varepsilon_i/n.$$

Since $\bar{x}$ is fixed

$$E(\bar{y}) = \alpha + \beta\bar{x} + E(\Sigma\ \varepsilon_i/n) = \alpha + \beta\bar{x} \quad [E(\varepsilon_i) = 0 \text{ for each } i]$$

and $\text{Var}(y) = \text{Var}(\Sigma\ \varepsilon_i/n) = (1/n^2)\text{Var}(\Sigma\ \varepsilon_i)$

$$= (1/n^2) \times n\sigma^2 \quad [\text{Var}(\varepsilon_i) = \sigma^2 \text{ for each } i].$$

But $a = \bar{y} - b\bar{x}$, where $\bar{x}$ is fixed

$$\Rightarrow E(a) = E(\bar{y}) - \bar{x}E(b) = \alpha + \beta\bar{x} - \bar{x}\beta.$$

Hence $E(a) = \alpha.$

In other words $\hat{\alpha} = a$ and a is an unbiased estimate of α.

It can be proved that $\bar{y}$ and b are independent (the proof is difficult) so that $a = \bar{y} - \bar{x}b$ is the difference between two independent variables: $\bar{y}$, whose variance is σ^2/n, and $\bar{x}b$, where the distribution of b has variance σ^2/S_{xx}.

Hence $\text{Var}(a) = \text{Var}(\bar{y}) + \bar{x}^2\ \text{Var}(b) = \sigma^2/n + \bar{x}^2\sigma^2/S_{xx}$

$$= \sigma^2(1/n + \bar{x}^2/S_{xx}) = \sigma^2(S_{xx} + n\bar{x}^2)/nS_{xx}$$

$$= \sigma^2\ \Sigma\ x^2/nS_{xx}.$$

Now the distribution of $\bar{y}$ is normal, as is the distribution of $\bar{x}b$, so that the distribution of $a - \alpha$ is

$$N(0, \sigma^2\Sigma\ x^2/nS_{xx})$$

and the distribution of a is

$$N[\alpha, \sigma^2\ \Sigma\ x^2/nS_{xx}].$$

21* Moment generating functions

During the course of this book you will often have come across the phrase: 'A proof of this result will be given later'. The reason for this is that in many cases the key to such a proof lies in the idea of a 'moment generating function' which has not, so far, been defined and which is the subject of this chapter.

The idea of a 'generating function' is, of course, not new. We met it in chapter 7, in connection with the discrete variable, X, whose probability generating function, $G(t)$, was defined as:

$$G(t) = p_0 t^0 + p_1 t^1 + p_2 t^2 \ldots p_n t^n \qquad [p_r = P(X = r)]$$

$$= \sum_{r=0}^{n} p_r t^r$$

$$= E(t^X).$$

The p.g.f. was found to be a very convenient auxiliary function for finding the mean and variance of discrete probability distributions, using:

$$G'(1) = \mu \quad \text{and} \quad G''(1) = \sigma^2 + \mu^2 - \mu.$$

The *moment generating function* is another auxiliary function which is defined for *both discrete and continuous* random variables. It, too, is useful for finding mean and variance, but its usefulness goes far beyond this. It is a powerful weapon in establishing some of the centrally important results about probability distributions. Later in this chapter we shall demonstrate this by proving the 'reproductive' property of the normal distribution, the normal distribution as a limit of the binomial, the normal as a limit of the Poisson distribution, and we will be able to indicate how the central limit theorem may be proved.

21.1 The moment generating function for discrete and continuous variables

The moment generating function, $M(t)$, of the random variable X is the expectation of e^{tX}, that is,

$$M(t) = E(e^{tX}).$$

If X is *discrete* and $P(X = x_r) = p_r, r = 0$ to n,

$$M(t) = \sum_{r=0}^{n} p_r \cdot e^{tx}r.$$

If X is *continuous* with density function $f(x)$, $-\infty < x < \infty$,

$$M(t) = \int_{-\infty}^{\infty} e^{tx} f(x)\,dx.$$

The usual abbreviation for moment generating function is m.g.f.

Examples of moment generating functions

1. The m.g.f. of the *binomial distribution B(n, p)*:
    Since the binomial probabilities are

$$P(X = r) = \binom{n}{r} q^{n-r}p^r \quad (r = 0, 1, 2, \ldots, n),$$

$$M(t) = q^n e^{0\cdot t} + \binom{n}{1} q^{n-1}pe^t + \binom{n}{2} q^{n-2}p^2 e^{2t} + \ldots$$

$$+ \binom{n}{r} q^{n-r}p^r e^{rt} + \ldots + p^n e^{nt}$$

$$= (q + pe^t)^n.$$

2. The m.g.f. of the *exponential distribution* with parameter λ:
    The density function for the exponential distribution is

$$f(x) = \begin{cases} \lambda e^{-\lambda x}, & x \geqslant 0 \\ 0 & , \quad \text{elsewhere,} \end{cases}$$

so that
$$M(t) = \int_0^{\infty} e^{tx}(\lambda e^{-\lambda x})\,dx = \lambda \int_0^{\infty} e^{-(\lambda - t)x}\,dx$$

$$= \frac{-\lambda}{\lambda - t}\left[e^{-(\lambda - t)x} \right]_0^{\infty} = \frac{\lambda}{\lambda - t}.$$

(As t is a 'dummy' variable we are free to assume that $t < \lambda$.)

It is important to realise that distributions like these are defined completely by their moment generating functions. There are, however, some distributions that do not have a moment generating function, the 't' distribution for example.

We shall now show how the m.g.f. is used to obtain the mean and variance.

21.2   Deriving mean and variance

Expanding e^{tX}, we have

$$M(t) = E(e^{tX}) = E(1 + tX + t^2 X^2/2! + t^3 X^3/3! + \ldots + t^r X^r/r! + \ldots)$$

$$= 1 + tE(X) + \frac{t^2}{2!} E(X^2) + \frac{t^3}{3!} E(X^3) + \ldots + \frac{t^r}{r!} E(X^r) \ldots$$

assuming that the result $E(X+Y+\ldots) = E(X)+E(Y)+\ldots$ is valid for an infinite series).

It can be seen that the coefficient of $t^r/r!$ is $E(X^r)$.

$E(X^r)$ is called the rth *moment of X about zero*, and in particular:

$E(X)$ is the first moment about zero (the mean, μ)
$E(X^2)$ is the second moment about zero.

Thus the m.g.f. is said to *generate* the moments of X about zero.

If, then, the m.g.f. can conveniently be expanded in ascending powers of t, this expansion can be used directly to find the mean and variance, as the following example shows.

EXAMPLE 1   Find the mean and variance of the exponential distribution with parameter λ.

Expanding the m.g.f. for the exponential distribution that we obtained earlier,

$$M(t) = \lambda(\lambda-t)^{-1} = (1-t/\lambda)^{-1}$$
$$= 1+t/\lambda+t^2/\lambda^2+\ldots$$
$$= 1+t(1/\lambda)+\frac{t^2}{2!}(2/\lambda^2)+\ldots.$$

Hence $E(X) = 1/\lambda$ and $E(X^2) = 2/\lambda^2$,

so that $\mu = 1/\lambda$ and $\sigma^2 = E(X^2)-\mu^2 = 2/\lambda^2-1/\lambda^2 = 1/\lambda^2$.

An alternative method

The expansion method is not always convenient: consider, for example the problem of expanding the m.g.f. of the binomial, $(q+pe^t)^n$ in ascending powers of t. In that case a different approach is needed.

Remembering that the key to using a *probability* generating function to find mean and variance was to differentiate the p.g.f. and use the results $G'(1) = \mu$ and $G''(1) = \sigma^2+\mu^2-\mu$, we shall now obtain analogous results for the m.g.f.

$$M(t) = 1+t\cdot E(X)+\frac{t^2}{2!}E(X^2)+\frac{t^3}{3!}E(X^3)+\ldots+\frac{t^r}{r!}E(X^r)+\ldots.$$

Differentiating w.r.t. t,

$$M'(t) = E(X)+tE(X^2)+\frac{t^2}{2!}E(X^3)+\ldots+\frac{t^{r-1}}{(r-1)!}E(X^r)+\ldots$$

(assuming the validity of differentiating term by term).
Putting $t = 0$, $M'(0) = E(X)$.
Differentiating again,

$$M''(t) = E(X^2)+t\cdot E(X^3)+\ldots+\frac{t^{r-2}}{(r-2)!}E(X^r)\ldots$$

and, putting $t = 0$, $M''(0) = E(X^2)$.

Successive differentiation, followed by substituting $t = 0$, leads to

$$M'''(0) = E(X^3),\ M''''(0) = E(X^4), \ldots, M^{(r)}(0) = E(X^r).$$

The 'higher' moments, $E(X^3)$, $E(X^4)$, etc. will be discussed later. Meanwhile we have the following useful results for finding mean and variance:

$$\mu = E(X) = M'(0)$$
$$\sigma^2 = E(X^2) - \mu^2 = M''(0) - [M'(0)]^2.$$

EXAMPLE 2   Find the mean and variance of the binomial distribution $B(n, p)$.

Differentiating $M(t) = (q + pe^t)^n$ twice gives

$$M'(t) = n(q + pe^t)^{n-1} \cdot pe^t$$
$$M''(t) = n(n-1)(q + pe^t)^{n-2} \cdot p^2 e^{2t} + n(q + pe^t)^{n-1} \cdot pe^t.$$

Hence $E(X) = M'(0) = n(q + p)p \Rightarrow \mu = np.$

$$E(X^2) = M''(0) = n(n-1)(q+p)p^2 + np(q+p)$$
$$= n(n-1)p^2 + np,$$

so that

$$\sigma^2 = E(X^2) - \mu^2 = (n^2 - n)p^2 + np - n^2 p^2 = np - np^2$$
$$= np(1-p) = npq.$$

EXAMPLE 3   Find the mean and variance of the exponential distribution with parameter λ.

In example 1 we used the expansion method for this distribution. The differentiation method is equally convenient.

$$M(t) = \lambda(\lambda - t)^{-1} \Rightarrow M'(t) = \lambda(\lambda - t)^{-2} \Rightarrow M''(t) = 2\lambda(\lambda - t)^{-3}.$$

Hence $E(X) = M'(0) = 1/\lambda \Rightarrow \mu = 1/\lambda.$

$$E(X^2) = M''(0) = 2/\lambda^2 \Rightarrow \sigma^2 = 2/\lambda^2 - 1/\lambda^2 = 1/\lambda^2.$$

If you now look back to page 185 where the mean and variance of this distribution were found by using

$$E(X) = \int_0^\infty x f(x)\, dx \quad \text{and} \quad E(X^2) = \int_0^\infty x^2 f(x)\, dx$$

you will be able to see that the use of the m.g.f. is considerably less laborious.

EXERCISE 21a

1. X is the score when an unbiased die is thrown. Find the moment generating function and use it to find the mean and variance of X.

2. The discrete random variable X has a geometric distribution, given by

$$P(X = r) = q^{r-1}p, \quad r = 1, 2, 3, \ldots$$

where $q = 1 - p$. Prove that the moment generating function is

$$pe^t/(1 - qe^t)$$

and hence show that the mean is $1/p$ and the variance is q/p^2.

3.   Prove that the moment generating function of a random variable X which is Poisson distributed with parameter a is

$$M(t) = e^{-a}e^{ae^t}.$$

Find $M'(t)$ and $M''(t)$ and deduce that the mean and variance are both equal to a. Find also $E(X^3)$.

4.   The discrete rectangular distribution is defined by

$$P(X = r) = 1/n \quad (r = 0, 1, 2, \ldots, n).$$

Prove that the moment generating function is given by

$$M(t) = (e^{nt} - 1)/n(e^t - 1).$$

5.   Prove that the moment generating function of the exponential distribution with density function

$$f(x) = \begin{cases} 3e^{-3x}, & x \geq 0 \\ 0 & , \quad \text{otherwise} \end{cases}$$

is $3/(3 - t)$.

By expanding this function and inspecting the coefficients of the relevant terms, write down the first three moments about zero. Deduce the variance.

6.   Find the moment generating function for the uniform continuous distribution

$$f(x) = \begin{cases} \frac{1}{2}, & 1 \leq x \leq 3 \\ 0, & \text{otherwise.} \end{cases}$$

By expanding this function in powers of t as far as t^2, find the variance.

7.   For the uniform distribution with density function

$$f(x) = \begin{cases} 1/2a, & -a \leq x \leq a \\ 0 & , \quad \text{otherwise} \end{cases}$$

prove that the moment generating function is

$$(e^{at} - e^{-at})/2at.$$

By expanding this function prove that the moments of odd order are zero and that

$$E(X^{2r}) = a^{2r}/(2r + 1).$$

8.   The two-parameter exponential distribution has density function

$$f(x) = \begin{cases} \lambda e^{-\lambda(x - a)}, & x \geq a \\ 0 & , \quad \text{elsewhere.} \end{cases}$$

Find $M(t)$, $M'(t)$ and $M''(t)$ and deduce $E(X)$ and $\text{Var}(X)$.

9.   A continuous random variable X has a frequency function $f(x)$ given by

$$f(x) = x/6 \quad , \quad \text{for } 0 \leq x \leq 3$$
$$f(x) = 2 - \tfrac{1}{2}x, \quad \text{for } 3 \leq x \leq 4$$
$$f(x) = 0 \quad , \quad \text{elsewhere.}$$

Show that the generating function for moments of X about the origin is

$$(1 - 4e^{3t} + 3e^{4t})/6t^2.$$

Hence find the mean and standard deviation. (O & C)

10. The continuous random variable X has density function

$$f(x) = \tfrac{1}{2}e^{-|x|}, \quad -\infty < x < \infty.$$

Show that its moment generating function is $(1 - t^2)^{-1}$, and derive its mean and variance.

11.★ A random variable X is said to have a geometric distribution with parameter α if

$$P(X = k) = (1 - \alpha)\alpha^{k-1} \quad (k = 1, 2, 3, \ldots),$$

where $0 < \alpha < 1$. For such a random variable, find:
(a) the probability that $X > k$;
(b) its moment generating function.
  Independent random variables U and V have geometric distributions with parameters β and γ respectively, and $W = \min(U, V)$ is the minimum of U and V. Write down an expression for the probability that $W = k$ in terms of the probabilities that U and V take certain values. Hence find the distribution of W, and deduce its moment generating function. (O)

21.3 Moments about the mean

The moments of a random variable X *about zero* have been defined as

$$E(X), E(X^2), E(X^3), \ldots, E(X^r)\ldots.$$

Moments of X *about the mean* are similarly defined as

$$E(X - \mu), E(X - \mu)^2, E(X - \mu)^3, \ldots, E(X - \mu)^r \ldots$$

where $E(X - \mu)^r$ is called the rth *moment about the mean*.
  It is convenient to distinguish between these two sets of moments by writing μ'_r for $E(X^r)$ and μ_r for $E(X - \mu)^r$, so that

$$\mu'_1 = E(X) = \mu \qquad \mu_1 = E(X - \mu) = 0 \quad [E(X) = \mu]$$

$$\mu'_2 = E(X^2) \qquad\qquad \mu_2 = E(X - \mu)^2 = \sigma^2$$

$$\mu'_3 = E(X^3) \qquad\qquad \mu_3 = E(X - \mu)^3$$

$$\mu'_4 = E(X^4) \qquad\qquad \mu_4 = E(X - \mu)^4 \qquad \text{etc.}$$

Just as μ_2 (σ^2) is a measure of the spread, other characteristics of a distribution are reflected in the values of μ_3 and μ_4.

Skewness

Clearly, if the distribution is symmetrical, to every positive value of $(X - \mu)$ there is an equal negative value so that μ_3 will be zero. (It is however possible to construct a distribution for which the converse is not true – see question 8 in exercise 21b). If the distribution is unsymmetrical and the right-hand 'tail' is longer than the left-hand one, then the deviations from the mean, $(X - \mu)$, to the right will outweigh the deviations to the left, and since each deviation is cubed, μ_3 will be positive. We say, in this case, that the distribution is skewed to the right. Similarly, if the distribution is skewed to the left, μ_3 will be negative. In other words the *skewness* of the distribution is reflected in the magnitude of μ_3.

In practice, skewness (which is a property of shape) is often measured by μ_3/σ^3, which is independent of the units in which X is measured. (Another measure is (mean − mode)/σ).

Kurtosis

As for μ_4, it can be shown that this moment reflects the degree to which the distribution is 'peaked', and the measure of 'peakedness', called *kurtosis*, is given by the non-dimensional formula μ_4/σ^4 when the concept is appropriate.

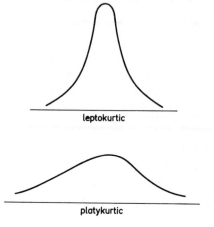

leptokurtic

platykurtic

The normal distribution has a kurtosis of 3. If a distribution has a kurtosis less than 3 it is called platykurtic; if more than 3, leptokurtic. The names are derived from Greek words for 'broad' and 'thin'.

Calculating moments about the mean

Method 1

The first method uses relations between moments about the mean and moments about zero.

The familiar result that $\sigma^2 = E(X-\mu)^2 = E(X^2)-\mu^2$ can now be written in the form

$$\mu_2 = \mu_2' - \mu^2. \tag{1}$$

There are similar relations for the higher moments:

$$\mu_3 = \mu_3' - 3\mu_2'\mu + 2\mu^3 \tag{2}$$

$$\mu_4 = \mu_4' - 4\mu_3'\mu + 6\mu_2'\mu^2 - 3\mu^4. \tag{3}$$

Relation (2) is proved as follows:

$$\mu_3 = E(X-\mu)^3 = E(X^3 - 3\mu X^2 + 3\mu^2 X - \mu^3)$$
$$= E(X^3) - 3\mu E(X^2) + 3\mu^2 E(X) - \mu^3$$
$$= \mu_3' - 3\mu\mu_2' + 3\mu^2\mu - \mu^3$$
$$= \mu_3' - 3\mu_2'\mu + 2\mu^3.$$

The proof of relation (3) follows similar lines.

In order to calculate μ_3, for example, the first step is to find μ_1' (or μ), μ_2', μ_3', either directly from the definition of $E(X)$, $E(X^2)$, $E(X^3)$ or by using the m.g.f. These values are then substituted in relation (2). Examples 4, 5 and 6 illustrate this approach.

EXAMPLE 4   Find μ_1, μ_2, and μ_3 for the following distribution:

X	-2	-1	0	1	2	3
p	0.04	0.10	0.30	0.30	0.20	0.06

$\mu'_1 = E(X) = \Sigma\, px = 0.7, \mu'_2 = E(X^2) = \Sigma\, px^2 = 1.90, \mu'_3 = E(X^3) = \Sigma\, px^3 = 3.10.$

Hence $\mu = \mu'_1 = 0.7$

$$\mu_2 = \mu'_2 - \mu^2 = 1.41 = \sigma^2.$$

$$\mu_3 = \mu'_3 - 3\mu'_2\mu + 2\mu^3$$

$$= 3.1 - 3(1.9)(0.7) + 2(0.7)^3$$

$$= -0.304.$$

The skewness $\mu_3/\sigma^3 = -0.182$ (skewed slightly to the left).

EXAMPLE 5   Find the first four moments about the mean for the distribution of the random variable X whose density function is

$$f(x) = \begin{cases} 6x(1-x), & 0 \leqslant x \leqslant 1 \\ 0 & , \quad \text{elsewhere.} \end{cases}$$

$$\mu = \int_0^1 6x^2(1-x)\,dx = 0.5, \qquad \mu'_2 = \int_0^1 6x^3(1-x)\,dx = 0.3$$

$$\mu'_3 = \int_0^1 6x^4(1-x)\,dx = 0.2, \qquad \mu'_4 = \int_0^1 6x^5(1-x)\,dx = 0.143.$$

Hence $\mu_2 = 0.3 - (0.5)^2 = 0.05$

$$\mu_3 = 0.2 - 3(0.3)(0.5) + 2(0.5)^3 = 0$$

$$\mu_4 = (0.143) - 4(0.2)(0.5) + 6(0.3)(0.5)^2 - 3(0.5)^4 = 0.0055.$$

EXAMPLE 6   Find the variance and skewness of the exponential distribution with parameter a.

The density function is

$$f(x) = \begin{cases} ae^{-ax}, & x \geqslant 0 \\ 0 & , \quad \text{otherwise,} \end{cases}$$

and the m.g.f., $M(t) = a(a-t)^{-1}$ [see example (3)].
Hence $M'(t) = a(a-t)^{-2}, M''(t) = 2a(a-t)^{-3}$ and $M'''(t) = 6a(a-t)^{-4}$.

$$\mu = M'(0) = 1/a, \mu'_2 = M''(0) = 2/a^2, \mu'_3 = M'''(0) = 6/a^3.$$

Using relations (1), (2) and (3)

$$\mu_2 = \mu'_2 - \mu^2 = 1/a^2 = \sigma^2$$

$$\mu_3 = \mu'_3 - 3\mu'_2\mu + 2\mu^3 = 6/a^3 - 3(2/a^2)(1/a) + 2(1/a)^3$$

$$= 2/a^3.$$

Thus the variance equals $1/a^2$ and the skewness equals

$$\mu_3/\sigma^3 = \frac{2/a^3}{1/a^3} = 2.$$

Method 2

A more direct way of calculating moments about the mean is to use the moment generating function of $X - \mu$.

Just as

$$M_x(t) = E(e^{tX}) = E\left[1 + tX + \frac{t^2}{2!}X^2 + \ldots\right]$$

$$= 1 + tE(X) + \frac{t^2}{2!}E(X^2) + \ldots$$

so that

$$M_x(t) = 1 + t\mu_1' + \frac{t^2}{2!}\mu_2' + \ldots,$$

if $Y = X - \mu$,

$$M_y(t) = E[e^{t(X-\mu)}] = E\left[1 + t(X-\mu) + \frac{t^2}{2!}(X-\mu)^2 + \ldots\right]$$

$$= 1 + tE(X-\mu) + \frac{t^2}{2!}E(X-\mu)^2 + \ldots$$

and

$$M_y(t) = 1 + t\mu_1 + \frac{t^2}{2!}\mu_2 + \ldots.$$

In other words, $M_y(t)$ generates moments about the mean.

Now $M_y(t)$ and $M_x(t)$ are related in the following way:

$$M_y(t) = e^{-\mu t}M_x(t)$$

since

$$M_y(t) = E(e^{tY}) = E[e^{t(X-\mu)}] = E[e^{tX} \cdot e^{-\mu t}] = e^{-\mu t}E(e^{tX}).$$

This means that $M_y(t)$ can be written down immediately once $M_x(t)$ and μ have been obtained.

$\mu_2, \mu_3, \mu_4, \ldots$ are then calculated, either by expanding $M_y(t)$ and picking out the coefficients of $t^2/2!$, $t^3/3!$, $t^4/4!, \ldots$ or by differentiating $M_y(t)$ successively and finding $M_y''(0)$, $M_y'''(0) \ldots$.

EXAMPLE 7   A discrete random variable X takes the values $2, 4, \ldots, 2n$ each with probability $1/n$.

Find the variance and the third and fourth moments about the mean.

The mean $\mu = \frac{1}{n}(2 + 4 + \ldots + 2n) = n + 1$.

$$M_x(t) = E(e^{tX}) = \frac{1}{n}(e^{2t} + e^{4t} + \ldots + e^{2nt}) = \frac{e^{2t}(e^{2nt} - 1)}{n(e^{2t} - 1)}.$$

Repeated differentiation would be laborious, so we find $M_y(t)$.

$$M_y(t) = e^{-\mu t} M_x(t) = e^{-(n+1)t} \times \frac{e^{2t}(e^{2nt}-1)}{n(e^{2t}-1)}$$

which simplifies to $\dfrac{e^{nt}-e^{-nt}}{n(e^t-e^{-t})}$.

This may conveniently be written as $\dfrac{\sinh nt}{n \sinh t}$.

Expanding in powers of t,

$$M_y(t) = \frac{1}{n}(nt+n^3t^3/3!+n^5t^5/5!+\ldots) \times t^{-1}[1+(t^2/3!+t^4/5!+\ldots)]^{-1}$$

which simplifies to

$$M_y(t) = 1+\frac{n^2-1}{6}t^2+\frac{3n^4-10n^2+7}{360}t^4+\ldots.$$

Hence $\sigma^2 = \mu_2 =$ the coefficient of $t^2/2! = (n^2-1)/3$

$$\mu_3 = 0$$

$$\mu_4 = \text{the coefficient of } t^4/4! = (3n^4-10n^2+7)/15.$$

EXERCISE 21b

1.   A die is biased in such a way that the probability of getting a particular score is proportional to that score. Find the second and third moments about the mean.

2.   Prove that for any probability distribution

$$\mu_4 = \mu'_4 - 4\mu'_3\mu + 6\mu'_2\mu^2 - 3\mu^4$$

where μ_r is a moment about the mean and μ'_r is a moment about zero.

3.   Find the second and third moments about the mean for the distribution of X whose density function is

$$f(x) = \begin{cases} \frac{3}{4}x(2-x), & 0 \leqslant x \leqslant 2 \\ 0 & , \quad \text{otherwise.} \end{cases}$$

4.   The random variable X has density function

$$f(x) = \begin{cases} \frac{4}{27}x^2(3-x), & 0 \leqslant x \leqslant 3 \\ 0 & , \quad \text{elsewhere.} \end{cases}$$

Find the second and third moments about the mean and derive the skewness.

5.   For the exponential distribution with parameter a, whose density function is given on page 407, prove that the moment generating function for moments about the origin is $a/(a-t)$. Find the mean and show that the moment generating function for moments about the mean is

$$ae^{-t/a}/(a-t).$$

Deduce that the 3rd moment about the mean is $2/a^3$.

6. Prove that, for a binomial distribution $B(n, p)$, the m.g.f. is

$$(q + pe^t)^n, \quad \text{where } q = 1 - p.$$

Hence show that the generating function for moments about the mean is

$$(qe^{-pt} + pe^{qt})^n.$$

Find the 3rd moment about the mean for this distribution.

7. A random variable X has probability density function

$$f(x) = \tfrac{1}{2}e^{-\frac{1}{2}x} \quad (x \geqslant 0).$$

Show that the moment generating function for moments about zero is $1/(1 - 2t)$, and derive the moment generating function for moments about the mean.
 Calculate the variance and third moment about the mean.

8. (a) The continuous random variable X has p.d.f.

$$f(x) = \begin{cases} 3x^2, & 0 \leqslant x \leqslant 1 \\ 0, & \text{elsewhere.} \end{cases}$$

Sketch the graph of $f(x)$. Repeat this for the p.d.f.

$$g(x) = \begin{cases} 3(1-x)^2, & 0 \leqslant x \leqslant 1 \\ 0, & \text{elsewhere.} \end{cases}$$

(b) For $r = 1, 2, \ldots$, the rth moment about the mean of a distribution is denoted by μ_r and defined by

$$\mu_r = E[(X - \mu)^r],$$

where μ is the mean of the distribution. Derive the results

$$\mu_1 = 0, \quad \mu_2 = E(X^2) - \mu^2, \quad \mu_3 = E(X^3) - 3\mu \cdot E(X^2) + 2\mu^3.$$

Hence obtain μ_3 for the two distributions whose p.d.fs. are given in (a).
 (c) μ_3 is often interpreted as giving a measure of the skewness of a distribution. Apply this interpretation to the two distributions in (a). By considering the discrete random variable X with $P(X = -3) = 1/10$, $P(X = -1) = \tfrac{1}{2}$, $P(X = 2) = 2/5$, show that it is *not* true that a distribution having $\mu_3 = 0$ is necessarily symmetric.   (MEI)

9. Show that, if μ_r' is the rth moment about the origin $(r = 2, 3, 4, \ldots)$ of a probability distribution of which μ_1' is the mean, then the third moment μ_3 about the mean is given by

$$\mu_3 = \mu_3' - 3\mu_2'\mu_1' + 2(\mu_1')^3.$$

The continuous random variable X takes all positive values with probability density function $(1 - \theta + \theta x)e^{-x}$. Find its moment generating function.
 Show that $\mu_r' = (1 + r\theta)(r!)$ and find μ_3 as a polynomial in θ.   (O & C)

10.★ The density function for a normal distribution $N(0, 1)$ is

$$f(x) = \frac{1}{\sqrt{(2\pi)}} e^{-\frac{1}{2}x^2} \quad -\infty < x < \infty.$$

If $Z = X^2$, prove that the m.g.f. for Z is $1/\sqrt{(1 - 2t)}$.

The substitution $x = \sqrt{(1 - 2t)} \cdot u$ will be helpful, and you may assume that

$$\frac{1}{\sqrt{2\pi}} \int_{-\infty}^{\infty} e^{-\frac{1}{2}x^2} \, dx = 1.$$

Prove that its mean is 1.
 If $Y = Z - \mu$, write down the m.g.f. of Y and hence find the variance. Find also the third moment about the mean.

21.4   Moment generating functions and 'reproductive' properties

Moment generating functions are particularly useful in establishing results about the distribution of the sum of two independent random variables with distributions of the same type. In earlier chapters, for example, we have often used (without proof) what is called the *reproductive* property of the normal distribution, which states that if X_1 and X_2 are independent random variables, each being normally distributed, then $X_1 + X_2$ is also normally distributed. Using the moment generation function we are now in a position to prove this and other reproductive properties, all of which depend on the following important theorem:

> The moment generating function of the sum of two independent random variables is the product of their moment generating functions.

Proof

If X_1 and X_2 are the two variables with m.g.fs. $M_1(t)$ and $M_2(t)$ respectively and $Y = X_1 + X_2$, then by definition $M_y(t)$, the m.g.f. of Y, is

$$M_y(t) = E(e^{tY}) = E[e^{t(X_1 + X_2)}] = E(e^{tX_1} \cdot e^{tX_2})$$
$$= E(e^{tX_1}) \cdot E(e^{tX_2}),$$

since e^{tX_1} and e^{tX_2} are independent.

Hence $M_y(t) = M_1(t) \cdot M_2(t)$.

A similar result was proved for probability generating functions on page 116, but it was of course only applicable to discrete variables.

The m.g.f. of the normal distribution and its reproductive property

For the reproductive property and other properties of the normal distribution we need its moment generating function.

Consider, first, the random variable Z (the standardised normal variable) with distribution $N(0, 1)$ and density function

$$\phi(z) = \frac{1}{\sqrt{(2\pi)}} e^{-\frac{1}{2}z^2}, \quad -\infty < z < \infty$$

$$M(t) = E(e^{tZ}) = \int_{-\infty}^{\infty} e^{tz} \left(\frac{1}{\sqrt{(2\pi)}} e^{-\frac{1}{2}z^2} \right) dz$$

$$= \frac{1}{\sqrt{(2\pi)}} \int_{-\infty}^{\infty} e^{tz - \frac{1}{2}z^2} dz = \frac{1}{\sqrt{(2\pi)}} \int_{-\infty}^{\infty} e^{-\frac{1}{2}(z^2 - 2tz)} dz$$

$$= \frac{1}{\sqrt{(2\pi)}} \int_{-\infty}^{\infty} e^{-\frac{1}{2}(z - t)^2} \cdot e^{\frac{1}{2}t^2} dz = \frac{1}{\sqrt{(2\pi)}} e^{\frac{1}{2}t^2} \int_{-\infty}^{\infty} e^{-\frac{1}{2}(z - t)^2} dz.$$

Making the substitution $u = z - t$,

$$M(t) = \frac{1}{\sqrt{(2\pi)}} e^{\frac{1}{2}t^2} \int_{-\infty}^{\infty} e^{-\frac{1}{2}u^2} du = e^{\frac{1}{2}t^2}. \quad \left(\frac{1}{\sqrt{(2\pi)}} \int_{-\infty}^{\infty} e^{-\frac{1}{2}u^2} du = 1 \right).$$

Hence   the m.g.f. for $N(0, 1)$ is $e^{\frac{1}{2}t^2}$.

For the general normal distribution, by a similar method,

the m.g.f. for $N(\mu, \sigma^2)$ is $e^{t\mu + \frac{1}{2}\sigma^2 t^2}$.

(For a proof of this result, see the Appendix, page 415.)

We shall now use the second of these results to prove the reproductive property, which, stated formally, is:

If X_1 and X_2 are independent random variables with distributions $N(\mu_1, \sigma_1^2)$ and $N(\mu_2, \sigma_2^2)$ and $Y = X_1 + X_2$, then the distribution of Y is $N(\mu_1 + \mu_2, \sigma_1^2 + \sigma_2^2)$.

Proof

Let the m.g.fs. of X_1, X_2 and Y be $M_1(t)$, $M_2(t)$ and $M_y(t)$ respectively.

Since $\qquad\qquad Y = X_1 + X_2, \qquad M_y(t) = M_1(t) \cdot M_2(t)$

$$= e^{\mu_1 t + \frac{1}{2}\sigma_1^2 t^2} \times e^{\mu_2 t + \frac{1}{2}\sigma_2^2 t^2}$$

$$= e^{(\mu_1 + \mu_2)t + \frac{1}{2}(\sigma_1^2 + \sigma_2^2)t^2},$$

which is the m.g.f. of $N(\mu_1 + \mu_2, \sigma_1^2 + \sigma_2^2)$.

It can be seen that we make the assumption that no two different distributions have the same moment generating function. In other words a moment generating function is characteristic of one and only one distribution.

EXAMPLE 8   The independent random variables X and Y each have uniform distribution in the interval $(0, 1)$ so that each has density function

$$f(x) = \begin{cases} 1, & 0 \leqslant x \leqslant 1 \\ 0, & \text{otherwise.} \end{cases}$$

If $Z = X + Y$, find its moment generating function.

W is a random variable with a triangular distribution given by the density function

$$f(w) = \begin{cases} w, & 0 \leqslant w \leqslant 1 \\ 2 - w, & 1 \leqslant w \leqslant 2. \end{cases}$$

By finding its moment generating function, show that the distribution of W is the same as that of Z.

$$M_x(t) = M_y(t) = \int_0^1 e^{tx} \cdot 1 \, dx = (e^t - 1)/t.$$

Hence $\quad M_z(t) = M_x(t) \cdot M_y(t) = (e^t - 1)^2/t^2.$

$$M_w(t) = \int_0^1 e^{tw} \cdot w \, dw + \int_1^2 e^{tw}(2 - w) \, dw,$$

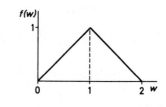

which, after integration by parts, gives

$$M_w(t) = (e^{2t} - 2e^t + 1)/t^2 = (e^t - 1)^2/t^2.$$

21.5   The m.g.f. of a linear function of a random variable

Much theoretical work with moment generating functions depends on the following important result, which connects the m.g.f. of a random variable X with the m.g.f. of a related variable Y, where

$$Y = aX + b \quad (a \text{ and } b \text{ constants}).$$

If X is a random variable with m.g.f. $M_x(t)$ and $Y = aX + b$ with m.g.f. $M_y(t)$, then

$$M_y(t) = e^{bt} M_x(at).$$

Proof

By definition

$$M_y(t) = E(e^{ty}) = E[e^{t(aX+b)}] = E(e^{atX} \cdot e^{bt})$$
$$= e^{bt} E(e^{atX}) = e^{bt} M_x(at).$$

With $a = 1$ and $b = -1$, we get the special case that if $Y = X - \mu$, $M_y(t) = e^{-\mu t} M_x(t)$, a result we have already used in calculating moments about the mean.

EXERCISE 21c

1.   Prove that the m.g.f. of a Poisson distribution with parameter a is
$$M(t) = e^{a(e^t - 1)}$$
and deduce that the distribution of the sum of two independent variables X_1 and X_2, each with Poisson distributions with parameters a_1 and a_2 respectively, is a Poisson distribution with parameter $a_1 + a_2$. (This is the reproductive property of the Poisson distribution, which was proved on page 155 using probability generating functions.)

2.   The m.g.f. of X, distributed binomially is $M(t) = (0.8 + 0.2e^t)^6$ (see question 6 in exercise 21b).
Find the m.g.f. of $Y = 2X + 3$ and use it to find the mean, variance and third moment about zero of Y.

3.   The moment generating function $M(t)$ of a normal distribution is $\exp(at + bt^2)$. Find the mean and variance of the distribution.
The independent random variables X_1, X_2 are each normal with mean μ and variance σ^2. Use moment generating functions to show that $Y = X_1 + X_2$ has a normal distribution, and derive its mean and variance.
Write down functions of X_1, X_2 only which have normal distributions with:
(a) mean 0 and variance σ^2;
(b) mean 1.4μ and variance σ^2.   (O)

4.   The continuous random variable X has an exponential distribution with parameter α, that is, its probability density function is given by
$$f(x) = \begin{cases} 0 & , \quad \text{for } x < 0 \\ \alpha e^{-\alpha x}, & \quad \text{for } x \geqslant 0. \end{cases}$$
Show that the moment generating function of X is given by
$$M_x(t) = \alpha/(\alpha - t), \quad (t < \alpha).$$
A continuous random variable Y has an exponential distribution with parameter β, where

$\beta > \alpha$, and the random variable Z is given by $Z = X + Y$. Assuming that X and Y are independent, find the moment generating function of Z.

Show that the random variable with probability density function given by

$$y(z) = \begin{cases} 0 & , \quad \text{for } z < 0, \\ \dfrac{\alpha\beta}{\beta - \alpha}\,(e^{-\alpha z} - e^{-\beta z}), & \text{for } z \geqslant 0 \end{cases}$$

has the same moment generating function as Z.

Find $E(Z)$ and $\text{Var}(Z)$.   (C)

5.★ A continuous random variable Z has probability density function given by

$$f(z) = \begin{cases} 0 & , \quad \text{for } z < 0 \\ \alpha e^{-\alpha z}, & \text{for } z \geqslant 0, \end{cases}$$

where α is a positive constant. Find the moment generating function $M_z(t)$ of Z.

Two independent continuous random variables X_1 and X_2 are each normally distributed with mean 0 and standard deviation 1.

Assuming that

$$\int_{-\infty}^{\infty} e^{-\frac{1}{2}x^2}\,dx = \sqrt{(2\pi)},$$

show that the moment generating function of X_1^2, $E(e^{tX_1^2})$, is

$$(1 - 2t)^{-\frac{1}{2}},$$

and find the moment generating function $M_y(t)$, of Y where $Y = X_1^2 + X_2^2$.

For what value of α does $M_y(t) = M_z(t)$?   (C)

6.★ The random variable X has p.d.f.

$$f(x) = \begin{cases} \lambda^{k+1}x^k e^{-\lambda x}/k!, & x > 0 \\ 0 & , \quad \text{elsewhere,} \end{cases}$$

where $\lambda > 0$ and k is a non-negative integer.

Show that the moment generating function of X is

$$[\lambda/(\lambda - t)]^{k+1}.$$

The random variable Y is the sum of n independent random variables each distributed as X. Find the moment generating function of Y, and hence find the mean and variance of Y.

Write down the p.d.f. of Y.   (MEI)

7.★ (a) The m.g.f. of a random variable U is defined by $M_u(t) = E(e^{tU})$. Prove from this definition that, if $V = aU + b$, where a and b are constants, the m.g.f. of V is $M_v(t) = e^{bt}M_u(at)$.

(b) The non-negative variable X has p.d.f. $f(x) = e^{-x}, x \geqslant 0$. Show that the m.g.f. of X is $M_x(t) = 1/(1-t)$ (provided $t < 1$), and deduce that the mean and variance of X are both unity.

The random variable Y is defined by $Y = X_1 + X_2 + \ldots + X_n$, where $X_1, X_2, \ldots, X_n$ are independent random variables each distributed as X. Write down the m.g.f. of Y and the mean and variance of Y.

The random variable $\bar{X}$ is defined by $\bar{X} = (X_1 + X_2 + \ldots + X_n)/n = Y/n$.

Use the result of part (a), taking $a = 1/n$ and $b = 0$, to show that the m.g.f. of $\bar{X}$ is $(1 - t/n)^{-n}$. Write down the mean and variance of $\bar{X}$.

The random variable Z is defined by $Z = \dfrac{\bar{X} - 1}{1/\sqrt{n}};$

this has mean 0 and variance 1 for any n, and is called the *standardised mean* of n independent realisations of X. Again, using the result of part (a), show that the m.g.f. of Z is

$$M_z(t) = e^{-t\sqrt{n}}(1 - t/\sqrt{n})^{-n}.$$

Show that

$$\ln M_z(t) = \frac{t^2}{2} + \frac{t^3}{3}n^{-\frac{1}{2}} + \frac{t^4}{4}n^{-1} + \dots,$$

given that the m.g.f. of the standard normal random variable is $e^{\frac{1}{2}t^2}$. What does this result lead you to believe about the distribution of Z as n becomes large?   (MEI)

21.6   Properties of the normal distribution

In the Appendix that follows, various properties of the normal distribution that have been assumed without proof in earlier chapters will be proved. Not all of these proofs are difficult, but as a rough guide, if you intend to look at some of them, they are, in order of accessibility:

1.   The Poisson distribution with parameter μ approximates to a normal distribution, $N(\mu, \mu)$ if μ is large.

2.   The binomial distribution $B(n, p)$ approximates to a normal distribution $N(np, npq)$ when n is large and p is not too small.

3.   The central limit theorem, which states that if $X_1, X_2, \dots, X_n$ are independent random variables, all having the same distribution, and $Y = X_1 + X_2 + \dots + X_n$, then the distribution of Y is approximately normal for large values of n, regardless of the nature of the distribution of the X_i.

21.7   Appendix

The moment generating function of $N(\mu, \sigma^2)$

Since the density function of $N(\mu, \sigma^2)$ is

$$\phi(x) = \frac{1}{\sigma\sqrt{(2\pi)}} e^{-\frac{1}{2}(x-\mu)^2/\sigma^2}, \quad -\infty < x < \infty,$$

$$M(t) = E(e^{tX}) = \frac{1}{\sigma\sqrt{(2\pi)}} \int_{-\infty}^{\infty} e^{tx} e^{-\frac{1}{2}(x-\mu)^2/\sigma^2} \, dx.$$

Substituting $(x - \mu)/\sigma = u$, so that $x = \sigma u + \mu$ and $dx = \sigma \, du$

$$M(t) = \frac{1}{\sqrt{(2\pi)}} \int_{-\infty}^{\infty} e^{t(\sigma u + \mu)} e^{-\frac{1}{2}u^2} \, du = \frac{e^{t\mu}}{\sqrt{(2\pi)}} \int_{-\infty}^{\infty} e^{-\frac{1}{2}(u^2 - 2\sigma tu)} \, du$$

$$= \frac{e^{t\mu}}{\sqrt{(2\pi)}} \int_{-\infty}^{\infty} e^{-\frac{1}{2}[(u - \sigma t)^2 - \sigma^2 t^2]} \, du$$

$$= \frac{e^{t\mu + \frac{1}{2}\sigma^2 t^2}}{\sqrt{(2\pi)}} \int_{-\infty}^{\infty} e^{-\frac{1}{2}(u - \sigma t)^2} \, du.$$

Making the further substitution $v = u - \sigma t$ and $dv = du$

$$M(t) = \frac{e^{t\mu + \frac{1}{2}\sigma^2 t^2}}{\sqrt{(2\pi)}} \int_{-\infty}^{\infty} e^{-\frac{1}{2}v^2} \, dv = e^{t\mu + \frac{1}{2}\sigma^2 t^2}$$

since

$$\frac{1}{\sqrt{2(\pi)}} \int_{-\infty}^{\infty} e^{-\frac{1}{2}v^2} \, dv = 1.$$

Putting $\mu = 0$ and $\sigma^2 = 1$ gives the m.g.f. of $N(0, 1) = e^{\frac{1}{2}t^2}$, a result that was proved on page 411. In the proofs that follow it is the m.g.f. of $N(0, 1)$ that is needed.

The Poisson distribution with parameter $\mu \to N(\mu, \mu)$ if μ is large

Proof

In exercise 21c, question 1 the m.g.f. of a random variable X whose distribution is Poisson with parameter μ is given as

$$M_x(t) = e^{\mu(e^t - 1)},$$

and if $Y = aX + b$ the m.g.f. of Y is $M_y(t) = e^{bt} M_x(at)$.
   Consider the random variable

$$Y = \frac{X - \mu}{\sqrt{\mu}} = X/\sqrt{\mu} - \sqrt{\mu}.$$

If we put $a = 1/\sqrt{\mu}$ and $b = -\sqrt{\mu}$,

$$M_y(t) = e^{-\sqrt{(\mu)}t} M(t/\sqrt{\mu}) = e^{-\sqrt{(\mu)}t} e^{\mu(e^{t/\sqrt{\mu}} - 1)}.$$

Writing $\sqrt{\mu} = N$, $\quad M_y(t) = e^{-Nt} e^{N^2(e^{t/N} - 1)}$.

But $N^2(e^{t/N} - 1) = N^2(t/N + t^2/2N^2 + t^3/6N^3 + \ldots)$

$$= Nt + t^2/2 + t^3/6N + \ldots$$

$$\approx Nt + t^2/2 \quad \text{if } N \text{ is large.}$$

Hence $M_y(t) \approx e^{-Nt} e^{Nt + t^2/2} \quad \text{if } \mu \text{ is large}$

$$= e^{\frac{1}{2}t^2}.$$

But this is the m.g.f. of $N(0, 1)$, and it follows that the distribution of

$$Y = \frac{X - \mu}{\sqrt{\mu}} \approx N(0, 1)$$

and therefore that the distribution of $X \approx N(\mu, \mu)$.

The normal distribution as the limit of $B(n, p)$ as $n \to \infty$

Proof

On page 401 we showed that the m.g.f. of a random variable X whose distribution is $B(n, p)$ is

$$M_x(t) = (q + pe^t)^n, \quad \text{where } q = 1 - p,$$

and if $Y = aX + b$ the m.g.f. of Y is $M_y(t) = e^{bt}M_x(at)$.

Consider the random variable

$$Y = \frac{X - np}{\sqrt{(npq)}} = X/\sqrt{(npq)} - \sqrt{(np/q)}.$$

If we put $a = 1/\sqrt{(npq)}$ and $b = -\sqrt{(np/q)}$

$$M_y(t) = e^{-\sqrt{(np/q)}t} M(t/\sqrt{(npq)}) = e^{-\sqrt{(np/q)}t}[q + pe^{t/\sqrt{(npq)}}]^n.$$

Hence $\quad \ln M_y(t) = -\sqrt{(np/q)}t + n\ln[q + pe^{t/\sqrt{(npq)}}].$

Writing $\sqrt{(npq)} = N$ and $\sqrt{(np/q)} = np/\sqrt{(npq)} = np/N$,

$$\ln M_y(t) = -\frac{np}{N}t + n\ln[q + pe^{t/N}]$$

$$= -\frac{np}{N}t + n\ln[q + p(1 + t/N + t^2/2N^2 + \therefore)]$$

$$= -\frac{np}{N}t + n\ln[1 + pt/N + pt^2/2N^2 + \ldots]$$

$$= -\frac{np}{N}t + n[(pt/N + pt^2/2N^2 + \ldots) - \tfrac{1}{2}(pt/N + pt^2/2N^2 + \ldots)^2 + \ldots]$$

$$\approx -\frac{np}{N}t + n[pt/N + pt^2/2N^2 - p^2t^2/2N^2] \quad \text{if } N \text{ is large}$$

$$= -\frac{np}{N}t + \frac{np}{N}t + \frac{np(1-p)t^2}{2N^2} = \frac{npqt^2}{2npq} = t^2/2.$$

Hence $\qquad\qquad M_y(t) \approx e^{\frac{1}{2}t^2}.$

But this is the m.g.f. of $N(0, 1)$, and it follows that the distribution of

$$Y = \frac{X - np}{\sqrt{(npq)}} \approx N(0, 1)$$

and therefore that the distribution of $X \approx N(np, npq)$ for large n.

The central limit theorem

If $X_1, X_2, \ldots, X_n$ are independent random variables, all having the same distribution and $Y = X_1 + X_2 + \ldots + X_n$, then the distribution of Y is approximately normal for large values of n, regardless of the nature of the distribution of the X_i.

Proof

Although this is not a rigorous proof, it will suggest the way such a proof could be approached. We start by stating the results that will be needed in the course of the proof:

1. If μ is the mean of the distribution of X then the m.g.f. of $(X - \mu)$, which we will

call $M(t)$, without a suffix, is

$$M(t) = 1 + t\mu_1 + \frac{t^2}{2!}\mu_2 + \frac{t^3}{3!}\mu_3 + \dots \quad \text{where } \mu_1, \mu_2, \dots$$

are moments about the mean.

Since $\mu_1 = 0$ and $\mu_2 = \sigma^2$, this may be written:

$$M(t) = 1 + \sigma^2 t^2/2! + \mu_3 t^3/3! + \dots.$$

2. If the m.g.f. of X is $M(t)$, the m.g.f. of bX is $M(bt)$ (see page 413).

3. If $Z = Z_1 + Z_2$, then $M_z(t) = M_{z_1}(t) \times M_{z_2}(t)$. [$Z_1, Z$, independent.]

4. The m.g.f. of $N(0,1)$ is $e^{\frac{1}{2}t^2}$.

Let the distribution of each X_i have mean μ and variance σ^2, so that the mean of Y is $n\mu$ and its variance is $n\sigma^2$.

We will call the m.g.f. of each $(X_i - \mu)$, $M(t)$, $(i = 1, 2, \dots, n)$ and the m.g.f. of $(Y - n\mu)$, $M_z(t)$.

Now $Y - n\mu = (X_1 - \mu) + (X_2 - \mu) + \dots + (X_n - \mu)$, and it follows that

$$M_z(t) = [M(t)]^n \quad \text{(see 3).}$$

Let $M_s(t)$ be the m.g.f. of the random variable $\dfrac{Y - n\mu}{\sigma\sqrt{n}}$.

Then, using (2) with $b = 1/\sigma\sqrt{n}$,

$$M_s(t) = M_z(t/\sigma\sqrt{n}) = [M(t/\sigma\sqrt{n})]^n.$$

$$\Rightarrow \ln M_s(t) = n\ln[M(t/\sigma\sqrt{n})].$$

Now $M(t) = 1 + \sigma^2 t^2/2 + \mu_3 t^3/6 + \dots$ [see (1)]

$$\Rightarrow M(t/\sigma\sqrt{n}) = 1 + \sigma^2(t^2/n\sigma^2)/2 + \mu_3(t^3/n\sqrt{(n)}\sigma^3)/6 + \dots$$

$$\approx 1 + t^2/2n, \quad \text{when } n \text{ is large.}$$

$$\Rightarrow \ln M_s(t) \approx n\ln(1 + t^2/2n) \approx n \cdot t^2/2n,$$

since $\ln(1 + x) \approx x$ for small x.

Hence $\ln M_s(t) \approx \frac{1}{2}t^2 \Rightarrow M_s(t) \approx e^{\frac{1}{2}t^2}$,

which is the m.g.f. of $N(0, 1)$.

It follows that the distribution of $(Y - n\mu)/\sigma\sqrt{n}$ is approximately $N(0, 1)$, and that the distribution of Y is approximately normal with mean $n\mu$ and variance $n\sigma^2$.

Note: This proof depends on the assumption that the distribution of the X_i *has* a moment generating function.

22* Continuous distributions: further topics

The first part of this chapter takes the theory of continuous probability distributions that was developed in chapter 10 a stage further.

22.1 The distribution of a function of a continuous random variable

The kind of problem dealt with in this section is this: given the density function $f(x)$ of a random variable X, to find the density function for some *function of X*.

If, for example, the distribution of V, the speed of a molecule of mass m moving in a gas is known to be given by the density function $f(v)$, how do we find the density function for the kinetic energy of the particle, $\frac{1}{2}mV^2$?

The method is illustrated in the following examples. In each case, if Y is the function of X, we first find the *distribution* function of Y, $F(y)$ and derive the *density* function $f(y)$ by differentiation, since as we saw on page 180,

$$\frac{dF(y)}{dy} = f(y).$$

EXAMPLE 1 The random variable X has a uniform distribution in the interval $1 \leqslant x \leqslant 4$. Find the density function of:

   (a) $Y = 2X + 3$,   (b) $Y = e^X$.

(a) The density function of X,

$$f(x) = \begin{cases} 1/3, & 1 \leqslant x \leqslant 4 \\ 0, & \text{elsewhere,} \end{cases}$$

and its distribution function

$$F(x) = \int_1^x \tfrac{1}{3}\, dt = \tfrac{1}{3}(x-1).$$

By definition, the distribution function of Y,

$$G(y) = P(Y \leqslant y)$$

$$= P(2X + 3 \leqslant y)$$

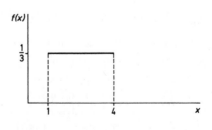

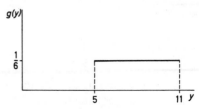

$$= P[X \leqslant \tfrac{1}{2}(y-3)]$$

$$= F[\tfrac{1}{2}(y-3)] = \tfrac{1}{3}[\tfrac{1}{2}(y-3)-1] = \frac{y}{6} - \frac{5}{6}.$$

Differentiating $G(y)$ w.r.t. y, the density function of Y

$$g(y) = G'(y) = \tfrac{1}{6} \quad \text{for } 5 \leqslant y \leqslant 11.$$

The two density functions are shown in the figures.

(b) As before, $F(x) = \tfrac{1}{3}(x-1)$ and

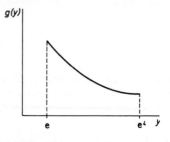

$$G(y) = P(Y \leqslant y)$$

$$= P(e^X \leqslant y)$$

$$= P(X \leqslant \ln y) = F(\ln y) = \tfrac{1}{3}(\ln y - 1).$$

Hence $g(y) = G'(y) = \dfrac{1}{3y} \quad \text{for } e \leqslant y \leqslant e^4.$

In the next example, $f(x)$ is not uniform.

EXAMPLE 2   The density function of X is

$$f(x) = \begin{cases} e^{-x}, & x > 0 \\ 0, & \text{elsewhere.} \end{cases}$$

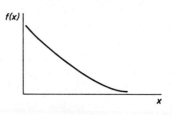

Find the density function of $Y = X^3$.

$$F(x) = \int_0^x e^{-t}\,dt = 1 - e^{-x}.$$

$$G(y) = P(Y \leqslant y)$$

$$= P(X^3 \leqslant y)$$

$$= P(X \leqslant y^{1/3}) = F(y^{1/3})$$

$$= 1 - e^{-y^{1/3}}$$

Hence $g(y) = G'(y) = \tfrac{1}{3}y^{-2/3} e^{-y^{1/3}} \quad \text{for } y > 0.$

In these examples the functions of X, $2X+3$, e^X, X^3, were all increasing functions. A little more care is needed when this is not so, as the next examples show.

EXAMPLE 3 X has a uniform distribution for $-2 \leqslant x \leqslant 2$. If Y is the area of the square of side X, find the density function of Y. Find also: (a) the mean area, (b) the probability that the area is greater than 2 square units.

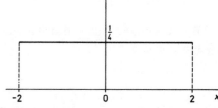

$$f(x) = \begin{cases} \tfrac{1}{4}, & -2 \leqslant x \leqslant 2 \\ 0, & \text{elsewhere} \end{cases}$$

and, as the diagram at the top of page 421 shows, y ranges from 0 to 4.

$$F(x) = \int_{-2}^{x} \tfrac{1}{4}\, dt = \tfrac{1}{4}(x+2).$$

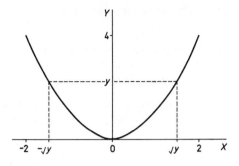

Since $Y = X^2$,

$$G(y) = P(Y \leqslant y)$$
$$= P(X^2 \leqslant y)$$
$$= P(-\sqrt{y} \leqslant X \leqslant \sqrt{y})$$

(see diagram)

$$= F(\sqrt{y}) - F(-\sqrt{y}) = \tfrac{1}{4}(\sqrt{y}+2) - \tfrac{1}{4}(-\sqrt{y}+2) = \tfrac{1}{2}\sqrt{y}.$$

Hence $g(y) = G'(y) = \tfrac{1}{4}y^{-\frac{1}{2}}$,   for $0 < y < 4$.

The mean area $= E(Y)$

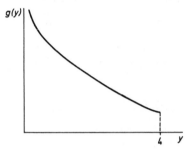

$$= \int_{0}^{4} y \cdot \tfrac{1}{4} y^{-\frac{1}{2}}\, dy$$

$$= \int_{0}^{4} \tfrac{1}{4} y^{\frac{1}{2}}\, dy$$

$$= \left[\tfrac{1}{6} y^{3/2}\right]_{0}^{4} = 4/3 \text{ square units.}$$

$$P(Y > 2) = 1 - \int_{0}^{2} \tfrac{1}{4} y^{-\frac{1}{2}}\, dy = 1 - \left[\tfrac{1}{2} y^{\frac{1}{2}}\right]_{0}^{2} = 0.29.$$

EXAMPLE 4   A point P is taken at random on a line of length 2 units and the distribution of $X = AP$ is uniform. Find the density function of Y where

$$Y = AP^2 + PB^2.$$

The density function of X is

$$f(x) = \begin{cases} \tfrac{1}{2}, & 0 \leqslant x \leqslant 2 \\ 0, & \text{elsewhere,} \end{cases}$$

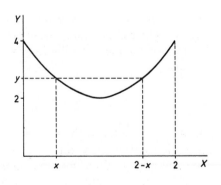

so that $F(x) = \tfrac{1}{2}x$.

Now $Y = X^2 + (2-X)^2$ and varies between 2 and 4.

$$G(y) = P(Y \leqslant y)$$

$$= P(x \leqslant X \leqslant 2-x) \quad \text{(see diagram)}$$

$$= F(2-x) - F(x)$$

$$= \tfrac{1}{2}(2-x) - \tfrac{1}{2}x = 1 - x.$$

But, since $y = x^2 + (2-x)^2$, $x = 1 - \tfrac{1}{2}\sqrt{(2y-4)}$, $(x < 1)$ so that

$$G(y) = \tfrac{1}{2}\sqrt{(2y-4)} \text{ and}$$

$$g(y) = \tfrac{1}{2}(2y-4)^{-\frac{1}{2}}, \quad 2 \leqslant y \leqslant 4.$$

22.2   The density function of X^2

The method used in example (3) is a special case of a general result that is of particular importance in connection with distributions derived from the normal distribution (see chapter 23). We shall therefore state and prove the general result.

If X is a continuous random variable that can assume both positive and negative values, and $Y = X^2$, then the distribution of Y has density function:

$$g(y) = \frac{1}{2\sqrt{y}} [f(\sqrt{y}) + f(-\sqrt{y})]$$

where $f(x)$ is the density function of X.

Proof

The distribution function of Y,

$$\begin{aligned} G(y) &= P(Y \leqslant y) \\ &= P(X^2 \leqslant y) \\ &= P(-\sqrt{y} \leqslant X \leqslant \sqrt{y}) \\ &= F(\sqrt{y}) - F(-\sqrt{y}), \end{aligned}$$

where $F(x)$ is the distribution function of X.

Hence
$$g(y) = G'(y)$$
$$= \tfrac{1}{2} y^{-\frac{1}{2}} F'(\sqrt{y}) + \tfrac{1}{2} y^{-\frac{1}{2}} F'(-\sqrt{y})$$
$$= \frac{1}{2\sqrt{y}} [f(\sqrt{y}) + f(-\sqrt{y})].$$

22.3   The Cauchy distribution

So far in this book it has been assumed that distribution models always have a finite variance. There is, however, one interesting distribution where this is not the case.

Consider first the t-distribution with one degree of freedom. Its density function (see page 293) is

$$f(t) = \frac{1}{\pi(1+t^2)}, \quad -\infty < t < \infty,$$

and it is so long-tailed that its variance is infinite. (You can easily verify this by integration.) From the point of view of sampling, this is clearly not much use, as we could not hope to draw any reasonable conclusions from a single sample of size 2 ($v = 2-1$), or two samples of sizes 1 and 2 ($v = 2+1-2$). From another standpoint, however, it is of considerable interest as being a special case of the *Cauchy distribution*.

Imagine a fireman is standing in front

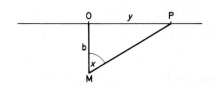

of a straight wall which is so long that it can be regarded as endless. His distance OM from the nearest point of the wall is b.

Out of boredom he spends his time spraying the wall at random with his fire-hose, and we will assume that he is sufficiently near for the narrow jet of water to strike the wall at points P which are always on the same horizontal line. Let OP $= y$, and suppose that the angle PMO (x) is uniformly distributed from (in theory) $-\pi/2$ to $\pi/2$. What is the distribution of Y?

The density function of X is

$$f(x) = \begin{cases} 1/\pi, & -\pi/2 < x < \pi/2 \\ 0 , & \text{elsewhere} \end{cases}$$

and its distribution function is

$$F(x) = \int_{-\pi/2}^{x} \frac{1}{\pi}\, dt = \frac{1}{\pi}\left(x+\frac{\pi}{2}\right).$$

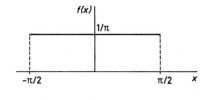

Now $Y = b \tan X$, and we find its distribution $G(y)$ and density function $g(y)$ in the usual way.

$$G(y) = P(Y \leqslant y) = P(b \tan X \leqslant y) = P\left(X \leqslant \tan^{-1}\frac{y}{b}\right)$$

$$= F\left(\tan^{-1}\frac{y}{b}\right) = \frac{1}{\pi}\left(\tan^{-1}\frac{y}{b}+\frac{\pi}{2}\right).$$

Hence $g(y) = G'(y) = \dfrac{1}{\pi}\left(\dfrac{b}{b^2+y^2}\right).$

Definition

The distribution of the random variable Y whose density function is

$$g(y) = \frac{b}{\pi(b^2+y^2)}, \quad -\infty < y < \infty,$$

is a Cauchy distribution with mean 0, and parameter b.

The general Cauchy distribution is the distribution of the random variable Y whose density function is

$$g(y) = \frac{b}{\pi[b^2+(y-\mu)^2]}, \quad -\infty < y < \infty.$$

The mean in the general case is μ.

EXERCISE 22a

1.   The random variable X is uniformly distributed in the interval $-1 \leqslant x \leqslant 1$. Find the probability density function of Y where:
(a) $Y = 3X+1$,   (b) $Y = e^{3x}$,   (c) $Y = 1/(X+2)$.

2.   If the random variable X is uniformly distributed over the interval $-3 \leqslant x \leqslant 3$, find the probability density function of Y where:
(a) $Y = X^2$,   (b) $Y = X^2+1$.

3.   The line AB is 10 cm long. A point P is taken at random on AB, that is the distribution of AP is uniform. Find the probability density function for Y, the area of the circle with radius AP.

What is the probability that the area of the circle will be greater than 10 cm²?

4.   The length of the side of a cube, X, is uniformly distributed between 2 and 3 cm. If Y is the volume of the cube, find its density function. Find also $E(Y)$ and $Var(Y)$.

5.   The density function of a random variable X is

$$f(x) = \begin{cases} 0 & , \quad x > 0 \\ 6x(1-x), & 0 \leqslant x \leqslant 1, \\ 0 & , \quad x > 1. \end{cases}$$

If $Y = X^2$, prove that the distribution function of Y is

$$F(y) = 3y - 2y^{3/2},$$

and deduce the density function of y.

6.   The line AB is of length $2a$. A point P is taken at random on AB. Find the expected value of the area of the rectangle with sides AP and PB.

Prove also that if the area of the rectangle is Y, the density function of Y is

$$1/2a\sqrt{(a^2 - y)}, \quad 0 \leqslant y < a^2.$$

7.   A measuring flask is so shaped that the volume V of the liquid in it is related to the height h of the surface of the liquid above the base of the flask by $V = \pi h^3/9$. The flask is used to measure volumes V of liquid which are uniformly distributed between 0 and 1 units. Find the probability density function of the distribution of the corresponding values of h and sketch the graph of this function. Calculate:
(a) the probability that h is less than 1 unit,
(b) the mean of h,
(c) the variance of h,
(d) the probability that h is less than the mean of h.
[Take $\sqrt[3]{(9/\pi)}$ as 1.42.]   (JMB)

8.   A child rides on a roundabout and his father waits for him at the point where he started. His journey may be regarded as a circular route of radius six metres and the father's position as a fixed point on the circle. When the roundabout stops, the shorter distance of the child from the father, measured along the circular path, is S metres. All points on the circle are equally likely stopping points, so that S is uniformly distributed between 0 and 6π. Find the mean and variance of S.

The direct linear distance of the child's stopping point from the father is D metres. Show that the probability density function of D is

$$f(d) = \frac{2}{\pi\sqrt{(144 - d^2)}}, \quad 0 \leqslant d \leqslant 12$$

and zero outside this range.

The father's voice can be heard at a distance of up to ten metres. Find to two decimal places the probability that the child can hear his father shout to him when the roundabout stops.   (JMB)

22.4   The distribution of the sum of two uniformly distributed random variables

When we introduced the uniform and triangular distributions (pages 182–3), it was stated that the first was a possible model for the distribution of errors in the measurements of drawn lines, and the second was a model for the distribution of the sum of the errors in measuring pairs of lines.

We are now in a position to establish the latter result, which we shall prove in the following form:

If X and Y are independent continuous random variables, each uniformly distributed between 0 and 1, and if $Z = X + Y$, then Z has a triangular distribution.

Proof

Let $f(z)$, $F(z)$ be the density and distribution functions of Z.

All possible pairs of values of X and Y can be represented by the coordinates (x, y) of points in a unit square.

For a particular value, z, of Z, all pairs satisfying $x + y = z$ are represented by points on the line AB.

Since the points may be expected to be spread uniformly over the square,

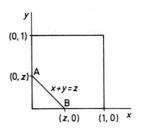

$$P(Z \leqslant z) \text{ is}$$

the fraction of area of the square to the left of AB.

If $0 \leqslant z \leqslant 1$:

$$P(Z \leqslant z) = \tfrac{1}{2}z^2,$$

that is $F(z) = \tfrac{1}{2}z^2$.

Hence $f(z) = F'(z) = z$.

If $1 \leqslant z \leqslant 2$:

$$P(Z \leqslant z) = 1 - \tfrac{1}{2}(2 - z)^2$$

that is $F(z) = -1 + 2z - \tfrac{1}{2}z^2$ and

$$f(z) = F'(z) = 2 - z.$$

Thus

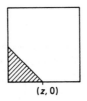 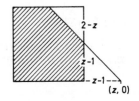

$$f(z) = \begin{cases} z & , \quad 0 \leqslant z \leqslant 1 \\ 2 - z, & 1 < z \leqslant 2 \\ 0 & , \quad \text{elsewhere,} \end{cases}$$

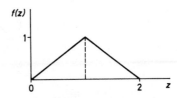

which is a triangular distribution.

Similarly, if X, Y are errors made in measuring lines to the nearest milli-metre, their distributions are uniform between -0.05 cm and 0.05 cm and Z (the sum of the errors) has density function

$$f(z) = \begin{cases} 100(z-0.1), & -0.1 \leqslant z \leqslant 0 \\ 100(0.1-z), & 0 < z \leqslant 0.1 \\ 0, & \text{elsewhere.} \end{cases}$$

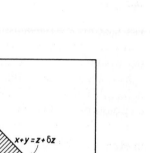

As an alternative argument:

If $0 \leqslant z \leqslant 1$:

$$P(z \leqslant Z \leqslant z+\delta z)$$

$$= \text{area between AB and CD}$$

$$= \tfrac{1}{2}(z+\delta z)^2 - \tfrac{1}{2}z^2 \approx z\,\delta z.$$

but, by definition,

$$P(z \leqslant Z \leqslant z+\delta z) \approx f(z)\,\delta z.$$

Hence $f(z) = z$.

In a similar way, if $1 \leqslant z \leqslant 2$:

$$P(z \leqslant Z \leqslant z+\delta z)$$

can be shown to be approximately

$$(2-z)\,\delta z$$

so that

$$f(z) = 2-z.$$

EXERCISE 22b

1. If X and Y are independent continuous random variables, each uniformly distributed between 0 and 1, and if $W = X - Y$, prove that W has the probability density function

$$f(w) = \begin{cases} 1-w, & 0 \leqslant w \leqslant 1 \\ 1+w, & -1 \leqslant w < 0 \\ 0, & \text{otherwise.} \end{cases}$$

(*Hint:* All pairs (x, y) satisfying $x-y = w$ are represented by points on the line AB.
   Find the shaded areas in the two cases:
(a) $0 \leqslant w \leqslant 1$,
(b) $-1 \leqslant w < 0$.

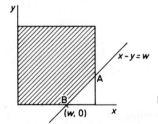

2.★ If X, Y and Z are three independent continuous random variables, each uniformly distributed between 0 and 1, and if $W = X + Y + Z$, prove that the distribution function for W is

$$F(w) = \begin{cases} w^3/6 & , \quad 0 \leqslant w < 1 \\ w^3/6 - \frac{1}{2}(w-1)^3, & 1 \leqslant w < 2 \\ 1 - (3-w)^3/6 & , \quad 2 \leqslant w \leqslant 3. \end{cases}$$

Hence show that W has a density function consisting of three parabolic arcs. Sketch the density function.

(*Hint*: All possible triplets can be represented by coordinates of points in a unit cube. For a particular value w of W all triplets satisfying $x + y + z = w$ lie on a plane. Find the volumes of the parts of the cube cut off by this plane for the three ranges of w given in the answer.)

22.5   The Poisson and exponential distributions

The heading to this section links together a discrete distribution (Poisson) and a continuous one (exponential) and it will be our purpose to use the Poisson distribution to show that the exponential distribution has some important applications that have not so far been considered. In order to do this we must first look again at the Poisson distribution and the way it is derived.

In chapter 9 it was arrived at as the limit of the binomial distribution $B(n, p)$ as $n \to \infty$, $p \to 0$ while $np = a$. The resulting probabilities of $0, 1, 2, \ldots$ occurrences of an event turned out to be

$$e^{-a}, ae^{-a}, \frac{a^2}{2!} e^{-a}, \ldots.$$

We shall now show that these probabilities may be arrived at by a quite different approach.

An alternative approach to the Poisson distribution

In the examples used in chapter 9 to illustrate the Poisson distribution typical events were:

        calls received at a telephone exchange in a given time;

        accidents reported at a police station in a given time.

Let us now examine the assumptions that were made about these events:
(a) the events are *random* and *independent* (in the sense that the number of events occurring in any interval of time is unaffected by what has happened in any other, non-overlapping interval;
(b) the events occur *uniformly* (in the sense that the mean number over a given period of time is constant, depending only on the length of the period).

In both examples, too, the events can be thought of as being represented by points on a *time axis*, and the random variable is the number of events occurring in a specified time interval.

When these assumptions seem to apply the Poisson distribution is likely to be the appropriate model. Further examples of situations where these assumptions are reasonable are:

>a Geiger counter, where the events are the arrival of atomic particles at the counter;

>a machine, where the events are breakdowns;

>a factory, where the events are industrial accidents;

>an airport, where the events are requests by aircraft to land.

What we shall now do is prove that the model for the distribution of events like this *is* the Poisson distribution, the *only* assumptions made being that the events (happening in *time*) are random, independent and uniform as defined above. (It will be shown later that the idea of events happening in time only may be extended.)

Consider the telephone exchange example again where the random variable is the number of calls received in a given time period.

Let λ = the mean number of calls per unit time. (In our example on pages 150–1 the mean number of calls per hour is 2, so that, if the unit of time is one minute, the value of λ would be 2/60.)

We shall think of the calls as being represented by points on a time axis and we shall be concerned with the random variable R, which is the number of calls received in a time interval of t.

Let $P_0(t)$ be the probability that there are no calls before time t.

Let δt be an interval small enough for the chance of more than one call arriving in it to be negligible.

The probability of a call arriving in this interval = $\lambda\,\delta t$, and the probability that there are no calls between times t and $t+\delta t$ = $1 - \lambda\,\delta t$ (since, in this small interval, a call either does or does not arrive.)

>$P_0(t+\delta t)$= the probability that there are no calls before time $t+\delta t$
>
>= (the probability that there are no calls before time t)
>
>× (the probability that there are no calls between times t and $t+\delta t$).
>
>= $P_0(t) \times (1 - \lambda\,\delta t)$.

Hence
$$\frac{P_0(t+\delta t) - P_0(t)}{\delta t} = -\lambda P_0(t),$$

and, letting δt tend to zero,
$$\frac{\mathrm{d}P_0(t)}{\mathrm{d}t} = -\lambda P_0.$$

The solution of this differential equation is
$$P_0(t) = k\mathrm{e}^{-\lambda t},$$

where k is an arbitrary constant. Now the probability that no calls arrive during a zero time interval is 1, so that $P_0(0) = 1$. Hence $k = 1$ and we have
$$P_0(t) = \mathrm{e}^{-\lambda t}.$$

Suppose now that $P_r(t)$ is the probability that exactly r calls occur up to time t. We derive a differential equation for $P_r(t)$ in a similar way.

First we find $P_r(t+\delta t)$, remembering that δt is small enough for us to be able to assume that between times t and $t+\delta t$ there is either no call or one call. This means that there can be r calls up to time $t+\delta t$ in two ways:

> either, r calls up to time t and no calls between t and $t+\delta t$,
>
> or, $r-1$ calls up to time t and one call between t and $t+\delta t$.

Hence
$$P_r(t+\delta t) = P_r(t)(1-\lambda\ \delta t) + P_{r-1}(t)\lambda\ \delta t$$

$$\Rightarrow \frac{P_r(t+\delta t)-P_r(t)}{\delta t} = \lambda P_{r-1}(t) - \lambda P_r(t).$$

Letting δt tend to zero,

$$\frac{dP_r(t)}{dt} = \lambda P_{r-1}(t) - \lambda P_r(t), \quad r = 1, 2, \ldots.$$

We now have a series of differential equations for different values of r. In order to see how to solve the general case, we will first look at the two special cases, $r = 1$ and $r = 2$.

Putting $r = 1$,
$$\frac{dP_1(t)}{dt} = \lambda P_0(t) - \lambda P_1(t)$$

or
$$\frac{dP_1(t)}{dt} + \lambda P_1(t) = \lambda e^{-\lambda t} \quad [\text{since } P_0(t) = e^{-\lambda t}].$$

This is solved by using the integrating factor $e^{\lambda t}$, and the solution is

$$e^{\lambda t} P_1(t) = \lambda t + k'.$$

The arbitrary constant k' is found by substituting $t = 0$, and using the fact that $P_1(0) = $ the probability of one call in zero time $= 0$. Hence $k' = 0$ and

$$P_1(t) = \lambda t e^{-\lambda t}.$$

Putting $r = 2$,
$$\frac{dP_2(t)}{dt} = \lambda P_1(t) - \lambda P_2(t)$$

or
$$\frac{dP_2(t)}{dt} + \lambda P_2(t) = \lambda^2 t e^{-\lambda t} \quad [\text{since } P_1(t) = \lambda t e^{-\lambda t}],$$

which, using the same integrating factor, has the solution

$$P_2(t) = \frac{(\lambda t)^2}{2} e^{-\lambda t}.$$

Using induction it is easily shown that the general solution is

$$P_r(t) = \frac{(\lambda t)^r}{r!} e^{-\lambda t}, \quad r = 0, 1, 2, \ldots.$$

These are, of course, the Poisson probabilities as defined on page 148 with parameter λt, the mean number of calls in time t.

At this stage you may be wondering whether this fairly lengthy piece of mathematics is necessary for understanding the Poisson process. The answer is that, while many elementary applications of the Poisson distribution depend only on knowing what the Poisson probabilities are, we shall shortly be exploring some extremely important extensions of the Poisson idea for which knowledge of this theoretical approach is helpful. Indeed, if you tackle the final section of this chapter on queueing, you will find much of the mathematics familiar if you have followed the arguments we have just been using.

Spatial examples of the Poisson process

In this theoretical approach, the Poisson probabilities involve the parameter λt, the mean number of events in an interval of *time t*. Clearly the mathematics will be similar if we regard 't' as a length or an area or a volume, and 'λ' as the mean number of 'events' in a specified length, area or volume. The assumptions of randomness and independence must be made, and the uniformity assumption is that the mean number of events in a given length, area or volume is constant and depends only on the length, area or volume (and not, in the case of area and volume, on its shape).

The description 'Poisson process' is given to any set of random variables whose distribution is Poisson with parameter depending on time, length, area or volume.

Spatial examples are:

A loom weaving a length of cotton, where the events are faults.
R = the number of flaws in a length L and the parameter $= \lambda L$.

Blood cells in suspension under a microscope, where the events are appearances of blood cells.
R = the number of blood cells visible in an area A and the parameter $= \lambda A$.

Stars in part of a galaxy, where the events are the appearances of stars.
R = the number of stars in volume V and the parameter $= \lambda V$.

In 'time' examples λ is called the expected *rate*.
In 'spatial' examples λ is called the expected *density*.

EXAMPLE 5   During a week of air attacks the authorities kept a record of the number of bombs that fell in 200 different equal areas of a city. The results were as follows:

Number of bombs per area	0	1	2	3	4	5	6 or more
Number of areas	46	71	48	23	9	3	0

Find the theoretical distribution of bombs, assuming a Poisson distribution.

(From these figures the mean number of bombs per area is 1.435 and the variance is 1.386, close enough to the mean to make the use of a Poisson distribution plausible.)

The assumptions are that the numbers of bombs appearing in non-overlapping areas of the city are random variables and that the probability of more than one bomb hitting a very small part of the city is zero.

The parameter used is 1.435 ($= \lambda A$, where A is the size of the areas in square units and λ is the expected density.)

The resulting theoretical distribution is:

$$48 \quad 68 \quad 49 \quad 24 \quad 8 \quad 2 \quad 1.$$

22.6   Exponential distributions associated with the Poisson process

We are now in a position to give two important examples, each connected with the Poisson process, where the exponential distribution is the appropriate model.

The distribution of 'waiting times' until an event occurs in a Poisson process

This will be illustrated by considering again the case where the event is the arrival of a call at a telephone exchange during a given time. The random variable, X, is however not the number of calls that arrive, but the *time*, measured from an arbitrary starting moment, that one would have to wait until the first call arrives.

Now we know that the number of calls arriving in a given time interval has a Poisson distribution. Suppose that the expected time rate is λ, so that the mean number of calls arriving from $t = 0$ to $t = x$ is λx.

Using the Poisson distribution, the probability that there are *no* calls until $t = x$ is

$$e^{-\lambda x},$$

so that $P(X > x) = e^{-\lambda x}$ and $P(X \leqslant x) = 1 - e^{-\lambda x}$.

This means that the distribution function for X is

$$F(x) = 1 - e^{-\lambda x}, \quad x \geqslant 0.$$

Differentiating, the density function for X is

$$f(x) = \lambda e^{-\lambda x}, \quad x \geqslant 0,$$

which is an *exponential distribution*.

Further examples of this kind are the distributions of waiting times until the first failure of an electrical component, or the first disintegration of a chemical substance. The theory can easily be generalised to the case where the random variable X is the *waiting time until k such events have occurred* (see Appendix on page 441).

The distribution of time intervals between successive events in a Poisson process

Using the same example as before, the random variable X is now the time between successive telephone calls.

As before, the mean number of calls from $t = 0$ to $t = x$ is λx, so that the probability that there are no calls from $t = 0$ to $t = x$ is

$$e^{-\lambda x}.$$

Let δx be an interval of time small enough for the chance of more than one call arriving in it to be negligible.

Now the probability that a call arrives in a small interval of length δx is $\lambda \, \delta x$, so that the probability that the first call after $t = 0$ (the 'preceding' call) occurs between $t = x$ and $t = x + \delta x$ is equal to the probability that there are *no* calls between $t = 0$ and $t = x$ and one *call* between $t = x$ and $t = x + \delta x$

$$= e^{-\lambda x} \times \lambda \; \delta x.$$

Hence, if X is the time from $t = 0$ to the *next* call,

$$P(x \leqslant X \leqslant x + \delta x) = \lambda e^{-\lambda x} \; \delta x.$$

Now, if the density function of X is $f(x)$, then, by definition,

$$P(x \leqslant X \leqslant x + \delta x) \approx f(x) \, \delta x,$$

so that in this case $f(x) = \lambda e^{-\lambda x}$, and the distribution of X is exponential.

EXERCISE 22c

1.   The number of particles emitted in t hours from a radioactive substance is kt and has a Poisson distribution. Show that the number of hours until the first particle is emitted has an exponential distribution with parameter k.
   If $k = 20$ find the probability that the time between successive emissions is: (a) greater than 5 minutes, (b) less than 1 minute.

2.   The number of accidents in a factory has a Poisson distribution with a mean 3 per week. Find the probability that the time between successive accidents will be more than 2 days. (There are 5 days in a working week.)

3.   The mean number of calls per minute made on a certain telephone exchange is two. Find: (a) the probability that in any given half-minute interval exactly two calls will be made, (b) the probability that there will be no call in the half-minute immediately following a call.

4.   In the presence of a steady stream of radiation of α-particles the probability that just n particles will hit the sensitive part of a Geiger counter in a time interval of length t is

$$(\lambda t)^n e^{-\lambda t} / n!,$$

where λ is a positive constant.
   The table gives the number of particles hitting a Geiger counter in 200 intervals of 15 s.

Number of particles	0	1	2	3	4	5	6	7	8	9	10	11	12	13 or more
Number of 15 s intervals	0	12	16	30	38	23	30	23	9	10	4	4	1	0

Estimate λ (in s^{-1}) and hence estimate:
(a) the probability that exactly 2 particles hit in an interval of 6 s;
(b) the expectation of the time that elapses before 2 particles have hit the counter;
(c) the probability that the time that elapses before 2 particles have hit the counter is more than 6 s but less than 12 s.   (MEI)

5.   The probability density function of the time T to the first breakdown of a machine A is $\alpha e^{-\alpha t}$. Given that the machine is working at time t, show that the probability of a breakdown in the small interval of time $(t, t + \delta t)$ is $\alpha \, \delta t$ to the first order of small quantities.
   Two other machines, B and C, have corresponding probability density functions $\beta e^{-\beta t}$ and $\gamma e^{-\gamma t}$ independently of each other and of A. The three machines are in use together and the time to the occurrence of the first breakdown is U. Show that the probability density function of U is

$$(\alpha + \beta + \gamma)e^{-(\alpha + \beta + \gamma)u}.$$ (C)

22.7   Queueing

The familiar experience of queueing, whether it is lining up to buy stamps at a post office or sitting in a doctor's waiting room, is the subject of an important branch of probability theory with extensive applications to telephone engineering. We shall look at what is called a *simple queueing process*, taking as our example a queue in a post office, where there is only a single position for serving customers, operating on a 'first come, first served' principle.

You will no doubt remember that in discussing the Poisson process we made certain assumptions: that the events we were concerned with were random and independent and that the influences producing the events were constant so that they occurred uniformly. We shall make similar assumptions here:

(a) that the probability of someone joining the queue in the short interval between times t and $t+\delta t$ is $\lambda\, \delta t$, and

(b) that the probability that in the short interval between times t and $t+\delta t$ the person at the head of the queue completes his business and leaves is $\mu\, \delta t$,

and that these probabilities are independent. It is also understood that δt is small enough for it to be safe to ignore the possibility that more than one person joins the queue (or leaves it) in the interval $(t, t+\delta t)$. (This is sometimes expressed by saying that the probability of more than one arrival or of more than one departure is $o(\delta t)$, which means that it is negligible compared with δt.)

Suppose $P_n(t)$ = the probability that there are n people in the queue at time t, including the one at the head being served. The question we now ask is this: under what circumstances will there still be n people in the queue after time $t+\delta t$? The answer is that there are three possibilities:

1.   there were n people in the queue at time t, and no one joined or left it in the interval $(t, t+\delta t)$. The probability of this happening is

$$P_n(t)\,(1-\lambda\, \delta t)\,(1-\mu\, \delta t);$$

2.   there were $n-1$ people in the queue at time t, one person joined, but no one left in the interval. The probability for this is

$$P_{n-1}(t)\,(\lambda t)\,(1-\mu\, \delta t);$$

3.   there were $n+1$ people in the queue at time t, one person left but no one joined in the interval. The probability for this is

$$P_{n+1}(t)\,(1-\lambda\, \delta t)\,(\mu\, \delta t).$$

Adding these probabilities,

$$P_n(t+\delta t) = P_n(t)\,(1-\lambda\, \delta t)\,(1-\mu\, \delta t)+P_{n-1}(t)\,(\lambda\, \delta t)\,(1-\mu\, \delta t)$$
$$+P_{n+1}(t)\,(1-\lambda\, \delta t)\,(\mu\, \delta t)$$

which may be written

$$P_n(t+\delta t)-P_n(t) \approx -(\lambda+\mu)P_n(t)\delta t+\lambda P_{n-1}(t)\delta t+\mu P_{n+1}(t)\delta t$$

[ignoring terms in $(\delta t)^2$].

Hence

$$\frac{P_n(t+\delta t)-P_n(t)}{\delta t} \approx -(\lambda+\mu)P_n(t)+\lambda P_{n-1}(t)+\mu P_{n+1}(t).$$

Letting δt tend to zero we arrive at the differential equations

$$\frac{\mathrm{d}P_n(t)}{\mathrm{d}t} = -(\lambda+\mu)P_n(t)+\lambda P_{n-1}(t)+\mu P_{n+1}(t). \tag{1}$$

which hold for $n \geqslant 1$.

For the case $n = 0$, the argument is similar. There will be none in the queue at time $t+\delta t$, if either:

(a) there were none at time t, and no one joined in the interval $(t, t+\delta t)$, or

(b) there was one at time t who left in the interval $(t, t+\delta t)$ and no one joined in that interval.

Hence

$$P_0(t+\delta t) = P_0(t)(1-\lambda\,\delta t)+P_1(t)(1-\lambda\,\delta t)(\mu\,\delta t)$$

which leads to the differential equation

$$\frac{\mathrm{d}P_0(t)}{\mathrm{d}t} = -\lambda P_0(t)+\mu P_1(t). \tag{2}$$

At this stage we are interested in the solution of these equations when the process has become *stationary*, that is to say when the probability that there are n in the queue *does not depend* on t. It can be proved, though not easily, that $P_n(t)$ does tend to a limit P_n as $t \to \infty$, so that a long way after the start the distribution *will* be stationary in the sense we have defined. (It is important to note that it is the probabilities that become stationary, not the numbers in the queue, which will obviously fluctuate.)

For the stationary state, then, we will rewrite equations (1) and (2), replacing $P_n(t)$ by P_n and equating $\mathrm{d}P_n(t)/\mathrm{d}t$ to zero:

$$(\lambda+\mu)P_n = \lambda P_{n-1}+\mu P_{n+1} \quad (n \geqslant 1) \tag{3}$$

$$\lambda P_0 = \mu P_1. \tag{4}$$

If you are not familiar with the general technique for solving a 'recurrence relation' like relation (3), the following method is a simple alternative. We write down (4), followed by (3) for $n = 1, 2, \ldots$:

From (4) $\qquad\qquad\qquad \lambda P_0 = \mu P_1$

From (3) $\qquad\qquad\qquad (\lambda+\mu)P_1 = \lambda P_0+\mu P_2$

$$\qquad\qquad\qquad\qquad (\lambda+\mu)P_2 = \lambda P_1+\mu P_3$$

$$\qquad\qquad\qquad\qquad\qquad \vdots \qquad\qquad \vdots$$

$$\qquad\qquad\qquad (\lambda+\mu)P_{n-1} = \lambda P_{n-2}+\mu P_n.$$

Adding these gives $\lambda P_{n-1} = \mu P_n$

so that
$$P_n = \left(\frac{\lambda}{\mu}\right)P_{n-1} = \left(\frac{\lambda}{\mu}\right)^2 P_{n-2} = \ldots = \left(\frac{\lambda}{\mu}\right)^n P_0.$$

The solution, therefore, is $P_n = P_0\left(\dfrac{\lambda}{\mu}\right)^n.$

Now if $P_0, P_1, P_2, \ldots$ give a probability distribution, it is necessary that

$$P_0 + P_1 + P_2 + \ldots = 1.$$

In other words $P_0\left[1 + \dfrac{\lambda}{\mu} + \left(\dfrac{\lambda}{\mu}\right)^2 + \ldots\right] = 1.$

The sum to infinity of this series only exists if $\dfrac{\lambda}{\mu} < 1$, in which case

$$\frac{P_0}{1 - (\lambda/\mu)} = 1 \quad \text{and} \quad P_0 = 1 - \frac{\lambda}{\mu}.$$

We have thus arrived at the solution we were looking for. The probability distribution is:

Number in queue	0	1	2	$\ldots$	n	$\ldots$	
Probabilities	k	$k\left(\dfrac{\lambda}{\mu}\right)$	$k\left(\dfrac{\lambda}{\mu}\right)^2$	$\ldots$	$k\left(\dfrac{\lambda}{\mu}\right)^n$	$\ldots$	$[k = 1 - (\lambda/\mu)]$

which is a *geometric distribution*.

Note: The more usual method of solving the recurrence relation (3) is to assume that there is a solution of the form $P_n = x^n$.
  Substituting in (3) gives $\mu x^{n+1} - (\lambda + \mu)x^n + \lambda x^{n-1} = 0$

$$\Rightarrow \mu x^2 - (\lambda + \mu)x + \lambda = 0 \Rightarrow (\mu x - \lambda)(x - 1) = 0 \Rightarrow x = \lambda/\mu \text{ or } 1.$$

Hence the general solution is $P_n = A(\lambda/\mu)^n + B$.
  Using (4), $P_1 = P_0(\lambda/\mu)$. Hence $B = 0$ and $A = P_0$.
  This simple queueing process was, in fact, first investigated in connection with telephone engineering, with incoming calls queueing to be connected to a single line and it has been developed to deal with the greater complexities that arise when there are many lines.
  We shall conclude this elementary approach to an important branch of statistical theory by considering some interesting practical considerations that are raised even in the simple process.
  The average number in the queue is easily shown to be $\lambda/(\mu - \lambda)$ which will be large if λ/μ is near 1. At the same time, the probability that there are none in the queue, $1 - \lambda/\mu$, will be small. Whether or not this is desirable will depend on the circumstances. If the queue consist of 'calls' waiting to be connected to a telephone line, it will be advisable to arrange for λ/μ *not* to be near 1 so that the mean queue size is not large. It will not matter if this results in P_0 being quite large, which means that there is a good chance of slack periods. A machine does not get impatient if it has no 'customers'. On the other hand, if the queue consists of patients waiting to

see a specialist in a hospital, it is likely that he will not want slack periods and may not mind if as a result the queue is large. In that case a value of λ/μ near 1 will be sought. (This whole matter will, of course be more complicated if an appointments system is used.)

EXERCISE 22d

1. For the simple queueing process described on pages 433–6:
(a) prove that the mean number of people in the queue is $\lambda/(\mu-\lambda)$;
(b) calculate the probability that there are more than four people in the queue when $\lambda = \frac{1}{2}\mu$.

2.   A queue is formed by motorists arriving at a petrol station with two pumps, each with one attendant. The probability that a motorist arrives in a short interval of time $(t, t+\delta t)$ is $\alpha\ \delta t+o(\delta t)$, the probability that no motorist arrives is $1-\alpha\ \delta t+o(\delta t)$, and the probability that two or more motorists is $o(\delta t)$, where $o(\delta t)$ denotes a quantity which is negligible compared with δt as $\delta t \to 0$. The probability that a motorist has his service completed in a time interval $(t, t+\delta t)$ is

$$\begin{cases} 0 & \text{if there are no motorists at the pumps,} \\ \beta\ \delta t & \text{if there is one motorist at the pumps,} \\ 2\beta\ \delta t & \text{if there are two motorists at the pumps.} \end{cases}$$

Denoting by $p_n(t)$ the probability that there are n motorists in the system at time t, including those being served, show that

$$p_n(t+\delta t) = [1-\alpha\ \delta t-2\beta\ \delta t+o(\delta t)]p_n(t)$$
$$+[\alpha\ \delta t+o(\delta t)]p_{n-1}(t)$$
$$+[2\beta\ \delta t+o(\delta t)]p_{n+1}(t)+o(\delta t), \quad n = 2, 3, \ldots.$$

Find the corresponding equations for the cases $n = 0, 1$, and hence find expressions for

$$\frac{dp_n(t)}{dt}, \quad n = 0, 1, 2, \ldots.$$

Determine and solve the equations for the stationary probabilities $p_n(t) = p_n$, independent of t, in terms of $\alpha/2\beta$. (MEI)

EXERCISE 22e (*miscellaneous*)

1.   The side X of a square is uniformly distributed from $x = 0$ to $x = 1$. Find the density function and distribution function for the area of the square.
Sketch their graphs and find the expected value and variance of the area.

2.   The random variable X is uniformly distributed over the interval $(0, 1)$. Find the probability density function of the random variable Y, where $Y = X^2 - 1$.

3.   The random variable X is uniformly distributed over the interval $(-1, 1)$. Find the density function for the random variable Y, where:
(a) $Y = \sin(\pi X/2)$,   (b) $Y = \cos(\pi X/2)$.

4.   Find the mean and the variance of the distribution whose probability density function is given by

$$f(x) = \begin{cases} x, & 0 \leqslant x < 1 \\ 2-x, & 1 \leqslant x \leqslant 2 \\ 0, & \text{otherwise.} \end{cases}$$

A rectangle is constructed with adjacent sides of lengths x cm and $(2-x)$ cm where x is a random value from the above distribution. Find the probability that the area of the rectangle will exceed 0.75 cm². Also find the expected value of the area of the rectangle.   (JMB)

5.   AB is a rod of length 20 cm. A point P is taken at random on AB, the points being uniformly distributed. If the stick is broken at P, find the expected value of $AP^2 + PB^2$.

If the ends A and B of the two parts are placed a distance $10\sqrt{3}$ cm apart on a table, so that APB forms a triangle, find the probability that the angle APB is greater than 90°.

6.   The continuous random variable, X, has probability density function

$$\tfrac{1}{2}e^x \quad (x \leqslant 0), \qquad \tfrac{1}{2}e^{-x} \quad (x \geqslant 0).$$

Find $P(X \leqslant x)$. By substituting $Y = X^2$, find $P(Y \leqslant y)$, and hence deduce the probability density function of Y.   (O)

7.   The length of the straight line AB is $2l$. A point P is selected at random on the line AB, so that the length of OP is uniformly distributed in the interval $(0, 2l)$. The random variable Y denotes the area of a rectangle having adjacent sides of lengths AP and PB. Find the probability density function of Y and show that

$$E(Y) = \tfrac{2}{3}l^2.$$ (C)

8.   The value of a rough diamond varies as the square of its weight. A diamond of value V is broken at random into two parts, the larger part being a fraction k of the total weight w. Show that Y, the total value of the two parts, is equal to $V(2k^2 - 2k + 1)$.

Deduce that Y lies between $\tfrac{1}{2}V$ and V.

Given that the fraction k is uniformly distributed between $\tfrac{1}{2}$ and 1, show that the probability density function of Y is

$$\frac{1}{V\sqrt{[(2y/V)-1]}}.$$

By using the probability density function of k, or otherwise, find the median value of Y and also the probability that Y is less than $\tfrac{3}{4}V$.   (JMB)

9.   A continuous random variable, X, has probability density function

$$4x^3 \quad \text{for } 0 \leqslant x \leqslant 1.$$

What is $P(X \leqslant x)$?
(a) Define a random variable $Y = X^4$, and by writing $x^4 = y$ and considering $P(X^4 \leqslant x^4)$, find $P(Y \leqslant y)$. Hence deduce the probability density function of Y.
(b) Define a random variable $Z = X^n (n > 0)$, and by writing $x^n = z$, find $P(Z \leqslant z)$. Hence deduce the probability density function of Z.

(O)

10.   The random variable X is rectangularly distributed in the interval $0 \leqslant x \leqslant \pi$. The random variable Y is a transformation of X given by

$$y = \tan(x/2 - \pi/4).$$

Show that the probability density function, $g(y)$ of Y is given by

$$g(y) = \frac{2}{\pi(1+y^2)}, \quad a \leqslant y \leqslant b,$$

$$g(y) = 0 \qquad , \quad \text{otherwise,}$$

and state the values of a and b.

Sketch the graph of $g(y)$ and state the mean value of Y.   (JMB)

11.   The random variables X_1 and X_2 are independent. The random variable Y is defined by the greater of X_1 and X_2. Explain why

$$P(Y \leqslant y) = P(X_1 \leqslant y) \cdot P(X_2 \leqslant y).$$

It is known that X_1 and X_2 both have rectangular distribution taking values in the interval 0 to a. Find the cumulative distribution function $G(y)$ of Y, and hence, or otherwise, find the probability density function of Y and the expectation of Y.   (C)

12.   The random variable X has the probability density function $f(x)$, where

$$f(x) = \begin{cases} a^{-1}, & 0 \leqslant x \leqslant a, \\ 0, & \text{otherwise.} \end{cases}$$

Determine the mean and variance of X.

A second random variable Y is independent of X, and has the same probability density function. Write down the values of the mean and variance of Z, where $Z = X + Y$.

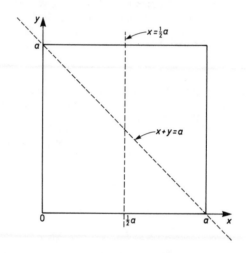

By considering the areas of suitable regions on the sketch, or otherwise, find the probabilities that:
(a) Z is less than a;
(b) Z is less than a and X is less than $\frac{1}{2}a$;
(c) Z is less than a, given that X is less than $\frac{1}{2}a$.   (MEI)

13.   The incidence of faults in a television set is such that the number of faults in a time interval of any fixed length T after the set is put into service has a Poisson distribution with mean λT, where λ is a constant. Let S denote the time up to the occurrence of the first fault. Show that

$$P(S \leqslant s) = 1 - e^{-\lambda s}$$

and deduce that the probability density function of S is $\lambda e^{-\lambda s}$.

Show that the probability density function of the time, U, up to the occurrence of the nth fault is given by

$$f(u) = \frac{\lambda(\lambda u)^{n-1} e^{-\lambda u}}{(n-1)!}.$$ (JMB)

14.   In the presence of a steady stream of radiation of α-particles the probability that just n particles will hit the sensitive part of a Geiger counter in a time interval of length t is

$$\frac{(\lambda t)^n}{n!} e^{-\lambda t},$$

where λ is a positive constant.

Denote by t_n the time that elapses before n particles have hit. Show that the probability that t_n is less than t is

$$1 - \left\{ 1 + \frac{\lambda t}{1!} + \frac{(\lambda t)^2}{2!} + \ldots + \frac{(\lambda t)^{n-1}}{(n-1)!} \right\} e^{-\lambda t},$$

and deduce that the probability density function of t_n is

$$\frac{\lambda(\lambda t)^{n-1}}{(n-1)!} e^{-\lambda t}, \quad (t \geqslant 0). \tag{MEI}$$

15.   In the manufacture of paper, faults may be regarded as occurring at random distances from the beginning of the roll. The random variable is the distance of a fault from the beginning of the roll, $\lambda \, \delta x$ is the probability of a fault in the small interval $(x, x + \delta x)$ and $P_0(x)$ is the probability that there are no faults in the interval $(0, x)$. Prove that

$$\frac{dP_0(x)}{dx} = -\lambda P_0(x).$$

Hence show that $P_0(x) = e^{-\lambda x}$.
The random variable Y is the distance, measured in metres, to the first fault. Obtain the probability density function of Y. The mean distance between faults is observed to be 10 metres. Estimate the value of λ. (C)

16.   The probability of an event E occurring in the small interval $(x, x + \delta x)$ of a continuous real variable x is $\lambda \, \delta x$. The probability that the event E occurs n times in the interval $(0, x)$ is $P_n(x)$.

Prove that $\dfrac{dP_n(x)}{dx} + \lambda P_n(x) = \lambda P_{n-1}(x), \quad (n \geqslant 1)$.

Given that $P_0(x) = e^{-\lambda x}$, prove, by induction or otherwise, that

$$P_n(x) = e^{-\lambda x} \frac{(\lambda x)^n}{n!}.$$

Obtain the probability density function of the value of x at which the first event E occurs. (C)

17.   In a laboratory there is a piece of apparatus which is so vital to keep running that as soon as it breaks down repairs begin immediately. If it is running satisfactorily at time t the probability of a breakdown in the small interval $(t, t + \delta t)$ is $\lambda \, \delta t$, whereas if it is being repaired at time t the probability that the repair will be completed in the interval $(t, t + \delta t)$ is $\mu \, \delta t$.
If $p(t)$ is the probability that the apparatus is under repair at time t, find a relation between $p(t + \delta t)$ and $p(t)$. Hence prove that

$$p(t) = \frac{\lambda}{\lambda + \mu} [1 - e^{-(\lambda + \mu)t}].$$

18.   An experiment consists of counting the number of times a certain event occurs in a time interval. It is known that the probability that the event occurs once in the interval between t and $t + \delta t$ is $\lambda \, \delta t$ if δt is small, where λ is a constant, and that the probability that it occurs more than once in that interval is negligible. Assuming that events occur independently, show that the probability of r events occurring between 0 and t is p_r, given by the Poisson distribution

$$p_r = (\lambda t)^r e^{-\lambda t}/r! \quad r = 0, 1, 2, \ldots.$$

Show that the probability that the first event occurs in the interval between t and $t + \delta t$ is $\lambda e^{-\lambda t} \, \delta t$ if δt is small, and hence find the expected time when the first event occurs.   (*contd.*)

A system can be in one of two states, A and B. The probability of a transition from one state to the other in the interval between t and $t+\delta t$ is $\mu\,\delta t$ if δt is small, where μ is a non-zero constant, and the probability of more than one transition in that interval is negligible. This probability is independent of previous transitions. Initially the system is in state A. Using the Poisson distribution, or otherwise, show that the probability that the system is in state A at time t is

$$\tfrac{1}{2}(1+e^{-2\mu t}).$$

(You may find it helpful to consider the power series for e^x+e^{-x}.)

It follows that as $t\to\infty$ the system is equally likely to be in state A or state B. Show that this result is independent of the initial state of the system.   (MEI)

19. The event E occurs in the small time interval $(t,t+\delta t)$ with probability $\lambda\,\delta t$. The probability that E occurs n times in the interval $(0,t)$ is P_n. Show that for $n\geqslant 1$

$$\frac{\mathrm{d}P_n}{\mathrm{d}t}=-\lambda P_n+\lambda P_{n-1}.$$

Obtain an expression for $\mathrm{d}P_0/\mathrm{d}t$.

Hence show that the probability generating function G, defined by

$$G=\sum_{n=0}^{\infty}P_n z^n,$$

satisfies, for a given value of z, the differential equation $\mathrm{d}G/\mathrm{d}t=\lambda(z-1)G$. Deduce an expression for G in terms of z, λ and t.   (C)

20. A single queue forms at a checkpoint, the probability of an individual joining the queue in the small time interval $(t,t+\delta t)$ being $m\,\delta t$. Checking procedures are such that the probability that the individual at the head of the queue leaves in $(t,t+\delta t)$ is $n\,\delta t$, where $n>m$.

In the steady state, the probability that there are r individuals in the queue is p_r. Show that

$$(m+n)p_r=mp_{r-1}+np_{r+1},\quad\text{for }r\geqslant 1$$

and obtain an equation connecting p_0 and p_1.

Hence show that $mp_r=np_{r+1}$ and evaluate p_r in terms of m, n and r.

Find also the mean number of individuals in the queue.   (C)

21. In a simplified model of traffic flow past a barrier, the probability that one car passes the barrier in a second is p while the probability that no cars pass the barrier is $1-p$. The probability that one more car arrives to join the queue in a second is k while the probability that no cars arrive is $1-k$. These events are independent of each other and of all previous and subsequent movements. If $P_n(t)$ is the probability that there are n cars in the queue at time t second, show that

$$P_n(t+1)=p(1-k)P_{n+1}(t)+(1-k-p+2kp)P_n(t)+k(1-p)P_{n-1}(t)$$

for all $n\geqslant 1$.

Find a similar formula for $P_0(t+1)$ assuming, that when there is no queue it is possible for a car to both arrive and pass the barrier in one second.

Verify that for suitable values of c and x that are independent of n and t, there is a time-independent solution of the form

$$P_n=cx^n$$

for all n, provided that $k<p$, and find c and x.   (O & C)

22.8   Appendix

The distribution of waiting times until k events have occurred in a Poisson process

Consider a Poisson process in which the events are failures of an electrical component and the mean number of failures in time t is λt.

Suppose a *breakdown* occurs if k or more failures occur, and let the random variable, X, be the waiting time to breakdown, measured from an arbitrary moment.

The mean number of failures from $t = 0$ to $t = x$ is λx, so that the probability that there are fewer than k failures in this time is

$$\sum_{r=0}^{k-1} e^{-\lambda x}(\lambda x)^r/r!$$

that is
$$P(X > x) = 1 - \sum_{r=0}^{k-1} e^{-\lambda x}(\lambda x)^r/r!$$

and
$$P(X \leqslant x) = \sum_{r=0}^{k-1} e^{-\lambda x}(\lambda x)^r/r!.$$

The distribution function for X is thus

$$F(x) = \sum_{r=0}^{k-1} e^{-\lambda x}(\lambda x)^r/r!.$$

Hence, by differentiation of $F(x)$, the density function is
$$f(x) = \frac{\lambda(\lambda x)^{k-1}}{(k-1)!} e^{-\lambda x}, \ x \geqslant 0.$$

23 Chi-squared tests: goodness of fit and contingency

Throughout this book we have stressed that in using a model for interpreting a set of observations we need some way of testing whether or not the model is a valid one. On page 152, for instance, starting with an observed frequency distribution giving the number X of bacteria in each of fifty 0.1 litre samples of a liquid:

X	0	1	2	3	4	5 or more
Observed frequency	23	14	9	3	1	0

and assuming that the model is Poisson, the parameter $\mu = 0.9$ was estimated from the sample. The theoretical frequencies for a Poisson distribution with this parameter were:

Expected frequency	20.33	18.30	8.23	2.47	0.56	0.11	(unrounded figures)

The problem is to find a method of testing whether the 'fit' between observed and expected frequencies is close enough to warrant the assumption that the sample comes from a Poisson distribution with mean 0.9.

23.1 Goodness of fit tests when parameters are given

Before dealing with an example of this kind – where the parameter is estimated from a sample – we shall look at a simpler example, where the parameters are *given*. Consider an experiment in which two supposedly unbiased coins are tossed 100 times and the number of heads is recorded. The results are shown below, together with the expected frequencies, obtained by assuming a binomial model $B(2, \frac{1}{2})$:

Number of heads	Observed frequencies (O_i)	Expected frequencies (E_i)
0	$o_1 = 22$	$e_1 = 25$
1	$o_2 = 47$	$e_2 = 50$
2	$o_3 = 31$	$e_3 = 25$
	$n = \Sigma\, e_i$	$= 100$

The observations are in 3 'cells' or classes.

442

Pearson's statistic

How are we to test whether the gaps between the o_i and e_i are small enough for us to be able to say, with reasonable confidence, that the binomial model $B(2, \frac{1}{2})$ is appropriate? The answer, in practice, is to use a statistic (which we will call, for the moment, X^2) which was devised by Karl Pearson, and is defined as follows:

$$X^2 = \sum_{i=1}^{3} \frac{(o_i - e_i)^2}{e_i}, \quad \text{for 3 cells, as in this example,}$$

and in the general case

$$X^2 = \sum_{i=1}^{k} \frac{(o_i - e_i)^2}{e_i}, \quad \text{for } k \text{ cells.}$$

On the face of it, this statistic is a sensible measure of 'closeness of fit'. If the fit is exact, $o_i - e_i$ is zero for each i so that $X^2 = 0$, and the larger the differences between the o_i and e_i, the greater will be the value of X^2. The differences $o_i - e_i$ are squared to get round the problem that the sum of all the differences will always be zero (this is the device used in defining variance), and we divide by e_i as a way of reflecting the fact that a small difference is relatively more important when the frequency itself is small.

Now if an experiment of this kind were repeated many times, and on each occasion the observations were classified into k cells, the resulting sets of observed frequencies would lead to different values of X^2. In other words we would have a *sampling distribution* of the statistic X^2. We therefore need to know what the distribution of X^2 is so that we can apply the usual kind of significance test at a suitable level of significance.

It turns out that none of the distributions we have met so far fill the bill and that we must use one of a family of distributions called χ^2 (or chi-squared, pronounced ky-squared).

23.2   The χ^2 distributions

The χ^2 distributions are

$$\chi_1^2, \chi_2^2, \ldots, \chi_\nu^2$$

where the suffix ν is called 'the number of degrees of freedom' (the term we used in connection with the t-distribution). The meaning of ν will be discussed later.

Each of these distributions has a different density function, which we shall not quote at this stage; the shapes of two of them are shown in the sketches below. This means that separate tables are needed to find the critical percentage points for hypothesis testing. As examples we give extracts from the tables (see page 520) for $\nu = 2$ and $\nu = 12$.

P	99	95	10	5	1	0.1
$\nu = 2$	0.0201	0.103	4.61	5.99	9.21	13.81
$\nu = 12$	3.57	5.23	18.55	21.03	26.22	32.91

χ^2_2

The shape of the χ^2 distribution when $v = 2$ is shown opposite.

When $P = 5$, $\chi^2_2 = 5.99$. This means that the shaded area is 5% of the total area under the curve, that is

$$P(\chi^2_2 \geqslant 5.99) = 5/100,$$

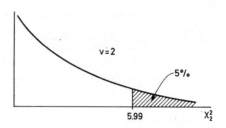

and this area is the critical region for a test at the 5% level.

χ^2_{12}

When $P = 1$, $\chi^2_{12} = 26.22$, that is

$$P(\chi^2_{12} \geqslant 26.22) = 1/100.$$

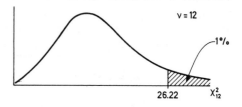

and the shaded area is the critical region for a test at the 1% level.

Some values of v are not included in the table, in which case linear interpolation should be used. Suppose, for instance, that the critical χ^2 value at the 5% level is needed for $v = 49$. We interpolate between the value for $v = 40$ (55.76) and the value for $v = 50$ (67.50), so that the value for $v = 49$ is:

$$55.76 + \frac{9}{10}(67.50 - 55.76) = 66.33.$$

The statistic X^2 and the χ^2 distributions

It can be shown that the distribution of the discrete variable X^2 is approximately χ^2, where the number of degrees of freedom, v, is *the number of cells minus the number of linear constraints on the expected frequencies.*

What this rather elaborate phrase means is best illustrated by our coin-tossing example. If you look back at page 442, you will see that when e_1 and e_2 have been calculated, e_3 is determined since $e_1 + e_2 + e_3$ has to equal 100, the sum of the observed frequencies. This relation $e_1 + e_2 + e_3 = 100$ is a 'linear constraint' imposed on our calculation of X^2, and we say, therefore, that in this case $v = 3 - 1 = 2$.

In general, then, when the parameter of the model is known

$$v = \text{the number of cells minus } 1 = k - 1.$$

We shall consider later how to interpret v when the parameter (or parameters) have to be estimated.

EXAMPLE 1   Test the goodness of fit between the results of the coin-tossing experiment on page 442 and the expected results if we assume that the model is $B(2,\frac{1}{2})$.

The figures to be compared, together with the calculation of X^2 are given in the table:

O_i	E_i	$(O_i - E_i)^2/E_i$
22	25	0.36
47	50	0.18
31	25	1.44
		$X^2 = 1.98$

The null hypothesis H_0, is that the model is $B(2,\frac{1}{2})$, and X^2 has a χ_2^2 distribution, since, as we saw above, $k = 3$ and $v = 2$.

Now the critical region at the 5% level for χ_2^2 starts at 5.99, whereas our sample value of $X^2 = 1.98$ (< 5.99). Hence H_0 is accepted at the 5% level.

A justification for using a χ^2 distribution in a 'goodness of fit' test

The application of a χ^2 distribution to testing 'goodness of fit' as in example 1 is easily handled, but before we go on to give further examples, something should be said about the origin of the χ^2 distributions themselves. So far we have just stated that they exist, without defining them. It is true that a complete definition, including the density function, is irrelevant if you just need to know how to use a χ^2 test, but in applying such a test certain procedures are adopted which cannot really be understood without an elementary knowledge of the definition. Moreover, with this knowledge, it will be possible to give some *justification* for the use of a χ^2 distribution as an approximation to the distribution of the statistic X^2 – an important matter that we have so far avoided. In order to do this we shall confine our attention to χ_1^2, the χ^2 distribution with one d.f. (a useful abbreviation for 'degree(s) of freedom').

The basic fact about χ_1^2 is that it is the distribution of the *square of a standardised normal variable*, and is defined formally as follows:

If Z is a random variable whose distribution is $N(0,1)$, then the distribution of Z^2 is called the χ^2 distribution with one degree of freedom.

With this definition it is not difficult to demonstrate that for a set of observed and expected frequencies given in *two* cells ($k = 2$), the distribution of X^2 is approximately χ_1^2 ($v = 2-1$).

Let us suppose that a traffic-monitoring survey is carried out at a fairly remote country crossroad. In a previous survey carried out over a long period of time, the probability distribution of the number of vehicles passing that point between 12 am and 1 pm had been given in the form shown here, where π was, of course, known.

Number of vehicles	Probability
Fewer than 3	π
3 or more	$1 - \pi$

In the new survey, spread over n hours (also between 12 am and 1 pm) the expected numbers of vehicles falling into these two categories (cells) would be $n\pi$ and $n(1-\pi)$ on the null hypothesis, H_0, that the same binomial model applies. We shall suppose that the observed numbers are Y and $n-Y$. This gives the table

Number	Observed (O)	Expected (E)
Fewer than 3	Y	$n\pi$
3 or more	$n-Y$	$n(1-\pi)$

so that

$$X^2 = \frac{\Sigma\,(O-E)^2}{E} = \frac{(Y-n\pi)^2}{n\pi} + \frac{[n-Y-n(1-\pi)]^2}{n(1-\pi)}$$

$$= \frac{(Y-n\pi)^2}{n\pi} + \frac{(n\pi-Y)^2}{n(1-\pi)} = \frac{(Y-n\pi)^2(1-\pi+\pi)}{n\pi(1-\pi)}$$

$$= \frac{(Y-n\pi)^2}{n\pi(1-\pi)} = \left[\frac{Y-n\pi}{\sqrt{[n\pi(1-\pi)]}}\right]^2.$$

Under H_0, the distribution of Y will be $B(n,\pi)$ and if neither π nor $1-\pi$ is too small, this distribution will be approximately normal, $N[n\pi, n\pi(1-\pi)]$. This means that

$$\frac{Y-n\pi}{\sqrt{[n\pi(1-\pi)]}}$$

is approximately a standardised normal variable and that, by definition, the distribution of X^2 is approximately χ_1^2. We may therefore use χ_1^2 for testing H_0.

A similar justification when there are more than two cells is much more complicated as the binomial model can no longer be used. You will have to take it for granted that, when the number of cells is k, the equivalent approximate distribution of X^2 is χ_v^2, where $v = k - 1$. (A full definition of χ_v^2 appears on page 462.)

One very important practical point emerges from this demonstration. The normal approximation to the binomial, on which the argument rests, is only valid for large values of n and values of π and $1-\pi$ that are not too small. The working rule is this:

The number in any expected frequency cell should not be less than five. If any expected frequency is too small, it must be combined with one or more of the cells above or below it.

EXAMPLE 2   The number of the colour strains red, yellow, white and blue in a particular flower are expected to appear in the proportions $4:12:5:4$ according to genetic theory. In 1600 plants the numbers of red, yellow, white and blue strains were 218, 821, 306 and 255 respectively. Are these results significantly different from the expected frequencies?

The table of observed and expected frequencies, is given below, together with the value of X^2, which we shall write as χ^2 from now on:

Strain	O	E	$(O-E)^2/E$
Red	218	256	5.64
Yellow	821	768	3.66
White	306	320	0.61
Blue	255	256	0.00
			$\chi^2 = \overline{9.91}$

The number of cells is 4, so that $v = 3$. Now the critical region for a 5% level test starts at $\chi_3^2 = 7.81$, and since $9.91 > 7.81$, we conclude that the null hypothesis that the proportions are $4:12:5:4$ must be rejected.

EXAMPLE 3   The numbers of road-traffic accidents in one area of a city that are reported to a police station seem to follow a Poisson distribution with mean 2 per day, a day being one of the 'working days' from Monday to Friday. Use a χ^2 distribution to test whether the following distribution, giving the daily number of accidents reported (X) on 200 successive days, fits the assumption of the stated Poisson distribution.

X	0	1	2	3	4	5	6 or more	
f	24	60	52	32	22	5	5	(Total 200)

The Poisson probabilities are $P(X = r) = e^{-2} \cdot 2^r/r!$ and the expected frequencies are $200 \times e^{-2} \cdot 2^r/r!$. These are shown in the following tables, in the second of which the last two cells have been combined, since the expected frequency of 3.22 is less than 5. Notice that we do not round off the expected frequencies to whole numbers,

X	O	E
0	24	27.07
1	60	54.13
2	52	54.13
3	32	36.09
4	22	18.04
5	5	7.22
6 or more	5	3.32

X	O	E	$(O-E)^2/E$
0	24	27.07	0.348
1	60	54.13	0.636
2	52	54.13	0.084
3	32	36.09	0.463
4	22	18.04	0.869
5 or more	10	10.54	0.028
			$\chi^2 = 2.428$

The number of degrees of freedom, $v = 6-1 = 5$, and the critical region at the 5% level starts at $\chi_5^2 = 11.07$. Since the value of χ^2 from the table is 2.428, the null hypothesis of a Poisson distribution with mean 2 is accepted at this level.

EXERCISE 23a

1.   The number of male children in a sample of 128 families, each of which contains four children, is given in the following table:

Number of sons	0	1	2	3	4
Frequency	14	28	50	24	12

Test the hypothesis that the model is $B(4, \frac{1}{2})$ using a χ^2 test with a 5% significance level.

2. If, in question 1, the frequency distribution had shown the results for 64 families as being:

Number of sons	0	1	2	3	4
Frequency	7	14	25	12	6

what would the result of carrying out the χ^2 test have been?

3. A paperback bookseller calculated that on average, out of every 20 books sold, six were science fiction, seven were thrillers, four were other fiction and three were non-fiction.
   After he had moved his shop to a different part of the town he found that the first 200 sales were as follows:

> science fiction 52   thrillers 80   other fiction 27   non-fiction 41.

On this evidence, using a χ^2 test at the 5 % level, determine whether he could conclude that the sales pattern was significantly different as a result of the move.
   After 400 sales the sales figures were:

> science fiction 110   thrillers 150   other fiction 68   non-fiction 72.

Using the same test comment on these results.   (C)

4. The number of minor accidents on any given day in a large factory is believed to have a Poisson distribution with mean 0.6. Over a period of 500 days the number of accidents was recorded with the following results:

Number of daily accidents	0	1	2	3 or more
Frequency	285	149	45	21

Use a χ^2 test at the 5 % significance level to test whether the given Poisson distribution is likely to be the appropriate model.

5. In an experiment, six dice are thrown together and the number of '6's is recorded. In 600 repetitions of the experiment the results were as follows:

Number of '6's	0	1	2	3	4 or more
Frequency	186	246	113	42	13

Test the hypothesis that the dice are unbiased.

6. A nurseryman is raising plants from seed and selling them to a retailer in batches of five plants. He expects that 25 % of the seeds will produce white-flowered plants. The retailer is asked to examine 100 batches and to report on the number of white-flowered plants in them. The data he produces are contained in the following table.

Number of white flowers	0	1	2	3	4	5
Number of batches	32	41	20	5	1	1

Apply a χ^2 test for goodness of fit between this distribution and the expected binomial distribution.

23.3   Goodness of fit tests when parameters have to be estimated

In the examples we have given so far, the parameters of the model with which the given distribution is compared have always been *given*. Suppose that this is not the case, as in the first example on page 442, where the theoretical Poisson frequencies were calculated using an *estimated* parameter. This will affect the number of degrees of freedom, which has been defined as the number of cells minus the number of linear constraints on the expected frequencies.

If a parameter has to be estimated from the observed sample frequencies, this introduces an extra linear constraint. Taking the introductory example, the parameter of the Poisson distribution is the mean, and equating the mean of the expected values to the mean of the observed values constitutes an additional linear constraint. Similarly, if two parameters have to be estimated in order to calculate the expected frequencies (the mean and standard deviation of a normal distribution, for instance), then two extra constraints are introduced. The effect of all this is that we can re-state the number of degrees of freedom as being *the number of cells minus one* (since the total is given) *minus one for each parameter to be estimated*, so that

$$v = k - 1 - r$$

where k = the number of cells and r = the number of parameters estimated.

We will now show how to apply this, starting with the introductory example.

EXAMPLE 4   The number, X, of bacteria in each of fifty 0.1 litre samples is given in the following table:

X	0	1	2	3	4	5 or more
f	23	14	9	3	1	0

Test the hypothesis that X has a Poisson distribution.

Since the mean of this distribution, $\Sigma fx/N = 45/50$, we take as our estimate of the population mean $\hat{\mu} = 0.9$.

The null hypothesis μ_0 to be tested is that the parent distribution is Poisson with parameter 0.9. An alternative hypothesis cannot be stated precisely as it would have to include various possibilities such as that the distribution is Poisson with a different parameter, or that it is not Poisson at all.

Under H_0 the expected frequencies are $50e^{-0.9}(0.9)^r/r!$, for $r = 0, 1, 2, 3, 4, 5$ or more, giving

$$20.33 \quad 18.30 \quad 8.23 \quad 2.47 \quad 0.56 \quad 0.11$$

The last three frequencies are each less than 5, and their sum is also less than 5. We therefore combine them with the one above, 8.23, to give:

X	O	E	$(O-E)^2/E$
0	23	20.33	0.35
1	14	18.30	1.01
2 or more	13	11.37	0.23
			$\chi^2 = 1.59$

There are 3 cells and 2 constraints (the sum must equal 50, and the mean is estimated) so that $v = 3 - 2 = 1$. Since the critical value of χ_1^2 at the 5% level is 3.84, H_0 is accepted.

EXAMPLE 5   Packets of chocolates are made up for distribution at a large children's party arranged by a charitable organisation. Each packet contains four chocolates, with a mixture of raspberry- and mint-flavoured centres. After the party, one of the organisers, out of interest, checks the contents of a sample of 50 packets that have been left over. She finds that the number of mint flavours in these packets is distributed as follows:

Number of mint-flavoured chocolates in a packet (X)	0	1	2	3	4
Number of packets (f)	12	15	10	9	4

Test the hypothesis that the distribution of the parent population is binomial with the same mean as the sample.

If the parent population is binomial it will be $B(4, \pi)$, where π is the probability of a chocolate being mint-flavoured. This means that the mean number of mint-flavoured chocolates in a packet of four will be 4π.

From the table, the observed mean is $\Sigma fx/N = 78/50 = 1.56$. Taking $\hat{\mu}$ as 1.56, $\hat{\pi}$ will be given by $4\hat{\pi} = 1.56 \Rightarrow \hat{\pi} = 0.39$.

Under H_0, the null hypothesis that the parent distribution is $B(4, 0.39)$, the expected frequencies,

$$50 \times \binom{4}{r}(0.39)^r(0.61)^{4-r}, \quad (r = 0, 1, 2, 3, 4)$$

are:

$$6.92 \quad 17.70 \quad 16.98 \quad 7.24 \quad 1.16$$

Combining the last two frequencies we get:

X	O	E	$(O-E)^2/E$
0	12	6.92	3.73
1	15	17.70	0.41
2	10	16.98	2.87
3 or 4	13	8.40	2.52
			$\chi^2 = 9.53$

With 4 cells and 2 constraints, $v = 2$. The critical value of χ_2^2 at the 5% level is 5.99, so that H_0 is rejected at this level.

23.4   Testing goodness of fit with a continuous distribution

Testing the fit between a set of grouped observations and the continuous probability distribution which is assumed to be the model is similar to the discrete case, but there are points of detail that need care, especially when the model concerned (the exponential and normal distributions, for instance) is defined over an infinite range of values of the random variable.

EXAMPLE 6   A monitoring post is established at a point on a narrow but busy road to time the gaps (to the nearest second) between successive vehicles passing the point. The results for 100 vehicles were:

Gap (X)	Up to 10	10–	20–	30–	40–	50–	60–	70–80
f	34	30	16	7	5	3	3	2

Is it reasonable to assume that the model is an exponential distribution?

The probability density function of an exponential distribution is

$$f(x) = \begin{cases} \lambda e^{-\lambda x}, & 0 < x < \infty \\ 0, & \text{otherwise,} \end{cases}$$

where the parameter λ and the mean, μ, of the distribution are connected by the relation $\mu = 1/\lambda$ (see page 185).

The mean, $\bar{x}$, of the given distribution is found to be 20, so that we estimate $\hat{\mu} = 20$ and $\hat{\lambda} = 1/20 = 0.05$.

In order to calculate the expected frequencies, which are derived from $P(0 \leqslant X < 10)$, $P(10 \leqslant X < 20), \ldots$ we need the distribution function $F(x)$.

Now in this case
$$F(x) = P(X < x) = \int_0^x f(t)\, dt = \int_0^x \lambda e^{-\lambda t}\, dt$$
$$= 1 - e^{-\lambda x} = 1 - e^{-0.05x}.$$

Using this formula, $P(0 \leqslant X < 10) = F(10)$, $P(10 \leqslant X < 20) = F(20) - F(10)$, and so on. The resulting probabilities and expected frequencies are laid out in a table:

Class	x Upper bound	F(x)	P	Expected frequency
0–			0.393	39.3
	10	0.393		
10–			0.239	23.9
	20	0.632		
20–			0.145	14.5
	30	0.777		
30–			0.088	8.8
	40	0.865		
40–			0.053	5.3
	50	0.918		
50–			0.032	3.2
	60	0.950		
60–			0.020	2.0
	70	0.970		
70–			0.030	3.0
	∞	1.000		

(In the model the upper bound of the last interval has to be ∞ even though there are no observed values above 80.)

With these expected frequencies we can make up the final table for calculating χ^2, combining the last three frequencies:

X	O	E	$(O-E)^2/E$
0–	34	39.3	0.71
10–	30	23.3	1.93
20–	16	14.5	0.16
30–	7	8.8	0.37
40–	5	5.3	0.02
50–	8	8.2	0.00
			$\chi^2 = \overline{3.14}$

The number of cells, $k = 6$ and a parameter has been estimated, so that $v = 6-2 = 4$. Since the critical region at 5% for χ_4^2 is 9.49, it seems reasonable to accept the exponential distribution as model.

EXAMPLE 7   Test the hypothesis that the following frequency distribution is consistent with the assumption that the distribution of X is normal. The data have been taken from the example on page 211.

X	< 10	10–	15–	20–	25–	30–	⩾ 35
f	1	3	7	14	20	9	4 Total 58

There is no information about the degree of accuracy to which the observations have been measured, so we take the class centres as 7.5, 12.5,..., assuming a class width of 5 for the first and last classes. Calculating the mean and variance in the usual way, $\bar{x} = 25.4$, and $\text{var}(x) = 43.14$. Estimates for the parent distribution are therefore $\hat{\mu} = 25.4$ and $\hat{\sigma} = \sqrt{43.14} = 6.57$, since the sample is a large one and there would be little point in using s^2. The expected frequencies were calculated on page 212.

$-\infty$–10	10–	15–	20–	25–	30–	35–∞
0.58	2.73	8.64	15.66	16.36	9.86	4.18

Combining the first three and the last two frequencies, we get

X	O	E	$(O-E)^2/E$
< 20	11	11.95	0.075
20–	14	15.66	0.176
25–	20	16.36	0.810
⩾ 30	13	14.04	0.077
			$\chi^2 = \overline{1.138}$

There are 4 cells and two parameters have been estimated, so that $v = 4-3 = 1$. The critical region for χ_1^2 starts at 3.84, and we conclude that it *is* reasonable to assume a normal distribution as model.

EXERCISE 23b

1.   A coded message consists of 10 000 letters. The message is divided up into 1000 groups of letters, each group containing 10 letters. The number of times that the letter W occurs in each group of ten letters is counted, with the following results:

Number of times that W occurs in a group	0	1	2	3	4	5 or more
Number of groups	362	361	202	65	10	0

Use a χ^2 test to determine whether or not a binomial distribution gives an appropriate model for the data. Give adequate details of your test and state clearly your conclusion.   (C)

2.   During the weaving of cloth, the thread sometimes breaks. 147 lengths of thread of equal length were observed during weaving and the table records the number of these threads for which the indicated number of breaks occurred.

Number of breaks per thread	0	1	2	3	4	5
Number of threads	48	46	30	12	9	2

Fit a Poisson distribution to the data and examine, at both a 5% and a 1% level of significance, whether the deviation between theory and experiment is significant.   (MEI)

3.   Most of the business done by a large firm of estate agents occurs during the months April to September. The records for 1984 showed that the number of house purchases completed during these months were:

Month	April	May	June	July	August	September
Number of house purchases	14	24	16	24	20	22

In allocating staff to deal with the work for 1985 the management worked on the assumption that no one of these months is likely to be busier than another. Do these figures suggest that the assumption is justified?

4.   A record was made of the number of visits made by each of the vehicles in an army unit to a repair depot, with these results:

Number of visits	0	1	2	3	4	5	6 or more	
Number of vehicles	295	190	53	5	5	2	0	(Total 550)

Fit a Poisson distribution with the same mean as the observed data and test for goodness of fit.   (MEI)

5.   A random sample of 36 values of the continuous variable X, gives the following distribution:

X	0–	1–	2–	3–	4–	5–6
f	0	2	7	8	9	10

It is suggested that X has density function given by

$$f(x) = \begin{cases} kx, & 0 \leqslant x \leqslant 6 \\ 0, & \text{elsewhere.} \end{cases}$$

Use a χ^2 test to determine whether this model is reasonable.

6.   The following frequency distribution gives a sample of the values of the continuous random variable X between 0 and 3.

X	0–	1–	2–3
f	30	42	28

Test whether this is compatible with a density function for X of

$$f(x) = \begin{cases} \frac{2}{9}x(3-x), & 0 \leqslant x \leqslant 3 \\ 0 & , \quad \text{elsewhere.} \end{cases}$$

7. Using a 5% significance level, test whether the following frequency distribution is consistent with the assumption of a normal model with mean 25 and variance 100:

x	f
$x \leqslant 5$	3
$5 < x \leqslant 10$	6
$10 < x \leqslant 15$	9
$15 < x \leqslant 20$	15
$20 < x \leqslant 25$	24
$25 < x \leqslant 30$	16
$30 < x \leqslant 35$	14
$35 < x \leqslant 40$	8
$40 < x \leqslant 45$	3
$x > 45$	2

8. The following frequency table has been compiled from data which are supposed to be a sample from an exponential distribution with parameter $\lambda = 0.5$. Test the assertion.

x	$0 \leqslant x < 1$	$1 \leqslant x < 2$	$2 \leqslant x < 3$	$3 \leqslant x < 4$	$4 \leqslant x < 5$	
f	36	23	17	8	9	
x	$5 \leqslant x < 6$	$6 \leqslant x < 7$	$7 \leqslant x < 8$	$8 \leqslant x < 9$	$9 \leqslant x$	
f	0	1	5	1	0	Total 100

(O)

9. The shape of the human head is the subject of an international project financed by the World Council for Health and Welfare. Observations were taken in many countries and the nose lengths, to the nearest millimetre, of 150 Italians are summarised below.

Nose lengths x (mm)	$-\infty < x \leqslant 44$	$45 \leqslant x \leqslant 47$	$48 \leqslant x \leqslant 50$
Frequency f	4	12	63
Nose lengths x (mm)	$51 \leqslant x \leqslant 53$	$54 \leqslant x \leqslant 56$	$57 \leqslant x < \infty$
Frequency f	59	10	2

Estimate the mean and standard deviation of the population from which these observations were taken. (For these calculations you should assume that the lower and upper classes have the same range as the other classes.) (See exercise 11e, question 4.)

Use the χ^2 distribution and a 1% level of significance to test the adequacy of the normal distribution as a model for these data. (AEB 1982)

10. Test for the goodness of fit of the following set of observations of values of the discrete random variable X with a geometric distribution for which

$$P(X = x) = p(1-p)^{x-1}, \quad x = 1, 2, \ldots \text{to } \infty.$$

The mean of a geometric distribution of this form is $1/p$.

X	1	2	3	4	5	6	7	8 or more
f	74	52	36	11	10	4	4	0

23.5 Contingency tables

To what extent is any tendency of members of a group of motorists to have accidents associated with the state of their eyesight? Are television viewing habits dependent on income or education? In a smallpox epidemic, do the numbers of people who recovered suggest that recovery depended on whether or not they had been vaccinated? It is to questions like these that we can often apply a χ^2 test, with the data set out in what are called 'contingency' tables. The word 'contingent' means 'dependent' and in all these examples what we want to know is whether it is reasonably safe to say that one characteristic (accident proneness, say) is, or is not, independent of another (eyesight). In this respect, the problem differs from the kind of question to which calculating a correlation coefficient gave an answer, where the relationship between sets of *measurements* is the subject.

Suppose three different drugs, A, B, and C are administered to three groups of people suffering from the same disease, and the numbers cured or not cured are as shown in the table that follows (a contingency table):

	Drug given			Total
	A	B	C	
Cured (X)	24	13	18	55
Not cured (Y)	26	27	42	95
Total	50	40	60	150

H_0, the null hypothesis, is that the three drugs are equally efficacious. On this hypothesis, the expected proportions cured would be the same for each drug and would equal 55/150, and the proportions not cured would be 95/150.

Under H_0, then, the expected numbers cured would be

$$50 \times \frac{55}{150}, \qquad 40 \times \frac{55}{150}, \qquad 60 \times \frac{55}{150}$$

and the numbers not cured would be

$$50 \times \frac{95}{150}, \qquad 40 \times \frac{95}{150}, \qquad 60 \times \frac{95}{150},$$

giving the table

	A	B	C	Total
X	18.33	14.67	22	55
Y	31.67	25.33	38	95
Total	50	40	60	150

We now combine the two tables, showing observed and expected numbers in each category: cured and took drug A, not cured and took drug A, and so on.

	O	E	$(O-E)^2/E$
X and A	24	18.33	1.754
Y and A	26	31.67	1.015
X and B	13	14.67	0.190
Y and B	27	25.33	0.110
X and C	18	22.00	0.727
Y and C	42	38.00	0.421
		$\chi^2 =$	4.217

To find the number of degrees of freedom we need to know how many *independent* calculations are necessary to determine the expected values for the overall table. The number of cells is 3×2, but how many linear constraints are there on the expected numbers? To answer this we note that the total numbers taking each drug (A, B and C) are given (50, 40 and 60) and that the total numbers cured or not cured are also given (55 and 95). The totals in each category for the expected numbers must be the same (as in the table).

The expected numbers in each cell are therefore subject to the following constraints:

$e_{11} + e_{21} = 50$

$e_{12} + e_{22} = 40$

$e_{13} + e_{23} = 60$

$e_{11} + e_{12} + e_{13} = 55$

$(e_{21} + e_{22} + e_{23} = 95)$.

	A	B	C	
X	e_{11}	e_{12}	e_{13}	55
Y	e_{21}	e_{22}	e_{23}	95
	50	40	60	

However, the relation in brackets can be derived from the other four (by adding the first three and subtracting the fourth), so that the number of linear constraints in this case is 4. Hence, using the rule stated on page 444, $v =$ the number of cells minus the number of linear constraints $= 3 \times 2 - 4 = 2$. For χ_2^2 the critical region at the 5% level starts at 5.99, and H_0 is therefore accepted.

The number of degrees of freedom for a contingency table

There is, in fact, an easier method of calculating v than by counting the number of linear constraints. It can be illustrated by looking at a 3×4 contingency table, that is, one with 3 rows and 4 columns.

The observed data are shown opposite.

ΣA = the sum of the observed
values in the first row,

ΣP = the sum of the observed
values in the first column,

and so on.

	P	Q	R	S	
A	o_{11}	o_{12}	o_{13}	o_{14}	ΣA
B	o_{21}	o_{22}	o_{23}	o_{24}	ΣB
C	o_{31}	o_{32}	o_{33}	o_{34}	ΣC
	ΣP	ΣQ	ΣR	ΣS	N

From this, the table for the expected data will be:

	P	Q	R	S
A	$\dfrac{\Sigma P \times \Sigma A}{N}$	$\dfrac{\Sigma Q \times \Sigma A}{N}$	$\dfrac{\Sigma R \times \Sigma A}{N}$	$\dfrac{\Sigma S \times \Sigma A}{N}$
B	$\dfrac{\Sigma P \times \Sigma B}{N}$	$\dfrac{\Sigma Q \times \Sigma B}{N}$	$\dfrac{\Sigma R \times \Sigma B}{N}$	$\dfrac{\Sigma S \times \Sigma B}{N}$
C	$\dfrac{\Sigma P \times \Sigma C}{N}$	$\dfrac{\Sigma Q \times \Sigma C}{N}$	$\dfrac{\Sigma R \times \Sigma C}{N}$	$\dfrac{\Sigma S \times \Sigma C}{N}$

This is the invariable pattern and should enable you to complete the table of expected values quickly from any contingency table.

If, now, we look at the diagram opposite, it can be seen that the number of *independent* expected values must equal the number within the dotted line.

Hence for a 3×4 table, $v = 2 \times 3$.

The general result is that:

e_{11}	e_{12}	e_{13}	e_{14}	ΣA
e_{21}	e_{22}	e_{23}	e_{24}	ΣB
e_{31}	e_{32}	e_{33}	e_{34}	ΣC
ΣP	ΣQ	ΣR	ΣS	

For an $h \times k$ table, the number of degrees of freedom,

$$v = (h-1)(k-1).$$

2×2 contingency tables

EXAMPLE 8   The table shows the results of testing men and women for colour blindness:

	Normal vision	Colour blind	
Men	30	30	60
Women	120	220	340
	150	250	400

From this sample, test the hypothesis that sex and vision are independent.

This problem can be dealt with as above with $v = 1 \times 1 = 1$. The statistic $\Sigma (O-E)^2/E$ gives $\chi^2 = 4.70$, and since the critical region at the 5% level starts at $\chi_1^2 = 3.84$, the null hypothesis is rejected. It is of interest, however, to treat it as a problem in comparing two proportions, using the method outlined on page 298.

If π_1 is the population proportion of men with normal vision and π_2 is the corresponding population proportion for women, the null hypothesis to be tested is:

$$H_0: \pi_1 - \pi_2 = 0 \quad \text{as against} \quad H_1: \pi_1 - \pi_2 \neq 0.$$

Under H_0, an unbiased estimate for π, the common proportion for men and women together is $150/400 = 0.375$, and the distribution of $p_1 - p_2$, where p_1 and p_2 are the sample proportions for men and women, is binomial, with mean 0 and variance

$$0.375(1-0.375)(1/60+1/340) = 0.004596 = (0.0678)^2.$$

Using the normal approximation $N(0, 0.004596)$, when $p_1 = 30/60$ and $p_2 = 120/340$, the value of the test statistic Z is

$$\frac{(0.5-0.353)-0}{0.0678} = 2.17.$$

This is in the critical region for a two-tail 5% test (> 1.96), and we conclude, as before, that the difference between the proportions of men and women with normal vision is significant.

Yates' correction for a 2 × 2 contingency table

The fact that $\Sigma (O-E)^2/E$ is a discrete variable and the χ^2 distribution is continuous, may lead to error when any of the expected numbers in the cells of a 2 × 2 contingency table are small (< 5, say). In that case it is advisable to use Yates' correction, which involves replacing $\Sigma (O-E)^2/E$ by

$$\Sigma (|O-E|-\tfrac{1}{2})^2/E.$$

This is a continuity correction, analogous to the one employed when the normal distribution is used as an approximation to the binomial distribution. The next example shows how it is used.

EXAMPLE 9   An investigation into sufferers from bronchitis among smokers and non-smokers yielded the following results:

	Bronchial	Not Bronchial	
Smokers	9	3	12
Non-smokers	15	33	48
	24	36	60

Test the hypothesis that smoking and being bronchial are independent.

Under the null hypothesis the expected figures are

	Bronchial	*Not bronchial*
Smokers	$24 \times \dfrac{12}{60} = 4.8$	$36 \times \dfrac{12}{60} = 7.2$
Non-smokers	$24 \times \dfrac{48}{60} = 19.2$	$36 \times \dfrac{48}{60} = 28.8$

which give the following table:

O	E	$O - E$	$\lvert O - E \rvert - \frac{1}{2}$	$(\lvert O - E \rvert - \frac{1}{2})^2 / E$
9	4.8	4.2	3.7	2.85
15	19.2	-4.2	3.7	1.90
3	7.2	-4.2	3.7	0.71
33	28.8	4.2	3.7	0.48
				$\chi^2 = \overline{5.94}$

Now for $v = 1$ the critical value of χ_1^2 is 3.84, and we conclude that the null hypothesis should be rejected at the 5 % level (though not at the 1 % level, for which the critical value is 6.63.)

Limitations of the χ^2 test for goodness of fit and contingency

You may well have realised by now that the χ^2 test has certain limitations. The first example on contingency we worked on (page 455) raises more questions than the test can begin to answer. How much, for instance, does the use of any particular drug contribute towards a cure?

Similarly, if you did question 3 in exercise 23a, you will have discovered that the answer yielded by a χ^2 test can change when the sample size is increased. This is liable to happen when the null hypothesis is likely to be only approximately true.

EXERCISE 23c

1. Groups of English and American students were graded by their performance in a mathematics test, with the following results:

	Grade		
	1	2	3
English	53	28	19
American	32	38	30

Test whether nationality and grade are associated.

2. In a survey, 200 people were divided into three groups A, B and C according to their weekly incomes. In each group the numbers who owned cars in three different price ranges P, Q and R were recorded. Use a χ^2 test to determine whether there is reason to associate weekly income with type of car.

		A	B	C
	P	10	28	54
Car	Q	16	13	10
	R	20	33	16

Income group

3. In the principality of Viewmania a survey of 200 families, known to be regular television viewers, was undertaken. They were asked which of the three channels they watched most during an average week. A summary of their replies is given in the following table, together with the region in which they lived.

		North	East	South	West
Channel	CCB1	29	16	42	23
watched	CCB2	6	11	26	7
most	VIT	15	3	12	10

Region

Find the expected frequencies on the hypothesis that there is no association between the channel watched most and the region. Use the χ^2 distribution and a 5% level of significance to test the above hypothesis.   (AEB 1980)

4. In an investigation into the effect of personality traits on the incidence of cancer, the following results were obtained on the frequencies of the various categories of patient.

	With cancer	Without cancer
With trait	8	3
Without trait	6	23

Analyse the table of results as a 2×2 contingency table.   (MEI)

5. An investigation into proneness to summer hay-fever amongst 80 people gave the following results:

	Yes	No
Men	5	22
Women	15	38

Hay-fever sufferer

Test whether these results provided evidence of an association between gender and a tendency to catch hay-fever.

6. Three hundred and fifty newspaper readers were categorised into types *A, B, C, D* according to their average weekly income. The numbers of each type who were regular buyers of four newspapers *W, X, Y, Z* were noted. The results are given in the table below:

		Newspaper			
		W	X	Y	Z
Reader type	A	34	41	18	37
	B	24	29	12	25
	C	13	20	18	29
	D	9	10	12	19

Test whether there is an association between type of reader and newspaper bought regularly.

7.   A number of children in three different regions of a small African state were found to be suffering from rickets, classified as severe, moderate, or minor. The figures were:

		Severe	Moderate	Minor
	Coastal	23	9	6
Region	Inland plain	21	4	3
	Hill country	34	24	17

Use a χ^2 distribution to test the hypothesis that there is no association between region and degree of suffering from rickets.

Summary of the rules for the number of degrees of freedom

Goodness of fit

If the parameters of the expected distribution are given

$$v = \text{the number of cells} - 1.$$

If the parameters of the expected distribution have to be estimated

$$v = \text{the number of cells} - 1 - r$$

where r is the number of parameters estimated.

Contingency tables

For an $h \times k$ table $v = (h-1)(k-1)$.

23.6★ Properties of the χ^2 distribution

We have been using the χ^2 distribution for the particular practical purpose of testing goodness of fit and contingency, but in fact it is a distribution with many important theoretical applications. The derivation of the 't' distribution, for instance, depends on χ^2, as do other distributions. In chapters 24 and 25 we shall be looking at two of these, both connected with tests of variance, and for this reason we will give here a rather fuller account of χ^2 than the brief description on pages 443–6.

As we have already said, the family of χ^2 distributions are derived from the distribution of squares of standardised normal variables.

If a random variable Z has a distribution $N(0, 1)$, that is, it is a standardised normal variable, then the distribution of the random variable Y, where

$$Y = Z^2$$

has a χ^2 distribution with 1 degree of freedom (χ_1^2).

Similarly, if Z_1 and Z_2 are independent random variables, each with distribution $N(0, 1)$, then the distribution of the random variable Y, where

$$Y = Z_1^2 + Z_2^2$$

has a χ^2 distribution with 2 degrees of freedom (χ_2^2).

The general definition of χ_n^2 is therefore this:

If $Z_1, Z_2, \ldots, Z_n$ are independent random variables, each with distribution $N(0, 1)$, then the distribution of the random variable Y, where

$$Y = Z_1^2 + Z_2^2 + \ldots + Z_n^2$$

has a χ^2 distribution with n degrees of freedom (χ_n^2).

The density function of a χ^2 distribution

The method of deriving the probability density function of χ_1^2 from the definition will be demonstrated shortly. The derivation of χ_n^2 is beyond the scope of this book and we shall just state the result without proof.

The density function of the random variable Y, where $Y = \sum_{i=1}^{n} Z_i^2$ is

$$f(y) = K_n y^{\frac{1}{2}n - 1} e^{-\frac{1}{2}y}, \quad 0 \leqslant y \leqslant \infty,$$

where the value of the constant K_n depends on n.

In one or two cases, K_n may be found quite easily by equating $\int_0^\infty f(y)\, dy$ to 1, but it is convenient to have the general result, which is:

$$K_n = \begin{cases} \dfrac{1}{(n-2)(n-4)\ldots} \cdot \left[\dfrac{1}{\sqrt{(2\pi)}}\right] & \text{if } n \text{ is odd,} \\[3ex] \dfrac{1}{(n-2)(n-4)\ldots} \cdot \left(\dfrac{1}{2}\right) & \text{if } n \text{ is even,} \end{cases}$$

where the product continues for as long as the terms remain positive.

Thus $K_1 = \dfrac{1}{\sqrt{(2\pi)}}$, $K_3 = \dfrac{1}{1} \cdot \dfrac{1}{\sqrt{(2\pi)}}$, $K_5 = \dfrac{1}{3.1} \cdot \dfrac{1}{\sqrt{(2\pi)}}, \ldots$

and $K_2 = \dfrac{1}{2}$, $K_4 = \dfrac{1}{2} \cdot \dfrac{1}{2}$, $K_6 = \dfrac{1}{4.2} \cdot \dfrac{1}{2}, \ldots$

The density functions of the first three χ^2 distributions are therefore:

$$\chi_1^2: \frac{1}{\sqrt{(2\pi)}} y^{-\frac{1}{2}} e^{-\frac{1}{2}y}, \qquad \chi_2^2: \tfrac{1}{2}e^{-\frac{1}{2}y}, \qquad \chi_3^2: \frac{1}{\sqrt{(2\pi)}} y^{\frac{1}{2}} e^{-\frac{1}{2}y}.$$

The shape of the χ^2 distribution

The figure below shows the shape of χ_n^2 for $n = 2, 4, 8$ and 12.

It can be seen that the curve is fairly symmetrical for $n = 12$, and indeed, when n is large the distribution is approximately normal. This follows from the central limit theorem, which states that the sum of a large number of independent random variables has an approximately normal distribution.

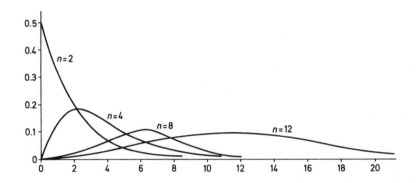

EXAMPLE 10   If the distribution of the random variable Y is χ_2^2, find:
(a) the mean and variance of Y,
(b) $P(Y > 3)$,
(c) the value of y such that $P(Y \leqslant y) = 0.99$. In other words, find the critical region for χ_2^2 at the 1% level.

(a) Since $f(y) = \tfrac{1}{2}e^{-\frac{1}{2}y}, 0 \leqslant y < \infty$,

$$\mu = E(Y) = \int_0^\infty y \cdot \tfrac{1}{2}e^{-\frac{1}{2}y} \, dy, \text{ which, after integration by parts,}$$

$$= 2.$$

Similarly $E(Y^2) = \displaystyle\int_0^\infty y^2 \cdot \tfrac{1}{2}e^{-\frac{1}{2}y} \, dy = 8$ (the details are omitted).

Hence $\sigma^2 = 8 - 2^2 = 4$.

(b) $P(Y > 3) = 1 - P(Y \leqslant 3) = 1 - \displaystyle\int_0^3 \tfrac{1}{2}e^{-\frac{1}{2}y} \, dy = 1 - \left[-e^{-\frac{1}{2}y} \right]_0^3$

$$= e^{-1.5} = 0.223.$$

(c) If $P(Y \leqslant y) = 0.99$, $\displaystyle\int_0^y \tfrac{1}{2}e^{-\frac{1}{2}y} \, dy = 0.99 \Rightarrow 1 - e^{-\frac{1}{2}y} = 0.99$

$$\Rightarrow y = 9.2 \quad \text{(which is the value for } \chi_2^2 \text{ given in the tables).}$$

EXAMPLE 11   Find the probability density function for the χ^2 distribution with one degree of freedom.

On page 422 it was shown that, if the density function for the continuous random variable X is $f(x)$ and $Y = X^2$, then the density function of Y is

$$g(y) = \frac{1}{2\sqrt{y}} [f(\sqrt{y}) + f(-\sqrt{y})].$$

Now χ_1^2 is the distribution of Y, where $Y = Z^2$ and Z, being a standardised normal variable has density function

$$\phi(z) = \frac{1}{\sqrt{(2\pi)}} e^{-\frac{1}{2}z^2}, \quad -\infty < z < \infty.$$

It follows that the density function of Y,

$$g(y) = \frac{1}{2\sqrt{y}} [\phi(\sqrt{y}) + \phi(-\sqrt{y})]$$

$$= \frac{1}{2\sqrt{y}} \times 2\phi(\sqrt{y}) \quad [\text{since } \phi(z) = \phi(-z)]$$

$$= \frac{1}{\sqrt{(2\pi)}} y^{-\frac{1}{2}} e^{-\frac{1}{2}y}, \quad 0 < y < \infty.$$

The moment generating function of the χ^2 distribution

Many of the theoretical applications of χ^2 depend on using its moment generating function, which also provides the easiest method of finding the mean, variance and higher moments. We shall now prove that the moment generating function for χ_n^2 is

$$M(t) = \frac{1}{(1-2t)^{\frac{1}{2}n}}.$$

Proof

Starting with χ_1^2, the distribution of Y when $Y = Z^2$ and Z has density function

$$\phi(z) = \frac{1}{\sqrt{(2\pi)}} e^{-\frac{1}{2}z^2}, \quad -\infty < z < \infty,$$

$$M_y(t) = E(e^{tY}) = E(e^{tZ^2})$$

$$= \int_{-\infty}^{\infty} e^{tz^2} \cdot \frac{1}{\sqrt{(2\pi)}} e^{-\frac{1}{2}z^2} \, dz = \frac{1}{\sqrt{(2\pi)}} \int_{-\infty}^{\infty} e^{-\frac{1}{2}z^2(1-2t)} \, dz.$$

Making the substitution $z\sqrt{(1-2t)} = u$, this integral becomes

$$\frac{1}{\sqrt{(1-2t)}} \cdot \frac{1}{\sqrt{(2\pi)}} \int_{-\infty}^{\infty} e^{-\frac{1}{2}u^2} \, du = \frac{1}{\sqrt{(1-2t)}} \times 1.$$

Hence the m.g.f. of χ_1^2 is $\dfrac{1}{\sqrt{(1-2t)}}$.

Now if $Y = Z_1^2 + Z_2^2$, its m.g.f. is the product of the m.g.fs. of Z_1^2 and Z_2^2, each of which is $1/\sqrt{(1-2t)}$.

It follows that the m.g.f. of χ_2^2 is

$$\left[\frac{1}{\sqrt{(1-2t)}}\right]^2 = \frac{1}{1-2t}.$$

The general result follows immediately.

The mean and variance of χ_n^2

Using $E(Y) = M'(0)$ and $E(Y^2) = M''(0)$ it is easily shown that

$$\text{the mean of } \chi_n^2 = n \text{ and the variance of } \chi_n^2 = 2n.$$

23.7   The sum of two χ^2 distributions

The last important property that will be needed in later work is this:

If Y_1 and Y_2 are independent random variables having χ^2 distributions with n_1 and n_2 degrees of freedom respectively, then the distribution of $Y_1 + Y_2$ is χ^2 with $n_1 + n_2$ degrees of freedom.

This follows immediately from the definition of χ^2, since $Y_1 + Y_2$ is clearly the sum of the squares of $n_1 + n_2$ independent standardised normal variables.

EXERCISE 23d

1.   The random variable Y has χ_2^2 distribution, given by the density function
$$f(y) = \tfrac{1}{2}e^{-\frac{1}{2}y}, \quad 0 \leqslant y < \infty.$$
If $P(Y \leqslant a) = 0.90$ and $P(Y \leqslant b) = 0.95$, find a and b, the critical 10% and 5% points of the distribution.

2.   The random variable Y has χ_4^2 distribution, given by the density function
$$f(y) = Kye^{-\frac{1}{2}y}, \quad 0 \leqslant y < \infty.$$
(a) Prove, by integration, that $K = \tfrac{1}{4}$.
(b) If $P(Y \leqslant y) = 0.9$, show that
$$(y+2)e^{-\frac{1}{2}y} = 0.2$$
and verify that $y = 7.78$ satisfies this equation approximately.

3.   Use the moment generating functions to prove that:
(a) the mean and variance of χ_1^2 are 1 and 2 respectively;
(b) the mean and variance of χ_n^2 are n and $2n$ respectively.

4.   The distribution of Y is χ_2^2. Use the moment generating function to show that the mean is 2 and the variance is 4.
Find also μ_3', the 3rd moment about zero and deduce μ_3, the 3rd moment about the mean, using the formula

$$\mu_3 = \mu'_3 - 3\mu\mu'_2 + 2\mu^3.$$

$[\mu'_2 = E(Y^2) = M''(0)$ and $\mu'_3 = E(Y^3) = M'''(0).]$

Hence find the skewness, as measured by the formula μ_3/σ^3.

5. The random variable Y whose distribution is χ_2^2 has density function

$$f(y) = \tfrac{1}{2}e^{-\frac{1}{2}y}, \quad 0 \leqslant y < \infty.$$

Using the definition $M(t) = E(e^{tY})$, prove by integration that

$$M(t) = 1/(1-2t).$$

6. If the random variables X_1 and X_2 have distributions $N(\mu, \sigma_1^2)$ and $N(\mu, \sigma_2^2)$ respectively and

$$Y = \frac{(X_1 - X_2)^2}{\sigma_1^2 + \sigma_2^2}$$

find the mean and variance of the distribution of Y.

EXERCISE 23e (miscellaneous)

1. (a) The following table shows the frequency distribution of the number of particles emitted by a radioactive source in each of 100 intervals of a second.

Number emitted	0	1	2	3	4	5	6	7	8
Frequency	20	19	29	13	14	4	0	0	1

Test whether these data are consistent with a Poisson distribution with mean 2.
   (b) The following table shows the frequency distribution of the number of accidents in a factory each week in 2 years each of 50 working weeks.

Number of accidents	0	1	2	3	4	5	6	7	8
Frequency	20	19	29	13	14	4	0	0	1

Fit a Poisson distribution to the data and test for goodness of fit.
   (c) Explain why in one of parts (a) and (b) the Poisson distribution is an acceptable fit and in the other it is not.   (O)

2. A six-sided die with faces numbered as usual from 1 to 6 was thrown 5 times and the number x of '6's was recorded. The experiment was repeated 200 times with the following results:

x	0	1	2	3	4	5
Frequency	66	82	40	10	2	0

On this evidence, would you consider the die to be biased? Fit a suitable distribution to the data and test and comment on the goodness of fit.   (MEI)

3. The table below shows the number of employees and the number of accidents occurring in a year in four factories:

Factory	A	B	C	D
Employees (thousands)	4	3	1	2
Accidents	22	13	11	17

Use χ^2 and a 5% level of significance to test the hypothesis that the expected number of accidents is proportional to the number of employees in each factory.   (O & C)

4.   To investigate the distribution of growth of runner beans, a sample distance along one row was chosen. The sample contained 500 beans whose lengths were recorded in the following frequency distribution:

Length (cm)	0–4	4–8	8–12	12–16	16–20	20–24	more than 24
Number of beans	31	72	140	159	76	15	7

Test the hypothesis that the appropriate model is a normal distribution with mean 12.0 cm and standard deviation 5.0 cm.

5.   The number of misprints per page of a newspaper were recorded for randomly selected pages as given in the table below.
(a) Find the mean and variance.
(b) Test the goodness of fit between the given distribution and a Poisson distribution.

Number of misprints per page	0	1	2	3	4	5	6	7	$\geqslant 8$
Number of pages	7	13	20	25	16	10	5	4	0

(MEI)

6.   The table below gives the cumulative frequency distribution for the masses of 100 eighteen-year-old female applicants to join the police force.

Mass (kg) *not greater than*	30	40	50	60	70	80	90	100
Cumulative frequency	0	3	18	43	66	84	96	100

On the basis of previous experience, it is believed that a normal distribution with mean 65 kg and standard deviation 15 kg gives a good model. Use a χ^2 test, at the 2.5% level, to determine whether or not this model provides a good fit for the given data. State how many degrees of freedom are appropriate for your test.   (C)

7.   Smallwoods Ltd. run a weekly football pools competition. One part of this involves a fixed-odds contest where the entrant has to forecast correctly the result of each of five given matches. In the event of a fully correct forecast, the entrant is paid out at odds of 100 to 1. During the last two years Miss Fortune has entered this fixed-odds contest 80 times. The table below summarises her results.

Number of matches correctly forecast per entry (x)	0	1	2	3	4	5
Number of entries with x correct forecasts (f)	8	19	25	22	5	1

(a) Find the frequencies of the number of matches correctly forecast per entry given by a binomial distribution having the same mean and total as the observed distribution.
(b) Use the χ^2 distribution and a 10% level of significance to test the adequacy of the binomial distribution as a model for these data.
(c) On the evidence before you, and assuming that the point of entering is to win money, would you advise Miss Fortune to continue with this competition and why?
(AEB 1981)

8.   A sample of 100 observations from the distribution of the continuous random variable X are given in the following table:

X	0–2	2–4	4–8	8–16
f	40	30	15	15

Test the hypothesis that the model for this distribution is exponential. The density function for an exponential distribution is

$$f(x) = \begin{cases} \lambda e^{-\lambda x}, & 0 \leqslant x < \infty, \\ 0 & , \quad \text{otherwise.} \end{cases}$$

9.   Three brands of sleeping pill, denoted by A, B and C, were tested for their effectiveness in helping people to sleep. From 1000 volunteers, 450 were chosen at random and given pill A, 350 were chosen at random and given pill B, and the remaining 200 were given pill C. After taking these pills, 200 of the volunteers got less sleep than usual, 300 got their usual amount of sleep, and 500 got more sleep than usual. On the assumption that the pills are identical in their effect, complete a table, similar in form to that below, to show, for each brand of pill, the expected number of volunteers getting less sleep, getting the usual amount of sleep and getting more sleep.
  The given table shows the observed number of volunteers in each category. Carry out a χ^2 test with four degrees of freedom to determine whether the observed results are consistent with the expected results. State your conclusion briefly.

Sleeping pill	Number of volunteers		
	Getting less sleep than usual	Getting usual amount of sleep	Getting more sleep than usual
A	82	104	264
B	65	112	173
C	53	84	63

(C)

10.★ (a) The proportion of viewers who watch a television play is P. The value of P is unknown, except that $0 \leqslant P \leqslant 1$. In a random sample of viewers, a people said that they had watched the play and b people said they had not, where a and b are large. Show that the value of χ^2 required to test the hypothesis that $P = p$ is given by

$$\frac{[a(1-p)-bp]^2}{(a+b)p(1-p)}.$$

  (b) Hence test the hypothesis that $P = 0.6$ when $a = 40$ and $b = 50$.
  (c) Show further that if $y = a-b$ and $x = a+b$ then, on testing the hypothesis that $P = 0.5$, the value obtained for χ^2 will be significant (at the 5% level) if, and only if, the point (x, y) lies outside a certain parabola whose equation should be given.   (O & C)

11.★ It is believed that the probability P that a lamp will survive for more than x months is

$$\exp(-ax^b)$$

where a and b are unknown contestants. Show that

$$\lg(-\lg P) = A + B \lg x,$$

and find A and B in terms of a and b.
  The lives of 100 such lamps, measured in months, are summarised in the table below.

Life	0–5	5–10	10–20	> 20
Number of lamps	15	34	37	14

Estimate a and b to 2 significant figures by plotting the sample values of $\lg(-\lg P)$ against $\lg x$.

(You may take $\lg 5 = 0.7$, $\lg 20 = 1.3$, $\lg \lg e = \bar{1}.64 = -0.36$.)

Use your graph to show that the numbers of lamps expected to fail are approximately:

Life	0–5	5–10	10–20	> 20
Expected numbers	16	29	42.5	12.5

Using these expected numbers, apply χ^2 with 1 degree of freedom to test the original hypothesis; explain briefly why 1 d.f. is appropriate. (O & C)

12.★ Show that, for the following 2 × 2 contingency table:

	X	Y
P	a	b
Q	c	d

$$\chi^2 = \frac{(ad - bc)^2 n}{(a+c)(a+b)(b+d)(c+d)} \quad \text{where } n = a+b+c+d.$$

13.★ In a contingency table with r rows and 2 columns, the observations in the ith row are a_i and $(n_i - a_i)$ for $i = 1, 2, \ldots, r$, and the column totals are A and $(N - A)$ respectively. Show that the value of χ^2 which tests the hypothesis that the expected values in the ith row are An_i/N and $(N - A)n_i/N$ for all i may be expressed in the form

$$\frac{N^2}{A(N-A)} \sum_{i=1}^{r} \frac{(a_i - An_i/N)^2}{n_i}.$$

Hence, or otherwise, show that it can be expressed as

$$\frac{N^2}{A(N-A)} \left[\left(\sum_{i=1}^{r} \frac{a_i^2}{n_i} \right) - \frac{A^2}{N} \right]. \tag{O & C}$$

14.★ In a survey of 100 voters in 1980, the number intending to vote for a certain political party was 40. In a similar survey of 100 voters in 1981, the number intending to vote for this party was x. Show that, ignoring Yates' correction, the value of the usual χ^2 statistic testing whether the probability of voting for this party is the same in both years is

$$200(40 - x)^2 / [(40 + x)(160 - x)]. \tag{O & C}$$

15.★ The probability density function of χ^2 with 4 degrees of freedom is

$$Cxe^{-x/2}, \quad (0 \leqslant x < \infty).$$

Find the value of C. If χ_0^2 satisfies

$$\int_0^{\chi_0^2} Cxe^{-x/2} \, dx = 0.95,$$

show that

$$(2\chi_0^2 + 4)\exp(-\tfrac{1}{2}\chi_0^2) = 0.2$$

and verify that an approximate solution of this equation is $\chi_0^2 = 9.49$. Explain the usefulness of this result.

A group of 300 people is classified according to their occupations and smoking habits while at work, the following table being drawn up:

	Cigarette smoker	Cigar/pipe smoker	Non-smoker
Manual worker	36	10	74
Office worker	23	16	61
Executive	22	25	33

Test at the 5% level of significance whether there is an association between occupation and smoking habits, and comment on your answer.   (MEI)

16.★ The random variable x is normally distributed with mean μ and variance σ^2. Show that, if $y = (x-\mu)^2/\sigma^2$, then the probability that $y \leqslant z$ is

$$(2\pi)^{-\frac{1}{2}} \int_0^z s^{-\frac{1}{2}} e^{-\frac{1}{2}s} \, ds.$$

Deduce that y is distributed like χ^2 with one degree of freedom. (The probability density function of χ^2 with n degrees of freedom is $Cz^{(\frac{1}{2}n-1)}e^{-\frac{1}{2}z}$, where C is a constant.)   (MEI)

17.★★ The random variable X has a χ^2 distribution with n degrees of freedom. By using the formulae for the density function of χ^2 with n, $n+2$ and $n+4$ degrees of freedom, or otherwise, show that the mean and variance of X are n and $2n$ respectively.   (O & C)

18.★★ Define the χ^2 distribution with n degrees of freedom.
(a) Prove that the mean of the distribution is n.
(b) Given that W_1, W_2 are independent χ^2 variables with n_1, n_2 degrees of freedom, show that $W_1 + W_2$ is a χ^2 variable with $n_1 + n_2$ degrees of freedom.
(c) The variable T is the corrected sum of squares

$$\sum_{i=1}^{k} (x_i - \bar{x})^2$$

for a sample of k values with mean $\bar{x}$ from the normal variable X, where X is the distribution $N(\mu, \sigma^2)$. The unbiased estimator for σ^2 is s^2 and it is given that T/σ^2 has the χ^2 distribution with $k-1$ degrees of freedom. Given the variance estimates from two independent samples of sizes k_1, k_2, show that the expression

$$\frac{(k_1-1)s_1^2 + (k_2-1)s_2^2}{k_1 + k_2 - 2}$$

is an unbiased estimator of σ^2 with $k_1 + k_2 - 2$ degrees of freedom.   (C)

19.★ A random sample of x observations $(x_1, x_2, \ldots, x_n)$ is taken from a normal distribution with zero mean and unknown variance σ^2 which is estimated by calculating s^2 where

$$s^2 = \frac{1}{n} \sum_{i=1}^{n} x_i^2.$$

By expressing x_i as σy_i or otherwise, find the expected value and variance of s^2.
(The relationship between the χ_1^2 distribution and the standard normal distribution with zero mean and unit variance should be clearly stated, but need not be proved.)   (O & C)

20.★ A random variable $Y = X^2$, where X is distributed normally with mean μ and unit variance. Show that the m.g.f. for the distribution of Y is

$$M_y(t) = \frac{\exp\{\mu^2 t(1-2t)^{-1}\}}{(1-2t)^{\frac{1}{2}}}, \quad |t| < \tfrac{1}{2}.$$ (MEI)

24* Variance tests: the *F*-distribution

24.1 Confidence interval for variance

A biologist doing field work on an island off the Scottish coast is able to collect 16 lizards of a particular species. In writing up an account of his findings he includes the mean length of the sample, $\bar{x} = 24.3$ cm (to the nearest millimetre) and the variance, var$(x) = 4.41$ cm^2, giving unbiased estimates of the population mean and variance as

$$\hat{\mu} = 24.3 \text{ cm}, \qquad \hat{\sigma}^2 = s^2 = 4.41 \times 16/15 = 4.70 \text{ cm}^2.$$

Suppose he wants to calculate 95 % confidence limits for μ and σ^2. There is no problem about the first of these, which are

$$\bar{x} \pm t \frac{s}{\sqrt{n}},$$

where $\bar{x} = 24.3$, $s = \sqrt{4.70}$, $n = 16$ and t, the $P = 5$ value of t_{15}, is 2.13. The confidence interval is $[23.15, 25.45]$ cm. This calculation is based, of course, on the very reasonable assumption that the parent distribution is normal.

What about the confidence limits for σ^2? So far the only information that has been established about the distribution of s^2 is that its mean is σ^2. In fact it turns out that we can conveniently use, not the distribution of s^2 itself, but the distribution of a related statistic

$$\frac{(n-1)s^2}{\sigma^2}.$$

In the Appendix on page 480 it is shown that:

If a random sample, size n, is taken from the normal distribution $N(\mu, \sigma^2)$ and $s^2 = \Sigma (X - \bar{X})^2/(n-1)$ (the unbiased estimator of σ^2), then the distribution of

$$\frac{(n-1)s^2}{\sigma^2}$$

is χ^2 with $n-1$ degrees of freedom. (Note that $(n-1)s^2/\sigma^2 = \Sigma (x - \bar{x})^2/\sigma^2$.)

Applying this to the lizard problem, where $n = 16$ and $s^2 = 4.70$, the distribution of

$$15 \times 4.70/\sigma^2$$

is χ^2 with 15 degrees of freedom. To find 95 % confidence limits in a skew distribution, it is usual to take an interval which gives $2\frac{1}{2}\%$ in each tail. From the tables, these are $27.49(P = 2.5)$ and $6.26(P = 97.5)$.

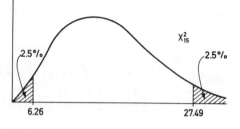

Hence

$$P[6.26 < 15 \times 4.70/\sigma^2 < 27.49] = 0.95$$

$$\Rightarrow P[6.26\sigma^2 < 70.5 < 27.49\sigma^2] = 0.95.$$

Rearranging, 95 % confidence limits for σ^2 are 70.5/27.49 and 70.5/6.26, giving a confidence interval of

$$[2.56, 11.26].$$

As this example shows, the confidence interval is wide for a small sample.

It can easily be shown that what we have said about the distribution of $(n-1)s^2/\sigma^2$ is true for $n = 2$, since if a sample of two observations from the distribution of X, a normal distribution $N(\mu, \sigma^2)$ are X_1 and X_2, then $\bar{X} = \frac{1}{2}(X_1 + X_2)$ and

$$\sum_{i=1}^{2} (X - \bar{X})^2 = [X_1 - \tfrac{1}{2}(X_1 + X_2)]^2 + [X_2 - \tfrac{1}{2}(X_1 + X_2)]^2 = \tfrac{1}{2}(X_1 - X_2)^2$$

so that $\quad s^2 = \frac{1}{2}(X_1 - X_2)^2/1$ and

$$(n-1)s^2/\sigma^2 = 1 \times (X_1 - X_2)^2/2\sigma^2 = [(X_1 - X_2)/\sqrt{(2)}\sigma]^2.$$

But the distribution of X_1 and of X_2 is $N(\mu, \sigma^2)$, so that the distribution of $X_1 - X_2$ is $N(0, 2\sigma^2)$.

It follows that $(n-1)s^2/\sigma^2$ is, in this case, the square of a standardised normal variable and its distribution is therefore χ_1^2.

Confidence limits for σ^2 when two samples are available

If two samples are available, one of size n_1 for which $s_1^2 = \Sigma (x_1 - \bar{x}_1)^2/(n_1 - 1)$, the other of size n_2, for which $s_2^2 = \Sigma (x_2 - \bar{x}_2)^2/(n_2 - 1)$, pooling this information leads to the following.

The distribution of $(n_1 - 1)s_1^2/\sigma^2$ is χ^2 with $n_1 - 1$ degrees of freedom, and the distribution of $(n_2 - 1)s_2^2/\sigma^2$ is χ^2 with $n_2 - 1$ degrees of freedom.

By the result on page 465, it follows that the distribution of

$$[(n_1 - 1)s_1^2 + (n_2 - 1)s_2^2]/\sigma^2$$

is χ^2 with $n_1 + n_2 - 2$ degrees of freedom, a result which may be extended for any number of samples.

EXAMPLE 1   The carbon content, measured in suitable units, of two samples of steel taken at different times were:
(a) 3.5   4.2   3.7   4.0   3.6
(b) 3.5   3.8   4.1   4.2   4.0   4.3   3.4.
Assuming a common variance σ^2, find a 90 % confidence interval for σ^2.

From the first sample, where $n_1 = 5$, s_1^2 is found to be 0.085. From the second sample, where $n_2 = 7$, s_2^2 is found to be 0.120.

The distribution of $(4s_1^2 + 6s_2^2)/\sigma^2$ or $1.06/\sigma^2$ is χ_{10}^2.

For a 90% symmetrical confidence interval we take the $P = 95$ and $P = 5$ values from the χ^2 table, which are 3.94 and 18.31 respectively.

Hence $P[3.94 < 1.06/\sigma^2 < 18.31] = 0.90$, which gives a confidence interval of

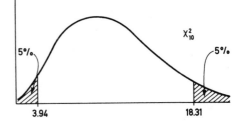

[0.06, 0.27]   to 2 dec. places.

EXERCISE 24a

1. A random sample of 30 from a normal population yields a variance of 11.5. Find: (a) a 95% confidence interval, (b) a 99% confidence interval for the population variance.

2. A random sample from a normal distribution gives the following values:

$$32 \quad 31 \quad 32 \quad 28 \quad 32 \quad 30 \quad 33 \quad 30.$$

Calculate a 90% confidence interval for the population variance.

3. Ten bags of fertiliser supplied to a garden centre were found to have the following masses (in kilograms):

$$2.57 \quad 2.05 \quad 1.65 \quad 2.62 \quad 2.44 \quad 1.48 \quad 2.31 \quad 1.58 \quad 2.60 \quad 1.85$$

Find a symmetrical 95% confidence interval for the population variance, assuming that the population distribution is normal.

4. Two samples are taken from the same normal distribution as follows:

Sample 1:	2.4	2.2	2.2	2.4	2.3	2.3	
Sample 2:	2.2	2.2	2.1	2.4	2.3	2.2	2.1

Calculate a 95% confidence interval for the population variance.

5. A random sample of 12 steel ingots was taken from a production line. The masses, in kilograms, of these ingots are given below.

$$24.8 \quad 30.8 \quad 28.1 \quad 24.8 \quad 27.4 \quad 22.1 \quad 24.7 \quad 27.3 \quad 27.5 \quad 27.8 \quad 23.9 \quad 23.2$$

Assuming that this sample came from an underlying normal population, calculate a 90% confidence interval for the variance of the population.   (AEB 1980)

24.2   The F-distribution

Let us look again at the biologist with his sample of lizards whose lengths he has measured. Suppose that on another island a colleague has collected data on another sample of lizards which seem to be of a similar type. How could they test whether it is reasonable to assume that as far as lengths are concerned, these samples could be regarded as coming from the same (normal) population? As a first step it will be necessary to test whether the underlying population *variances* are the same, since unless this is so it is not possible to test whether or not the means differ significantly (see page 281).

Such a test of variances depends on a distribution called the *F*-distribution which must now be described.

The *F*-distribution involves two χ^2 distributions and is defined as follows:

If X_1 and X_2 are two independent random variables whose distributions are χ^2 with v_1 and v_2 degrees of freedom respectively, and F is the ratio

$$F = \frac{X_1/v_1}{X_2/v_2}$$

then the distribution of this ratio is called the *F*-distribution and is defined for values of f from 0 to ∞.

(The density function of F is given in the Appendix on page 481.)

Since F has two parameters, v_1 and v_2, the distribution for a particular pair is written

$$F(v_1, v_2) \text{ for short.}$$

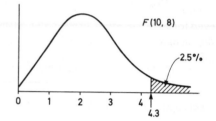

The sketch opposite shows a typical *F* curve for $v_1 > 4$.

The fact that there are two parameters means that there has to be a separate table for each critical percentage point: 5%, 2.5%, 1% and 0.1%. (See pages 521–4). The tables work like this. Suppose we want the critical 2.5% point for $F(10, 8)$. From the table, this is 4.30 (see diagram opposite).

This means that

$$P[F(10, 8) > 4.30] = 0.025.$$

Note: 1. For values of v_1 and v_2 not given in a table, linear interpolation may be used, except when $v_1 > 12$ or $v_2 > 40$.

2. The critical values of F at the other end of the percentage scale, 95%, 97.5%, 99%, 99.9% are not given, and we shall not have much occasion to use them. If they are wanted, the following relation may be used:

$$F_{(100-\alpha)\%}(v_1, v_2) = 1/F_{\alpha\%}(v_2, v_1).$$

This follows from the definition of F, which implies that if the distribution of Y is $F(v_2, v_1)$ then the distribution of $1/Y$ is $F(v_1, v_2)$.

Now
$$P[Y > F_{5\%}(v_2, v_1)] = 0.05$$

$$\Rightarrow P[1/Y < 1/F_{5\%}(v_2, v_1)] = 0.05$$

$$\Rightarrow P[1/Y > 1/F_{5\%}(v_2, v_1)] = 0.95.$$

which means that the critical 95 % point for $1/Y$ is $1/F_{5\%}(v_2, v_1)$. In other words, the critical 95 % value of $F(v_1, v_2)$ is the reciprocal of the critical 5 % value of $F(v_2, v_1)$.

24.3   Testing the ratio of two variances

The procedure for testing whether, on the evidence of two independent samples, it is reasonable to suppose that the variances of their underlying populations are the same is as follows.

If the two estimates of the population variance are s_1^2 and s_2^2, based on samples of size n_1 and n_2 respectively, we calculate the *variance ratio*

$$s_1^2/s_2^2 \quad \text{where } s_1^2 \text{ is the } greater \text{ of the two.}$$

The null hypothesis is that there *is* a common population variance. If it is true, we would expect that this ratio will not differ greatly from 1 for large samples, and that even for small samples it will not be significantly greater than 1, taking into account the sizes of the samples. The reason we use a ratio for this test (rather than a difference) is that, conveniently, under the null hypothesis,

$$\text{the distribution of } s_1^2/s_2^2 \quad \text{is} \quad F(v_1, v_2)$$

where

$$v_1 = n_1 - 1 \quad \text{and} \quad v_2 = n_2 - 1.$$

Proof

Suppose the two population variances are σ_1^2 and σ_2^2. As we saw on page 471, the distribution of $(n_1 - 1)s_1^2/\sigma_1^2$ is χ^2 with $n_1 - 1$ degrees of freedom.

Similarly the distribution of $(n_2 - 1)s_2^2/\sigma_2^2$ is χ^2 with $n_2 - 1$ degrees of freedom.

By definition of the F-distribution, under the null hypothesis $\sigma_1^2 = \sigma_2^2 (= \sigma^2)$, the distribution of

$$\frac{(s_1^2/\sigma^2)}{(s_2^2/\sigma^2)}$$

is therefore $F(n_1 - 1, n_2 - 1)$, that is s_1^2/s_2^2 has distribution $F(n_1 - 1, n_2 - 1)$.

We are now ready to use this test in an example.

EXAMPLE 2   The biologist mentioned earlier estimated the variance of the underlying population for his sample of 16 as 4.70. Suppose that his colleague produced an estimate of 5.40, based on a sample of 13. Do these estimates differ significantly?

If the hypothetical population variances are σ_1^2 and σ_2^2, the hypotheses to be tested are

$$H_0 : \sigma_1^2 = \sigma_2^2 \quad \text{as against} \quad H_1 : \sigma_1^2 \neq \sigma_2^2,$$

and we shall use a 5 % significance level.

Taking s_1^2 as 5.40, with $n_1 = 13$ and s_2^2 as 4.70 with $n_2 = 16$, the variance ratio is

$$5.4/4.7 = 1.15$$

and the appropriate F-distribution is $F(12, 15)$.

Now, although the test is a two-tail one, since we put the larger value of s^2 in the numerator the variance ratio *can only be significantly different from 1 by being on the right-hand side of the distribution curve.*

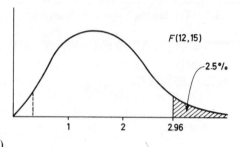

This means that we are only concerned to discover whether or not it is in the *right-hand critical region*, corresponding to $P = 2.5\%$ (the right-hand half of a 5% two-tail critical region).

The critical, $P = 2.5\%$ value of $F(12, 15)$ is 2.96, and since $1.15 < 2.96$, H_0 is accepted and we conclude that it is reasonable to assume a common variance.

It will be noticed that the tables for the F-distribution include values for $v_1 = \infty$ and $v_2 = \infty$. These are used when we wish to compare a genuine sample variance with a hypothetical population variance. The latter is then regarded as a sample of infinite size. The following example (which can also be done by using the distribution of $(n-1)s^2/\sigma^2$) will illustrate the method.

EXAMPLE 3   A sample of 20 items has sample variance 4.2. Can it be regarded as coming from a large population with variance 10.2?

Method 1

Treating this as a problem of testing the ratio of two estimated variances,

$$s_2^2 = 4.20 \times 20/19 = 4.42 \quad \text{and} \quad s_1^2 = 10.2,$$

from samples of size 20 and ∞ respectively; the variance ratio is $10.2/4.42 = 2.30$, and the appropriate F-distribution is $F(\infty, 19)$.

The critical $P = 2.5$ value of $F(\infty, 19)$ is 2.13, so that the hypothesis of a common variance is rejected at the 5% level.

Method 2

Under the null hypothesis $H_0: \sigma^2 = 10.2$, the distribution of $(n-1)s^2/\sigma^2$ is χ^2 with 19 degrees of freedom.

From the χ^2 table, 95% of the distribution lies between the $P = 2.5$ and $P = 97.5$ points, which are 32.85 and 8.91 respectively. Now when $s^2 = 4.42$ and $\sigma^2 = 10.2$, the value of $(n-1)s^2/\sigma^2$ is 8.23. Since this is outside the limits, H_0 is rejected.

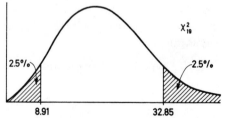

EXAMPLE 4   The following table gives the reaction times in seconds of two men, X and Y, exposed to identical stimuli. Eight readings are available for X and 10 for Y.

X	0.52	0.59	0.56	0.52	0.58	0.54	0.59	0.54		
Y	0.48	0.51	0.47	0.49	0.51	0.53	0.53	0.54	0.56	0.54

Test the assumption that the distributions for the times of the two men are the same.

From these figures, $\bar{x} = 0.555$ and $s_x^2 = 8.57 \times 10^{-4}$; $n_x = 8$

$$\bar{y} = 0.516 \text{ and } s_y^2 = 8.49 \times 10^{-4}; n_y = 10.$$

Two hypotheses have to be tested:
(a) that the population variances are not significantly different;
(b) that the difference between the means is not significant.

Hypothesis (b) may only be tested if hypothesis (a) is shown to be acceptable.

For (a) the hypotheses to be tested are

$$H_0: \sigma_x^2 = \sigma_y^2 \quad \text{as against} \quad H_1: \sigma_x^2 \neq \sigma_y^2.$$

The variance ratio $= 8.57/8.49 = 1.01$. This is less than the $P = 2.5$ value of $F(7,9)$, which is 4.20, so that H_0 is accepted at the 5% level.

We now use the method on page 281 to test the hypotheses for (b), which are

$$H_0: \mu_x = \mu_y \quad \text{as against} \quad H_1: \mu_x \neq \mu_y.$$

The estimated common variance

$$s^2 = \frac{(n_x - 1)s_x^2 + (n_y - 1)s_y^2}{n_x + n_y - 2}$$

$$= (7 \times 8.57 + 9 \times 8.49) \times 10^{-4}/16$$

$$= 8.53 \times 10^{-4}.$$

Under H_0 the distribution of

$$\frac{(\bar{X} - \bar{Y}) - 0}{s\sqrt{(1/n_x + 1/n_y)}}$$

is 't' with $n_1 + n_2 - 2$ degrees of freedom.

The value of this statistic for these samples is

$$\frac{0.039}{0.0138} = 2.82$$

and since the critical value of t at the 5% level for 16 degrees of freedom is 2.10, H_0 is rejected. We conclude that the difference between the means is significant and that therefore the distributions of the men's reaction times are not the same.

EXERCISE 24b

1.   A sample of 5 gave a sample variance of 10.5 and a sample of 9 gave a sample variance of 4.5. Are the estimated population variances significantly different?

2.   Samples *A* and *B* yield the following results:

A	2.41	2.38	2.37	2.42	2.35	2.38
B	2.32	2.36	2.38	2.33	2.38	

Test the hypothesis that these samples could come from populations with the same variance.

3.   A sample of 14 gives a sample variance of 2.97. Could this be from a large population of variance 9.6? (Use two methods: (a) *F*-distribution, (b) χ^2 distribution.)

4.   Fourteen boys on the Arts side of a sixth form took a reasoning test. Their mean score was 27.56 with standard deviation 10.58. Twenty-nine boys on the science side took the same test. Their mean score was 31.72 with standard deviation 13.36. Use the *F*-test to determine whether the difference between the variances is significant. If it is not, use the *t*-test to determine whether the difference between the mean scores is significant.

EXERCISE 24c (miscellaneous)

1.   Over a long period of driving to work, a man had found that the standard deviation of his journey time was 9.3 minutes. After the opening of a new stretch of motorway, he recorded the following journey times (in minutes) for a random sample of 10 journeys:

$$37.4 \quad 48.6 \quad 39.9 \quad 61.2 \quad 50.3$$
$$44.7 \quad 46.6 \quad 51.2 \quad 38.3 \quad 45.9.$$

Test at the 5% level of significance whether there has been a change in the variability of journey time, and provide a two-sided 90% confidence interval for the new variance.

(MEI)

2.   The masses (in grams) of two samples of eggs, one sample from the nests of species *A*, and one sample from the nests of a related species *B*, are given below:

Species A	12.3	13.7	10.4	11.4	14.9	12.6	
Species B	15.7	10.3	12.6	14.5	12.6	13.8	11.9

Test whether there is a significant difference between the two samples as regards variability and as regards mean mass. (Assume that both samples come from normal distributions.)

3.   A sample of nine values of a normal variable *X* yields the following data: $\Sigma x = 108.0$ and $\Sigma (x - \bar{x})^2 = 30.6$.
A sample of 17 values of a normal variable *Y* yields $\Sigma y = 171.7$ and $\Sigma (y - \bar{y})^2 = 108.6$.
Show that it is a reasonable assumption that the samples come from the same population and find 95% confidence limits for the estimate of the population variance based on the two samples combined.

(C)

4.   The random variable S^2 is defined by

$$S^2 = \sum_{i=1}^{n} (X_i - \bar{X})^2 / (n-1),$$

where $X_1, X_2, \ldots, X_n$ are independent identically distributed random variables and

$$\bar{X} = \sum_{i=1}^{n} X_i.$$

State the distribution of S^2 for the case where the common distribution of the X_i is $N(\mu, \sigma^2)$.

The management of an airline are concerned about the day-to-day variability in the number of passengers using a certain flight. Over a long period, the standard deviation of the number of passengers had been found to be 18.4. In an attempt to reduce this, a simplified fares structure has been introduced, since when the numbers of passengers on this flight on 10 randomly chosen days were

$$87 \quad 64 \quad 52 \quad 73 \quad 81 \quad 65 \quad 73 \quad 80 \quad 49 \quad 70.$$

Does this constitute evidence, significant at the 10% level, that the variance has been reduced? Provide a two-sided 95% confidence interval for the new variance. (You may assume an underlying normal population.) (MEI)

5. In order to test the effectiveness of a new rust-proofing treatment for steel, the times until rust started to appear were recorded for two samples, one being given the standard treatment and the other the new treatment. The figures below give the times (in units of 100 days):

New treatment	22.0	23.9	20.9	23.8	25.0	24.0
	21.7	22.8	23.1	23.5	23.0	
Standard treatment	23.2	22.0	22.2	21.2	21.6	21.9
	22.9	22.8				

(a) Test the hypothesis that the population variances for the two samples are the same.
(b) Test whether the mean time for the new treatment is significantly greater than the mean time for the standard treatment.
(Use a 5% significance level for both tests.)

6. Samples A and B are taken from normally distributed variables X and Y respectively. The following data are calculated from the samples.

Sample A: number in sample $= 8$; $\Sigma x = 56$; $\Sigma x^2 = 1078$.
Sample B: number in sample $= 12$; $\Sigma y = 168$; $\Sigma y^2 = 2605$.

Hypothesis 1 states that X has variance 50.
Hypothesis 2 states that Y has variance 50.
Hypothesis 3 states that X and Y have the same variance.

Test each hypothesis at the 5% level of significance and comment on the results. (C)

7. The sulphur contents in p.p.m. (= parts per million) of 10 samples of a liquid are determined by a new method, giving the results $x_1, x_2, \ldots, x_{10}$. Find 98% confidence limits for the standard deviation of the population given that

$$\sum_{i=1}^{10} x_i = 2900, \qquad \sum_{i=1}^{10} x_i^2 = 847\,000.$$

Using the old method the estimate for the standard deviation was 55 p.p.m., based on 20 samples. Does the new method give a smaller standard deviation than the old? Give reasons for your answer. (O & C)

8.★ Use the definition of the F-distribution on page 481 to determine the constant K when $v_1 = 2$ and $v_2 = v$.

9.★ (a) Show that, if $F(\lambda, v)$ is a random variable with an F-distribution having λ and v degrees of freedom, then
$$P[F(2, v) < f] = 1 - [v/(v + 2f)]^{\frac{1}{2}v}.$$

(b) A random sample of 5 observations $x_i(i = 1,\ldots,5)$ is drawn from a normally distributed population whose variance σ^2 is unknown. Find a one-sided upper 95% confidence limit for σ^2 in terms of X where

$$X = \Sigma\ (x_i - \bar{x})^2 \quad \text{and} \quad \bar{x} = \Sigma\ x_i/5.$$

[The probability density function of the F-distribution is given on page 481.]   (O & C)

10.★ Show that, if $C(\lambda)$ is a random variable with a χ^2 distribution having λ degrees of freedom, then

$$P[C(4) > c] = (1 + \tfrac{1}{2}c)e^{-\frac{1}{2}c}.$$

A random sample of 5 observations x_i $(i = 1,\ldots,5)$ is drawn from a normally distributed population whose variance σ^2 is unknown, and the statistic X is calculated, where

$$X = \sum_i\ (x_i - \bar{x})^2 \quad \text{and} \quad \bar{x} = \sum_i x_i/5.$$

If X is less than a fixed value b, the hypothesis H_0 that $\sigma^2 = 3$ will be accepted; if $X > b$, the alternative hypothesis $\sigma^2 > 3$ will be accepted. Show that, if the probability of rejecting H_0 when it is true is 0.01, then $b = 40$ approximately.
For this value of b, find the probability $P(s)$ of rejecting H_0 when $\sigma^2 = s$. Sketch the curve $P(s)$ in the (s, P) plane and show its asymptote.
[The probability density function of the χ^2 distribution is given on page 462.]

(O & C)

11.★ The random variable X is normally distributed with mean 28 and variance 25. A sample of 10 values believed to come from this distribution gives $\Sigma\ x = 270$ and $\Sigma\ x^2 = 7740$. Show that the variance estimate given by the sample does not lie in the critical region of a two-tail test at the 5% significance level of the hypothesis that the variance is 25.
The power of a test is defined as $1 - P$ (type II error). Find approximately the power of this test when the sample in fact comes from a population with variance 40.

$P(\%)$	40	35	30	25	20	15	10	5	2.5	1	0.5	0.1
$F_{9,\infty}$	1.13	1.18	1.22	1.27	1.34	1.46	1.63	1.88	2.12	2.41	2.62	3.12
$F_{\infty,9}$	1.45	1.49	1.51	1.53	1.62	1.81	2.16	2.71	3.33	4.31	5.19	7.81

[For type II errors, see page 318.]   (C)

24.4   Appendix

The distribution of $(n-1)s^2/\sigma^2$

If a random sample, size n, is taken from the normal distribution $N(\mu, \sigma^2)$ and $s^2 = \Sigma\ (X - \bar{X})^2/(n-1)$, then the distribution of

$$\frac{(n-1)s^2}{\sigma^2} \text{ is } \chi^2 \text{ with } n-1 \text{ d.f.}$$

Proof

On page 34, exercise 3b, question 6, we showed that if $X_1, X_2, \ldots, X_n$ is a set of observations and a is any constant, then

$$\sum_{i=1}^{n}\ (X_i - a)^2 = \sum_{i=1}^{n}\ (X_i - \bar{X})^2 + n(\bar{X} - a)^2.$$

Dropping the suffix i and replacing a by μ,

$$\Sigma\,(X-\mu)^2 = \Sigma\,(X-\bar{X})^2 + n(\bar{X}-\mu)^2$$

$$\Rightarrow \Sigma\,\frac{(X-\mu)^2}{\sigma^2} = \Sigma\,\frac{(X-\bar{X})^2}{\sigma^2} + \frac{n(\bar{X}-\mu)^2}{\sigma^2}$$

$$\Rightarrow \Sigma\left(\frac{X-\mu}{\sigma}\right)^2 = \frac{(n-1)s^2}{\sigma^2} + \left(\frac{\bar{X}-\mu}{\sigma/\sqrt{n}}\right)^2 \qquad [\Sigma\,(X-\bar{X})^2 = (n-1)s^2].$$

Now the distribution of X is $N(\mu,\sigma^2)$, so that $(X-\mu)/\sigma$ is a standardised normal variable.

Similarly the distribution of $\bar{X}$ is $N(\mu,\sigma^2/n)$, so that $(\bar{X}-\mu)/(\sigma/\sqrt{n})$ is also a standardised normal variable.

It follows that the distribution of $\Sigma\,[(X-\mu)/\sigma]^2$, the sum of the squares of n standardised variables is

$$\chi^2 \text{ with } n \text{ degrees of freedom.}$$

Similarly, the distribution of

$$\left(\frac{\bar{X}-\mu}{\sigma/\sqrt{n}}\right)^2,$$

the square of a standardised normal variable is

$$\chi^2 \text{ with one degree of freedom.}$$

Assuming that

$$\frac{(n-1)s^2}{\sigma^2} \quad \text{and} \quad \left(\frac{\bar{X}-\mu}{\sigma/\sqrt{n}}\right)^2$$

are independently distributed, it follows that

$$\frac{(n-1)s^2}{\sigma^2} \text{ has a } \chi^2 \text{ distribution with } n-1 \text{ d.f.,}$$

(using the property of χ^2 on page 465.)
(The independence is, in fact, not easy to prove.)

The probability density function of the F-distribution

The distribution of the random variable F is defined by the density function

$$y = K\,\frac{f^{\frac{1}{2}v_1-1}}{(v_1\,f+v_2)^{\frac{1}{2}(v_1+v_2)}}, \qquad 0 < f < \infty,$$

where K is a constant depending on v_1 and v_2, chosen so that

$$\int_0^\infty y\,df = 1.$$

25* Analysis of variance

25.1 Single-factor analysis of variance

In earlier chapters we devoted a good deal of space to the problem of comparing two means. In this chapter we shall outline a method of comparing several means simultaneously, using the technique known as analysis of variance (commonly written as ANOVA for short).

Consider the following problem. Three machines produce metal rods. The diameters of a sample of five rods from each machine, measured in tenths of a millimetre are given in the following table, together with the mean of each sample.

	Machine		
	1	2	3
	47	55	54
	53	54	50
	49	58	51
	50	61	51
	46	52	49
Means	49	56	51

On the evidence of these observations it seems possible that the machines are performing differently. We want, therefore, to test the null hypothesis

$$H_0: \mu_1 = \mu_2 = \mu_3$$

where μ_1, μ_2, μ_3 are the means of the parent populations of the output of the separate machines, which are assumed to be normal and to have a common variance σ^2.

The null hypothesis amounts to saying that the samples may be regarded as coming from a common population and that the division of these 15 observations into three samples is an artificial one.

Now H_0 could be tested by taking the sample means in pairs and using a t-test for each pair, which, for a large number of samples would be laborious. In any case, the method we are about to describe enables us to tell immediately whether H_0 is likely to be true. If it turns out not to be true, we can then consider using t-tests.

The question to be answered, then, is whether the differences between the sample

482

means is a consequence of differences between the population means, or whether chance fluctuations in the performance of each machine may be responsible.

In order to decide between these alternatives we need some way of testing whether the *variability* between the sample means is large in comparison with what might be expected from the operation of chance fluctuations alone. The first stage, therefore, is to calculate the obvious measure of variability between the means, the *variance of the sample means*.

Notation

The notation used here and in subsequent work is:

Number of samples $= r$; Number in each sample $= n$;
Number in total sample $= N = rn$.
The value in the ith row and jth column is x_{ij}.

$$\bar{x}_j = \text{the mean of the } j\text{th sample}$$

$$m = \text{the overall mean} = \Sigma \, \bar{x}_j/r = \Sigma \, x_{ij}/N.$$

In this case,

$r = 3$ $n = 5$ $N = 15$

$\bar{x}_1 = 49$ $\bar{x}_2 = 56$ $\bar{x}_3 = 51$

$m = 52.$

i \ j	1	2	3
1	47	55	54
2	53	54	50
3	49	58	51
4	50	61	51
5	46	52	49

The between-samples estimate s_b^2

The mean of the sample means $= m = 52$, so that the variance of the sample means

$$= \Sigma_j \frac{(\bar{x}_j - m)^2}{r} = [(49-52)^2 + (56-52)^2 + (51-52)^2] \div 3 = 26/3.$$

As usual the estimated variance of the population of sample means, which we shall call $s_{\bar{x}}^2$ is obtained by multiplying this value by $r/(r-1)$, giving

$$s_{\bar{x}}^2 = \frac{26}{3} \times \frac{3}{2} = 13.$$

Now this can be used to estimate σ^2, the variance of the population from which the total sample of 15 is presumed to come. We know that the variance of the distribution of the means of samples of size n is σ^2/n, so that it seems reasonable to say that

$$\frac{\hat{\sigma}^2}{n} = s_{\bar{x}}^2$$

or, writing s^2 for $\hat{\sigma}^2$,

$$s^2 = ns_{\bar{x}}^2 = 5 \times 13 = 65.$$

(This step will be justified later.)

This estimate of σ^2, arrived at by using the variations *between* the sample means is called the *between-samples* estimate, and is written s_b^2.

Since it was obtained by calculating $\Sigma (\bar{x}_j - m)^2$, dividing this by $r-1$, and multiplying the result by n,

$$s_b^2 = \frac{n \sum_j (\bar{x}_j - m)^2}{r-1} \qquad (j = 1, 2, 3; r = 3).$$

The question we now have to answer is whether this estimate for σ^2 is comparable with the value that would be expected if chance fluctuations only were responsible for the difference between the means. In other words we need *another* estimate of σ^2, using only these chance fluctuations. Now it might seem that the obvious candidate for this second estimate would be one based on the variance of the whole sample of 15. In fact, for reasons that will appear later, this turns out not to be independent of s_b^2. The estimate we use for comparison is based on the variations *within* the individual samples, that is to say on the variances of the samples.

The within-samples estimate s_w^2

With the usual notation, the formula for estimating σ^2 when three samples are available (see page 239) is

$$s^2 = \frac{\sum_{i=1}^{5} (x_{i1} - \bar{x}_1)^2 + \sum_{i=1}^{5} (x_{i2} - \bar{x}_2)^2 + \sum_{i=1}^{5} (x_{i3} - \bar{x}_3)^2}{n_1 + n_2 + n_3 - 3}.$$

$$\sum_{i=1}^{5} (x_{i1} - \bar{x}_1)^2 = (47 - 49)^2 + (53 - 49)^2 + (49 - 49)^2 + (50 - 49)^2 + (46 - 49)^2$$

$$= 30.$$

Similarly $\sum_{i=1}^{5} (x_{i2} - \bar{x}_2)^2 = 50$ and $\sum_{i=2}^{5} (x_{i3} - \bar{x}_3)^2 = 14.$

With $n_1 = n_2 = n_3 = 5$, the pooled sample estimate of σ^2 is

$$s^2 = \frac{30 + 50 + 14}{15 - 3} = 7.83.$$

This estimate of σ^2, using variations *within* the samples is called the *within-samples* estimate and is written s_w^2.

It can be seen that an equivalent, and more concise, expression for this estimate is

$$s_w^2 = \frac{\sum\limits_{j} \left[\sum\limits_{j} (x_{ij} - \bar{x}_j)^2 \right]}{N - r} \quad (i = 1, 2, 3, 4, 5; j = 1, 2, 3,; N = 15; r = 3).$$

To sum up, then we now have two estimates of σ^2:

$$s_b^2 = 65, \text{ based on } r - 1 \, (= 2) \text{ degrees of freedom}$$

and

$$s_w^2 = 7.83, \text{ based on } r(n-1) \text{ or } N - r \, (= 12) \text{ degrees of freedom,}$$

since, for the calculation of each of the r sample variances there were $n - 1$ degrees of freedom.

In order to decide whether these are significantly different estimates we use the F-ratio in the form

$$F = \frac{s_b^2}{s_w^2} \quad (s_b^2 \text{ always in the numerator}).$$

Here $F = 65/7.83 = 8.30$, based on 2 and 12 degrees of freedom, which, if H_0 is true, should not be significantly *greater* than 1.

From the F-distribution table, the critical 5% ($P = 5$) value for $F(2, 12)$ is 3.89. As $8.30 > 3.89$ we conclude that H_0 is rejected at the 5% level and that the differences between the means as reflected in the between-samples estimate of σ^2 is too large to be accounted for by chance fluctuations.

Note: Remember that in using the F-ratio for comparing two estimates of σ^2 in general (see page 475)

$$F = \frac{s_1^2}{s_2^2}$$

with the larger estimate in the numerator, the reason being that the F-table only gives the right-hand critical values for F. However, either of the two estimates might have turned out to be the greater, so that the test is a two-tail one, and we use $P = 2.5$, *to give the right-hand critical region for a 5% level test.*

The situation is different when we are comparing s_b^2 and s_w^2; s_b^2 is always the numerator. If H_0 is true (and the samples are large) s_b^2 will be near 1 (and could even be a little less than 1). If H_0 is not true, the spread of the sample means will result in an increase in s_b^2 without a corresponding change in s_w^2. This means that s_b^2 will always overestimate σ^2 when H_0 is false, so that the F-ratio can only range from near 1 to a larger value. A *one-tail test* is therefore appropriate.

The Anova table

The method we have just described may be summarised, using the following notation:

For r samples, each of size n, where $rn = N$,

$$x_{ij} = \text{value in the } i\text{th row}$$

and jth column,

$$\bar{x}_j = \text{mean of } j\text{th sample},$$

$$m = \text{overall mean} = \Sigma\, x_{ij}/N$$

$$= \Sigma\, \bar{x}_j/r.$$

$\diagdown$ $^{j}_{i}$	1	2	...	j	...	r
1	.	.	.	.	.	.
2	.	.	.	.	.	.
i	.	.	.	x_{ij}	.	.
⋮	.	.	.	.	.	.
n	.	.	.	.	.	.
				$\bar{x}_j$		

The two estimates of σ^2 are:

$$s_b^2 = \frac{n\,\sum\limits_{j}\,(\bar{x}_j - m)^2}{r-1} \qquad \text{(between-samples estimate)}$$

$$s_w^2 = \frac{\sum\limits_{j}\left[\sum\limits_{i}(x_{ij} - \bar{x}_j)^2\right]}{N-r} \qquad \text{(within-samples estimate)}.$$

It is customary to give the name 'sum of squares' to the numerators of these expressions, with abbreviations SS_b and SS_w respectively, so that

$$\text{for between-samples variations } SS_b = n\sum_{j}(\bar{x}_j - m)^2$$

$$\text{for within-samples variations } SS_w = \sum_{j}\left[\sum_{i}(x_{ij} - \bar{x}_j)^2\right].$$

It is also possible to calculate a sum of squares for the *total* sample, that is to say the sum of the squared deviations of each observation x_{ij} from the overall mean m:

$$\text{for total variation } SS_t = \sum_{j}\sum_{i}(x_{ij} - m)^2.$$

It can be proved (see Appendix, page 509) that

$$SS_t = SS_b + SS_w,$$

in other words, the total variation consists of two parts,

1. the variation due to classification of the observations into samples,

2. the variation due to chance fluctuations after (1) has been eliminated.

The number of degrees of freedom associated with SS_t is, of course, $N-1$.

Calculation of SS_t is not strictly necessary but, as we shall see it is often useful. In our example

$$SS_t = (47-52)^2 + (53-52)^2 + \ldots + (49-52)^2 = 224,$$

with $15-1 = 14$ degrees of freedom.

These sums of squares, together with their degrees of freedom are usually displayed in a table as follows:

Source of variation	Sum of squares	d.f.	Mean sum of squares (variance)
Between samples	SS_b	$r-1$	s_b^2
Within samples	SS_w	$N-r$	s_w^2
Total	SS_t	$N-1$	—

The name 'mean sum of squares' is given to the estimates of σ^2 obtained by dividing the second column by the third column. A proof that s_b^2 and s_w^2 calculated in this way do give unbiased estimates of σ^2 is given in the Appendix on pages 510–11. Notice that an estimate based on SS_t is not included in the table, as it is not independent of s_b^2.

The calculations in our example will, therefore, appear like this:

Source of variation	Sum of squares	d.f.	Mean sum of squares (variance)
Between samples	130	2	$s_b^2 = 65$
Within samples	94	12	$s_w^2 = 7.83$
Total	224	14	—

The variance ratio $F = s_b^2/s_w^2$ and the distribution used for testing the significance of this ratio is $F(r-1, N-r)$, with, for a 5 % level test, the critical value given by $P = 5$ (one-tail test).

This procedure is called *single-factor analysis of variance*, 'analysis of variance' because the total variation has been analysed into two parts, SS_b and SS_w, and 'single-factor' since the sample in which an observation is found depends on a single criterion – in our example, the machine from which it is taken.

25.2   Alternative forms for the sums of squares

After this rather lengthy account of the method we shall work a further example, this time in a concise form, and use more convenient alternative forms for the sums of squares. These derive from the familiar relation:

$$\Sigma (x-\bar{x})^2 = \Sigma x^2 - n\bar{x}^2,$$

so that $SS_b = n\left(\sum_{j=1}^{r} \bar{x}_j^2 - rm^2 \right)$

$$SS_w = \sum_{j=1}^{r} \left[\sum_{i=1}^{n} x_{ij}^2 - n\bar{x}_j^2 \right]$$

$$SS_t = \sum_{j=1}^{r} \sum_{i=1}^{n} x_{ij}^2 - Nm^2.$$

In addition, the calculations may be simplified by coding the figures. This makes no difference to the result since we are only concerned to test the significance of the *ratio* of two variances.

The relation $SS_t = SS_b + SS_w$ will also be used when it is more convenient to calculate SS_t and SS_b and to derive SS_w by subtraction.

(A further simplified procedure will be described later. It is the method used by statisticians generally, but you should use the method we have been describing initially as it will give you a better understanding of what is involved in the process.)

EXAMPLE 1   The table below gives the tensile strength (in appropriate units) of samples of five pieces of wire turned out by four machines. Assuming a common variance, test whether there is evidence of variability in the mean tensile strength of the wire produced by the machines.

	Sample		
1	2	3	4
9.3	9.2	9.5	9.8
9.2	9.2	9.4	9.6
9.4	9.3	9.7	9.6
9.1	9.4	9.6	9.3
9.5	9.3	9.4	9.2

The hypotheses to be tested are:

$H_0: \mu_1 = \mu_2 = \mu_3 = \mu_4$ H_1: at least two of the means are not equal where μ_1, μ_2, μ_3, μ_4, are the underlying population means for the machines.

In the table on the right the observations have been coded and the values of

$$\Sigma x_{ij}, \bar{x}_j \quad \text{and} \quad \Sigma x_{ij}^2$$

have been computed for each sample.

$r = 4$ $n = 5$ $N = 20$

$m = (3 + 2.8 + 5.2 + 5)/4$

$= 4.$

i \ j	1	2	3	4
1	3	2	5	8
2	2	2	4	6
3	4	3	7	6
4	1	4	6	3
5	5	3	4	2
Σx_{ij}	15	14	26	25
$\bar{x}_j$	3	2.8	5.2	5
Σx_{ij}^2	55	42	142	149

The easiest sums of squares to calculate are SS_b and SS_t:

$$SS_b = n(\Sigma \bar{x}_j^2 - rm^2) = 5(3^2 + 2.8^2 + 5.2^2 + 5^2 - 4 \times 4^2) = 24.4$$

$$\text{(degrees of freedom } r - 1 = 4 - 1 = 3).$$

$$SS_t = \sum_j \sum_i x_{ij}^2 - Nm^2 = (55 + 42 + 142 + 149) - 20 \times 4^2 = 68$$

$$\text{(degrees of freedom } N - 1 = 20 - 1 = 19).$$

Hence

$$SS_w = SS_t - SS_b = 68 - 24.4 = 43.6$$

$$\text{(degrees of freedom } N - r = 20 - 4 = 16).$$

With these figures the Anova table is:

Source of variation	Sum of squares	d.f.	Mean sum of squares (variance)
Between samples	24.4	3	8.13
Within samples	43.6	16	2.73
Total	68	19	—

The variance ratio $F = 8.13/2.73 = 2.98$, and since the critical $(P = 5)$ value of $F(3, 16)$ is 3.24, H_0 is accepted at the 5% level.

Note: The direct calculation of SS_w is marginally less easy:

$$\sum_j \left[\sum_i x_{ij}^2 - n\bar{x}_j^2 \right] = (5 - 5 \times 3^2) + (42 - 5 \times 2.8^2)$$

$$+ (142 - 5 \times 5.2^2) + (149 - 5 \times 5^2) = 43.6.$$

In the next example, where the observations are presented in summary form, we shall show how to proceed when the null hypothesis is rejected.

EXAMPLE 2 From each of four very large student departments in a university, 50 students were given a test in manual dexterity. The results were as follows, where X was the individual score:

	Department			
	A	B	C	D
Σx	3400	3700	3500	3400
Σx^2	237250	281000	248200	236200

Are the department means significantly different?

If the samples from A, B, C and D are samples $j = 1, 2, 3$ and 4 respectively, and the individual scores are x_{ij} ($i = 1$ to 50, $j = 1$ to 4) then, with the usual notation:

$$H_0: \mu_1 = \mu_2 = \mu_3 = \mu_4 \quad H_1: \text{at least two of the means are not equal}$$

$$r = 4, n = 50, N = 200.$$

The sample means are

$$\bar{x}_1 = 3400/50 = 68, \quad \bar{x}_2 = 3700/50 = 74$$

$$\bar{x}_3 = 3500/50 = 70, \quad \bar{x}_4 = 3400/50 = 68,$$

and $m = \Sigma\, \bar{x}_j/4 = 70$.

$$SS_b = n(\Sigma\, \bar{x}_j^2 - rm^2) = 50(68^2 + 74^2 + 70^2 + 68^2 - 4 \times 70^2)$$

$$= 1200 \quad (r - 1 = 3 \,\text{d.f.})$$

$$SS_t = \Sigma\, x_{ij}^2 - Nm^2 \; = (237\,250 + 281\,000 + 248\,200 + 236\,200) - 200 \times 70^2$$

$$= 22\,650 \quad (N - 1 = 199 \,\text{d.f.}).$$

Hence $SS_w = 22\,650 - 1200 = 21\,450 \quad (N - r = 196 \,\text{d.f.}).$

The Anova table is therefore:

Source of variation	Sum of squares	d.f.	Mean sum of squares (variance)
Between samples	1200	3	400
Within samples	21450	196	109.4
Total	22650	199	—

The variance ratio $F = 400/109.4 = 3.66$, and since the critical 5% value of $F(3, 196)$ is less than 2.68, we conclude that the population means are *not* all the same and that at least two of them are different.

Commonsense would suggest that μ_1 and μ_4 differ from μ_2, and that possibly they differ from μ_3. In order to test these we use the t-test between pairs of sample means.

The estimated common variance to be used is s_w^2, since, as we pointed out on page 487, this reflects the residual variation due to chance fluctuations after the between-samples variation SS_b has been eliminated. For this reason it is the *best estimate*.

In this case $s^2 = 109.4 \Rightarrow s = 10.46$, calculated with 196 d.f.

For the first two samples

$$H_0 : \mu_1 = \mu_2 \quad H_1 : \mu_1 \neq \mu_2.$$

Since each sample contains 50 scores the test statistic

$$t = \frac{\bar{x}_1 - \bar{x}_2}{s\sqrt{\left(\dfrac{1}{50} + \dfrac{1}{50}\right)}} = \frac{68 - 74}{10.46\sqrt{0.04}} = -2.87.$$

The critical value of t for a 5% level two-tail test when $\nu = 196$ is between 1.96 and 1.98, so that H_0 is rejected.

Similarly for the 2nd and 3rd samples $t = 1.91$, and for the 3rd and 4th samples $t = 0.96$. In both cases we conclude that the means do not differ. For the 1st and 4th samples $t = 0$.

To sum up: at the 5% level μ_1 and μ_4 *are* significantly different from μ_2, but none of the other pairs differ significantly.

Note: To obtain confidence limits for σ^2, use the fact that the distribution of SS_w/σ^2 is χ^2 with $N-r$ degrees of freedom (see page 472).

25.3 Samples of different sizes

Although it is preferable to use samples of equal size this is not always possible. For instance, tests on groups of students over a period of time, as in example 2, may end with results from unequal samples owing to the unavailability of some students due to illness or absence. In these circumstances the procedure has to be modified slightly.

Suppose that, with the usual notation,

$$n_j = \text{the size of sample } j \quad (j = 1, 2, \dots, r)$$

then the overall mean

$$m = \frac{\Sigma n_j \bar{x}_j}{N} = \frac{\Sigma \Sigma x_{ij}}{N}, \quad (N = \Sigma n_j).$$

The modified expressions for the sums of squares are then:

$$SS_b = \sum_{j=1}^{r} n_j(\bar{x}_j - m)^2$$

$$SS_w = \sum_{j=1}^{r} \left[\sum_{i=1}^{n_j} (x_{ij} - \bar{x}_j)^2 \right]$$

$$SS_t = \sum_{j=1}^{r} \left[\sum_{i=1}^{n_j} (x_{ij} - m^2) \right] = \sum_j \sum_i x_{ij}^2 - Nm^2 \quad \text{(as before)}.$$

These results may be proved by making minor modifications to the proof for samples of equal size.

(Alternative forms for SS_b and SS_w are not given here. A better method, using short cuts, will be described shortly.

EXAMPLE 3   Samples of four different varieties of corn gave the following yields in kilograms.

	Variety		
1	2	3	4
30	39	28	36
32	27	25	36
34	27		36
			34

Are the mean yields significantly different at the 1% level?

The results, coded by subtracting 30, are shown in the table at the top of page 492, with the values of Σx_{ij}, $\bar{x}_j$ and Σx_{ij}^2 calculated, using the usual notation.

i \ j	1	2	3	4
1	0	9	−2	6
2	2	−3	−5	6
3	4	−3		6
4				4
Σx_{ij}	6	3	−7	22
$\bar{x}_j$	2	1	−3.5	5.5
Σx_{ij}^2	20	99	29	124

$r = 4$ $N = 12$

$n_1 = 3,\ n_2 = 3,\ n_3 = 2,\ n_4 = 4$

$m = (6+3-7+22)/12 = 2$

$H_0: \mu_1 = \mu_2 = \mu_3 = \mu_4$ H_1: at least two means differ.

$$SS_b = \Sigma\, n_j(\bar{x}_j - m)^2$$
$$= 3 \times (2-2)^2 + 3 \times (1-2)^2 + 2 \times (-3.5-2)^2 + 4 \times (5.5-2)^2$$
$$= 112.5 \quad (\text{d.f. } r-1 = 3).$$
$$SS_t = \Sigma\,\Sigma\, x_{ij}^2 - Nm^2$$
$$= 20+99+29+124-12 \times 2^2 = 224 \quad (\text{d.f. } N-1 = 11).$$

Hence $SS_w = SS_t - SS_b = 224 - 112.5 = 111.5$ (d.f. $N-r = 8$).

The Anova table is:

Source of variation	Sum of squares	d.f.	Mean sum of squares (variance)
Between samples	112.5	3	37.50
Within samples	111.5	8	13.94
Total	224	11	—

The variance ratio $= 37.50/13.94 = 2.69$, and since the critical 1% value of $F(3,8) = 7.59$, H_0 is accepted.

EXERCISE 25a

These exercises may all be done using a simplified procedure (see exercise 25b).

1.   The heights of 12 plants, tabulated according to their positions A, B, C in a plot of ground are given, coded, in the following table. Is the difference between the mean heights significant?

A	B	C
2	0	5
3	0	4
0	1	3
2	−1	3

2.   The following table shows the marks gained by a sample of five pupils, each from three unstreamed classes in a school, in a mathematics test. Are the sample means significantly different?

	Class		
	A	B	C
Marks	76	77	76
	77	74	76
	78	75	75
	79	75	76
	75	74	77

3.   From each of four suburban areas of a city five families were sampled and their annual income, x, (in units of £1000, to the nearest £1000) was recorded. The table shows the results. Test the hypothesis that the population means for the four areas are the same.

	Area			
	A	B	C	D
Σx	35	25	30	30
Σx^2	255	131	182	198

4.   Three diets were given to samples of three animals each. The resulting increase of mass (in kilograms) is shown below.

	Diet		
	P	Q	R
Increase in mass	2.20	2.18	2.15
	2.17	2.19	2.13
	2.19	2.18	2.14

   Use a 5% level test to show that the means of the mass increases for the three diets are unlikely to be equal.
   Test also whether two of these means may be equal, or whether all three are unequal. (Use a 5% level test).

5.   Five samples of ore from each of 3 different veins are analysed for sulphur. The results x (expressed in tenths of a percent) are summarised below.

Vein	A	B	C
Σx	10	15	20
Σx^2	23	47	83

   Assume that these are random samples from three independent normally distributed random variables X, Y, Z which have the same variance, but whose means are unknown.
   (a) Estimate the means of X, Y, Z and their common variance.
   (b) Use a suitable significance test to show that the means of X, Y, Z are unlikely to be all equal.
   (c) Find the standard error of the difference between the estimates of any two of these means.

Use a 1% level of significance to test whether these three means are all unequal or whether two of them might be equal. (O & C)

6.   Eastside Area Health Authority has a policy whereby any patient admitted to a hospital with a suspected coronary heart attack is automatically placed in the intensive care unit. The table below gives the number of hours spent in intensive care by such patients at five hospitals in the area.

		Hospital		
A	B	C	D	E
30	42	65	67	70
25	57	46	58	63
12	47	55	81	80
23	30	27		
16				

Use a one-factor analysis of variance to test, at the 1% level of significance, for differences between hospitals.

Write down an estimate, with its standard error, of the mean time spent in intensive care by this type of patient in hospital C. (AEB 1980)

7.   From each of 3 machines making rods, a number n is taken at random and the amounts x by which their diameters exceed 8 mm are measured in hundredths of a millimetre. The results are summarised in the table below.

Machine	A	B	C
n	7	9	4
Σx	205	212	135
Σx^2	6200	5140	4700

Find the sample mean diameter for each machine in millimetres. Assuming these are random samples from three normally distributed populations which have the same variance:
(a) estimate the common standard deviation in millimetres,
(b) give an upper 95% confidence limit for the common standard deviation,
(c) test the hypothesis that the three population means are equal. (O & C)

25.4   Simplified procedure for computing the sums of squares

The most convenient procedure for computing sums of squares is one which uses simplified forms for SS_t, SS_w and SS_b, and a simplified notation.

Referring to the table of values on the right, where the numbers in each sample are equal, we will use the following standard notation:

T_j = the sum of all items in sample j, so that $\bar{x}_j = T_j/n$.

T = the sum of *all* items in the *whole sample* $= \sum_j \sum_i x_{ij}$ so that $m = T/N$, where $N = rn$.

		Sample			
1	2	3	...	j	...	r
x_{11}	x_{12}	x_{13}	.	x_{1j}	.	x_{1r}
x_{21}	x_{22}	x_{23}	.	x_{2j}	.	x_{2r}
$\vdots$	$\vdots$	$\vdots$		$\vdots$		$\vdots$
x_{i1}	x_{i2}	x_{i3}	.	x_{ij}	.	x_{ir}
$\vdots$	$\vdots$	$\vdots$		$\vdots$		$\vdots$
x_{n1}	x_{n2}	x_{n3}	.	x_{nj}	.	x_{nr}
T_1	T_2	T_3	.	T_j	.	T_r

S = sum of the squares of all items in the whole sample = $\sum_j \sum_i x_{ij}^2$.

Since, for a sample of size n, with mean $\bar{x}$,

$$\sum (x - \bar{x})^2 = \sum x^2 - n\bar{x}^2,$$

it follows that:

$$SS_t = \sum_j \sum_i (x_{ij} - m)^2 = \sum_j \sum_i x_{ij}^2 - Nm^2 = S - T^2/N.$$

Considering the sum of squares for sample j,

$$\sum_i (x_{ij} - \bar{x}_j)^2 = \sum_i x_{ij}^2 - n\bar{x}_j^2 = \sum_i x_{ij}^2 - n(T_j/n)^2.$$

Hence, summing for $j = 1$ to r,

$$SS_w = \sum_j \sum_i (x_{ij} - \bar{x}_j)^2 = \sum_j \sum_i x_{ij}^2 - \sum_j T_j^2/n = S - \sum_j T_j^2/n.$$

By subtraction,

$$SS_b = SS_t - SS_w = (S - T^2/N) - \left(S - \sum_j T_j^2/n\right) = \sum_j (T_j^2/n) - T^2/N.$$

It can be seen that these expressions for the sums of squares are considerably simpler than the ones we have been using up to now, and that, for example, the means of samples no longer have to be calculated.

The Anova table now looks like this:

Source of variation	Sum of squares	d.f.	Mean sum of squares (variance)
Between samples	$\sum_j T_j^2/n - T^2/N$	$r-1$	s_b^2
Within samples	$S - \sum_j T_j^2/n$	$N-r$	s_w^2
Total	$S - T^2/N$	$N-1$	—

where $T = \sum x_{ij}$, $S = \sum x_{ij}^2$, T_j = sum of sample j.

EXAMPLE 4   We shall re-work example 1 on page 488, using this simplified procedure.

In the table on the next page, the values of

$$T, \quad S \quad \text{and} \quad \sum T_j^2$$

have been computed.

$$r = 4, \ n = 5, \ N = 20.$$

i \ j	1	2	3	4	
1	3	2	5	8	
2	2	2	4	6	
3	4	3	7	6	
4	1	4	6	3	
5	5	3	4	2	
					Total
T_j	15	14	26	25	$T = 80$
T_j^2	225	196	676	625	$\Sigma\, T_j^2 = 1722$
$\Sigma\, x_{ij}^2$	55	42	142	149	$S = 388$

$SS_b = \Sigma\, T_j^2/n - T^2/N = 1722/5 - 80^2/20 = 24.4$ (d.f. $r - 1 = 3$).

$SS_w = S - \Sigma\, T_j^2/n \quad = 388 - 1722/5 \quad = 43.6$ (d.f. $N - r = 16$).

$SS_t = S - T^2/N \quad\quad = 388 - 80^2/20 \quad = 68$ (d.f. $N - 1 = 19$).

(As each of these sums of squares is easy to calculate, it is advisable to compute all three, to provide a check.)

With these figures the Anova table is:

Source of variation	Sum of squares	d.f.	Mean sum of squares (variance)
Between samples	24.4	3	8.13
Within samples	43.6	16	2.73
Total	68	19	—

The variance ratio $8.13/2.73 = 2.98$, and since the critical $(P = 5)$ value of $F(3, 16)$ is 3.24, H_0 is accepted at the 5% level.

For samples of *different* sizes the Anova table is:

Source of variation	Sum of squares	d.f.	Mean sum of squares (variance)
Between samples	$\sum_j (T_j^2/n_j) - T^2/N$	$r - 1$	s_b^2
Within samples	$S - \sum_j (T_j^2/n_j)$	$N - r$	s_w^2
Total	$S - T^2/N$	$N - 1$	—

The proof follows the same course as the proof for samples of the same size.

EXAMPLE 5   We shall now re-work example 3 on page 491, using this simplified procedure.

In the table the values of

$$T, S \text{ and each } T_j^2$$

have been computed.

$$r = 4, N = 12$$

$$n_1 = 3, n_2 = 3, n_3 = 2, n_4 = 4.$$

i \ j	1	2	3	4	
1	0	9	−2	6	
2	2	−3	−5	6	
3	4	−3		6	
4				4	
					Total
T_j	6	3	−7	22	$T = 24$
T_j^2	36	9	49	484	
Σx_{ij}^2	20	99	29	124	$S = 272$

Hence $\Sigma (T_j^2/n_j) = 36/3 + 9/3 + 49/2 + 484/4 = 160.5$

$$SS_b = \Sigma (T_j^2/n_j) - T^2/N = 160.5 - 24^2/12 = 112.5 \quad \text{(d.f. } r-1 = 3\text{)}.$$

$$SS_w = S - \Sigma (T_j^2/n_j) \qquad = 272 - 160.5 \qquad = 111.5 \quad \text{(d.f. } N-r = 8\text{)}.$$

$$SS_t = S - T^2/N \qquad = 272 - 24^2/12 \quad = 224 \qquad \text{(d.f. } N-1 = 11\text{)}.$$

The Anova table is:

Source of variation	Sum of squares	d.f.	Mean sum of squares (variance)
Between samples	112.5	3	37.50
Within samples	111.5	8	13.94
Total	224	11	—

The variance ratio $= 37.50/13.94 = 2.69$, and since the critical 1% value of $F(3, 8) = 7.59$, H_0 is accepted at this level.

EXERCISE 25b

1. Do question 1 of exercise 25a using the simplified procedure.

2. Do question 2 of exercise 25a using the simplified procedure.

3. Do question 3 of exercise 25a using the simplified procedure.

4. Do question 4 of exercise 25a using the simplified procedure.

5. Do question 5 of exercise 25a using the simplified procedure.

6. Do question 6 of exercise 25a using the simplified procedure.

7. Do question 7 of exercise 25a using the simplified procedure.

8.   Each of three groups of four plots of land were given different fertilising treatment. The yield in each plot, measured in suitable units, is shown in the following table. Is there any significant difference in the mean yields for the three fertilisers?

	Fertiliser		
	A	B	C
Yield	9	10	6
	8	10	8
	8	12	5
	7	8	5

9.   Five machines each dispensing soap powder in nominal quantities of 415 g produce the following quantities when five samples are taken from each:

1	2	3	4	5
413	421	406	421	429
417	425	416	399	417
415	429	414	408	418
426	412	413	416	421
418	420	408	413	420

Are the mean quantities produced by the machines significantly different at the 5% level?

10.   Four African countries take the same GCE examination. n pupils are selected at random from pupils taking the mathematics examination and their marks are recorded. The results are summarised in the following table. Test the hypothesis that the population means for the four countries are the same.

	Country			
	A	B	C	D
n	40	110	20	50
Σx	2040	5852	952	2300
Σx^2	111880	330900	49720	117050

11.   From each of six machines producing electrical components samples are tested for their length of life, in hours. The results are shown below. Use a one-factor analysis of variance to test for differences between the machines.

		Machine			
1	2	3	4	5	6
648	647	645	646	646	645
646	645	643	644	645	644
648	645	646	645	643	645
	641	642	645	641	
			648		

12.   An experiment was conducted to study the effects of various diets on pigs. A total of 24 similar pigs were selected and randomly allocated to one of the five groups such that the control group, which was fed a normal diet, had 8 pigs and each of the other groups, to which the new diets were given, had 4 pigs each. After a fixed time the gains in mass, in kilograms, of the pigs were measured. Unfortunately by this time two pigs had died, one of which was on diet A and one which was on diet C. The gains in mass of the remaining pigs are recorded below.

Diets	Gains in mass (kg)
Normal	23.1, 9.8, 15.5, 22.6, 14.6, 11.2, 15.7, 10.5
A	21.9, 13.2, 19.7
B	16.5, 22.8, 18.3, 31.0
C	30.9, 21.9, 29.8
D	21.0, 25.4, 21.5, 21.2

Use a one-factor analysis of variance to test, at the 5 % level of significance, for differences between diets.

What further information would you require about the dead pigs and how might this affect the conclusions of your analysis?   (AEB 1982)

13.   As part of a stylistic analysis of the plays written by Shakespeare at different stages of his career, samples of eight lines were taken at random from each of three plays, and the number of words per line of verse counted. With these data, determine whether the mean numbers of words in a line are significantly different.

> Bless you, fair dame! I am not to you known,
> Though in your state of honour I am perfect.
> I doubt, some danger does approach you nearly:
> If you will take a homely man's advice,
> Be not found here; hence with your little ones.
> To fright you thus, methinks, I am too savage;
> To do worse to you were fell cruelty,
> Which is too nigh your person. Heaven preserve you!
>
> Macbeth, Act IV, Scene 2

> Madam, you have bereft me of all words,
> Only my blood speaks to you in my veins:
> And there is such confusion in my powers,
> As, after some oration fairly spoke
> By a beloved prince, there doth appear
> Among the buzzing pleased multitude;

Where every something, being blent together,
Turns to a wild of nothing, save of joy ....

Merchant of Venice, Act III, Scene 2

There be some sports are painful; but their labour
Delight in them sets off: some kinds of baseness
Are nobly undergone; and most poor matters
Point to rich ends. This my mean task would be
As heavy to me, as 'tis odious: but
The mistress which I serve, quickens what's dead,
And makes my labours pleasures: O, she is
Ten times more gentle, than her father's crabbed ....

Tempest, Act III, Scene 1

(Count "'tis' and 'what's' as one word.)

25.5   Two-factor analysis of variance

So far we have been dealing with observations which have been classified into samples according to a single criterion. In the first example, on page 482, the criterion was the particular machine from which a rod came; in example 2 (page 489) it was the university department to which the student belonged; in example 3 (page 491) it was the kind of corn which was being grown. A researcher may, however, want to investigate observations which are classified in a way which involves two criteria.

Let us suppose that the figures on page 482 arise from having five different operators, each working three different machines. One observation is available from each combination of machine and operator.

		Machine		
		1	2	3
	1	47	55	54
	2	53	54	50
Operator	3	49	58	51
	4	50	61	51
	5	46	52	49

We now have two factors that may affect performance and our analysis will be aimed at detecting possible differences between:
(a) the means of three samples of five items, due to variations in the machines, and
(b) the means of five samples of three items, due to variations in the operators.
The null hypothesis H_0 is in two parts:

1. that there is no difference between the underlying population means of the columns,

2. that there is no difference between the underlying population means of the rows.

As in single-factor analysis, we develop the theory for the general case shown in the following table:

Factor A

		1	2	...	j	...	r
	1	x_{11}	x_{12}	.	x_{1j}	.	x_{1r}
	2	x_{21}	x_{22}	.	x_{2j}	.	x_{2r}
	⋮	⋮	⋮	⋮	⋮	⋮	⋮
Factor B	⋮	⋮	⋮	⋮	⋮	⋮	⋮
	i	x_{i1}	x_{i2}	.	x_{ij}	.	x_{ir}
	⋮	⋮	⋮	⋮	⋮	⋮	⋮
	⋮	⋮	⋮	⋮	⋮	⋮	⋮
	n	x_{n1}	x_{n2}	.	x_{nj}	.	x_{nr}

We shall write $\bar{x}_j$ as the mean of column j and $\bar{x}_i$ as the mean of row i.

m is the mean of the total sample, and of both the column means and row means, that is,

$$m = \sum_i \sum_j x_{ij}/N = \sum_j \bar{x}_j/r = \sum_i \bar{x}_i/n$$

where $N = rn$.

Now on pages 486–7 we stated that the sum of the squared deviations from the mean of the whole sample (the total variation) could be analysed into two parts, the variation due to classification into samples (the between-samples variation) and the variation due to chance fluctuations within the sample (the within-samples variation), which is the residual variation after the first has been eliminated. This was expressed by the relation

$$SS_t = SS_b + SS_w.$$

In two-factor analysis there is a similar relation, which includes a *between-rows* variation. With the following notation:

SS_t = the total variation (as before),

SS_c = the between-columns variation (which takes the place of the between-samples variation)

SS_r = the between-rows variation,

SS_e = the residual variation that is left after SS_c and SS_r have been eliminated,

the relation is

$$SS_t = SS_c + SS_r + SS_e.$$

The proof of this result is similar to the single-factor case but is considerably more complicated. We shall therefore omit it. The sums of squares just mentioned are defined as follows.

If $SS_t = \sum_j \sum_i (x_{ij} - m)^2$ (as before)

$SS_c = n \sum_j (\bar{x}_j - m)^2$ (previously the form for SS_b)

$SS_r = r \sum_i (\bar{x}_i - m)^2$

then the form for the residual variation can be shown to be

$$SS_e = \sum_j \left[\sum_i (x_{ij} - \bar{x}_j - \bar{x}_i + m)^2 \right].$$

As with single-factor analysis, the procedure is to use SS_c, SS_r and (if convenient) SS_e to derive estimates of σ^2 by dividing each by the appropriate number of degrees of freedom. Thus:

1.   for the total sample the number of degrees of freedom is $N - 1$;

2.   for the *between-columns* estimate the r means have one constraint on them, since their sum must be rm. Similarly for the *between-rows* estimate, the sum of the n means must be nm. The numbers of degrees of freedom for these are therefore $r - 1$ and $n - 1$ respectively;

3.   for the *residual* estimate there are $(n - 1)(r - 1)$ since the mean of each row and column is given.

The tabulated results are therefore:

Source of variation	Sum of squares	d.f.	Mean sum of squares (variance)
Between columns	SS_c	$r - 1$	s_c^2
Between rows	SS_r	$n - 1$	s_r^2
Residual	SS_e	$(n-1)(r-1)$	s_e^2
Total	SS_t	$N - 1$	—

The sums of squares are best computed using a simplified procedure with the following notation:

$$T = \text{sum of all terms in the whole sample} = \sum_i \sum_j x_{ij}$$

$$S = \text{sum of the squares of all terms} = \sum_i \sum_j x_{ij}^2$$

$$C_j = \text{sum of items in column } j$$

$$R_i = \text{sum of items in row } i.$$

The simplified expressions for SS_c and SS_r have already been found (see page 495 with minor modifications) and SS_e is obtained by subtraction from SS_t. Hence the Anova table is:

Source of variation	Sum of squares	d.f.	Mean sum of squares (variance)
Between columns	$\sum_j C_j^2/n - T^2/N$	$r-1$	s_c^2
Between rows	$\sum_i R_i^2/r - T^2/N$	$n-1$	s_r^2
Residual	$S - \sum_j C_j^2/n - \sum_i R_i^2/r + T^2/n$	$(n-1)(r-1)$	s_e^2
Total	$S - T^2/N$	$N-1$	

To test the null hypothesis, the between-columns and between-rows estimates for σ^2 are compared separately with the residual estimate, using the F-test. If either is significantly different from the residual estimate, that part of the null hypothesis is rejected.

(If the numbers of items in rows and columns are *not uniform* it is easily shown that, in the above table

$$\sum_j C_j^2/n \quad \text{becomes} \quad \sum_j C_j^2/n_j \quad \text{and} \quad \sum_i R_i^2/r \quad \text{becomes} \quad \sum_i R_i^2/r_i.)$$

EXAMPLE 6 For the results given on page 482, carry out an analysis of variance and test for differences between machines and between operators.

The figures, coded by subtracting 50, are given in the table below, together with computations of the values of C_j, R_j, C_j^2, R_j^2, S and T.

		Machine					$\sum x_{ij}^2$
i \ j		1	2	3	Totals		
Operator	1	-3	5	4	$R_1 = 6$	$R_1^2 = 36$	50
	2	3	4	0	$R_2 = 7$	$R_2^2 = 49$	25
	3	-1	8	1	$R_3 = 8$	$R_3^2 = 64$	66
	4	0	11	1	$R_4 = 12$	$R_4^2 = 144$	122
	5	-4	2	-1	$R_5 = -3$	$R_5^2 = 9$	21
Totals		$C_1 = -5$	$C_2 = 30$	$C_3 = 5$	$T = 30$	$\sum_i R_i^2 = 302$	$S = 284$
		$C_1^2 = 25$	$C_2^2 = 900$	$C_3^2 = 25$	$\sum_j C_j^2 = 950$		
$\sum_i x_{ij}^2$		35	230	19	$S = 284$		

The hypotheses to be tested are:

H_0: (a) there is no difference between the underlying population means of the columns;

(b) there is no difference between the underlying population means of the rows.

H_1: At least two of the means in each case are equal.

From the figures in the table, with $r = 3, n = 5, N = 15$

$$SS_c = \sum_j C_j^2/n - T^2/N = 950/5 - 900/15 = 130 \qquad \text{(d.f. } r-1 = 2).$$

$$SS_r = \sum_i R_i^2/r - T^2/N = 302/3 - 900/15 = 40.67 \quad \text{(d.f. } n-1 = 4).$$

$$SS_t = S - T^2/N \qquad = 284 - 900/15 \quad = 224 \qquad \text{(d.f. } N-1 = 14).$$

$$SS_e = S - \sum_j C_j^2/n - \sum_i R_i^2/r + T^2/N$$

$$= 284 - 950/5 - 302/3 + 900/15 = 53.33 \quad \text{(d.f. } (n-1)(r-1) = 8).$$

(SS_e can either be calculated, as here, and used as a check, or calculated from the other sums of squares by subtraction.)

The Anova table is thus:

Source of variation	Sum of squares	d.f.	Mean sum of squares (variance)
Between columns	130	2	65
Between rows	40.67	4	10.17
Residual	53.33	8	6.67
Total	224	14	—

The variance ratio for comparing the between-columns and residual estimates of σ^2 is $65/6.67 = 9.75$, which is greater than the 5% critical value of $F(2,8)$, 4.46. We conclude that the column means are not equal and that the machines are not operating uniformly.

The variance ratio for comparing the between-rows and residual estimates of σ^2 is $10.17/6.67 = 1.52$, which is less than the 5% critical value of $F(4,8)$, 3.84. We conclude that there is no strong evidence that the operators differ.

Note: In this account of two-factor analysis, two matters have been passed over. The first is that the table of observations necessarily includes only a single value in each category, so that in our example, we only had one value for operator 2 working machine 3. Although we shall not pursue the point here, you should be aware that there is a way of allowing for this difficulty by calculating *predicted* values to replace each of these single observations.

The other matter that has been ignored is that it is quite possible for the two factors to *interact*. A particular operator might, for instance, work differently on different machines. In order to make an allowance for the possibility of interaction a more complicated method of analysis is used. This, too, will not be described here.

EXERCISE 25c

1.   Three men are employed packing boxes. A check on the number packed by each between 10 and 11 am, between 2 and 3 pm and between 4 and 5 pm gave the following results.

		10–11	2–3	4–5
	1	31	29	33
Man	2	36	35	36
	3	32	26	29

Test for differences between men and between hours.

2.   Four kinds of paint are tested for weathering in each of three coastal towns by exposing panels with each kind in each town for six months. The conditions of the panels were then given the following 'scores':

		Paint			
		A	*B*	*C*	*D*
	1	69	68	72	77
Town	2	72	73	76	79
	3	67	70	73	71

Test for differences between paints and between towns.

3.   Four hospitals each test five different treatments on patients suffering from a certain disease. The number of days taken for patients to recover for each hospital and treatment is recorded in the following table:

		Treatment				
		1	2	3	4	5
	1	16	30	11	51	21
	2	23	20	15	53	24
Hospital	3	23	39	15	55	28
	4	11	36	2	46	21

Test for differences between hospitals and between treatments.

4.   Four brands of petrol were compared on three makes of car. The table below gives the distances in kilometres travelled under identical conditions on the same amount of petrol.

		Brand of petrol			
		1	2	3	4
	A	32	39	27	20
Car	*B*	23	30	39	45
	C	27	34	57	63

Carry out an analysis of variance and test for differences between brands of petrol and between makes of car.

EXERCISE 25d (*miscellaneous*)

1.   Tests were carried out by a chemist to measure the percentage of a certain impurity in a chemical solution, using five different methods on samples of four. The percentage of impurity was recorded as follows.

		Method		
1	2	3	4	5
31	35	37	30	37
28	29	34	31	38
31	30	34	31	38
27	31	31	27	34

Is there any significant difference between the mean percentages obtained by the different methods. (Use a 5% significance level.)

2.   It was planned to record the effect of a particular kind of diet on three different kinds of rabbit by measuring the percentage increase in weight of five of each kind after a month's trial. Owing to illness, data for a complete sample of five was not available for each type of rabbit. The incomplete data giving percentage increase in weight is given below.

	Type of rabbit	
X	Y	Z
5	8	9
7	7	9
4	7	11
4		7
5		

Use a 5% significance level test to find whether the sample means differ significantly.

3.   Samples of six leaves were taken from a tree in each of four Caribbean islands, and their lengths (in centimetres) were measured, with the following results.

	Island		
A	B	C	D
16.8	19.8	18.0	17.4
17.4	16.8	18.4	16.0
18.0	18.2	17.8	17.2
16.8	17.0	17.2	16.2
17.0	18.2	17.0	16.4
17.2	17.6	18.4	16.4

Use these data to show that it is unlikely that the trees can be regarded as being of the same type, and use a *t*-test to examine the differences between the mean lengths.

4.   In a horticultural experiment, three varieties of tomato plant are grown. The number n of plants of each variety and the yield x (in kilograms) of each plant are summarised in the table following:

Variety	n	Σx	Σx^2
Standard	8	95	1160
New A	6	92	1430
New B	6	76	1000

Assuming that the plant yields are normally distributed about means μ_s, μ_a, μ_b respectively and have the same variance σ^2 in all three varieties:
(a) estimate μ_s, μ_a, μ_b,
(b) estimate σ^2,
(c) use the F-distribution to test the hypothesis $\mu_s = \mu_a = \mu_b$,
(d) find two-sided 95 % confidence limits for μ_a.

(O & C)

5. Three different kinds of tyre were tried out by four taxi drivers each using the same model of taxi. The lives of the tyres (in days) were found to be

Taxi driver	Tyre		
	A	B	C
1	2100	2500	2000
2	1800	2200	1800
3	1200	1900	1600
4	1500	2100	1500

Test for differences between drivers and makes of tyre.

6. After completing a six-month typing course with the Speedy-fingers Institute four women had their typing speed measured in words per minute, on each of five different kinds of work. The results are given in the table below.

	Kinds of work				
	Legal document	Business letter	Numerical tabulation	Prose from manuscript	Prose from typescript
Mrs A	40	47	42	45	53
Mrs B	34	32	14	36	44
Miss C	33	40	31	48	44
Mrs D	24	26	25	27	45

Carry out an analysis of variance and test, at the 5 % level of significance, for differences between women and between kinds of work.

Subsequently it transpired that Mrs A and Miss C used electric typewriters, while Mrs B and Mrs D used manual typewriters. Does this information affect your conclusions and why? (AEB 1981)

7.   Random samples $x_1, x_2, \ldots, x_a$; $y_1, y_2, \ldots, y_b$; $z_1, z_2, \ldots, z_c$ are taken from three normal distributions which have means μ_1, μ_2, μ_3 respectively and a common variance σ^2.
Statistics A, B and C are calculated, where

$$A = \sum_1^a (x_i - \bar{x})^2 + \sum_i^b (y_i - \bar{y})^2 + \sum_1^c (z_i - \bar{z})^2,$$

$$B = a(\bar{x} - g)^2 + b(\bar{y} - g)^2 + c(\bar{z} - g)^2,$$

$$C = \sum_1^a (x_i - g)^2 + \sum_1^b (y_i - g)^2 + \sum_1^c (z_i - g)^2,$$

$$\bar{x} = \left(\sum_1^a x_i\right) \bigg/ a, \quad \bar{y} = \left(\sum_1^b y_i\right) \bigg/ b, \quad \bar{z} = \left(\sum_1^c z_i\right) \bigg/ c, \quad g = \left(\sum_1^a x_i + \sum_1^b y_i + \sum_1^c z_i\right) \bigg/ (a+b+c).$$

Show that $A + B = C$. Explain how the ratio B/A can be used to test the null hypothesis $\mu_1 = \mu_2 = \mu_3$.

Make this test if

$$\Sigma x^2 = 5800, \qquad \Sigma y^2 = 8100, \qquad \Sigma z^2 = 13\,600$$

$$\Sigma x = 240, \qquad \Sigma y = 310, \qquad \Sigma z = 520$$

$$a = 10 \qquad\qquad b = 12 \qquad\qquad c = 20.$$

Estimate σ^2.

(O & C)

25.6   Appendix

$SS_t = SS_b + SS_w$

Proof

The table below gives the values of the variable in the samples. x_{ij} is the ith item in the jth sample:

			Sample			
1	2	3	...	j	...	r
x_{11}	x_{12}	x_{13}	.	x_{1j}	.	x_{1r}
x_{21}	x_{22}	x_{23}	:	x_{2j}	:	x_{2r}
⋮	⋮	⋮	⋮	⋮	⋮	⋮
x_{i1}	x_{i2}	x_{i3}	:	x_{ij}	:	x_{ir}
⋮	⋮	⋮	⋮	⋮	⋮	⋮
x_{n1}	x_{n2}	x_{n3}	.	x_{nj}	.	x_{nr}

We shall write $\bar{x}_j$ as the mean of sample j, and m as the mean of these sample means, so that

$$m = \frac{1}{r} \sum_j \bar{x}_j \quad (j = 1 \text{ to } r).$$

It can be seen that m is also the mean of the total sample since

$$\bar{x}_j = \frac{1}{n} \sum_i x_{ij} \quad (i = 1 \text{ to } n)$$

so that $m = \frac{1}{r} \sum_j \bar{x}_j = \frac{1}{nr} \sum_j \sum_i x_{ij} \quad (j = 1 \text{ to } r; i = 1 \text{ to } n).$

Now we know (see page 34) that for any distribution, if the mean of a sample size n is $\bar{x}$, its variance $(1/n) \sum (x - \bar{x})^2$ can be written in the form

$$\frac{1}{n} \sum (x - a)^2 - (\bar{x} - a)^2,$$

where a is an arbitrary number, and hence that

$$\sum (x - \bar{x})^2 = \sum (x - a)^2 - n(\bar{x} - a)^2. \tag{1}$$

We shall use this result to express the sum of the squared deviations from the mean of the *whole* sample,

$$SS_t = \sum_j \sum_i (x_{ij} - m)^2 \quad (j = 1 \text{ to } r, i = 1 \text{ to } n)$$

as the sum of two distinct parts.

Consider sample 1.

The sum of the squared deviations from the mean $\bar{x}_1$ of this sample

$$= \sum_i (x_{i1} - \bar{x}_1)^2 \quad (i = 1 \text{ to } n).$$

Using (1), with $a = m$,

$$\sum_i (x_{i1} - \bar{x}_i)^2 = \sum_i (x_{i1} - m)^2 - n(\bar{x}_1 - m)^2$$

$$\Rightarrow \sum_i (x_{i1} - m)^2 = \sum_i (x_{i1} - \bar{x}_1)^2 + n(\bar{x}_1 - m)^2.$$

Similarly for sample 2,

$$\sum_i (x_{i2} - m)^2 = \sum_i (x_{i2} - \bar{x}_2)^2 + n(\bar{x}_2 - m)^2$$

and in general, for sample j,

$$\sum_i (x_{ij} - m)^2 = \sum_i (x_{ij} - \bar{x}_j)^2 + n(\bar{x}_j - m)^2 \quad (i = 1 \text{ to } n).$$

Summing over all the samples

$$\sum_j \sum_i (x_{ij} - m)^2 = \sum_j \sum_i (x_{ij} - \bar{x}_j)^2 + n \sum_j (\bar{x}_j - m)^2 \quad (j = 1 \text{ to } r, \quad i = 1 \text{ to } n)$$

or $SS_t = SS_w + SS_b.$ \hfill (2)

The first term on the right-hand side of this relation is the sum of the squared deviations from the mean *within* each sample. The second is $n \times$ the sum of the squared deviations *between* the means of samples.

In other words the sum of the squared deviations from the mean for the whole sample has been split up into two parts, reflecting:

(a) the variability due to classification into samples, which is called the '*between sample*' variation (the second term), and

(b) the variation due to chance fluctuation within the samples, that is, the residual variation after (a) has been eliminated. This is called the '*within sample*' variation (the first term).

Derivation of s_w^2 and s_b^2 from SS_w and SS_b

We now show that, on the assumption of a common variance σ^2, the two sums of squares may be used to obtain unbiased estimates of σ^2.

Using the within-sample variation

Consider sample 1, size n with mean $\bar{x}_1$.

$$E\left[\sum_i (x_{i1} - \bar{x}_1)^2\right] = (n-1)\sigma^2 \quad \text{(see page 238)}.$$

Similarly, for sample 2

$$E\left[\sum_i (x_{i2} - \bar{x}_2)^2\right] = (n-1)\sigma^2$$

and in general, for sample j,

$$E\left[\sum_i (x_{ij} - \bar{x}_j)^2\right] = (n-1)\sigma^2.$$

Summing for $j = 1$ to r,

$$E\left[\sum_j \sum_i (x_{ij} - \bar{x}_j)^2\right] = r(n-1)\sigma^2$$

$$\Rightarrow E(SS_w) = r(n-1)\sigma^2 \Rightarrow E\left[\frac{SS_w}{r(n-1)}\right] = \sigma^2.$$

Hence $\quad s_w^2 = \dfrac{SS_w}{r(n-1)} \quad$ is an unbiased estimate of σ^2.

Using the between-sample variation

Consider the whole sample

$$E\left[\sum_j \sum_i (x_{ij} - m)^2\right] = (N-1)\sigma^2$$

From relation (2)

$$n \sum_j (\bar{x}_j - m)^2 = \sum_j \sum_i (x_{ij} - m)^2 - \sum_j \sum_i (x_{ij} - \bar{x}_j)^2.$$

Hence

$$nE[\Sigma \ (\bar{x}_j - m)^2] = E\left[\sum_j \sum_i (x_{ij} - m)^2\right] - E\left[\sum_j \sum_i (x_{ij} - \bar{x}_j)^2\right]$$

$$= (N-1)\sigma^2 - r(n-1)\sigma^2 \quad \text{(see above)}$$

$$= \sigma^2(N-1-rn+r)$$

$$= (r-1)\sigma^2, \quad \text{since } rn = N.$$

$$\Rightarrow E(SS_b) = (r-1)\sigma^2 \Rightarrow E\left[\frac{SS_b}{r-1}\right] = \sigma^2.$$

Hence $\quad s_b^2 = \dfrac{SS_b}{r-1} \quad$ is an unbiased estimate of σ^2.

Miscellaneous exercises in probability

1. A machine contains four components, A, B, C and D, each of which, independently of the others, may fail with probabilities 0.1, 0.3, 0.4 and 0.2 respectively. Show that there is a probability of 0.88 that at least one of B and C will not fail.

The machine will fail to operate if A fails or if D fails or if both B and C fail, but otherwise it will operate. Calculate the probability that the machine will operate.

A second machine also contains the four components A, B, C and D, with the same failure probabilities as above, except that if either B or C fails the probability of D failing is increased to 0.4. Assuming that this machine will fail to operate under the same conditions as those for the first machine, calculate the probability that the second machine will operate.   (JMB)

2. A and B play a game as follows: A draws and keeps a card from a shuffled pack numbered 1 to 6. B then draws a card from the remaining five. The winner is the one with the card possessing the greater number. Determine whether there is an advantage in drawing first.

If the rules are altered so that A returns his card to the pack after noting its value and, also, if A and B draw the same card all the cards are shuffled and they draw again, find:
(a) the probability that A wins on the first draw,
(b) the probability that A wins on the rth draw,
(c) the probability that A wins, if the game can continue for as long as is necessary.   (SUJB)

3. The probabilities of A, B or C winning a certain game in which all three take part are 0.5, 0.3 and 0.2 respectively. A match is won by the player who first wins two games. Find the probability that A will win a match involving all three players.

When the players are joined by a fourth player, D, the probabilities of A, B or C winning a game in which all four take part are reduced to 0.3, 0.2 and 0.1 respectively. A match is played with all four players taking part in each game; again, the player who first wins two games wins the match. Find the probabilities that D wins in fewer than: (a) four games, (b) five games, (c) six games.   (JMB)

4. (a) In a school, 40% of the boys have fair hair, 25% have blue eyes; 15% have both fair hair and blue eyes. A boy is selected at random.
(i)   Given that he has fair hair, find the probability that he has blue eyes.
(ii) Given that he has blue eyes, find the probability that he does not have fair hair.
(iii) Find the probability that he has neither fair hair nor blue eyes.
   (b) Two distinct digits are selected from the digits 1 to 9 inclusive. Given that their sum is even, find the probability that both numbers are odd.   (L)

5. Show that the total number of random samples of size r that can be drawn from a population of size n is
$$\frac{n!}{r!(n-r)!}.$$

In a class of 10 boys and 10 girls there are 5 children with blue eyes. A random sample of 4 children is taken. Find the probabilities that in this sample there are exactly:
(a) 2 boys,   (b) 2 children with blue eyes.

Half the children living in a large city are boys, and one quarter of the children have blue eyes. A random sample of 4 children is taken. Estimate the probabilities that in this sample there are exactly:
(c) 2 boys,   (d) 2 children with blue eyes.   (MEI)

6.   A snooker champion is engaged to play two opponents A and B alternately until one of them wins. The first game is against A; find the probability that B wins before A when P (champion wins against A) = 19/20 and P (champion wins against B) = 15/16. It may be assumed that the results of all matches are statistically independent.   (MEI)

7.   Of the group of pupils studying at A-level in schools in a certain area, 56 % are boys and 44 % are girls. The probability that a boy of this group is studying chemistry is 1/5 and the probability that a girl of this group is studying chemistry is 1/11.
(a) Find the probability that a pupil selected at random from this group is a girl studying chemistry.
(b) Find the probability that a pupil selected at random from this group is not studying chemistry.
(c) Find the probability that a chemistry pupil selected at random from this group is male.
    (You may leave your answers as fractions in their lowest terms.)   (O & C)

8.   (a) A motorist regularly makes a certain journey which involves three traffic lights A, B, C. He reckons that the probabilities of being delayed at these lights are respectively 0.6, 0.75 and 0.8 and that the delays at the points are independent. Calculate the probabilities of 0, 1, 2, 3 delays on the journey and also the likely total number of delays in a period of 20 journeys.
    (b) Four of the numbers 0, 1, 2, 3, 4, 5, 6, 7, 8, 9 are chosen independently from a random number table. This implies that each choice is unaffected by the others, and that duplication can occur. Calculate in decimal form the following probabilities:
(i)   that all the numbers lie in the set $S = \{n: 3 \leqslant n \leqslant 7\}$;
(ii)   that 2 is chosen just once and all the others exceed 2;
(iii) that 8 is chosen just once and all the others are less than 8;
(iv)   that 2 and 8 are chosen once each and the remaining numbers lie between them in value.
    (MEI)

9.   Out of 50 patients being treated at a clinic for a severe allergy, 10 are chosen at random to receive a new dietary treatment as opposed to the standard drug treatment given to the other 40 patients. It is known that the probability of a cure with the standard treatment is 0.6, whereas the probability of a cure with the new treatment is 0.9. Sometime later, one of these patients returns to the clinic to thank the doctors for his cure. What is the probability that he had been given the new treatment?
    (AEB 1982)

10.   (a) Show that $P(A \cap B \,|\, C) = P(A \,|\, B \cap C) P(B \,|\, C)$ provided that $P(B \cap C) \neq 0$.
    (b) Independent events A and B are such that A is twice as likely to occur as B which, in turn, is three times as likely to occur as that neither A nor B occurs. Find the probabilities of A and B.   (O)

11.   A doctor is certain that a patient has the symptoms of disease A but knows that diseases B and C show similar symptoms. In the general population these diseases are found in the ratios 5:1:2. The doctor decides to use a new test and makes 4 independent trials of this test. Find the probability that the patient has disease A if 3 of the trials give positive results and 1 gives a negative one. It is known that the test reacts positively to diseases A, B and C with probabilities 0.9, 0.5 and 0.3 respectively.   (MEI)

12. Show that

$$P(A \cup B \cup C) = P(A) + P(B) + P(C) - P(A \cap B) - P(B \cap C) - P(C \cap A) + P(A \cap B \cap C).$$

[You may use the result $P(D \cup E) = P(D) + P(E) - P(D \cap E)$ without proof.]

Alan throws three coins into a fountain, Brian throws two, and Colin throws one. David selects three of these six coins at random and gives one to each of Alan, Brian and Colin. What are the probabilities that: (a) Alan gets one of his own back; (b) Alan and Brian each get one of their own coins back; (c) at least one person gets one of his own coins back? (O)

13. An amethyst die and a beryl die are rolled. The score obtained, S, is defined as the larger of the two numbers shown by the dice if they show different numbers, and as 0 if they show the same number. Determine the probability distribution of S and show that the expected value of S is 35/9.

The events A, B, C and D are defined as follows:

A: The amethyst die shows a 2;
B: The beryl die shows a 2;
C: The score S is greater than 3;
D: $A \cup B$.

Determine $P(C|A)$, $P(C|D)$ and $P(D|C)$. (C)

14. In a maternity hospital n babies are put to sleep in one room after feeding and the number who are still asleep after one hour is recorded. It is suggested that the probability that this number is r is

$$\frac{2(n-r+1)}{(n+1)(n+2)}.$$

Show that this suggestion is consistent with the requirements of probability theory.

Adopting this suggestion, find the probability that at least m of the babies are still asleep after one hour, and deduce that if n is large the probability that at least half of the babies are still asleep is approximately $\frac{1}{4}$.

Find the smallest value of n for which the probability of there being at least one baby awake after one hour is greater than 0.95. (MEI)

15. Coloured balls are distributed in three identical boxes in the following way:

	Box 1	Box 2	Box 3
Red	4	3	2
White	1	4	3
Blue	3	3	5

A box is selected at random and a ball is taken out of it. If the ball turns out to be red, what is the probability that Box 2 was selected?

16. (a) A, B and C represents 3 events. If $A \cap B$ is the event that both A and B occur and $P(B|A)$ is the probability that B occurs given that A has already occurred, show that

$$P(A \cap B) = P(A)P(B|A).$$

Deduce, or show otherwise, that

$$P(A \cap B \cap C) = P(A)P(B|A)P(C|A \cap B).$$

(b) An athlete aims to measure his fitness by subjecting himself to a sequence of 3 physical tests, the completion of each test in a specified time being classed by him as a 'pass'. The probability that he passes the first test in the sequence is p, but the probability of passing any

subsequent test is half the probability of passing the immediately preceding test. Show that, if the probability of passing all 3 tests is $1/216$, the value of p is $1/3$. Hence find the probabilities:
(i) that he fails all the tests;
(ii) that he passes exactly 2 of the 3 tests.   (MEI)

17.   In a game, three cubical dice are thrown by a player who attempts to throw the same number on all three. What is the chance of the player: (a) throwing the same number on all three? (b) throwing the same number on just two?
   If the first throw results in just two dice showing the same number, then the third is thrown again. If no two dice show the same number, then all are thrown again. The player then comes to the end of his turn. What is the chance of the player succeeding in throwing three identical numbers in a complete turn?
   What is the chance that all numbers are different at the end of a turn?   (O & C)

18.   Define the conditional probability of the event A given the event B.
   An electrically heated hot water service is operated by two relay switches, A and B whose probabilities of failure, at the moment the service is switched on, are

$$P_A = 0.1, \qquad P_B = 0.05.$$

Assuming the switches operate independently of each other, calculate the probability that at least one of the switches will fail at this moment.
   In practice the switches do not operate independently. It has been observed that the failure of switch A depends on switch B and that the conditional probability of A failing given that B has not failed is 0.105. Calculate the probability that both of the switches will fail, at the moment the service is switched on, if the individual probabilities of failure are as before.   (MEI)

19.   One day a man buys two match boxes each containing 50 matches. He carries both in his pocket and every time he wants a match he chooses, with equal probability, one of the boxes and takes a match from there. One day he finds, for the first time that the box he has chosen is empty. Show that the probability that the other box contains r matches when this occurs is u_r, where

$$u_r = \frac{(100-r)!}{50!(50-r)!} \, (\tfrac{1}{2})^{100-r} \quad (r = 0, 1, \dots, 50).$$

Given that $u_0 = 0.0796$ to four places of decimals, determine the probabilities that the other box contains: (a) exactly 3 matches, (b) not more than 3 matches.   (MEI)

20.   A meteorologist has found that, in a certain district over a long period, the probability of the weather changing, from wet one day to dry the next, or vice versa, is $1/3$. If p_n is the probability that day n is dry, explain carefully why (for $n \geqslant 1$)

$$p_{n+1} = \tfrac{2}{3}p_n + \tfrac{1}{3}(1 - p_n).$$

Given that day 1 is dry, prove by induction that

$$p_n = \tfrac{1}{2}[1 + (\tfrac{1}{3})^{n-1}].$$ (SMP)

21.   A game is played with two dice thrown simultaneously. A symbol is moved, through a distance equal to the sum of the scores shown on the dice, on a track as shown in the figure:

A	B	C	D	E	F	G	H				

   A player's symbol is situated at A. Landing on E at the first throw of the dice is denoted by E_1; landing on H at the second throw by H_2, and similarly for other results. He continues to throw until he reaches or passes the twelfth square. Calculate:
(a) $P(B_1)$, $P(G_1)$ and $P(B_1 \cup G_1)$,
(b) $P(G_2)$ and $P(G)$, where G denotes landing on G at any throw.
                                                                                   (L)

22. A random sample of 6 articles is taken from a large consignment and tested in two independent stages. The probability that an article will pass either stage is q. All six articles are first tested at stage 1; provided five or more pass, those which pass are retested at stage 2. The consignment is accepted if there is no more than one failure at each stage.

Find polynomials in q for
(a) the probability that stage 2 of the test will be required,
(b) the number of articles expected to undergo stage 2,
(c) the probability $P(q)$ of accepting the consignment.

Show that $dP/dq = 0$ when $q = 1$ and find $P(0.9)$ and $P(0.8)$. Sketch the graph of $P(q)$ for $0 \leqslant q \leqslant 1$.   (O & C)

23. A bag contains one white ball and three blue balls, which are identical apart from colour. Three gamblers play a game which consist of a series of 'rounds'. In each round the gamblers are blindfold and simultaneously draw one ball each from the bag. Any gambler who draws the white ball is eliminated and the other two continue. After each round all the balls are put back in the bag. The game finishes when the white ball is drawn for the second time.
(a) Show that the probability that exactly two players remain in the game after one round is 3/4, and that the probability that exactly two players remain after two rounds is 9/16.
(b) Find the probability that exactly two players remain in the game after r rounds $(r = 1, 2, 3, \ldots)$.
(c) Hence show that the probability that the game finishes after exactly r rounds is
$$3.2^{-r} - 6.4^{-r} \quad (r = 2, 3, 4, \ldots).$$
(d) Find, using the binomial expansion of $(1-x)^{-2}$, the expected length of the game.
                                                                                     (MEI)

24. A chess match between two grandmasters A and B is won by whoever first wins a total of 2 games. A's chances of winning, drawing or losing any particular game are p, q, r respectively. The games are independent and $p + q + r = 1$. Show that the probability that A wins the match after $(n+1)$ games $(n \geqslant 1)$ is
$$p^2[nq^{n-1} + n(n-1)rq^{n-2}].$$

By considering suitable operations on the infinite geometric series $(q + q^2 + \ldots)$, or otherwise, show that the probability that A wins the match is
$$p^2(p + 3r)/(p + r)^3.$$

Find the probability that there is no winner and comment on your result.   (MEI)

25. In a competition, two players A and B are playing a tennis match. Players serve in alternate games. The probability of A winning a game in which he serves is a. The probability of B winning a game in which he serves is b. The score in a match is 6 games each, with A about to serve. A player wins the match when he has won two games more than his opponent. Show that the probability that A wins the match is
$$\frac{a(1-b)}{1-\lambda},$$
where $\lambda = ab + (1-a)(1-b)$.

In another competition, the same two players are playing. The score is 6 games each and A is about to serve. The rules of this competition are the same as those of the previous competition with the additional rule that, should the score reach 8 games each, the match will then be decided by a special 'tie-break' game. The probability that A wins such a 'tie-break' game is $a/(a+b)$. Show that the probability that A goes on (from the score of six games each) to win this match is
$$a(1-b)(1+\lambda) + \lambda^2 a/(a+b).$$ (C)

Tables

Critical values for the Wilcoxon signed-rank test (T)

P	1	2.5	5
$n =$ 5			0
6		0	2
7	0	2	3
8	1	3	5
9	3	5	8
10	5	8	10
11	7	10	13
12	9	13	17
13	12	17	21
14	15	21	25
15	19	25	30
16	23	29	35
17	27	34	41
18	32	40	47
19	37	46	53
20	43	52	60
21	49	58	67
22	55	65	75
23	62	73	83
24	69	81	91
25	76	89	100
26	84	98	110
27	92	107	119
28	101	116	130
29	110	126	140
30	120	137	151

(Taken from *New Cambridge Elementary Statistical Tables*, Lindley and Scott 1984.)

The standard normal distribution

Values of z, the standard normal variable, from 0.0 by steps of 0.01 to 3.9, showing the cumulative probability up to z. (Probability correct to 4 decimal places.)

z	0.00	0.01	0.02	0.03	0.04	0.05	0.06	0.07	0.08	0.09
0.0	.5000	.5040	.5080	.5120	.5160	.5199	.5239	.5279	.5319	.5359
.1	.5398	.5438	.5478	.5517	.5557	.5596	.5636	.5675	.5714	.5753
.2	.5793	.5832	.5871	.5910	.5948	.5987	.6026	.6064	.6103	.6141
.3	.6179	.6217	.6255	.6293	.6331	.6368	.6406	.6443	.6480	.6517
.4	.6554	.6591	.6628	.6664	.6700	.6736	.6772	.6808	.6844	.6879
.5	.6915	.6950	.6985	.7019	.7054	.7088	.7123	.7157	.7190	.7224
.6	.7257	.7291	.7324	.7357	.7389	.7422	.7454	.7486	.7517	.7549
.7	.7580	.7611	.7642	.7673	.7704	.7734	.7764	.7794	.7823	.7852
.8	.7881	.7910	.7939	.7967	.7995	.8023	.8051	.8078	.8106	.8133
.9	.8159	.8186	.8212	.8238	.8264	.8289	.8315	.8340	.8365	.8389
1.0	.8413	.8438	.8461	.8485	.8508	.8531	.8554	.8577	.8599	.8621
.1	.8643	.8665	.8686	.8708	.8729	.8749	.8770	.8790	.8810	.8830
.2	.8849	.8869	.8888	.8907	.8925	.8944	.8962	.8980	.8997	.9015
.3	.9032	.9049	.9066	.9082	.9099	.9115	.9131	.9147	.9162	.9177
.4	.9192	.9207	.9222	.9236	.9251	.9265	.9279	.9292	.9306	.9319
.5	.9332	.9345	.9357	.9370	.9382	.9394	.9406	.9418	.9429	.9441
.6	.9452	.9463	.9474	.9484	.9495	.9505	.9515	.9525	.9535	.9545
.7	.9554	.9564	.9573	.9582	.9591	.9599	.9608	.9616	.9625	.9633
.8	.9641	.9649	.9656	.9664	.9671	.9678	.9686	.9693	.9699	.9706
.9	.9713	.9719	.9726	.9732	.9738	.9744	.9750	.9756	.9761	.9767
2.0	.9772	.9778	.9783	.9788	.9793	.9798	.9803	.9808	.9812	.9817
.1	.9821	.9826	.9830	.9834	.9838	.9842	.9846	.9850	.9854	.9857
.2	.9861	.9864	.9868	.9871	.9875	.9878	.9881	.9884	.9887	.9890
.3	.9893	.9896	.9898	.9901	.9904	.9906	.9909	.9911	.9913	.9916
.4	.9918	.9920	.9922	.9925	.9927	.9929	.9931	.9932	.9934	.9936
.5	.9938	.9940	.9941	.9943	.9945	.9946	.9948	.9949	.9951	.9952
.6	.9953	.9955	.9956	.9957	.9959	.9960	.9961	.9962	.9963	.9964
.7	.9965	.9966	.9967	.9968	.9969	.9970	.9971	.9972	.9973	.9974
.8	.9974	.9975	.9976	.9977	.9977	.9978	.9979	.9979	.9980	.9981
.9	.9981	.9982	.9982	.9983	.9984	.9984	.9985	.9985	.9986	.9986
3.0	.9987	.9987	.9987	.9988	.9988	.9989	.9989	.9989	.9990	.9990
.1	.9990	.9991	.9991	.9991	.9992	.9992	.9992	.9992	.9993	.9993
.2	.9993	.9993	.9994	.9994	.9994	.9994	.9994	.9995	.9995	.9995
.3	.9995	.9995	.9995	.9996	.9996	.9996	.9996	.9996	.9996	.9997
.4	.9997	.9997	.9997	.9997	.9997	.9997	.9997	.9997	.9997	.9998
.5	.9998	.9998	.9998	.9998	.9998	.9998	.9998	.9998	.9998	.9998
.6	.9998	.9998	.9999	.9999	·9999	.9999	.9999	.9999	.9999	.9999
.7	.9999	.9999	.9999	.9999	.9999	.9999	.9999	.9999	.9999	.9999
.8	.9999	.9999	.9999	.9999	.9999	.9999	.9999	.9999	.9999	.9999
.9	1.0000									

The curve is $N(0, 1)$, the standard normal variable. The table entry is the shaded area $\Phi(z) = P(Z < z)$. For example, when $z = 1.96$ the shaded area is 0.9750.

(Taken from *New Cambridge Elementary Statistical Tables*, Lindley and Scott 1984.)

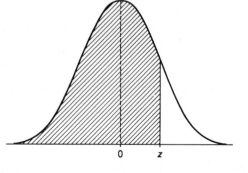

Percentage points of the *t*-distribution

P	20	10	5	2	1	0.2	0.1
$v = 1$	3.08	6.31	12.7	31.8	63.7	318	637
2	1.89	2.92	4.30	6.96	9.93	22.3	31.6
3	1.64	2.35	3.18	4.54	5.84	10.2	12.9
4	1.53	2.13	2.78	3.75	4.60	7.18	8.61
5	1.48	2.02	2.57	3.36	4.03	5.89	6.87
6	1.44	1.94	2.45	3.14	3.71	5.21	5.96
7	1.41	1.89	2.36	3.00	3.50	4.78	5.41
8	1.40	1.86	2.31	2.90	3.36	4.50	5.04
9	1.38	1.83	2.26	2.82	3.25	4.29	4.78
10	1.37	1.81	2.23	2.76	3.17	4.15	4.58
11	1.36	1.80	2.20	2.72	3.11	4.02	4.44
12	1.36	1.78	2.18	2.68	3.05	3.93	4.32
13	1.35	1.77	2.16	2.65	3.01	3.85	4.22
14	1.35	1.76	2.14	2.62	2.98	3.79	4.14
15	1.34	1.75	2.13	2.60	2.95	3.73	4.07
20	1.33	1.72	2.09	2.53	2.85	3.55	3.85
30	1.31	1.70	2.04	2.46	2.75	3.38	3.64
40	1.30	1.68	2.02	2.42	2.70	3.31	3.55
50	1.30	1.68	2.01	2.40	2.68	3.26	3.50
60	1.30	1.67	2.00	2.39	2.66	3.23	3.46
∞	1.28	1.64	1.96	2.33	2.58	3.09	3.29

The values for $v = \infty$ are those from a normal probability function, which is the limiting form for large v.

If X is a random variable with probability density function that of t with v degrees of freedom then $P/100$ is the probability that $|X| \geqslant t_P$. It is *tabulated for two-tail tests*. The probability that $x \geqslant t_P$ is half that given at the head of the table.

Taken from *SMP Advanced Tables*, 1979.)

Percentage points of the χ^2-distribution

P	99.5	99	97.5	95	10	5	2.5	1	0.5	0.1
$v=1$	0.0^4393	0.0^3157	0.0^3982	0.00393	2.71	3.84	5.02	6.63	7.88	10.83
2	0.0100	0.0201	0.0506	0.103	4.61	5.99	7.38	9.21	10.60	13.81
3	0.0717	0.115	0.216	0.352	6.25	7.81	9.35	11.34	12.84	16.27
4	0.207	0.297	0.484	0.711	7.78	9.49	11.14	13.28	14.86	18.47
5	0.412	0.554	0.831	1.15	9.24	11.07	12.83	15.09	16.75	20.52
6	0.676	0.872	1.24	1.64	10.64	12.59	14.45	16.81	18.55	22.46
7	0.989	1.24	1.69	2.17	12.02	14.07	16.01	18.48	20.28	24.32
8	1.34	1.65	2.18	2.73	13.36	15.51	17.53	20.09	21.95	26.12
9	1.73	2.09	2.70	3.33	14.68	16.92	19.02	21.67	23.59	27.88
10	2.16	2.56	3.25	3.94	15.99	18.31	20.48	23.21	25.19	29.59
11	2.60	3.05	3.82	4.57	17.28	19.68	21.92	24.73	26.76	31.26
12	3.07	3.57	4.40	5.23	18.55	21.03	23.34	26.22	28.30	32.91
13	3.57	4.11	5.01	5.89	19.81	22.36	24.74	27.69	29.82	34.53
14	4.07	4.66	5.63	6.57	21.06	23.68	26.12	29.14	31.32	36.12
15	4.60	5.23	6.26	7.26	22.31	25.00	27.49	30.58	32.80	37.70
16	5.14	5.81	6.91	7.96	23.54	26.30	28.85	32.00	34.27	39.25
17	5.70	6.41	7.56	8.67	24.77	27.59	30.19	33.41	35.72	40.79
18	6.26	7.01	8.23	9.39	25.99	28.87	31.53	34.81	37.16	42.31
19	6.84	7.63	8.91	10.12	27.20	30.14	32.85	36.19	38.58	43.82
20	7.43	8.26	9.59	10.85	28.41	31.41	34.17	37.57	40.00	45.31
21	8.03	8.90	10.28	11.59	29.62	32.67	35.48	38.93	41.40	46.80
22	8.64	9.54	10.98	12.34	30.81	33.92	36.78	40.29	42.80	48.27
23	9.26	10.20	11.69	13.09	32.01	35.17	38.08	41.64	44.18	49.73
24	9.89	10.86	12.40	13.85	33.20	36.42	39.36	42.98	45.56	51.18
25	10.52	11.52	13.12	14.61	34.38	37.65	40.65	44.31	46.93	52.62
26	11.16	12.20	13.84	15.38	35.56	38.89	41.92	45.64	48.29	54.05
27	11.81	12.88	14.57	16.15	36.74	40.11	43.19	46.96	49.64	55.48
28	12.46	13.56	15.31	16.93	37.92	41.34	44.46	48.28	50.99	56.89
29	13.12	14.26	16.05	17.71	39.09	42.56	45.72	49.59	52.34	58.30
30	13.79	14.95	16.79	18.49	40.26	43.77	46.98	50.89	53.67	59.70
40	20.71	22.16	24.43	26.51	51.81	55.76	59.34	63.69	66.77	73.40
50	27.99	29.71	32.36	34.76	63.17	67.50	71.42	76.15	79.49	86.66
60	35.53	37.48	40.48	43.19	74.40	79.08	83.30	88.38	91.95	99.61
70	43.28	45.44	48.76	51.74	85.53	90.53	95.02	100.4	104.2	112.3
80	51.17	53.54	57.15	60.39	96.58	101.9	106.6	112.3	116.3	124.8
90	59.20	61.75	65.65	69.13	107.6	113.1	118.1	124.1	128.3	137.2
100	67.33	70.06	74.22	77.93	118.5	124.3	129.6	135.8	140.2	149.4

If X is a variable distributed as χ^2 with v degrees of freedom, $P/100$ is the probability that $X \geqslant \chi_v^2(P)$. For $v < 100$, linear interpolation in v is adequate. For $v > 100$, $\sqrt{(2\chi^2)}$ is approximately normally distributed with mean $\sqrt{(2v-1)}$ and unit variance.

(Taken from *New Cambridge Elementary Statistical Tables*, Lindley and Scott 1984.)

5 per cent points of the F-distribution

$v_1 =$	1	2	3	4	5	6	7	8	10	12	24	∞
$v_2 = 1$	161.4	199.5	215.7	224.6	230.2	234.0	236.8	238.9	241.9	243.9	249.0	254.3
2	18.5	19.0	19.2	19.2	19.3	19.3	19.4	19.4	19.4	19.4	19.5	19.5
3	10.13	9.55	9.28	9.12	9.01	8.94	8.89	8.85	8.79	8.74	8.64	8.53
4	7.71	6.94	6.59	6.39	6.26	6.16	6.09	6.04	5.96	5.91	5.77	5.63
5	6.61	5.79	5.41	5.19	5.05	4.95	4.88	4.82	4.74	4.68	4.53	4.36
6	5.99	5.14	4.76	4.53	4.39	4.28	4.21	4.15	4.06	4.00	3.84	3.67
7	5.59	4.74	4.35	4.12	3.97	3.87	3.79	3.73	3.64	3.57	3.41	3.23
8	5.32	4.46	4.07	3.84	3.69	3.58	3.50	3.44	3.35	3.28	3.12	2.93
9	5.12	4.26	3.86	3.63	3.48	3.37	3.29	3.23	3.14	3.07	2.90	2.71
10	4.96	4.10	3.71	3.48	3.33	3.22	3.14	3.07	2.98	2.91	2.74	2.54
11	4.84	3.98	3.59	3.36	3.20	3.09	3.01	2.95	2.85	2.79	2.61	2.40
12	4.75	3.89	3.49	3.26	3.11	3.00	2.91	2.85	2.75	2.69	2.51	2.30
13	4.67	3.81	3.41	3.18	3.03	2.92	2.83	2.77	2.67	2.60	2.42	2.21
14	4.60	3.74	3.34	3.11	2.96	2.85	2.76	2.70	2.60	2.53	2.35	2.13
15	4.54	3.68	3.29	3.06	2.90	2.79	2.71	2.64	2.54	2.48	2.29	2.07
16	4.49	3.63	3.24	3.01	2.85	2.74	2.66	2.59	2.49	2.42	2.24	2.01
17	4.45	3.59	3.20	2.96	2.81	2.70	2.61	2.55	2.45	2.38	2.19	1.96
18	4.41	3.55	3.16	2.93	2.77	2.66	2.58	2.51	2.41	2.34	2.15	1.92
19	4.38	3.52	3.13	2.90	2.74	2.63	2.54	2.48	2.38	2.31	2.11	1.88
20	4.35	3.49	3.10	2.87	2.71	2.60	2.51	2.45	2.35	2.28	2.08	1.84
21	4.32	3.47	3.07	2.84	2.68	2.57	2.49	2.42	2.32	2.25	2.05	1.81
22	4.30	3.44	3.05	2.82	2.66	2.55	2.46	2.40	2.30	2.23	2.03	1.78
23	4.28	3.42	3.03	2.80	2.64	2.53	2.44	2.37	2.27	2.20	2.00	1.76
24	4.26	3.40	3.01	2.78	2.62	2.51	2.42	2.36	2.25	2.18	1.98	1.73
25	4.24	3.39	2.99	2.76	2.60	2.49	2.40	2.34	2.24	2.16	1.96	1.71
26	4.23	3.37	2.98	2.74	2.59	2.47	2.39	2.32	2.22	2.15	1.95	1.69
27	4.21	3.35	2.96	2.73	2.57	2.46	2.37	2.31	2.20	2.13	1.93	1.67
28	4.20	3.34	2.95	2.71	2.56	2.45	2.36	2.29	2.19	2.12	1.91	1.65
29	4.18	3.33	2.93	2.70	2.55	2.43	2.35	2.28	2.18	2.10	1.90	1.64
30	4.17	3.32	2.92	2.69	2.53	2.42	2.33	2.27	2.16	2.09	1.89	1.62
32	4.15	3.29	2.90	2.67	2.51	2.40	2.31	2.24	2.14	2.07	1.86	1.59
34	4.13	3.28	2.88	2.65	2.49	2.38	2.29	2.23	2.12	2.05	1.84	1.57
36	4.11	3.26	2.87	2.63	2.48	2.36	2.28	2.21	2.11	2.03	1.82	1.55
38	4.10	3.24	2.85	2.62	2.46	2.35	2.26	2.19	2.09	2.02	1.81	1.53
40	4.08	3.23	2.84	2.61	2.45	2.34	2.25	2.18	2.08	2.00	1.79	1.51
60	4.00	3.15	2.76	2.53	2.37	2.25	2.17	2.10	1.99	1.92	1.70	1.39
120	3.92	3.07	2.68	2.45	2.29	2.18	2.09	2.02	1.91	1.83	1.61	1.25
∞	3.84	3.00	2.60	2.37	2.21	2.10	2.01	1.94	1.83	1.75	1.52	1.00

(Taken from *New Cambridge Elementary Statistical Tables*, Lindley and Scott 1984.)

$2\frac{1}{2}$ per cent points of the F-distribution

$v_1 =$	1	2	3	4	5	6	7	8	10	12	24	∞
$v_2 = 1$	648	800	864	900	922	937	948	957	969	977	997	1018
2	38.5	39.0	39.2	39.2	39.3	39.3	39.4	39.4	39.4	39.4	39.5	39.5
3	17.4	16.0	15.4	15.1	14.9	14.7	14.6	14.5	14.4	14.3	14.1	13.9
4	12.22	10.65	9.98	9.60	9.36	9.20	9.07	8.98	8.84	8.75	8.51	8.26
5	10.01	8.43	7.76	7.39	7.15	6.98	6.85	6.76	6.62	6.52	6.28	6.02
6	8.81	7.26	6.60	6.23	5.99	5.82	5.70	5.60	5.46	5.37	5.12	4.85
7	8.07	6.54	5.89	5.52	5.29	5.12	4.99	4.90	4.76	4.67	4.42	4.14
8	7.57	6.06	5.42	5.05	4.82	4.65	4.53	4.43	4.30	4.20	3.95	3.67
9	7.21	5.71	5.08	4.72	4.48	4.32	4.20	4.10	3.96	3.87	3.61	3.33
10	6.94	5.46	4.83	4.47	4.24	4.07	3.95	3.85	3.72	3.62	3.37	3.08
11	6.72	5.26	4.63	4.28	4.04	3.88	3.76	3.66	3.53	3.43	3.17	2.88
12	6.55	5.10	4.47	4.12	3.89	3.73	3.61	3.51	3.37	3.28	3.02	2.72
13	6.41	4.97	4.35	4.00	3.77	3.60	3.48	3.39	3.25	3.15	2.89	2.60
14	6.30	4.86	4.24	3.89	3.66	3.50	3.38	3.29	3.15	3.05	2.79	2.49
15	6.20	4.76	4.15	3.80	3.58	3.41	3.29	3.20	3.06	2.96	2.70	2.40
16	6.12	4.69	4.08	3.73	3.50	3.34	3.22	3.12	2.99	2.89	2.63	2.32
17	6.04	4.62	4.01	3.66	3.44	3.28	3.16	3.06	2.92	2.82	2.56	2.25
18	5.98	4.56	3.95	3.61	3.38	3.22	3.10	3.01	2.87	2.77	2.50	2.19
19	5.92	4.51	3.90	3.56	3.33	3.17	3.05	2.96	2.82	2.72	2.45	2.13
20	5.87	4.46	3.86	3.51	3.29	3.13	3.01	2.91	2.77	2.68	2.41	2.09
21	5.83	4.42	3.82	3.48	3.25	3.09	2.97	2.87	2.73	2.64	2.37	2.04
22	5.79	4.38	3.78	3.44	3.22	3.05	2.93	2.84	2.70	2.60	2.33	2.00
23	5.75	4.35	3.75	3.41	3.18	3.02	2.90	2.81	2.67	2.57	2.30	1.97
24	5.72	4.32	3.72	3.38	3.15	2.99	2.87	2.78	2.64	2.54	2.27	1.94
25	5.69	4.29	3.69	3.35	3.13	2.97	2.85	2.75	2.61	2.51	2.24	1.91
26	5.66	4.27	3.67	3.33	3.10	2.94	2.82	2.73	2.59	2.49	2.22	1.88
27	5.63	4.24	3.65	3.31	3.08	2.92	2.80	2.71	2.57	2.47	2.19	1.85
28	5.61	4.22	3.63	3.29	3.06	2.90	2.78	2.69	2.55	2.45	2.17	1.83
29	5.59	4.20	3.61	3.27	3.04	2.88	2.76	2.67	2.53	2.43	2.15	1.81
30	5.57	4.18	3.59	3.25	3.03	2.87	2.75	2.65	2.51	2.41	2.14	1.79
32	5.53	4.15	3.56	3.22	3.00	2.84	2.72	2.62	2.48	2.38	2.10	1.75
34	5.50	4.12	3.53	3.19	2.97	2.81	2.69	2.59	2.45	2.35	2.08	1.72
36	5.47	4.09	3.51	3.17	2.94	2.79	2.66	2.57	2.43	2.33	2.05	1.69
38	5.45	4.07	3.48	3.15	2.92	2.76	2.64	2.55	2.41	2.31	2.03	1.66
40	5.42	4.05	3.46	3.13	2.90	2.74	2.62	2.53	2.39	2.29	2.01	1.64
60	5.29	3.93	3.34	3.01	2.79	2.63	2.51	2.41	2.27	2.17	1.88	1.48
120	5.15	3.80	3.23	2.89	2.67	2.52	2.39	2.30	2.16	2.05	1.76	1.31
∞	5.02	3.69	3.12	2.79	2.57	2.41	2.29	2.19	2.05	1.94	1.64	1.00

(Taken from *New Cambridge Elementary Statistical Tables*, Lindley and Scott 1984.)

1 per cent points of the *F*-distribution

$v_1 =$	1	2	3	4	5	6	7	8	10	12	24	∞
$v_2 = 1$	4052	5000	5403	5625	5764	5859	5928	5981	6056	6106	6235	6366
2	98.5	99.0	99.2	99.2	99.3	99.3	99.4	99.4	99.4	99.4	99.5	99.5
3	34.1	30.8	29.5	28.7	28.2	27.9	27.7	27.5	27.2	27.1	26.6	26.1
4	21.2	18.0	16.7	16.0	15.5	15.2	15.0	14.8	14.5	14.4	13.9	13.5
5	16.26	13.27	12.06	11.39	10.97	10.67	10.46	10.29	10.05	9.89	9.47	9.02
6	13.74	10.92	9.78	9.15	8.75	8.47	8.26	8.10	7.87	7.72	7.31	6.88
7	12.25	9.55	8.45	7.85	7.46	7.19	6.99	6.84	6.62	6.47	6.07	5.65
8	11.26	8.65	7.59	7.01	6.63	6.37	6.18	6.03	5.81	5.67	5.28	4.86
9	10.56	8.02	6.99	6.42	6.06	5.80	5.61	5.47	5.26	5.11	4.73	4.31
10	10.04	7.56	6.55	5.99	5.64	5.39	5.20	5.06	4.85	4.71	4.33	3.91
11	9.65	7.21	6.22	5.67	5.32	5.07	4.89	4.74	4.54	4.40	4.02	3.60
12	9.33	6.93	5.95	5.41	5.06	4.82	4.64	4.50	4.30	4.16	3.78	3.36
13	9.07	6.70	5.74	5.21	4.86	4.62	4.44	4.30	4.10	3.96	3.59	3.17
14	8.86	6.51	5.56	5.04	4.70	4.46	4.28	4.14	3.94	3.80	3.43	3.00
15	8.68	6.36	5.42	4.89	4.56	4.32	4.14	4.00	3.80	3.67	3.29	2.87
16	8.53	6.23	5.29	4.77	4.44	4.20	4.03	3.89	3.69	3.55	3.18	2.75
17	8.40	6.11	5.18	4.67	4.34	4.10	3.93	3.79	3.59	3.46	3.08	2.65
18	8.29	6.01	5.09	4.58	4.25	4.01	3.84	3.71	3.51	3.37	3.00	2.57
19	8.18	5.93	5.01	4.50	4.17	3.94	3.77	3.63	3.43	3.30	2.92	2.49
20	8.10	5.85	4.94	4.43	4.10	3.87	3.70	3.56	3.37	3.23	2.86	2.42
21	8.02	5.78	4.87	4.37	4.04	3.81	3.64	3.51	3.31	3.17	2.80	2.36
22	7.95	5.72	4.82	4.31	3.99	3.76	3.59	3.45	3.26	3.12	2.75	2.31
23	7.88	5.66	4.76	4.26	3.94	3.71	3.54	3.41	3.21	3.07	2.70	2.26
24	7.82	5.61	4.72	4.22	3.90	3.67	3.50	3.36	3.17	3.03	2.66	2.21
25	7.77	5.57	4.68	4.18	3.86	3.63	3.46	3.32	3.13	2.99	2.62	2.17
26	7.72	5.53	4.64	4.14	3.82	3.59	3.42	3.29	3.09	2.96	2.58	2.13
27	7.68	5.49	4.60	4.11	3.78	3.56	3.39	3.26	3.06	2.93	2.55	2.10
28	7.64	5.45	4.57	4.07	3.75	3.53	3.36	3.23	3.03	2.90	2.52	2.06
29	7.60	5.42	4.54	4.04	3.73	3.50	3.33	3.20	3.00	2.87	2.49	2.03
30	7.56	5.39	4.51	4.02	3.70	3.47	3.30	3.17	2.98	2.84	2.47	2.01
32	7.50	5.34	4.46	3.97	3.65	3.43	3.26	3.13	2.93	2.80	2.42	1.96
34	7.45	5.29	4.42	3.93	3.61	3.39	3.22	3.09	2.90	2.76	2.38	1.91
36	7.40	5.25	4.38	3.89	3.58	3.35	3.18	3.05	2.86	2.72	2.35	1.87
38	7.35	5.21	4.34	3.86	3.54	3.32	3.15	3.02	2.83	2.69	2.32	1.84
40	7.31	5.18	4.31	3.83	3.51	3.29	3.12	2.99	2.80	2.66	2.29	1.80
60	7.08	4.98	4.13	3.65	3.34	3.12	2.95	2.82	2.63	2.50	2.12	1.60
120	6.85	4.79	3.95	3.48	3.17	2.96	2.79	2.66	2.47	2.34	1.95	1.38
∞	6.63	4.61	3.78	3.32	3.02	2.80	2.64	2.51	2.32	2.18	1.79	1.00

(Taken from *New Cambridge Elementary Statistical Tables*, Lindley and Scott 1984.)

0.1 per cent points of the *F*-distribution

$v =$	1	2	3	4	5	6	7	8	10	12	24	∞
$v_2 = 1^*$	4053	5000	5404	5625	5764	5859	5929	5981	6056	6107	6235	6366*
2	998.5	999.0	999.2	999.2	999.3	999.3	999.4	999.4	999.4	999.4	999.5	999.5
3	167.0	148.5	141.1	137.1	134.6	132.8	131.5	130.6	129.2	128.3	125.9	123.5
4	74.14	61.25	56.18	53.44	51.71	50.53	49.66	49.00	48.05	47.41	45.77	44.05
5	47.18	37.12	33.20	31.09	29.75	28.83	28.16	27.65	26.92	26.42	25.14	23.79
6	35.51	27.00	23.70	21.92	20.80	20.03	19.46	19.03	18.41	17.99	16.90	15.75
7	29.25	21.69	18.77	17.20	16.21	15.52	15.02	14.63	14.08	13.71	12.73	11.70
8	25.42	18.49	15.83	14.39	13.48	12.86	12.40	12.05	11.54	11.19	10.30	9.34
9	22.86	16.39	13.90	12.56	11.71	11.13	10.69	10.37	9.87	9.57	8.72	7.81
10	21.04	14.91	12.55	11.28	10.48	9.93	9.52	9.20	8.74	8.44	7.64	6.76
11	19.69	13.81	11.56	10.35	9.58	9.05	8.66	8.35	7.92	7.63	6.85	6.00
12	18.64	12.97	10.80	9.63	8.89	8.38	8.00	7.71	7.29	7.00	6.25	5.42
13	17.82	12.31	10.21	9.07	8.35	7.86	7.49	7.21	6.80	6.52	5.78	4.97
14	17.14	11.78	9.73	8.62	7.92	7.44	7.08	6.80	6.40	6.13	5.41	4.60
15	16.59	11.34	9.34	8.25	7.57	7.09	6.74	6.47	6.08	5.81	5.10	4.31
16	16.12	10.97	9.01	7.94	7.27	6.80	6.46	6.19	5.81	5.55	4.85	4.06
17	15.72	10.66	8.73	7.68	7.02	6.56	6.22	5.96	5.58	5.32	4.63	3.85
18	15.38	10.39	8.49	7.46	6.81	6.35	6.02	5.76	5.39	5.13	4.45	3.67
19	15.08	10.16	8.28	7.27	6.62	6.18	5.85	5.59	5.22	4.97	4.29	3.51
20	14.82	9.95	8.10	7.10	6.46	6.02	5.69	5.44	5.08	4.82	4.15	3.38
21	14.59	9.77	7.94	6.95	6.32	5.88	5.56	5.31	4.95	4.70	4.03	3.26
22	14.38	9.61	7.80	6.81	6.19	5.76	5.44	5.19	4.83	4.58	3.92	3.15
23	14.19	9.47	7.67	6.70	6.08	5.65	5.33	5.09	4.73	4.48	3.82	3.05
24	14.03	9.34	7.55	6.59	5.98	5.55	5.23	4.99	4.64	4.39	3.74	2.97
25	13.88	9.22	7.45	6.49	5.89	5.46	5.15	4.91	4.56	4.31	3.66	2.89
26	13.74	9.12	7.36	6.41	5.80	5.38	5.07	4.83	4.48	4.24	3.59	2.82
27	13.61	9.02	7.27	6.33	5.73	5.31	5.00	4.76	4.41	4.17	3.52	2.75
28	13.50	8.93	7.19	6.25	5.66	5.24	4.93	4.69	4.35	4.11	3.46	2.69
29	13.39	8.85	7.12	6.19	5.59	5.18	4.87	4.64	4.29	4.05	3.41	2.64
30	13.29	8.77	7.05	6.12	5.53	5.12	4.82	4.58	4.24	4.00	3.36	2.59
32	13.12	8.64	6.94	6.01	5.43	5.02	4.72	4.48	4.14	3.91	3.27	2.50
34	12.97	8.52	6.83	5.92	5.34	4.93	4.63	4.40	4.06	3.83	3.19	2.42
36	12.83	8.42	6.74	5.84	5.26	4.86	4.56	4.33	3.99	3.76	3.12	2.35
38	12.71	8.33	6.66	5.76	5.19	4.79	4.49	4.26	3.93	3.70	3.06	2.29
40	12.61	8.25	6.59	5.70	5.13	4.73	4.44	4.21	3.87	3.64	3.01	2.23
60	11.97	7.77	6.17	5.31	4.76	4.37	4.09	3.86	3.54	3.32	2.69	1.89
120	11.38	7.32	5.78	4.95	4.42	4.04	3.77	3.55	3.24	3.02	2.40	1.54
∞	10.83	6.91	5.42	4.62	4.10	3.74	3.47	3.27	2.96	2.74	2.13	1.00

* Entries for $v_2 = 1$ must be multiplied by 100.

(Taken from *New Cambridge Elementary Statistical Tables*, Lindley and Scott 1984.)

Answers

All answers given here are supplied by the author. The relevant Examining Boards accept no responsibility whatsoever for their accuracy.

Exercise 2b (page 16)

1. 2.375   2. 2.67   3. 49.2   4. 1.56; 438   5. B   6. 10.3
7. 4.24 tonnes; 12   8. 4.49%   9. 52.42 kg (assuming masses to nearest kg)
10. (a) 16   (b) 15.8   11. 16.1   12. £4.55; £4.61   13. Powell 5.25; Waugh 4.15
14. Thomas 4.03; Keneally 4.91

Exercise 2c (page 21)

1. (a) 6; 6   (b) 80; 80   2. 2; 3   3. 49; 49   4. 4.8   5. 42.9 mm
6. 10.6   7. 4.27 tonnes   8. 43.5 mm

Exercise 2d (page 23)

1. (a) $3\Sigma x + 4\Sigma y$ (b) $4\Sigma x - 5n$ (c) $3\Sigma + n$ (d) $2\Sigma x + cn$ (e) $\Sigma x + 2\Sigma y + 3\Sigma z + kn$
2. (a) $\Sigma x^2 + 10\Sigma x + 25n$ (b) $a^2\Sigma x^2 + 2ab\Sigma x + b^2 n$
3. (a) $\Sigma f_i x_i - a\Sigma f_i$ (b) $\Sigma f_i x_i^2 - 2a\Sigma f_i x_i + a^2\Sigma f_i$

Exercise 2e (page 24)

1. 149.49 cm   2. 19.4   3. 24.4   4. 30–39; 40 yrs
5. (a) 66.3 sh.   (b) 67.1 sh.; 27   6. (a) 26   (b) 25.8
7. (b) 4.40 mm   (c) 3.45; 3.95   (d) 4.54 mm   8. 78 mm   9. 86.6; 44.1; 188

Exercise 3a (page 29)

1. 5.27; 6.16   2. A: 25, 2.74; B: 25, 1.35; B
3. A: 15, 2.83 mins; B: 15, 1.89 mins; B   4. 2; 1.43   5. 3; 1.37   6. 5; 2.39
7. LC: s.d. = 1.44; CB: s.d. = 1.76; LC   8. 20; 6.32
9. 15; 8.94 mins; taking s.d. = 9: 49, 48, 42, 27, 9, 0, 0

Exercise 3b (page 33)

1. 4.8; 3.06   2. 6.88; 3.26   3. 1.32; 1.16; 1.34   4. 3.65; 1.51
5. 1.88, 1.29; 2.08, 1.32   7. 2.5, 1.12; 2.5, 0.79   8. 1.33; 0.94   9. 5.97; 818.06
10. 30.57; 4.41   11. 17.4; 4.00   12. 11.7; 2.17

Exercise 3c (page 40)

1. 149.6; 1.86 cm   2. 140.7 lb; 22.4 lb   3. 129.3 mm; 13.8 mm   4. 53.5; 405
5. 2.92   6. 17.85; 5.57   7. 70.43 kg; 71.0 kg^2 8. 2822.7; 7665
9. 11.17 cm^2 10.

number	0	1	2	3	4	5	6	7	8	9
f	4	3	11	10	15	19	13	8	7	10

; 4.9, 2.39

Exercise 3d (page 44)

1. 21   **2.** (a) 0.9 (b) 1.00   **3.** 2   **4.** (a) 46 (47) (b) 21; 72 (c) 51

Exercise 3e (page 45)

1. (a) 15–19 (b) 20.5 (d) 8
2. 2 m 38 s, 1 m 55 s (or 1 m 56 s); 2 m 16 s (or 2 m 18 s); LQ: 1 m 24 s (or 1 m 25.5 s);
UQ: 2 m 56 s (or 2 m 59 s)   **3.** (a) 13.3 mm (b) 13.29(5) mm (c) 0.102 mm
4. 6.59; LQ: 4.00; UQ: 9.25
5. (a)

Age	1	2	3	4	5	6	7	8	9
f	4	8	12	15	19	14	11	10	7

(b) 5.15; 2.14 yrs
6. (a) 0.785 (b) 4.4(4) cm   **7.** 1.32 kg, 1.08 kg^2, 1.04 kg; 1.3 kg   **8.** 51; 12.4
9. (a) £16 130 (c) £11 905; £18 545   **10.** 17.3 %; £72.4, £16
11. 15 m 8.5 s, 0.64 m^2, 0.8 m (48 s); 15 m 7 s
12. A = − 1, B = 236, C = 1728, D = − 288, E = 8642; 49.04 yrs, 14.72 yrs; d = 12.37

Exercise 4a (page 54)

1. (a) 1/2 (b) 1/2   **2.** 7/50   **3.** 1/22   **4.** (a) 1/13 (b) 1/4 (c) 5/13
5. 19/250   **6.** 1/2; 1/2   **7.** 37/100   **8.** (a) 3/8 (b) 1/4 (c) 1/2
9. 2/9; 5/18; 7/12   **10.** 1/7

Exercise 4b (page 57)

1. (a) 11!/7! (b) 30!/(25!5!)   **2.** (a) 2520 (b) 360 (c) 45 360 (d) 302 400
3. 17 280   **4.** 118   **5.** 43 680   **6.** 19 958 400   **7.** 7
8. (a) $n(n-1)$ (b) $n+1$ (c) $(n+1)n$

Exercise 4c (page 59)

1. 4845   **2.** 120   **3.** 20   **4.** 350   **5.** 24 502 500 (4950^2)   **6.** n^2
7. 6   **8.** 5

Exercise 4d (page 60)

1. 1/143 (0.007)   **2.** (a) 1/170 (0.0059) (b) 143/340 (0.421)
3. 132/323 (0.409); 9372/15 504 (0.604)   **4.** (a) 14/2907 (0.0048) (b) 7/969 (0.0072)
5. 114/1771 (0.064)   **6.** 4367/4368   **7.** (a) 5/204 (b) 65/816
8. 3243/10 829 (0.299)   **9.** (a) 11/850 (b) 26/221

Exercise 4e (page 61)

1. 7/64   **2.** 3/5   **3.** (a) 39 916 800 (b) 34 560   **4.** 99
5. (a) $6! \times 2^6 = 46 080$ (b) $12! - 2^6 \times 6! = 478 955 520$ **6.** 2520, 36
7. (a) 360 (b) 1296; 360   **8.** 70   **9.** 42   **10.** (a) 126 (b) 60
11. (a) 1/2584 (0.00039) (b) 35/969 (0.036) (c) 105/646 (0.162) (d) 33/646 (0.051)
12. (a) 1/1428 (0.0007) (b) 75/442 (0.170)   **13.** 120; (a) 1/10 (b) 7/10   **14.** 9/10
15. (a) 2/5; 7/15 (b) 4/9   **16.** (a) 5/203 (b) 108/203   **17.** (a) 1/3 (b) 5/36
18. (a) 0.00264 (b) 0.0179

Exercise 5a (page 67)

1. (a) 1/13 (b) 1/4 (c) 4/13 (d) 2/13   **2.** (a) 1/2 (b) 1/5 (c) 3/5   **3.** 2/11
4. (a) 0.0130 (72/5525) (b) 0.0132 (73/5525)   **5.** (a) 0.719 (b) 0.281
6. (a) 0.993 (2982/3003) (b) 0.426 (1281/3003)   **8.** (a) 0.33 (b) 0.67 (c) 0.08
9. 0.17   **11.** (a) 1/6 (b) 13/18 (c) 1/9 (d) 7/9
12. (a) 1/6 (b) 2/3 (c) 1/18 (d) 7/9 (e) 1/3 (f) 1/9 (g) 7/18

Exercise 5b (page 72)

1. (a) 6/11 (b) 1/22 (c) 9/22; 7/44   **2.** (a) 1/221 (b) 1/1326

13/28

4. $2/5; 3/10; 1/5; 1/10$ 5. $6/13$ 6. $2/9; 16/81$ 7. $45/812$
8. $1/15; 1/2; (13/24)$ $1/2$ 9. $1/2$ 10. (a) $3/8$ (b) $2/5$ (c) $1/2$
11. (a) 0.02 (b) 0.45 12. (a) $1/9$ (b) $5/9$ (c) $1/3$ 13. $1/3$
14. (a) $2/5$ (b) $1/2$ 15. (a) $1/7$ (b) $2/7$

Exercise 5c (page 76)

1. $0.02; 0.17; 0.45; 0.36$ 2. $115/147$ 3. (a) $27/2197$ (0.0123) (b) $120/169$
4. $98/125$; (a) $27/500$ (b) $473/500$ (c) $27/125$ 5. (a) $1/27$ (b) $98/243$ (c) $28/343$
6. (b) $1/8$ (c) $1/2$ (d) $1/4$ 8. Yes   9. (a) 0.1 (b) 0.07 (c) 0.73 10. No
11. Yes   12. No

Exercise 5d (page 80)

1. $1/3$ 2. (a) $1/3; 1/4; 5/12$ (b) $1/9$ 3. 0.265 4. 0.025

Exercise 5e (page 80)

1. (a) $1/170$ (b) $5/272$ (c) $15/136$ 2. (a) 0.02 (b) 0.78 (c) 0.76 (d) $1/30$
3. 0.2 (a) 0.32 (b) 0.18 (c) $0.5; 0.64$ 4. (a) 0.0138 ($1080/5^7$) (b) $1/2; 288$
5. (a) (i) $1/6$ (ii) $5/6$ (b) (i) $1/6$ (ii) $5/6$ 6. $3/10; 0; 1/15; 1/20$; No
7. (a) $2/49$ (b) $20/49$ (c) $8/49$ (d) $5/49$ 8. $5/192$
9. (a) (i) 0.1 (ii) 0.1 (iii) 0.01 (b) (i) 0.06 (ii) 0.19 (iii) 0.02
(c) In (a) independent, in (b) not   10. (a) $2/7$ (b) $1/7$ (c) $1/35$ 11. A and B
12. (a) $\binom{13}{3}\binom{13}{4}\binom{26}{6} \Big/ \binom{52}{13}$ (b) 0.355 (c) 0.920
13. (a) $1/2$ (b) $11/36$ (c) $1/6$ (d) $23/36$ (e) $6/11$ 14. (a) 0.3 (b) 0.3 (c) 0.5
15. (a) $3/8$ (b) $7/36$ (c) $29/36$ 16. (a) 0.35 (b) $4/13$ 17. 0.624
19. 6.3×10^{-12} 20. 5.47×10^{-4} 21. 4.12×10^{-6}
22. (a) $1/21$ (b) $2/7$ (c) $4/9$ (d) $5/42$
23. 0.257; p_2 is not the probability of at least one of four exclusive events, the probability of one of which occurring is p_1.

Exercise 6a (page 90)

1. (a) 2.5 (b) -0.4 2. $161/36$ 3. $35/18$ 4. $13/3$ 5. $-8/9p$
6. Neither   7. $3p$ 8. (a) 5.5 (b) 2.6 9. (a) 8 (b) 2.5

Exercise 6b (page 94)

1. (i) $3; 5\frac{1}{2}$ (ii) $5; 11.4$ (iii) $-2.66; 4.15$
2. (i) $2; 1\frac{1}{2}$ (ii) $3; 2.4$ (iii) $-0.83; 3.46$ 3. $35/6$ 4. $20/9$ 5. $6; 16/3$
6. $3/4$

Exercise 6c (page 97)

1. (a) $154.5; 33.75$ (b) $50.5; 3.75$ (c) $70.33; 6.67$
2. (a) 4μ (b) $2\mu - 6$ (c) $\frac{9}{16}\sigma^2$ (d) $4\sigma^2$ (e) σ^2 3. $1.6; 10$
4. $3.75, 0.963; 23.75, 4.815$ 5. The novel

Exercise 6d (page 102)

1.
	X	
Y	$15/32$	$5/32$
	$9/32$	$3/32$

2. (a)

1	2	3
$5/32$	$18/32$	$9/32$

(b)

0	-1	-2
$15/32$	$14/32$	$3/32$

(c)

0	1	2
$8/32$	$15/32$	$9/32$

3. (a) $3/4$ (b) $11/8$ (c) $3/16$ (d) $15/64$ (e) $17/8$ (f) $-5/8$ (g) $27/64$ (h) $27/64$
4. (a) 1.6 (b) 1.7 (c) 3.3 (d) 0.85
5.

X	-1	0	1
p	0.2	0.5	0.3

Y	0	1	2
p	0.1	0.7	0.2

(a) 0.1 (b) 1.1 (c) 0.11

6. (a) 1.0 (b) 1.0   (c) 0.4   (d) 0.4   (e) 2.0   (f) 1.2

7. Y
$$\begin{array}{|cccc}
\multicolumn{4}{c}{X} \\
\hline
1/64 & 3/32 & 3/16 & 1/8 \\
3/64 & 3/16 & 3/16 & 0 \\
3/64 & 3/32 & 0 & 0 \\
1/64 & 0 & 0 & 0
\end{array}$$
(a) $1\frac{1}{2}$ (b) 3/4   (c) $2\frac{1}{4}$ (d) 3/4   (e) 9/16   (f) 9/16

Exercise 6e (page 105)

1. (a) 103   (b) 14.5   (c) 7   (d) 14.5   (e) 213   (f) 81.7   (g) -34 (h) 88.5   (i) 131
(j) 18.8   (k) 63   (l) 31.7   **2.** 7/2, 35/12; 21/2, 35/4   **3.** 56.5, 2.92; 58, 3.48
4. $3; -4$ **5.** $\mu, \sigma^2/5; \mu, \sigma^2/n$

Exercise 6f (page 106)

1. (a) 7.94; 7.89   (b) 4/9   **2.** (a)
$$\begin{array}{c|cccccc}
X & 0 & 1 & 2 & 3 & 4 & 5 \\
\hline
p & 3/18 & 5/18 & 4/18 & 3/18 & 2/18 & 1/18
\end{array}$$
(b) 1.94; 2.05

3.
$$\begin{array}{c|cccc}
X & 1 & 2 & 3 & 4 \\
\hline
p & 5/8 & 15/56 & 5/56 & 1/56
\end{array}$$
$1\frac{1}{2}; 15/28$ **4.** 1/6; 1/3; 1/2   **5.** 6.2; 3.78

6. 7/8, 31/56; 7/8, 2/7, 1/4; No   **7.** (a) 6; 2   (b) 0; 2

$$\begin{array}{c|ccccccccc}
XY & 1 & 2 & 3 & 4 & 6 & 8 & 9 & 12 & 16 \\
\hline
p & 0.01 & 0.04 & 0.06 & 0.12 & 0.12 & 0.16 & 0.09 & 0.24 & 0.16
\end{array}$$
$E(XY) = 9$

8. 0, 1/2;
$$\begin{array}{ccccc}
-2 & -1 & 0 & 1 & 2 \\
\hline
1/16 & 1/4 & 3/8 & 1/4 & 1/16
\end{array}$$
$0, 1$; same as $X_1 - X_2$

9. (b) 3/2, 1; 0; 55/128

10. (a)
$$\begin{array}{c|ccccccccccc}
X & 2 & 3 & 4 & 5 & 6 & 7 & 8 & 9 & 10 & 11 & 12 \\
\hline
p & 1/36 & 2/36 & 3/36 & 4/36 & 5/36 & 6/36 & 5/36 & 4/36 & 3/36 & 2/36 & 1/36
\end{array}$$
(b)
$$\begin{array}{c|ccccccccc}
Y & 1 & 2 & 3 & 4 & 5 & 6 & 8 & 9 & 10 \\
\hline
p & 1/36 & 2/36 & 2/36 & 3/36 & 2/36 & 4/36 & 2/36 & 1/36 & 2/36
\end{array}$$
$$\begin{array}{c|ccccccccc}
Y & 12 & 15 & 16 & 18 & 20 & 24 & 25 & 30 & 36 \\
\hline
p & 4/36 & 2/36 & 1/36 & 2/36 & 2/36 & 2/36 & 1/36 & 2/36 & 1/36
\end{array}$$
(c) 1/3   (d) No: $P(Y = 12) \neq P(Y = 12 \mid X = 7)$
11. 5/3, 5/9; 7/3, 5/9   **12.** (a) 3/4   (b) 1/4096   **13.** $1 - 2A(1 - 2^{-M})$
14. 4/5; 1/5   **15.** (b) $1/p$; $-\pounds0.16$; $\pounds1.95$

Exercise 7a (page 115)

1. $1/2 + 3t/8 + t^2/8$; 5/8; 31/64   **2.** 3.0; 2.4   **3.** $3\frac{1}{2}$; 35/12   **4.** 3/2; 3/4
5. $3t/5 + 3t^2/10 + t^3/10$; $1\frac{1}{2}$, 9/20; $3t/(5 - 2t)$ **6.** $t/2 + t^2/4 + t^3/8 + \ldots$

Exercise 7b (page 118)

1. $(0.6 + 0.4t)^4$; 1.6; 0.96   **2.** $(t + t^2 + t^3 + t^4 + t^5 + t^6)^2/36$; 7
3. $(1 + t + 2t^2 + 2t^3)/6$; 11/6; 41/36   **4.** $(1/2 + t/6 + t^3/6 + t^5/6)^3$; 9/2; 3n/2
5. 19/4; 15/16

Exercise 7c (page 120)

1. $(t + t^2 + t^3 + t^4 + t^5 + t^6 + t^7 + t^8 + t^9 + 4t^{10})/13$; 85/13   **3.** $na(1 - a)$
5. $(4/5)^{r-1}(1/5)$; (a) $t/(5 - 4t)$ (b) 5   (c) 20   (d) $(4/5)^5$ (e) 4
6. $(1 + t)/2$; $(2 + t)/3$; 11/6; 17/36   **7.** (a) 1/6   (b) 1/6   (c) 2/3; 10/3
8. 21/2, 113/12; 5/48   **10.** $(t + t^{-1})/256$; 119/128   **11.** $2^{-n}(t^2 + t^{-1})^n$; n/2; 9n/4
12. $(1 - t^6)^2/[36t^5(1 - t)^2]$

Exercise 8a (page 127)

1. (a) 0.033 (b) 0.297 (c) 0.995   **2.** (a) 0.614 (b) 0.386   **3.** (a) 0.078 (b) 0.096
4. (b)   **5.** 1/6; 0.0032   **6.** (a) 0.114 (b) 0.886   **8.** 0.494
9. (a) 6 (b) 0.913   **10.** (a) 0.323 (b) £1.49   **11.** (a) 0.942 (b) 542
12. 11/16; (a) 0.0676 (b) 0.9991   **13.** 0.895; 0.277
14. (a) $45p^2(1-p)^8$ (b) $1 - (1-p)^{10} - 10p(1-p)^9 [= P]$; $10P(1-P)^9$

Exercise 8b (page 132)

1. 243/1024, 405/1024, 270/1024, 90/1024, 15/1024, 1/1024; 5/4; 15/16   **2.** 50; 0.3
3. 5, 2.5; 0.377   **4.** 8; 2.19; (a) 0.126 (b) 0.016   **5.** 1; 63.3, 84.4, 42.2, 9.4, 0.8
6. 1.0, 7.7, 23.0, 34.6, 25.9, 7.8
7. 0.2; 10.7, 26.8, 30.2, 20.1, 8.8, 2.6, 0.5, 0.1, 0.0, 0.0, 0.0; 0.322
8. 16.5, 42.4, 45.4, 25.9, 8.3, 1.4, 0.1   **10.** 3   **11.** 5/3; 10/9

Exercise 8c (page 137)

1. 0.073   **2.** (a) 0.116 (b) 0.039; 6   **3.** $1/9$; 6.8×10^{-5} **4.** 25; 0.625
5. 4   **7.** (a) 0.0643 (b) 0.0651   **8.** 50; 0.011

Exercise 8d (page 138)

1. (a) 0.073 (b) 0.053 (c) 0.270   **2.** 0.514   **3.** (a) 0.017 (b) 0.299 (c) 0.020
4. (a) 0.678 (b) 0.322; 1/15   **5.** (a) 5.30×10^{-11} (b) $1 - 1.18 \times 10^{-9}$
6. (a) 5 (b) 0.672   **7.** (a) 0.130 (b) 0.035; 0.270
8. 237.9, 374.9, 246.2, 86.2, 17.0, 1.8, 0.1
9. 2; 6, 1/3; 17.56, 52.67, 65.84, 43.90, 16.46, 3.29, 0.27   **10.** 0.046; 4; 2.4
11. (a) 243, 405, 270, 90, 15, 1 (b) 1.25, 0.968 (c) 1/64 (d) 2.5; 2
12. $(n-k+1)p/[k(1-p)]$; r greatest integer m such that $m \leqslant p(n+1)$
13. 0.081; $1 - (35/36)^n$ **14.** (a) 0.032 (b) 0.035 (c) 0.0013 (d) 0.04
15. 0.322; 50p; 50; 21   **16.** (a) *Method A*: $(1-p)^{10} + 10p(1-p)^9 + 45p^2(1-p)^8$
Method B: $(1-p)^5 + 5p(1-p)^9 + 25p^2(1-p)^8$

	$p = 0.2$	$p = 0.5$
(b) P(accept) A	0.678	0.055
P(accept) B	0.630	0.061

(c) Use *A*.

17. 2   **18.** 0.04; 0.022; £761
19. $P(\text{number} = 0, 1, 2, 3) = \frac{1}{3}q^2(1+2q), \frac{2}{3}pq(1+3q), \frac{1}{3}p^2(1+6q), \frac{2}{3}p^3$; mean $8p/3$
20. $n^2 - 4n(n-1)p(1-p)$ **21.** $(1-p)^9[1 + 9p(1-p)^8]$; 12.49

Exercise 9a (page 149)

1. (a) 0.082 (b) 0.205 (c) 0.257 (d) 0.544 (e) 0.891   **3.** 0.161; 0.163
4. (a) 0.204 (b) 0.027 (c) 0.0025   **5.** 0.054   **6.** 0.135, 0.271, 0.271, 0.180, 0.090
7. (a) 1.61 (b) 0.056   **8.** 4/3   **9.** 2   **10.** (b) 0.487
11. (a) 0.0003 (b) 0.011   **12.** (a) 0.089 (b) 0.938; 0.020

Exercise 9b (page 152)

1. (a) 0.223 (b) 0.809 (c) 0.050 (d) 0.423   **2.** (a) 0.082 (b) 0.257 (c) 0.084
3. 0.918   **4.** 0.122   **5.** 1.45; (a) 0.060 (b) 0.330
6. 202.2, 161.8, 64.7, 17.2, 3.5, 0.6   **7.** 108.6, 66.3, 20.2, 4.1, 0.6   **8.** 8

Exercise 9c (page 156)

1. mean = 1.4(4); variance = 1.3(9)   **2.** (a) 0.197 (b) 0.377
3. (a) 0.062 (b) 0.642; 1.67×10^{-5}
4. (a) 0.100 (b) 0.050 (c) 0.168. Most likely results: 1–0, 2–0, 1–1, 2–1 (*A* first).

Exercise 9d (page 157)

1. 0.833   **2.** 0.371; £60.37   **3.** (a) 0.5; 121.3, 60.7, 15.2, 2.5, 0.3, 0   (b) 0.93
4. (a) 0.082   (b) 0.242   **5.** (a) 0.983   (b) 0.184   (c) 0.199   **6.** 0.5; 0.910
7. (a) 0.300   (b) 0.301; 0.343; 0.70   **8.** (a) 0.419   (b) $1 - 5.27 \times 10^{-7}$

9. (a) 2.22   (b)

0.155	0.266	0.255	0.165	0.100	0.040	0.010	0.010
0.135	0.271	0.271	0.180	0.090	0.036	0.012	0.003

; 0.195

10. (a) $(1 - e^{-3})^{10}$ (b) $\binom{10}{8} e^{-6}(1 - e^{-3})^{8}$ **11.** 0.045   **12.** 3; 0.223; 0.988
13. $r \times P(Y = r) = P(Y = r - 1)$; (a) 0.05   (b) 0.162   (c) 0.60
14. (a) 0.191   (b) 0.018   (c) 0.264; 2   **15.** (a) 0.368   (b) 0.053   (c) 4
16. (a) 0.601   (b) 0.908; 0.775   **17.** $e^{-a}a^{t}/t!$ where $a = 0.4(s + c)$; 345
18. (a) 0.135   (b) 0.271   (c) 0.029   (d) 2   (e) 8.19

19. (a) $e^{-\mu}\mu^{r+k}\binom{r+k}{r}p^{r}(1-p)^{k}/(r+k)!$; Poisson with mean μp (b) 0.050   (c) 0.916

21. $B[n, \alpha/(\alpha + \beta)]$; (a) 0.0067   (b) 0.173   (c) 0.166
22. (a) $E(X) = \lambda$; $\mathrm{Var}(X) = \lambda$ (b) symmetrical and bell-shaped
24. $e^{-\lambda}\lambda^{x}/x![1 - e^{-\lambda}(1 + \lambda)]$; $x = 2, 3, 4, \ldots$.

Exercise 10a (page 170)

1. (a) 1/4   (b) 1/2   **2.** 1/64; 81/256   **3.** 1; 2/3   **4.** 0.156
5. (a) 2×10^{-4} (b) 0.263   (c) 0.191   **6.** 1/6; 13/24   **7.** $\pi/2$; 3/4   **8.** 0.68
9. (a) 11/16   (b) 29/40   **10.** 3/10; 3   **11.** 5/6; 1/3

Exercise 10b (page 176)

1. 1; 1/5; 1; 1   **2.** 12; 2/5; 1/25; 1/3   **3.** 1/4; 8/5; $8^{\frac{1}{4}}$; 8/75; 0.697   **4.** 6; 12/5
5. 4/27; 9/5; 0.475   **6.** (a) 12   (b) 0.6; 0.04   (c) 0.475   **7.** $(a+b)/2$; $(b-a)^2/12$
8. (a) 3/2   (b) 1.648; 0.284   (c) 2/5   (d) 6/5; 2
9. $C_1 = 3$, $\mu_A = 0.25$, $\sigma_A = 0.1936$; $C_2 = 12$, $\mu_B = 0.6$, $\sigma_B = 0.2$; (a) 0.125   (b) 0.3125
10. (a) 3/16   (c) 2   **11.** $2/n^2$; $n/3$; $n^2/18$; 7.76   **12.** 2; 0.124
13. (a) 2/5   (b) 2.6   (c) 1.5   **14.** 0.6; 0.2; 0.166   **15.** (a) $-3/16$ (b) 19/80   (d) 2

Exercise 10c (page 181)

1. (a) $x/20$ (b) $(x+3)/4$ **2.** $(3x^2 - x^3)/4$
3. (a) 3/52; $(x^{3/2} - 1)/26$ (b) 1/2; $\frac{1}{2}(\sin x + 1)$ (c) 2; $1 - e^{-2x}$
4. $x^2/2$ $(0 \leqslant x < 1)$, $-1 + 2x - x^2/2$ $(1 \leqslant x \leqslant 2)$ **5.** $(x^3 + 3x + 4)/8$
6. $3x^2/2 - x^3$ $(0 \leqslant x < 1)$, $1 - 1/x$ $(x \geqslant 1)$; 1/2, 1/8
7. (a) 1/4   (b) $2e^{2x}$ (c) $3x^2/8$ (d) $(x - x^2)^{-\frac{1}{2}}/\pi$

Exercise 10d (page 183)

1. 3/2; 25/12; (a) 4/5   (b) $1/\sqrt{3}$ **3.** 1/2   **4.** (a) 3/5   (b) 11/15
5. x, $(0 \leqslant x < 1)$, $2 - x$, $(1 \leqslant x \leqslant 2)$; (a) 1/6   (b) $2 - \sqrt{2}$
6. (a) 1.67×10^{-3} (b) $50(x + 0.1)^2$, $(-0.1 \leqslant x < 0)$, $1 - 50(0.1 - x)^2$, $(0 \leqslant x \leqslant 0.1)$
(c) 0.045   **7.** $(x + a)^2/2a^2$, $(-a \leqslant x < 0)$, $1 - (a - x)^2/2a^2$, $(0 < x \leqslant a)$; 5/32

Exercise 10e (page 186)

1. (a) 4; 16   (b) 0.393; 0.063   (c) 1.15   **2.** 1/5; 0.132
3. (a) 0.135   (b) 0.078   (c) 0.905   **4.** 0.0253   **5.** e^{-k}; 0.693   **6.** 0.223

Exercise 10f (page 188)

1. (a) 0.4   (b) 19/15; 1.25   (d) 0.207   **2.** 4/5; 4/25; £8   **3.** 12; 3/5; 2/3; 0.225
4. (a) 2/75   (b) 70/9   (c) 12/25; Money bond
5. 1/2; $\pi/2$; $\pi^2/4 - 2$; $\pi/2$; $\pi/3$; $2\pi/3$; 0.632

6. $1/8; 1.5; 0.15; 1.59$ **7.** (b) $\sqrt{3/5}$

8. (a) $f(x) = \frac{1}{4}, (1 \leqslant x < 3), \frac{1}{2}, (3 \leqslant x \leqslant 4)$ (c) $\sqrt{37/48}$ (d) $5/16$ **9.** $2; 3/4; 1/5$

10. $8/15; 11/225; 1/\sqrt{2}$ **11.** $-1; 2.08;$ (a) $1/4$ (b) $3/4; 37/64$ **12.** (a) $1.2; 3$

13. (a) 0.689 (b) 0.525 **14.** (a) 0.62 (b) 0.38 **15.** $0.135; 0.171$

16. $\lambda; 100; 10\,000; 0.607; 0.368$ **17.** $1500; 1.8 \times 10^5; 0.159$

18. $k;$ (a) 0.232 (b) $4.3; 19$ (c) 0.103 **19.** $1/\alpha; 1/\alpha;$ process A if $C > 0.11k$

20. $(1-k)(1+t)/[1+(1-k)t]; (1-k)(1+2t)/[1+(2-k)t]$

21. (b) $\theta/(\theta+3); \theta(\theta+1)/[(\theta+3)(\theta+4)]$ (c) $3\theta/[(\theta+3)^2(\theta+4)]$ (d) $1/5$

23. $20^{-\frac{1}{2}}, x/(2-20^{-\frac{1}{2}}) \leqslant \lambda \leqslant 20^{\frac{1}{2}}x$

24. $\lambda\theta\,e^{-\lambda(x-2)}; \theta(2+1/\lambda); 2\theta(2+2/\lambda+1/\lambda^2) - \theta^2(2+1/\lambda)^2$

25. $1/(m-n); \dfrac{mn}{m-n}(x^{n-1}-x^{m-1}); mn/(m+1)(n+1); (1-1/\sqrt{2})^{1/n}; 1+1/\sqrt{2}$

26. $\lambda^2; e^{-2}[(3-2\sqrt{2})e^{2\sqrt{2}} - (3+2\sqrt{2})e^{-2\sqrt{2}}]$

27. $(\beta+1)(\beta+2); 2/(\beta+3); 2(\beta+1)/[(\beta+3)^2(\beta+4)]; 1-(1-x)^{\beta+1}[1+x(\beta+1)]$

28. (a) $3/4a; 9/80a^2$ (b) $10; 60$

29. $1-e^{-kx}(kx+1), (0 \leqslant x \leqslant L), 1-e^{-kx}(kL+1)$ (a) $(x > L)$ (b) $1/k, L$

Exercise 11a (page 201)

1. (a) 0.947 (b) 0.826 (c) 0.071 (d) 0.359 (e) 0.466 (f) 0.136 (g) 0.910 (h) 0.972 (i) 0.653

2. (a) 0.92 (b) 0.33 (c) -0.05 (d) -1.08

3. (a) 1.55 (b) 2.05 (c) -0.13 (d) -1.64 (e) -1.04 **4.** $\pm 1.64(5)$

Exercise 11b (page 202)

1. (a) 0.952 (b) 0.010 (c) 0.952 (d) 0.657 **2.** (a) 0.067 (b) 0.974

3. (a) 0.258 (b) 0.137 (c) 0.183 **4.** (a) 28.0% (b) 17.1% **5.** 24 **6.** 242.3

7. $1.56\,\text{lb}$ **8.** (a) 5% (b) $103.26\,\text{mm}$ **9.** $104.2; 38.9$ **10.** $0.51\,\text{g}; 56.32\,\text{g}$

11. 0.709 **12.** $53\,\text{m}; 10\,\text{m}; 0646$ **13.** 781 gallons; $0.788; 0.042$

14. (a) 0.076 (b) 0.847 (c) $0.676; 1.313\,\text{m}$

Exercise 11c (page 205)

1. (a) 0.933 (b) 0.885 **2.** (a) 0.291 (b) $0.013; 0.789$ **3.** 0.522 **4.** 0.247

5. 15.9% **6.** (a) 0.674 (b) 0.139 **7.** (a) $N(195, 36)$ (b) $N(125, 36)$

8. 0.996 **9.** $23.2\,\text{h}; 0.89\,\text{h}$ (a) 9% (b) 18%

10. (a) $N(24, 81)$ (b) $N(16, 64)$ (c) $N(61, 72)$ (d) $N(2, 160)$ (e) $N(11.5, 3.25)$

11. 0.900

Exercise 11d (page 209)

1. (a) 0.029 (b) 0.018 (c) 0.011 **2.** (a) 0.932 (b) 0.912 **3.** (a) 0.922 (b) 0.028

4. 0.088 **5.** (a) 0.916 (b) 0.870 **6.** (a) 0 (b) 0 (c) 0.743

7. $0.000, 0.001, 0.015, 0.073, 0.161;$ No   **8.** $16/37, 0.323$ **9.** 24

10. (a) 0.794 (b) 0.588 **11.** 46

12.   Each airline should put on 2583 or more seats; (a) 0.010 (b) 0.075

Exercise 11e (page 213)

1. $0.4, 2.5, 9.8, 21.8, 28.9, 22.7, 10.6, 2.9, 0.5$ **2.** $60.8; 13.5$ (approx.)

3. $3.5, 10.2, 20.2, 28.4, 26.9, 18.1, 8.3, 2.6$

4. $50.3\,\text{mm}; 2.65\,\text{mm}; 2.1, 19.7, 57.6, 53.7, 15.4, 1.5$

Exercise 11f (page 216)

1. (a) 25.8% (b) 13.7% (c) 18.3% **2.** $9.4\,\text{cm}; 0.67$

3. (a) 0.076 (b) $0.223; 0.432$ **4.** (a) $113.3; 14.2$ (b) 50.7 **5.** 0.37

6. (a) $1097; 844$ (b) 63.5 (c) 715 **7.** $0.075\,\text{mm}$ **8.** $1166; 1050$

9. (a) 0.994 (b) 4730 (c) 0.106 **10.** (a) $0.960; 0.788$ (b) 0.857 (c) 99%

11. 0.567, 0.116, 0.159, 0.159; 12.7p   **12.** (a) 0.023   (b) 0.319
13. (a) 0.029   (b) 0.332; 0.242   **14.** 0.023; 0.851
15. (a) 0.245   (b) 0.659   (c) 0.409   **16.** 0.246; 0.080; 0.018
17. (a) 0.649   (b) 0.965   **18.** (a) 0.315   (b) 0.307; Worse; 0.524
19. (a) $5(0.6)^4(0.4)$ (b) $100(0.6)^{99}(0.4)$; 0.982   **20.** 0.095; 0.0004; 271   **21.** 0.176
22. 69.95; 1.26; 142; 176; 177
23. (a) 0.0125, 0.1, 0.33, 0.54, 0.86   (c) 64.3, 11 (approx.)   (d) 78.4
24. (a) 0.866   (b) $(225 - 4k)$ cm; 36.25 cm^2 (d) 0.866
25. (a) 48.37   (b) 0.888   (c) 10, 41; 0.94
26. 0.082; $W = 60 - X - Z$; 10; 16; 0.006
27. (a) 0.159   (b) 0.317; 1; 3; 3; 5   (c) 0.274   (d) 0.549
28. (a) 0.824   (b) 0.059; 1/4; 1/6   **29.** 0.65; 0.4 (approx.)
30. 8.70; 0.91; 9.30; 8.10; 0.076 (approx.)   **31.** 0; 3; 4/5; 49/25
32. $\exp\{-\frac{1}{2}[(x-3)/1.5]^2\} \div [1.5\sqrt{2\pi}\,\Phi(2)]$

Exercise 12a (page 232)

1. 100; 10/3; $\sqrt{10/3}$ **2.** 25   **3.** 5/2; 3/4; 5/2; 3/8   **4.** 7/2; 35/12; 7/2; 35/24
5. 5/2; 1/4

Exercise 12b (page 236)

1. 43; 138.9   **2.** 2.92; 0.030   **3.** 2.185; 0.946   **4.** $100/n$ **5.** 5.9; 8.1

Exercise 12c (page 240)

1. (b)   **2.** The first

Exercise 12d (page 245)

1. 0.067; 0.023   **2.** 0.131   **3.** 0.789   **4.** 0.669   **5.** 0.930   **6.** 0.007
7. 0.915   **8.** (a) 60; 1.067   (b) 20; 1.067   **9.** 0.674
10. (a) $N(90, 8/9)$ (b) $N(10, 8/9)$ (c) $N(220, 52/9)$
11. 4 (or 5)   **12.** 0.006; 0.933   **13.** $1/3; \sqrt{2/63}$; 0.1875 (3/16); 0.937

Exercise 12e (page 246)

1. $1/2n$ **2.** (a) 0.5   (b) 0.159   **3.** $n > 96.04$ or $n = 97$
4. 5/2; 5/4 (a) 5/2; 5/12   (b) 5/32; 11/32; 11/32; 7/8   **5.** 0.364; 2.05; 0.031
7. (a) 0.843   (b) 0.933   (c) 0.308 or 0.309   **8.** 5; 0.98
9. $\bar{X} \simeq N(\mu, \sigma^2/n); \Sigma(X_i - \bar{X})^2/(n-1); \hat{C} = 2\pi\bar{R}; \hat{C}$ is consistent for $C = 2\pi r$, by considering
Var$(\hat{C})$. $E(T) = \pi r^2 + \pi\sigma^2/n$, biased; consider $T' = T - \pi S^2/n$, $S^2 = \Sigma(R_i - \bar{R})^2/(n-1)$
10. 2; 0.7; 0.65. Takes into account the deviations within the samples.
11. $E(X) = \theta/(\theta+1)$; Var$(X) = \theta/[(\theta+2)(\theta+1)^2]$; $E(\bar{X}) = \theta/(\theta+1)$;
Var$(\bar{X}) = \theta/[n(\theta+2)(\theta+1)^2]$; $b(\theta) = -\theta^2/(\theta+1)$ **12.** $\sigma_1^2/n_1 + \sigma_2^2/n_2$; 30; 40

Exercise 13a (page 253)

1. $z = 2.11$, sig.   **2.** $z = -1.33$, not sig.
3. $z = 1.77$, (a) not sig.   (b) sig.   **4.** $z = 2.78$, sig.   **5.** $\bar{x} = 103.9$, $z = 1.78$, not sig.
6. $z = -2.5$, (a) sig.   (b) sig.   **7.** $z = 2.19$, (a) not sig.   (b) sig.
8. $z = 3.67$, highly sig.

Exercise 13b (page 256)

1. $z = 1.77$, not sig.   **2.** $z = 4.08$, highly sig.   **3.** $z = 2.12$, sig.
4. $\bar{x}_1 = 19.77$, $\bar{x}_2 = 20.086$, $z = 2.21$, not sig.

Exercise 13c (page 259)

1. $z = 2.49$, justified   **2.** $z = 2.42$, sig. evidence

3. (a) $z = -2.77$, reject H_0 at 1%   (b) $z = 3.10$, reject H_0 at 1%
4. (a) $z = 2.28$, sig.   (b) $z = 1.46$, not sig.; not sig.
5. 3.97; 3.51; 3.56; (a) $z = 2.08$, reject H_0 at 5%   (b) $z = -2.35$, reject H_0 at 5%

Exercise 13d (page 262)

1. $z = 2.07$, reject H_0 **2.** $z = 3.87$, sig. evidence   **3.** $z = 1.49$, not sig.

Exercise 13e (page 266)

1. (a) $(138.6, 145.4)$ (b) $(139.1, 144.9)$ **2.** $(25.5, 26.5)$ **3.** $(71.0, 76.2)$
4. $(74.2, 74.8)$ **5.** 26.1; 7.66; $(25.0, 27.2)$ **6.** $(90.5, 92.2)$

Exercise 13f (page 268)

1. (a) $(0.07, 1.93)$ (b) $(-0.32, 2.32)$ **2.** $(1.3, 6.7)$; $(1.2, 6.8)$

Exercise 13g (page 270)

1. $z = 2.4$, sig. at 5%   **2.** $(0.973, 0.987)$ **3.** 53.4; 413.3; $(47.9, 59.1)$
4. 48.75; 1.91; $(48.5, 49.0)$; $z = 2.6$, reject H_0 **5.** $z = 2.74$, reject H_0
6. $z = 2.09$, (a) sig.   (b) not sig.   **7.** $(2.6, 4.2)$; $z = 1.83$, accept H_0
8. $z = 1.48$, accept H_0 **9.** $602\,g$; $0.65\,g$; $15\,050\,g$; $3.25\,g$; $z = 1.66$, (a) not sig.   (b) sig.
10. $z = 2.19$, sig.   **11.** $(2.60, 3.12)$; $z = 2.95$, sig.
12. $(2.32, 4.68)$; $(3.43, 6.57)$; (a) $z = 2.5$ (b) $z = 1.5$ **13.** $z = 3.80$, highly sig.
14. (a) 0.029 (b) 0.193 (c) 0.038; $(100.3, 100.7)$
15. 27.33; $(26.74, 27.86)$; 2.4; $z = 1.96$, reject the theory
16. $z = 1.81$, accept claim; $z = 1.6$, reject claim
17. $(0.03, 1.07)$; (a) $z = 0.29$, accept H_0 (b) $z = 2.08$, reject H_0
18. 62; (a) 0.682 (b) 0.01 **19.** $\mu_1 - \mu_2$, $\sigma_1^2/2n + \sigma_2^2/n$; 0.83; 65
20. $(m' - 0.219\sigma, m' + 0.219\sigma)$ **21.** $(188, 192)$; $(13.0, 17.0)$

Exercise 14a (page 280)

1. $t = 3$, reject H_0 **2.** (a) $t = -0.88$, accept H_0 (b) $t = -2.65$, reject H_0
3. $t = -1.51$, no sig. evidence   **4.** $t = -3.85$, (a) yes   (b) yes
5. $\bar{x} = 42.68$; $t = -1.02$, he buys (mean breaking force not sig. less than 49)
6. $t = -2.83$, sig.

Exercise 14b (page 283)

1. $t = 2.45$, evidence sig. at 5%, not at 1%
2. $t = 2.41$, reject H_0 at 5%, accept at 1% (two-tail test)   **3.** $t = 3.86$, sig. greater
4. $t = 1.91$, no sig. difference   **5.** $t = 1.50$, accept H_0
6. $t = -0.76$, accept H_0 (no reduction in weight) at 5%

Exercise 14c (page 286)

1. $(67.1, 98.9)$ **2.** $(9.16, 11.22)$ **3.** $(1.125, 1.129)$ **4.** $(8.94, 43.34)$
5. $(-2.74, 5.96)$

Exercise 14d (page 288)

1. $z = 2.4$, sig.   **2.** $z = 1.16$, not sig.
3. (a) $(14.12, 14.28)$ (b) $z = 1.71$, consistent at 5%   **4.** 0.006; $z = -2.74$, sig.
5. (a) (i) $(42, 50)$ (ii) $(35.2, 50.8)$ (iii) $(41.1, 48.5)$ (b) $z = 2.44$, sig.
6. 28.375; 8.984; (a) $z = 2.17$, reject H_0 (b) $t = 2.10$, accept H_0; (b) is preferable; $(26.73, 30.02)$ **7.** $(50.06, 54.58)$
8. $t = -2.80$, radioactivity sig. less for weathered rocks at 1% level; $(4.55, 50.89)$
9. (a) 0.683 (b) $2.92 < \bar{x} < 3.08$ **10.** $t = 3.10$, sig. at 5%, not sig. at 1%
11. 64; 452.84 **12.** $(0.973, 0.987)$ **13.** $(-0.03, 2.97)$; $z = 1.92$, not sig.

14. $t = 2.69$, sig. at 5%, not sig. at 1%
15. (a) $z = 14.6$, highly sig.   (b) $t = 3.22$, sig. at 5%
16. $t = 0.026$, not sig.; (1.617, 1.798)
17. $t = 2.54$, sig. at 5%, not sig. at 1%; (0.010, 0.176)   **18.** $t = -3.29$, reject H_0
19. (a) (i) interval narrows   (ii) interval widens   (iii) interval widens   (b) (47.6, 51.4)
20. (6.04, 9.96); $z = 8$, highly sig.; $(-0.79, 3.83)$; $z = 1.29$, not sig.
21. 110; 90; 330; $t = 1.2$, sig. evidence   **22.**   0.03; (9.22, 9.58); 0.208; (9.21, 9.59)

Exercise 15a (page 301)

1. Accept H_0 **2.** Not sig.   **3.** Sig. at 1%, not at 5%   **4.** Not sig. at 5%
5. $z = -2.45$, sig. at 5%   **6.** $z = 1.17$, not sig. at 5%   **7.** $z = 2.39$, sig. at 5%
8. (a) First store $z = 2.18$, unlikely; second store $z = 0.59$, could be   (b) $z = 0.97$
9. $z = 1.92$, not sig. at 5%; $z = -2.60$, sig. at 5%   **10.** (0.22, 0.44)
11. (0.12, 0.28)   **12.** (0.615, 0.825); 7/9; $z = 2.25$, sig. difference   **13.** (0.07, 0.15)
14. (a) $z = -8.98$, highly sig.   (b) $z = 2.5$, sig. at 5%   (c) $z = 1.07$, not sig. at 5%

Exercise 15b (page 305)

1. (a) $X \leqslant 1$ (or $X \geqslant 7$)   (b) $X = 0$ and $X = 8$
2. (a) $X \leqslant 2$ (or $X \geqslant 9$)   (b) $X \leqslant 1$ (or $X \geqslant 10$)   **3.** $X = 7$, accept H_0
4. $X = 5$, not sig.   **5.** $X = 6$, no sig. evidence   **6.** $z = 1.87$, accept H_0
7. $z = 2.01$, reject H_0 **8.** $X = 2$, accept H_0 **9.** $z = 2.41$, not sig.

Exercise 15c (page 309)

1. $z = -0.83$, accept H_0 at 5%   **2.** $z = -1.77$, accept H_0 at 5%
3. $T = 13.5$, accept H_0 **4.** $T = 9$, accept H_0 **5.** $T = 3.5$, accept H_0

Exercise 15d (page 312)

1. 1/35, 1/35, 2/35, 4/35; (a) $W = 6$ (b) $W \leqslant 7$
2. $W = 7$, accept claim (critical region $W \leqslant 7$)   **3.** $W = 87.5$, $z = 1.02$, accept H_0
4. $W = 98.5$, $z = 0.49$, no evidence for difference

Exercise 15e (page 312)

1. $z = 1.76$, not sig. different at 5%   **2.** $z = 1.04$; (a) No   (b) Not sig. lower
3. $z = 8.9$; 0.9 is an understatement of the law
4. (a) (0.325, 0.407)   (b) (0.317, 0.415)   **5.** $z = 1.6$, sig. evidence; (0.094, 0.110)
6. $z = -3.5$, not reasonable; (0.23, 0.41)   **7.** $z = 1.48$, accept H_0; 0.56
8. $z = -1.93$, tends to confirm the claim, at 5%   **9.** $X = 6$, accept H_0
10. $X = 2$, reject H_0
11. (a) $X \geqslant 10$ and $X \leqslant 2$ (b) $X \geqslant 10$ (or $X \leqslant 2$); $X = 9$, no sig. evidence
12. $z = 3.2$, yes   **13.** $T = 12.5$, reject H_0 **14.** $T = 13$, no sig. difference
15. $T = 116$, $z = -6.6$, yes   **16.** $W = 18$, no sig. difference

Exercise 16a (page 324)

1. (a) $\alpha = 0.159$; $\beta = 0.067$ (b) 1.64; 0.740   (c) 87; 0.5
2. (a) $z = 1.36$, accept H_0 (b) 0.239   **3.** 0.055; 0.7
4. 0.053; Accept $p = 0.01$ if fewer than 5 faulty; 0.28   **5.** 0.348; 0.234
6. (a) $p^4 + (1-p)^4$ (b) $7p^6(1-p) + p^7 + 7(1-p)^6 p + (1-p)^7$; Procedure 2
7. 0.014; 0.238
8. (a) 0.264   (b) 0.092; $P(\text{reject } H_B \mid H_B \text{ true}) = 0.003$ if $n = 2$;
$P(\text{reject } H_B \mid H_B \text{ true}) = 0.00008$ if $n = 3$; $P(\text{reject } H_A \mid H_A \text{ true}) = 0.80$
9. 7/2, 35/12; 3.9, 2.49; 0.121; 594   **10.** 204.7; 0.004   **11.** (a) $30 - 3.29/\sqrt{n}$ (b) 9
12. (0.1, 0.806), (0.2, 0.386), (0.3, 0.125), (0.4, 0.030), (0.5, 0.005)

Exercise 17a (page 330)

2. $(b-a)/(a+b)$ 3. $3/5$ 4. r/n 5. $2/5$ 6. $(a+b+c)/(a+2b+3c)$
7. $3/23$ 8. $(a-c)/(a+b+c)$
9. (c) $a/p - b/(2-p) - c/(1-p) + d/(1+p) = 0$ (e) 0.12

Exercise 17b (page 334)

1. 0.347 2. $(x_1 + x_2 + x_3 + x_4 + x_5)/5$

Exercise 17c (page 334)

1. (b) $2np^2 + (b+2c)p - (2a+b) = 0$ (c) 0.22 2. 0.06
3. (a) $\sqrt{(1-p)}$ (b) $p/2$ (c) $p^2/8$; $p/2\sqrt{(1-p)}$; $1/2$
4. $\alpha^2[1/(\alpha^2 - x_1^2) + 1/(\alpha^2 - x_2^2)] = 3$ 5. $2\lambda^2 - \lambda - 4 = 0$; $\lambda = 1.69$; $\mu = 19.53$
6. $e^{-\mu} = 1 - \mu/3$; $\mu = 2.8$ (say) giving $\mu = 2.82$ (or equivalent)   7. $\text{Max}(x_1, x_2, \ldots, x_n)$

Exercise 18a (page 343)

1. (a) 0.882 (b) 0.284 (c) -0.903 (d) -0.736 2. 0; $y = x^2 - 4$ 3. 0.944
4. 0.544 5. 0.898 6. 0.934 7. 0.024

Exercise 18b (page 346)

1. 0.688 2. -0.603 3. 0.783

Exercise 18c (page 349)

1. 0.786; -0.048 2. -0.517 3. 0.488; 0.830 4. -0.879 5. 0.150

Exercise 18d (page 351)

1. $t = 6.3$, sig.   2. $t = 1.74$, (a) not sig.   (b) not sig.   3. $t = 8.67$, sig.
4. $t = 0.08$, not sig.   5. $t = 4.56$, (a) sig.   (b) sig.

Exercise 18e (page 352)

1. (a) 0.143 (b) 0.152 2. 0.235
3. (a) Straight line with negative gradient   (b) Monotonically increasing curve
(c) -0.924 (d) -0.900 4. -0.858
5. $r_{AB} = 0.727$, $r_{BC} = 0.566$, $r_{CA} = 0.790$; A and C 6. (a) 0.79 (b) 0.98
7. -0.036, no sig. correlation   8. 0.392 9. (a) ± 1 (b) 1 (c) -1

Exercise 19a (page 360)

1. $y = 0.47 + 1.63x$ 2. $y = 14.01 - 1.55x$ 3. $y = 6.19 + 7x$ 4. $y = 0.36 + 0.065x$

Exercise 19b (page 362)

1. $y = 56.3 + 0.43x$ 2. $y = 0.97 + 0.98x$; 45.1 3. $y = 94.40 - 1.45x$
4. $y = 105.15 - 0.51x$; 105 5. $W = 40.39 + 0.36T$; 60.5 6. $y = 0.045 + 0.021x$
7. $y = 29.32 + 0.017x$ 8. $y = 32.82 + 0.221x$; 75.7

Exercise 19c (page 366)

1. $x = 1.23y - 1.18$ 2. $y = 0.476x + 20.59$; $x = 1.036y + 7.35$; 53.0; 48.8
3. $y = 0.35x + 32.75$; $x = 0.065y + 110.27$; 67.8; 115.8

Exercise 19d (page 366)

1. $N = 47.67 - 2.92P$; -0.86 2. (a) -180.4 (b) -0.93 (c) $y = 75 - 0.75x$
3. 0.99; $y = 31.64 + 4.55x$; 81.7 ml   4. (a) $y = 0.74 + 0.29x$ (b) 0.97
5. $y = 0.58 + 3.03 \times 10^{-4}x$; 1.19; $y = 0.785 + 2 \times 10^{-4}x$; 1.185

6. (b) (40, 414)   (c) $y = 132.8 + 7.03x$; 273.4   **7.** $x = 0.3, y = 0.6$ **8.**   84.68 min
9. $H = 1.4 + 0.7T$; 53.9   **10.**   (a) $y = 0.49x - 2.40x$ (b) 0.997
11.   (a) $y = 48.4 + 2.75x$ (b) 0.79   **12.**   4.62; 157.3; 179.3
13. $F = 0.94I - 6.37$; $F = 20.75$ **14.**   12.5; 1.10

Exercise 20a (page 380)

1. -0.035 **2.**   32; 7; 35%   **3.** -0.404 **4.**   0; 28.8 cm^2 **7.** $2\sqrt{6}/5$
9.   0.979; 0; No   **10.**   No; Yes

Exercise 20b (page 385)

1.   0.63; (0.554, 0.706)   **2.**   (0.550, 0.710)   **3.**   21.92; (21.72, 22.12)

Exercise 20c (page 389)

1. $t = 2.14$, sig.   **2.** $t = -2.34$, sig. at 5%   **3.**   0.396   **4.** $z = 1.29$, Yes
5.   (0.59, 0.91)   **6.**   (a) $z = 1.2$, not sig.   (b) $z = 1.4$, not sig. at 5%
7. $z = 0.61$, accept H_0 **8.**   (0.26, 0.76)

Exercise 20d (page 392)

1. $\hat{\alpha} = 1.75$; $\hat{\beta} = 2.82$ **2.** $\hat{\alpha} = 33.42$; $\hat{\beta} = -1.95$; $\hat{\gamma} = -8.79$
3. $\hat{\alpha} = 9.95$; $\hat{\beta} = 2.99$; $\hat{\gamma} = 0.34$ **4.** $\hat{\alpha} = 11.68$; $\hat{\beta} = 0.17$; $\hat{\gamma} = 1.30$

Exercise 20e (page 393)

1.   (a) 7/2   (b) 455/12   (c) 0   (d) 35/12   **2.**   (a) No   (c) $p = 1/6$; $q = 1/12$
3.   0.78; (a) $t = 3.6$, reject H_0 at 5%   (b) $z = -1.1$, accept H_0 at 5%
4.   0.150; $t = 1.50$, theory supported; $z = 2.69$, No   **5.**   0.959; (0.990, 0.831)
6. $z = 0.90$, accept H_0
7.   (a) 31; 73; 5.196; 9.726   (b) 0.768   (c) $y = 28.45 + 1.44x$ (d) 0.424
8. $a = 1.46$; $b = 0.76$; $c = -2.25$ **9.** $\hat{\alpha} = 0.036$; $\hat{\beta} = 0.074$; $\hat{\mu} = 0.735$
10. $\sigma^2 [\Sigma a_i^2 - (\Sigma a_i b_i)^2 / \Sigma b_i^2]$ **12.**   (b) $\hat{\alpha} = \Sigma y_i / n - \hat{\beta} \Sigma x_i / n$; $\hat{y} = \Sigma z_i / n - \Sigma y_i / n$
13. $\beta, \sigma^2 / \Sigma (x_i - \bar{x})^2$ **14.** $a\mu_1 + b\mu_2$; $a^2 \sigma_1^2 + b^2 \sigma_2^2 + 2ab\rho$; $\sigma_1^2 = \sigma_2^2$

Exercise 21a (page 403)

1. $(e^t + e^{2t} + e^{3t} + e^{4t} + e^{5t} + e^{6t})/6$; 7/2; 35/12   **3.** $a + 3a^2 + a^3$ **5.**   1/3, 2/9, 2/9; 1/9
6. $(e^{3t} - e^t)/2t$; 1/3   **8.** $M(t) = e^{at}/(\lambda - t)$; $a + 1/\lambda$; $1/\lambda^2$ **9.**   7/3; $\sqrt{13/18}$ **10.**   0; 2
11.   (a) α^k (b) $(1-\alpha)e^t/(1-\alpha e^t)$ for $t < -1n\alpha$; geometric, $(1-\beta\gamma)e^t/(1-\beta\gamma e^t)$ for $t < -1n\beta\gamma$

Exercise 21b (page 409)

1.   20/9; $-52/27$ **3.**   1/5; 0   **4.**   0.36; -0.062; -2.86 **5.**   1/a
6. $npq(q-p)$ **7.** $e^{-2t}/(1-2t)$; 4; 16   **8.**   (b) $-1/160$; 1/160
9. $(1-t+\theta t)/(1-t)^2$; $2 + 6\theta - 6\theta^2 + 2\theta^3$ **10.**   2; 8

Exercise 21c (page 413)

2. $e^{3t}(0.8 + 0.2e^{2t})^6$; 5.4; 3.84; 224.28   **3.** a; $2b$; 2μ; $2\sigma^2$; $(X_1 - X_2)/\sqrt{2}$; $(3X_1 + 4X_2)/5$
4. $\alpha\beta/[(\alpha - t)(\beta - t)]$; $1/\alpha + 1/\beta$; $1/\alpha^2 + 1/\beta^2$
5. $M_z(t) = \alpha/(\alpha - t)$ for $\alpha > t$; $M_y(t) = (1 - 2t)^{-1}$; $\alpha = 1/2$
6. $[\lambda/(\lambda - t)]^{nk+n}$; $(nk+n)/\lambda$; $(nk+n)/\lambda^2$; $\lambda^{nk+n} x^{nk+n-1} e^{-\lambda x}/(nk+n-1)!$ $(x > 0)$
7.   (b) $(1-t)^{-n}$; n; n; 1, $1/n$; Tends to a normal distribution.

Exercise 22a (page 423)

1.   (a) 1/6, $-2 \leqslant y \leqslant 4$ (b) $1/6y$, $e^{-3} \leqslant y \leqslant e^3$ (c) $1/2y^2$, $1/3 \leqslant y \leqslant 1$
2.   (a) $1/6\sqrt{y}$, $0 \leqslant y \leqslant 9$ (b) $1/6\sqrt{y-1}$, $0 \leqslant y \leqslant 10$
3. $\sqrt{1/\pi y}/20$, $0 < y \leqslant 100\pi$; 0.82   **4.** $y^{-2/3}/3$, $8 \leqslant y \leqslant 27$; 65/4; 30.1

5. $3-3\sqrt{y}, 0 \leqslant y \leqslant 1$ **6.** $2a^2/3$

7. $\pi h^2/3, 0 \leqslant h \leqslant (9/\pi)^{1/3}$; (a) 0.35   (b) 1.06   (c) 0.076   (d) 0.42   **8.** $3\pi; 3\pi^2; 0.63$

Exercise 22c (page 432)

1. (a) 0.19   (b) 0.28   **2.** 0.30   **3.** (a) 0.18   (b) 0.63

4. 1/3; (a) 0.27   (b) 6 s   (c) 0.31

Exercise 22d (page 436)

1. (b) 1/32

2. $dp_0/dt = -\alpha p_0 + \beta p_1$; $dp_1/dt = \alpha p_0 - (\alpha+\beta)p_1 + 2\beta p_2$; $p_n = 2[(1-k)/(1+k)]k^n$ where $k = \alpha/2\beta$; $dp_n/dt = \alpha p_{n-1} - (\alpha+2\beta)p_n + 2\beta p_{n+1}$

Exercise 22e (page 436)

1. $1/2\sqrt{y}, \sqrt{y}, 0 < y \leqslant 1$; 1/3; 4/45   **2.** $1/2\sqrt{(y+1)}, -1 < y \leqslant 0$

3. (a) $1/\pi\sqrt{(1-y^2)}, -1 < y < 1$ (b) $2/\pi\sqrt{(1-y^2)}, 0 \leqslant y < 1$ **4.** 1; 1/6; 3/4; 5/6

5. $800/3; 1/\sqrt{2}$ **6.** $\frac{1}{2}e^x (x \leqslant 0); 1 - \frac{1}{2}e^{-x} (x \geqslant 0); e^{-\sqrt{y}}/2\sqrt{y}$

7. $1/2l\sqrt{(l^2-y)}, 0 \leqslant y < l^2$ **8.** $3V/8; 1/\sqrt{2}$

9. (a) $1 (0 \leqslant y \leqslant 1)$, i.e. uniform   (b) $(4/n)z^{4/(n-1)}$ **10.** $a = -1; b = 1; 0$

11. $y^2/a^2; 2y/a^2, 0 < y \leqslant a; 2a/3$ **12.** $a/2; a^2/12; a, a^2/6$; (a) 1/2   (b) 3/8   (c) 3/4

15. $\lambda e^{-\lambda y}, y \geqslant 0; 0.1$ **16.** $\lambda e^{-\lambda x}, x \geqslant 0$

17. $p(t+\delta t) = p(t)(1-\mu\delta t) + [1-p(t)]\lambda\delta t$ **18.** Expected time to first event $= 1/\lambda$

19. $dP_0/dt = -\lambda P_0$; $G = e^{\lambda(z-1)t^2/2}$ **20.** $mp_0 = np_1$; $p_r = (1-m/n)(m/n)^r, m/(n-m)$

21. $P_0(t+1) = (1-k+kp)P_0(t) + (p-kp)P_1(t)$; $c = (p-k)/p(1-k)$; $x = k(1-p)/p(1-k)$

Exercise 23a (page 447)

1. $\chi^2 = 9.08$, accept H_0 (4 d.f.)   **2.** $\chi^2 = 0.29$, accept H_0 (2 d.f.)

3. $\chi^2 = 10.75$, sig. different; $\chi^2 = 5.75$, not sig. different (3 d.f.)

4. $\chi^2 = 9.97$, unlikely (3 d.f.)   **5.** $\chi^2 = 16.30$, reject H_0 (4 d.f.)

6. $\chi^2 = 5.56$, accept H_0 (3 d.f.)

Exercise 23b (page 452)

1. $\hat{p} = 0.1, \chi^2 = 4.28$, accept H_0 at 5% (3 d.f.)

2. $\hat{\mu} = 1.279, \chi^2 = 7.56$, not sig. at either (3 d.f.)   **3.** $\chi^2 = 4.40$, justified (5 d.f.)

4. $\hat{\mu} = 0.62, \chi^2 = 0.73$, accept H_0 (2 d.f.)   **5.** $\chi^2 = 0.23$, reasonable (3 d.f.)

6. $\chi^2 = 1.59$, compatible (2 d.f.)   **7.** $\chi^2 = 3.20$, accept H_0 (7 d.f.)

8. $\chi^2 = 3.55$, accept assertion at 5% (5 d.f.)

9. $50.3; 2.657; \chi^2 = 4.41$, accept model (1 d.f.)

10. $\hat{p} = 0.442; \chi^2 = 7.23$, accept H_0 at 5% (4 d.f.)

Exercise 23c (page 459)

1. $\chi^2 = 9.17$, evidence of association at 5% (2 d.f.)

2. $\chi^2 = 30.94$, evidence of association at 5% (4 d.f.)   **3.** $\chi^2 = 13.45$, reject H_0 (6 d.f.)

4. $\chi^2 = 7.34$, evidence of association at 5% and 1% (1 d.f.)

5. $\chi^2 = 0.47$, no evidence of association at 5% (1 d.f.)

6. $\chi^2 = 12.3$, no evidence of association at 5% (9 d.f.)

7. $\chi^2 = 7.84$, accept H_0 (4 d.f.)

Exercise 23d (page 465)

1. $a = 4.605; b = 5.99$ **4.** $\mu_3' = 48; \mu_3 = 16$; skewness $= 2$ **6.** 1; 2

Exercise 23e (page 466)

1. (a) $\chi^2 = 9.87$, consistent at 5% (5 d.f.)   (b) $\chi^2 = 9.87$, not acceptable at 5% (4 d.f.)

2. $\chi^2 = 7.92$, evidence of bias at 5% (3 d.f.); $\hat{\mu} = 0.2$, $\chi^2 = 0.04$, a suspiciously low value. Since $P(\chi^2 < 0.10) = 0.05$, the results may have been fiddled.
3. $\chi^2 = 7.29$, accept H_0 at 5% (3 d.f.)　**4.** $\chi^2 = 3.84$, accept H_0 at 5% (5 d.f.)
5. (a) $3;3$　(b) $\chi^2 = 0.64$, good fit at 5% (4 d.f.)　**6.** $\chi^2 = 1.75$, good fit (4 d.f.)
7. (b) $\chi^2 = 1.73$, adequate (3 d.f.)　(c) Yes: $P(\text{win}) = 1/80$ (data), $P(\text{win}) = 0.01024$ (model); in either case 100 to 1 means long-run profit.
8. $\hat{\lambda} = 1/4$; $\chi^2 = 4.68$, accept H_0 at 5% (2 d.f.)
9. $\chi^2 = 42.95$, not consistent at 5%, so kind of pill affects sleep (4 d.f.)
10. (b) $\chi^2 = 9.07$, reject H_0 at 5%　(c) parabola is $y^2 = 3.84x$
11. $A = \lg a + \lg \lg e$; $B = b$; $a = 0.010\ (\pm 0.001)$; $b = 1.8\ (\pm 0.05)$; $\chi^2 = 1.82$, accept H_0
15. $C = 1/4$; Shows how χ^2 tables can be obtained; $\chi^2 = 20.16$, sig. association (4 d.f.)
19. σ^2; $2\sigma^4/n$

Exercise 24a (page 473)

1. (a) $(7.6, 21.5)$　(b) $(6.6, 26.3)$　**2.** $(1.3, 8.3)$　**3.** $(0.096, 0.674)$
4. $(0.005, 0.028)$　**5.** $(3.5, 15.0)$

Exercise 24b (page 477)

1. V.R. $= 2.59$, not sig. at 5%　**2.** V.R. $= 1.16$, accept H_0 at 5%
3. (a) V.R. $= 3.00$　(b) $(n-1)s^2/\sigma^2 = 4.33$; reject H_0 at 5%
4. V.R. $= 1.53$, difference not sig. at 5%; $t = 1.0$, difference not sig. at 5% (4 d.f.)

Exercise 24c (page 478)

1. V.R. $= 1.71$, no sig. change; $(26.8, 136.4)$
2. (a) V.R. $= 1.23$, difference not sig. at 5%　(b) $t = 0.95$, no sig. difference at 5%
3. V.R. $= 1.78$, $t = 1.91$ (24 d.f.); $(3.5, 11.2)$
4. $[\sigma^2/(n-1)]\chi^2$, with $(n-1)$ d.f.; $(n-1)s^2/\sigma^2 = 3.99$, sig. evidence; $(71, 500)$
5. (a) V.R. $= 2.9$, accept H_0　(b) $t = 1.8$, sig. greater
6. V.R. $= 1.96$, accept Hyp. 1; V.R. $= 2.17$, accept Hyp. 2; V.R. $= 4.26$, reject Hyp. 3
7. $(277, 2871)$; V.R. $= 4.54$, evidence for smaller s.d. at 5%　**8.** $K = v^{\frac{1}{2}v+1}$
9. (b) $1.41X$　**10.** $P(s) = (1 + 20/s)\,e^{-20/s}$　**11.** 0.21

Exercise 25a (page 492)

1. V.R. $= 13.34$, sig. at 5%　**2.** V.R. $= 3.33$, not sig. different at 5%
3. V.R. $= 1.48$, accept H_0 at 5%　**4.** V.R. $= 16.73$; Two may be equal.
5. (a) $2; 3; 4; 0.667$　(b) V.R. $= 7.5$　(c) 0.516; Two might be equal.
6. V.R. $= 12.8$, sig.; $\hat{\mu}_c = 48.25$ h, s.e. $= 5.64$ h
7. 8.293 mm, 8.236 mm, 8.338 mm; (a) 0.0535　(b) 0.075　(c) V.R. $= 5.56$, reject H_0 at 5%

Exercise 25b (page 498)

1.–7. (See Exercise 25a, questions 1–7.)
8. V.R. $= 9.0$, sig. at 5%　**9.** V.R. $= 3.49$, sig.　**10.** V.R. $= 3.32$, reject H_0 at 5%
11. V.R. $= 1.78$, sig. at 5%　**12.** V.R. $= 4.01$, sig.　**13.** V.R. $= 4.12$, sig. at 5%

Exercise 25c (page 505)

1. V.R.(hours) $= 3.10$, not sig.; V.R.(men) $= 13.45$, sig. at 5%
2. V.R.(paint) $= 7.70$, sig.; V.R.(town) $= 7.22$, sig. at 5%
3. V.R.(treatment) $= 38.41$, sig.; V.R.(hospitals) $= 2.73$, not sig. at 5%
4. V.R.(brands) $= 0.93$, not sig.; V.R.(cars) $= 1.65$, not sig. at 5%

Exercise 25d (page 506)

1. V.R. $= 8.12$, sig.　**2.** V.R. $= 11.12$, sig.
3. V.R. $= 5.45$, sig. difference between B and D, and between C and D.
4. (a) $11.88; 15.33; 12.67$　(b) 5.21　(c) V.R. $= 4.13$, reject H_0 at 5%　(d) $(13.37, 17.30)$

5. V.R.(tyres) = 21.94, sig. at 5%; V.R.(drivers) = 15.80, sig. at 5%
6. V.R.(work) = 6.91, sig.; V.R.(women) = 9.39, sig.   **7.** V.R. = 2.65, not sig. at 5%; 5.43

Miscellaneous exercises in probability (page 512)

1. 0.58; 0.3   **2.** No advantage; (a) 5/12   (b) $(1/6^{r-1})(5/12)$ (c) 0.5
3. 0.59; (a) 0.352   (b) 0.458   (c) 0.481   **4.** (a) (i) 0.375   (ii) 0.4   (iii) 0.5   (b) 5/8
5. (a) 135/323   (b) 70/323   (c) 3/8   (d) 27/128   **6.** 19/35
7. (a) 1/25   (b) 106/125   (c) 14/19
8. (a) 0.02, 0.17, 0.45, 0.36; 43   (b) (i) 0.0625   (ii) 0.1372   (iii) 0.2048   (iv) 0.015
9. 3/11   **10.** (b) $(5-\sqrt{7})/3$; $(5-\sqrt{7})/6$ **11.** 0.784
12. (a) 1/2   (b) 1/5   (c) 41/60   **13.** 1/2; 1/36; 1/4
14. $(n-m+1)(n-m+2)/(n+1)(n+2)$; 5   **15.** 1/3   **16.** (b) (i) 55/108   (ii) 1/12
17. (a) 1/36   (b) 15/36; 73/648; 25/81   **18.** 0.145; 0.00025
19. (a) 0.077   (b) 0.315   **21.** (a) 0; 5/36; 5/36   (b) 5/648 (0.008); 6841/46 656 (0.147)
22. (a) $6q^5 - 5q^6$ (b) $30q^5 - 24q^6$ (c) $30q^9 - 54q^{10} + 30q^{11} - 5q^{12}$; 0.796; 0.462
23. (b) $3(2^{-r}) - 3(4^{-3})$ (d) 10/3
24. P(no winner) = 0 since there must be a winner with an unlimited number of games (unless $p = r = 0$)

Index